TOPICS IN INTEGRAL CALCULUS

AF529340

TOPICS IN INTEGRAL CALCULUS

By

BANSI LAL

FIREWALL MEDIA
(An Imprint of Laxmi Publications Pvt.Ltd.)
An ISO 9001:2008 Company
BENGALURU • CHENNAI • COCHIN • GUWAHATI • HYDERABAD
JALANDHAR • KOLKATA • LUCKNOW • MUMBAI • RANCHI • NEW DELHI
BOSTON (USA) • ACCRA (GHANA) • NAIROBI (KENYA)

TOPICS IN INTEGRAL CALCULUS

© by Laxmi Publications (P) Ltd.
All rights reserved including those of translation into other languages. In accordance with the Copyright (Amendment) Act, 2012, no part of this publication may be reproduced, stored in a retrieval system, or transmitted in any form or by any means, electronic, mechanical, photocopying, recording or otherwise. Any such act or scanning, uploading, and or electronic sharing of any part of this book without the permission of the publisher constitutes unlawful piracy and theft of the copyright holder's intellectual property. If you would like to use material from the book (other than for review purposes), prior written permission must be obtained from the publishers.

Printed and bound in India
Typeset at Goswami Associates, Delhi
Seventh Edition: 2017
ISBN 978-81-318-0002-7

Limits of Liability/Disclaimer of Warranty: The publisher and the author make no representation or warranties with respect to the accuracy or completeness of the contents of this work and specifically disclaim all warranties. The advice, strategies, and activities contained herein may not be suitable for every situation. In performing activities adult supervision must be sought. Likewise, common sense and care are essential to the conduct of any and all activities, whether described in this book or otherwise. Neither the publisher nor the author shall be liable or assumes any responsibility for any injuries or damages arising here from. The fact that an organization or Website if referred to in this work as a citation and/or a potential source of further information does not mean that the author or the publisher endorses the information the organization or Website may provide or recommendations it may make. Further, readers must be aware that the Internet Websites listed in this work may have changed or disappeared between when this work was written and when it is read.

All trademarks, logos or any other mark such as Vibgyor, USP, Amanda, Golden Bells, Firewall Media, Mercury, Trinity, Laxmi appearing in this work are trademarks and intellectual property owned by or licensed to Laxmi Publications, its subsidiaries or affiliates. Notwithstanding this disclaimer, all other names and marks mentioned in this work are the trade names, trademarks or service marks of their respective owners.

Branches

Bengaluru	080-26 75 69 30	
Chennai	044-24 34 47 26,	24 35 95 07
Cochin	0484-237 70 04,	405 13 03
Guwahati	0361-254 36 69,	251 38 81
Hyderabad	040-27 55 53 83,	27 55 53 93
Jalandhar	0181-222 12 72	
Kolkata	033-22 27 43 84	
Lucknow	0522-220 99 16	
Mumbai	022-24 91 54 15,	24 92 78 69
Ranchi	0651-220 44 64	

PUBLISHED IN INDIA BY

FIREWALL MEDIA
(An Imprint of Laxmi Publications Pvt.Ltd.)

An ISO 9001:2008 Company
113, GOLDEN HOUSE, DARYAGANJ,
NEW DELHI - 110002, INDIA
Telephone : 91-11-4353 2500, 4353 2501
Fax : 91-11-2325 2572, 4353 2528
www.laxmipublications.com info@laxmipublications.com

C—
Printed at:

CONTENTS

PREFACE TO THE SEVENTH EDITION

The very rapid sale of the first five heavy editions and a roaring demand for the sixth make me believe that the book has supplied long-felt need and has really been found useful. I take this opportunity of thanking the numerous readers, from India and abroad, for the unsolicited letters of appreciation.

The book has been thoroughly revised, recast at some place and brought up to date. Older examples of lesser importance have been deleted or in a few cases replaced by modern examples of greater importance. It is hoped that the book, in its present form will be found more useful to the examinees for whom it is meant.

AUTHOR

PREFACE TO THE FIRST EDITION

For some time past my own students, friend-Professors and other numerous readers of my books have been pressing me hard to write book on Integral Calculus based on my own method of teaching in the class-room. This book has, therefore, been written with the following.

Distinguishing Features :

(1) It *just* covers the University syllabus and though as such, *other* things (however important they might appear from *other* point of view) have been carefully excluded, the book is fairly complete within its limits.

(2) Written primarily from the *point of view of an examinee* attempts to present the subject in an *'easy-to-understand'* and *'easy-to-reproduce'* form. For this reason a step which is likely to present difficulty to the beginner, has been explained in a footnote under the heading **"How to write this step"** or **"Why this step".**

(3) Every Article which is likely to be set in the Examination has been stated in the form in which it usually occurs in the question paper and has been done in the manner in which it ought to be answered there.

(4) In view of the heavy syllabus in Mathematics and the limited time at the disposal of the teacher in the class-room, *only typically* questions *just sufiicient* to give the student an adequate practice in the formulae or methods, have been given in a set after an Article or two. Great care has been taken to include in a gradual manner almost all the standard questions on the subject. References to Examinations have been invariably given so that the student may realize their relative importance. Hints to difficult questions have been given to encourage the student to come prepared from home and thus quicken the progress of the subject in the class-room. *For the more ambitious and voracious reader*, however, a set of Miscellaneous Examples, not to form *necessarily* a part of the class-work has been appended to each chapter to try his skill. For him is intended all the matter marked with a double asterisk (**).

(5) All Articles and Examples which are important from the *Examination* point of view have been printed in bold type. This will help the student in his judicious choice of them when the time at his disposal is limited.

(6) Whenever possible the method has been summarized into a rule, and, where necessary, an **Aid to memory** has been provided.

(7) A chapter-wise list of all the important results and formulae to be remembered by the student has been prefixed to the book for ready reference and occasional review.

In short no pains have been sparedto make the book useful or all types of students—ordinary, average and brilliant.

—AUTHOR

IMPORTANT RESULTS AND FORMULAE

[To be remembered]

CHAPTER I

1. I. $\int x^n\,dx = \frac{x^{n+1}}{n+1}$. $(n \neq -1)$ | II. $\int \frac{1}{x}\,dx = \log x$.

III. $\int a^x\,dx = \frac{a^x}{\log a}$. | IV. $\int e^x\,dx = e^x$.

V. $\int \sin x\,dx = -\cos x$. | VI. $\int \cos x\,dx = \sin x$.

VII. $\int \sec^2 x\,dx = \tan x$. | VIII. $\int \operatorname{cosec}^2 x\,dx = -\cot x$

IX. $\int \sec x \tan x\,dx = \sec x$. | X. $\int \operatorname{cosec} x \cot x\,dx = -\operatorname{cosec} x$. (Art. 5)

2. XI. $\int \sinh x\,dx = \cosh x$. | XII. $\int \cosh x\,dx = \sinh x$.

CHAPTER II

1. XIII. $\int \frac{1}{\sqrt{a^2 - x^2}}\,dx = \sin^{-1}\frac{x}{a}$.

XIV. $\int \frac{1}{a^2 + x^2}\,dx = \frac{1}{a}\tan^{-1}\frac{x}{a}$.

XV. $\int \frac{1}{x\sqrt{x^2 - a^2}}\,dx = \frac{1}{a}\sec^{-1}\frac{x}{a}$. (Art. 13)

2. XVI. $\int \frac{1}{\sqrt{a^2 + x^2}}\,dx = \sinh^{-1}\frac{x}{a}$.

XVII. $\int \frac{1}{\sqrt{x^2 - a^2}}\,dx = \cosh^{-1}\frac{x}{a}$. (Art. 15)

3. An important rule. *If the integral of a function of x is known then if x is multiplied by a constant, and to the product is added another constant, the integral is of the same form but is divided by the coefficient of x.* (Art. 16)

4. $\int [f(x)]^n f'(x)\,dx = \frac{[f(x)]^{n+1}}{n+1}$. $(n \neq -1)$ (Art. 17)

5. *The integral of a fraction whose numerator is the differential coefficient of the denominator is log (denominator).* (Art. 18)

6. XVIII. $\int \tan x\,dx = -\log \cos x$.

XIX. $\int \cot x\,dx = \log \sin x$. (Art. 19)

7. XX. $\int \sec x\,dx = \log \tan\left(\frac{\pi}{4} + \frac{x}{2}\right)$.

XXI. $\int \operatorname{cosec} x\, dx = \log \tan \frac{x}{2}.$ (Art. 20)

CHAPTER III

1. Rule of integration by parts.

Integral of product of two functions

= 1st function × integral of 2nd

– integral of [diff. coeff. of 1st × integral of 2nd]. (Art. 23)

[**Rule to choose the first function rightly in the rule of integration by parts :** If one factor of the product is a power of x, take it as the first function except when the other factor is an inverse circular (or hyperbolic) function or a logarithm in which case take that factor as the first function. (Note, Art. 23)]

2. XXII. $\int \sqrt{a^2 - x^2}\, dx = \frac{x\sqrt{a^2 - x^2}}{2} + \frac{a^2}{2} \sin^{-1} \frac{x}{a}.$

XXIII. $\int \sqrt{a^2 + x^2}\, dx = \frac{x\sqrt{a^2 + x^2}}{2} + \frac{a^2}{2} \sinh^{-1} \frac{x}{a}.$

XXIV. $\int \sqrt{x^2 - a^2}\, dx = \frac{x\sqrt{x^2 - a^2}}{2} - \frac{a^2}{2} \cosh^{-1} \frac{x}{a}.$ (Art. 24)

3. Rule to integrate $\int [f(x) + f'(x)]\, e^x\, dx$: Split the integral into two, and integrate only the first integral by parts. (Art 25)

4. XXV. $\int e^{ax} \sin bx\, dx = \frac{1}{a^2 + b^2} e^{ax} (a \sin bx - b \cos bx)$

$$= \frac{1}{\sqrt{a^2 + b^2}} e^{ax} \sin\left(bx - \tan^{-1} \frac{b}{a}\right).$$

XXVI. $\int e^{ax} \cos bx\, dx = \frac{1}{a^2 + b^2} e^{ax} (a \cos bx + b \sin bx)$

$$= \frac{1}{\sqrt{a^2 + b^2}} e^{ax} \cos\left(bx - \tan^{-1} \frac{b}{a}\right).$$ (Art. 27)

$$\int e^{-ax} \sin bx\, dx = -\frac{1}{\sqrt{a^2 + b^2}} e^{-ax} \sin\left(bx + \tan^{-1} \frac{b}{a}\right).$$

$$\int e^{-ax} \cos bx\, dx = -\frac{1}{\sqrt{a^2 + b^2}} e^{-ax} \cos\left(bx + \tan^{-1} \frac{b}{a}\right).$$ (Ex. 7, Art. 27)

5. General rule of integration by parts.

Integral of product of two functions

= 1st function × integral of 2nd – go on differentiating 1st, integrating 2nd, signs alternately +ve and – ve, ... (Art. 28)

CHAPTER IV

XXVII. $\int \frac{1}{x^2 - a^2}\, dx = \frac{1}{2a} \log \frac{x-a}{x+a}.\ (x > a).$

XXVIII. $\int \frac{1}{a^2 - x^2}\, dx = \frac{1}{2a} \log \frac{a+x}{a-x}.\ (x < a).$ (Art. 33)

CHAPTER V

1. Four standard cases.

Case I. Rule to integrate $\int \frac{1}{\text{linear}\sqrt{\text{linear}}}\, dx$:

Put $\sqrt{\text{linear}} = z$.

Case II. Rule to integrate $\int \frac{1}{\text{linear}\sqrt{\text{quadratic}}}\, dx$:

Put linear $= \frac{1}{z}$.

[**Extension. Rule to integrate** $\int \frac{1}{(\text{linear})^2\sqrt{\text{quadratic}}}\, dx$:

Put linear $= \frac{1}{z}$.]

Case III. Rule to integrate $\int \frac{1}{\text{quadratic}\sqrt{\text{linear}}}\, dx$:

Put $\sqrt{\text{linear}} = z$.

Case IV. Rule to integrate

$$\int \frac{1}{\text{pure quadratic}\sqrt{\text{pure quadratic}}}\, dx :$$

Put $x = \frac{1}{z}$, and then in the resulting integral, put

$$\sqrt{\text{pure quadratic}} = u.$$ (Arts. 43–46)

[**Extension.** If in any of the above four cases, instead of 1 in the numerator, there is a polynomial in x, whose degree is equal to, or greater than, that of the rational factor of the denominator, first divide the numerator by this rational factor, and then use the rule where necessary.]

Note 1. Rule to integrate

$$\int \frac{x}{\text{pure quadratic}\sqrt{\text{pure quadratic}}}\, dx :$$

Put $\sqrt{\text{pure quadratic}} = z$. (Art. 47)

****Note 2. Rule to integrate**

$$\int \frac{1}{\text{quadratic}\sqrt{\text{quadratic}}}\, dx, \left[\text{or} \int \frac{\text{linear}}{\text{quadratic}\sqrt{\text{quadratic}}}\, dx\right]$$

if the quadratic not under the square root can be resolved into real linear factors :

Resolve $\dfrac{1}{\text{quadratic}}\left[\text{or } \dfrac{\text{linear}}{\text{quadratic}}\right]$ into partial fractions, and split the integral into two, each of which is of the form

$$\int \frac{1}{\text{linear}\sqrt{\text{quadratic}}}\,dx.$$ (Ex. 5, end, Art. 46)

2. Rule to integrate $\int x^m (a + bx^n)^p\, dx$:

Case I. If p is a positive integer, expand.

Case II. If p is a fraction $= \dfrac{r}{s}$ (say) (where r and s are integers and s positive) then, if $\dfrac{m+1}{n}$ is a positive or negative integer, or 0, put $a + bx^n = z^s$, and then expand, or resolve into partial fractions, as the case may require.

Case III. If p is a fraction $= \dfrac{r}{s}$ (say) (where r and s are integers and s positive) then, if $\dfrac{m+1}{n} + p$ is a positive or negative integer, or 0, take x^n outside the binomial, and put $ax^{-n} + b = z^s$, and then expand, resolve into partial fractions, as the case may require. (Art. 47)

3. Rule to integrate an irrational algebraic function by rationalizing by an algebraic substitution :

Case I. If only one radical $\sqrt{\text{linear}}$ occurs in the integrand, at linear $= z^2$.

Generalization. If only one radical $(\text{linear})^{\frac{1}{n}}$ occurs in the integrand, put linear $= z^n$.

Case II. If fractional powers of only one linear occur in the integrand, put linear $= z^l$, where l is the L.C.M. of the denominators the fractional indices.

****Case III.** If only one radical $\sqrt{\text{quadratic}}$ occurs in the integrand, then

(*i*) If $\sqrt{x^2 + ax + b}$ occurs in the integrand, put $\sqrt{x^2 + ax + b} = z - x$,

(*ii*) If $\sqrt{-x^2 + ax + b}$ occurs in the integrand,

and $\sqrt{-x^2 + ax + b} = \sqrt{(x - \alpha)(\beta - x)}$,

put $\sqrt{\dfrac{\text{one factor}}{\text{other factor}}}$, *i.e.*, $\sqrt{\dfrac{x-\alpha}{\beta - x}} = z.$ (Arts. 48–50)

4. Rule to integrate an irrational algebraic function by rationalizing by a trigonometric substitution :

(*i*) If $\sqrt{a^2 - x^2}$ occurs in the integrand, put $x = a \sin \theta$.

(*ii*) If $\sqrt{a^2 + x^2}$ occurs in the integrand, put $x = a \tan \theta$.

(*iii*) If $\sqrt{x^2 - a^2}$ occurs in the integrand, put $x = a \sec \theta$. (Art. 51)

CHAPTER VI

1. Rule to integrate $\int \sin^n x\, dx$ or $\int \cos^n x\, dx$, where n is a positive integer :

Case I. (*i*) If the index of $\sin x$ is a positive odd integer, put $\cos x = z$.

(*ii*) If the index of $\cos x$ is a positive odd integer, put $\sin x = z$. (Art. 52)

Case II. If the index of $\sin x$ or $\cos x$ is a positive even integer, use the multiple-angle method (Art. 53) or the reduction formula (Art. 69-I).

2. Rule to integrate $\int \sin^p x \cos^q x\, dx$:

Case I. If either p or q is a positive odd integer, whatever the other may be, then

(*i*) If the index of $\sin x$ is a positive odd integer, put $\cos x = z$.

(*ii*) If the index of $\cos x$ is a positive odd integer, put $\sin x = z$. (Art. 55)

Case II. If both p and q are positive even integers, use the multiple-angle method (Art. 55) or the reduction formula (solved Ex. 1, Art. 68).

Case III. If $p + q$ is a negative even integer (p, q are not necessarily integers), express the integrand as the product of the powers of $\tan x$ and $\sec x$, and put $\tan x = z$. (Art. 56)

3. Rule to integrate $\int \sec^n x\, dx$ or $\int \text{cosec}^n x\, dx$, where n is a positive integer :

Case I. (*i*) If the index of $\sec x$ is a positive even integer, put $\tan x = z$.

(*ii*) If the index of $\text{cosec}\, x$ is a positive even integer, put $\cot x = z$. (Art. 57)

Case II. If the index of $\sec x$ or $\text{cosec}\, x$ is a positive odd integer, use the reduction formula (Arts. 74, 75).

4. If $a + b$ is positive, then

(*i*) if $a^2 > b^2$, $\displaystyle\int \frac{1}{a + b\cos x}\, dx = \frac{1}{\sqrt{a^2 - b^2}} \cos^{-1} \frac{b + a\cos x}{a + b\cos x}$;

(*ii*) if $a^2 < b^2$, $\displaystyle\int \frac{1}{a + b\cos x}\, dx = \frac{1}{\sqrt{b^2 - a^2}} \cosh^{-1} \frac{b + a\cos x}{a + b\cos x}$. (Art. 58)

CHAPTER VII

1. Rule of "smaller index + 1" to connect $\int x^m (a + bx^n)^p\, dx$ with a given integral of the same type :

(*i*) Let $P = x^{\lambda + 1} (a + bx^n)^{\mu + 1}$, where λ and μ are the smaller indices of x and $a + bx^n$ respectively in the two expressions whose integrals are to be connected.

(*ii*) Find $\dfrac{dP}{dx}$, and rearrange it as a linear function of the two expressions whose integrals are to be connected.

(*iii*) Integrate both sides w.r.t. x, transpose, and solve for the integral given to be connected. (Art. 67)

Note. Important. *The rule of "smaller index + 1" is also applicable to connect* $\int \sin^p x \cos^q x\, dx$ *with a given integral of the same type.*

(*Here λ and μ are the smaller indices of sin x and cos x respectively in the two expressions whose integrals are to be connected.*) (Art. 68)

2. Rule to write down the value of $\int_0^{\frac{\pi}{2}} \sin^n x\, dx$ or $\int_0^{\frac{\pi}{2}} \cos^n x\, dx$, where n is a positive integer :

$$\int_0^{\frac{\pi}{2}} \sin^n x\, dx \text{ or } \int_0^{\frac{\pi}{2}} \cos^n x\, dx = \frac{(n-1)\times \text{go on diminishing by } 2 \ldots}{n \times \text{go on diminishing by } 2 \ldots} \cdot \frac{\pi}{2}$$

only if n is a positive even integer (otherwise no $\frac{\pi}{2}$), each series of factors being continued so long as the factors are positive (*i.e.*, omitting zero and negative factors). (Arts. 69-II, 70)

3. Rule to write down the value of $\int_0^{\frac{\pi}{2}} \sin^p x \cos^q x\, dx$, where p, q are positive integers :

$$\int_0^{\frac{\pi}{2}} \sin^p x \cos^q x\, dx$$

$$= \frac{(p-1)\times \text{go on diminishing by } 2 \ldots \times (q-1)\times \text{go on diminishing by } 2 \ldots}{(p+q)\times \text{go on diminishing by } 2 \ldots} \cdot \frac{\pi}{2}$$

only if both p and q are positive even integers (otherwise no $\frac{\pi}{2}$), each series of factors being continued so long as the factors are positive (*i.e.*, omitting zero and negative factors).

(Art. 71)

4. Rule to integrate $\int \tan^n x\, dx$ or $\int \cot^n x\, dx$, where n is a positive integer :

Use the reduction formula. (Note 3, Art. 72; Note 3, Art. 73)

CHAPTER VIII

1. I. $\int_a^b f(x)\, dx = \int_a^b f(z)\, dz$. II. $\int_a^b f(x)\, dx = -\int_a^b f(x)\, dx$.

III. $\int_a^b f(x)\,dx = \int_a^c f(x)\,dx + \int_c^b f(x)\,dx.$ IV. $\int_0^a f(x)\,dx = \int_0^a f(a-x)\,dx.$

(*i*) If $f(a-x) = f(x)$, then $\int_0^a f(x)\,dx = 2\int_0^{\frac{a}{2}} f(x)\,dx.$

(*ii*) If $f(a-x) = -f(x)$, then $\int_0^a f(x)\,dx = 0.$ (Art. 78)

3. $\int_0^{\frac{\pi}{2}} \log \sin x\,dx = -\frac{\pi}{2}\log 2.$ (Solved Ex. 4, theorem VI, Art. 78)

4. (*i*) **Def.** $\int_a^b f(x)\,dx = \underset{h\to 0}{\text{Lt}}\ h[f(a) + f(a+h) + f(a+2h) + \ldots + f(a + {}_{n-1}h)]$

where $nh = b - a$. (Art. 81)

(*ii*) **Def. (Useful for $\int_a^b x^m\,dx$.)

$$\int_a^b f(x)\,dx = \underset{r\to 1}{\text{Lt}}\ (r-1)\,[f(a).a + f(ar).ar + f(ar^2).ar^2 + \ldots + f(ar^{n-1}).ar^{n-1}$$

where $r^n = \frac{b}{a}$. (Art. 83)

CHAPTER IX

Arc formulae.

Equation of curve	*Formula for arc s*	
(1) $y = f(x)$.	$s = \int \sqrt{1+\left(\frac{dy}{dx}\right)^2}\,dx.$	
(2) $x = f(y)$.	$s = \int \sqrt{1+\left(\frac{dx}{dy}\right)^2}\,dy.$	(Art. 88)
(3) $x = f(t)$, $y = \phi(t)$.	$s = \int \sqrt{\left(\frac{dx}{dt}\right)^2+\left(\frac{dy}{dt}\right)^2}\,dt.$	(Art. 89)
(4) $r = f(\theta)$.	$s = \int \sqrt{r^2+\left(\frac{dr}{d\theta}\right)^2}\,d\theta.$	
(5) $\theta = f(r)$.	$s = \int \sqrt{1+r^2\left(\frac{d\theta}{dr}\right)^2}\,dr.$	(Art. 91)
**(6) $p = f(r)$.	$s = \int \frac{r}{\sqrt{r^2-p^2}}\,dr.$	(Ex. 6, Art. 91)

The *limits of integration*, in each case, are the values of the independent variable at the ends of the arc whose length is required.

CHAPTER X

1. Area formulae.

Area bounded by	*Formula for area A*
1. the curve, the x-axis and two ordinates.	$A = \int y\, dx$. (Art. 96-I)
2. the curve, the y-axis and two abscissae.	$A = \int x\, dy$. (Art. 96-II)
3. the curve and two radii vectores.	$A = \int \frac{1}{2} r^2\, d\theta$. (Art. 98)
4. the closed curve $x = f(t)$, $y = \phi(t)$, or a sector of the curve.	$A = \int \frac{1}{2}(x\, dy - y dx)$ $= \int \frac{1}{2}\left(x \frac{dy}{dt} - y \frac{dx}{dt}\right) dt$. (Art. 100)

The *limits of integration,* in each case, are the values of the independent variable at the ends of the bounding arc.

2. Simpson's rule. ** $\int_a^b y\, dx$ is approximately

$$= \frac{h}{3}[(y_1 + y_{2n+1}) + 2(y_3 + y_5 + \ldots + y_{2n-1}) + 4(y_2 + y_4 + \ldots + y_{2n})],$$

where $h = \dfrac{b-a}{2n}$, and $y_1, y_2, y_3, \ldots$ are the values of y when $x = a, a + h, a + 2h, \ldots$.

Practical form of Simpson's rule.

$\int_a^b y\, dx$ is approx. $= \dfrac{h}{3}[X + 2O + 4E]$,

where $h = \dfrac{b-a}{2n}$,

X = sum of extreme ordinates (*i.e.*, first and last ordinates),
O = sum of other odd ordinates (*i.e.*, omitting first and last ordinates),
E = sum of even ordinates. (Art. 102 and Note)

CHAPTER XI

1. Volume formulae.

Volume of solid generated by the area by	*Formula for volume V*
(1) revolution about the x-axis.	$V = \int \pi y^2\, dx.$ (Art. 103-I)
(2) revolution about the y-axis.	$V = \int \pi x^2\, dy.$ (Art. 103-II)
**(3) revolution about any axis.	$V = \int \pi\, PM^2\, d\,(OM)$, where PM is the $\perp$ from any pt. P of the curve on the axis, and O a fixed pt. in the axis. (Art. 103-III)
(4) revolution about the initial line.	$V = \int \frac{2}{3} \pi r^3 \sin\theta\, d\theta.$ (Art. 106)

The *limits of integration*, in each case, are the values of the independent variable at the ends of the revolving arc.

2. Surface formulae.

Curved surface of solid generated by the area by	*Formula for curved surface S*
1. revolution about the x-axis.	$S = \int 2\pi y\, ds.$ (Art. 104-I)
2. revolution about the y-axis.	$S = \int 2\pi x\, ds.$ (Art. 104-II)
**(3) revolution about any axis.	$S = \int 2\pi\, PM\, ds$, where PM is the $\perp$ from any pt. P of the curve on the axis. (Art. 104-III)

The *limits of integration*, in each case, are the values of the independent variable at the ends of the revolving arc.

3. Pappus' (or Guldin's) theorems. If a closed curve revolves about an axis in its plane, which does not cross the curve, then

(I) volume of the solid generated

= area of curve × length of path of centroid of area. (Art. 105-I)

(II) curved surface of the solid generated

= length of arc of curve × length of path of centroid of arc. (Art. 105-II)

CHAPTER XII

Five standard cases of differential equations of the first order and first degree.

Case I. Variables separable. Rule to solve an equation in which the variables are separable, *i.e.*, $\frac{dy}{dx}$ = XY, where X is a function of x only and Y a function of y only :

(*i*) Multiply thro' out by dx.

(*ii*) Separate the variables x and y (*i.e.*, get all the functions of x and dx to one side, and all the functions of y and dy to the other).

(*iii*) Integrate both sides, and add an arbitrary constant on one side. (Art. 109)

Case II. Homogeneous equation. Rule to solve a homogeneous equation in x and y, *i.e.*, $\frac{dy}{dx}=\frac{f_1(x, y)}{f_2(x, y)}$, where $f_1(x, y)$, $f_2(x, y)$ are homogeneous functions of the same degree in x and y :

Put $y = vx$, and separate the variables v and x. (Art. 110, (*b*))

Case III.

I. Equation reducible to homogeneous.

Rule to solve the equation $\frac{dy}{dx}=\frac{ax+by+c}{a'x+b'y+c'}$, when $\frac{a}{a'}\neq\frac{b}{b'}$:

(*i*) Put $x = x' + h$, $y = y' + k$.

(*ii*) Put the constant terms (*i.e.*, terms not involving x' and y') in the numerator and denominator of R.H.S. equal to zero, and find the values of h, k.

(*iii*) The equation is now homogeneous in x' and y'.

II. Rule to solve the equation $\frac{dy}{dx}=\frac{ax+by+c}{a'x+b'y+c'}$, when $\frac{a}{a'}=\frac{b}{b'}$:

Put $ax + by = z$, and separate the variables z and x. (Art. 111)

Case IV. Linear equation. The solution of the linear equation, $\frac{dy}{dx}$ + Py = Q, where P, Q are functions of x, is

$$ye^{\int P dx} = \int Q e^{\int P dx}\, dx + c. \qquad \text{(Art. 112, (b))}$$

Case V. Equation reducible to linear. Rule to solve the equation $\frac{dy}{dx}$ + Py = Qy^n, where P, Q are functions of x :

(*i*) Divide throughout by y^n.

(*ii*) Put $y^{1-n} = z$.

(*iii*) The equation is now linear in z. (Art. 113)

CHAPTER XIII

1. Rule to solve the equation

$$\frac{d^n y}{dx^n} + P_1 \frac{d^{n-1} y}{dx^{n-1}} + P_2 \frac{d^{n-2} y}{dx^{n-2}} + \ldots + p_n y = 0,$$

where $P_1, P_2, \ldots, P_n$ are constants :

(*i*) Write the equation in the symbolic form

$$\left[\text{by putting } \frac{d}{dx} = D, \frac{d^2}{dx^2} = D^2, \ldots, \frac{d^n}{dx^n} = D^n, \textit{i.e.,}\right]$$

$$(D^n + P_1 D^{n-1} + P_2 D^{n-2} + \ldots + P_n)y = 0.$$

(*ii*) Write down the auxiliary equation (A.E.) [by equating to zero the symbolic coefficient of y, *i.e.,*]

$$D^n + P_1 D^{n-1} + P_2 D^{n-2} + \ldots + P_n = 0,$$

and solve for D as if it were an ordinary algebraic quantity.

(*iii*) From the roots of the A.E. write down the corresponding part of the complete solution (C.S.) as follows :

Roots of A.E.	*Corresponding part of C.*
(*a*) (*i*) One real and different root m_1.	$c_1 e^{m_1 x}$.
(*ii*) Two real and equal roots m_1, m_1.	$e^{m_1 x}(c_1 + c_2 x)$.
(*iii*) Three real and equal roots m_1, m_1, m_1.	$e^{m_1 x}(c_1 + c_2 x + c_3 x^2)$.
And so on.	
(*b*) (*i*) One pair of imaginary and different roots of $\alpha \pm i\beta$.	$e^{\alpha x}(c_1 \cos \beta x + c_2 \sin \beta x)$.
(*ii*) Two pairs of imaginary and equal roots $\alpha \pm i\beta, \alpha \pm i\beta$.	$e^{\alpha x}[\cos \beta x . (c_1 + c_2 x) + \sin \beta x . (c_3 + c_4 x)$.
And so on.	

2. Five standard cases of particular integrals.

Case I. Rule to evaluate $\frac{1}{f(D)} e^{ax}$: Put $D = a$. $[f(a) \neq 0]$

Case II. Rule to evaluate $\frac{1}{f(D^2)}$ **cos ax** : Put $D^2 = -a^2$. $[f(-a^2) \neq 0]$

Case III. Rule to evaluate $\frac{1}{f(D^2)}$ **sin ax** : Put $D^2 = -a^2$. $[f(-a^2) \neq 0]$

Case of failure. Rule to find the particular integral in case failure [*i.e.,* when by using the usual rule for finding the particular integral ((2), cases I-III), we get zero in the denominator]:

Case I′. $\dfrac{1}{f(D)} e^{ax} = x \dfrac{1}{\dfrac{d}{dD}[f(D)]} e^{ax}$, when $f(a) = 0$.

Case II′. $\dfrac{1}{f(D^2)} \cos ax = x \dfrac{1}{\dfrac{d}{dD}[f(D^2)]} \cos ax$, when $f(-a^2) = 0$.

Case III′. $\dfrac{1}{f(D^2)} \sin ax = x \dfrac{1}{\dfrac{d}{dD}[f(D^2)]} \sin ax$, when $f(-a^2) = 0$.

$$\left[i.e., x \frac{1}{\text{diff. coeff. of denom. w.r.t. D}} e^{ax} \text{ or } \cos ax \text{ or } \sin ax \text{ as the case may be.} \right]$$

(Art. 123)

Note. *If by using the above rule, again we get zero in the denominator, we repeat the rule, and so on.*

Case IV. Rule to evaluate $\dfrac{1}{f(D)} x^m$, where m is a positive integer : Expand $\dfrac{1}{f(D)}$ in ascending powers of D by the Binomial theorem, taking the expansion of $[1 + \phi(D)]^{-1}$ as far as the term in x^m, and operate on x^m by each term of the expansion of $\dfrac{1}{f(D)}$.

[☞ *The following case V may be omitted by the students of the Panjab University.*]

****Case V. Rule to evaluate $\dfrac{1}{f(D)} (e^{ax} V)$, where V is any function of x :**

$$\frac{1}{f(D)} (e^{ax} V) = e^{ax} \frac{1}{f(D + a)} V.$$

[In words : Take e^{ax} out from under the operator, and in the operator change D to $(D + a)$.] (Art. 121)

3. Rule to solve the equation

$$\frac{d^n y}{dx^n} + P_1 \frac{d^{n-1} y}{dx^{n-1}} + P_2 \frac{d^{n-2} y}{dx^{n-2}} + \ldots + P_n y = Q,$$

where $P_1, P_2, \ldots, P_n$ are constants, and Q is a function of x :

(*i*) Write the equation in the symbolic form

$$\left[\text{by putting } \frac{d}{dx} = D, \frac{d^2}{dx^2} = D^2, \ldots, \frac{d^n}{dx^n} = D^n, i.e., \right]$$

$$(D^n + P_1 D^{n-1} + P_2 D^{n-2} + \ldots + P_n) y = Q.$$

(*ii*) Write down the auxiliary equation (A.E.) [by equating to zero the symbolic coefficient of y, *i.e.*,]

$$D^n + P_1 D^{n-1} + P_2 D^{n-2} + \dots + P_n = 0$$

and solve for D as if it were an ordinary algebraic quantity.

(*iii*) From the roots of the A.E. write down the corresponding part of the complementary function (C.F.) by the same rule as for writing down the corresponding part of the complete solution (C.S.) if the R.H.S. of the given equation were zero (instead of Q).

(*iv*) From step (*i*), find the particular integral

$$\text{P.I.} = \frac{1}{D^n + P_1 D^{n-1} + P_2 D^{n-2} + \dots + P_n} Q$$

$$= \frac{1}{f(D)} Q. \text{ Evaluate by the usual rule.} \qquad \text{(See (2) above.)}$$

(*v*) Then the C.S. is $y = \text{C.F.} + \text{P.I.}$

CHAPTER I

INTEGRALS

SECTION I
INDEFINITE INTEGRAL

1. What is Integral Calculus ?

"Integral Calculus is the inverse (*i.e.*, opposite) of Differential Calculus."

The primary problem of Differential Calculus is :

Given a function ; to find its differential coefficient.

But the primary problem of Integral Calculus is the

Converse : *Given the differential coefficient of a function ; to form the function.*

2. An integral. Definition and notation.

If $\frac{d}{dx}[\phi(x)] = f(x)$, then $\phi(x)$ is called **an integral** of f and is written $\int$* $f(x)\, dx$ [read "integral of $f(x)\, dx$"].

Thus if $\frac{d}{dx}[\phi(x)] = f(x)$ ...(1) [*Diff. coeff. form*]

then $\int f(x)\, dx = \phi(x)$ ...(2) [*Integral form*]

Note. *Method to write a differential coefficient (1) in integral form (2) :*

Remove $\frac{d}{dx}$ *from the L.H.S* †. *and insert* $\int dx$ *on the R.H.S., thus getting*

$$\phi(x) = \int f(x)\, dx.$$

Now interchange sides, i.e., write the R.H.S. on the left and L.H.S. on the right, thus getting

$$\int f(x)\, dx = \phi(x).$$

---

*Historically, the sign $\int$ is a distorted *S*, the first letter of the word because an integral was originally considered as the limit of *a Sum*. In fact Integral Calculus was discovered in an attempt to find the area of a curve by dividing this area into a very large number of very small parts called *elements,* and find the limit of the *sum* of these elements when each element $\to 0$ and number $\to \infty$.

†The letters "*L.H.S.*" stand, as usual, for "*Left Hand Side*" and "*R.H.S.*" for "*Right Hand Side*".

Example. *Write down the integral of 0.*

$$\int 0\, dx = c^*,$$

where c is an arbitrary† constant. $\left[(\text{Art. } 2) \because \dfrac{d}{dx}(c) = 0\right]$

3. Addition of an arbitrary constant.

If $$\frac{\mathbf{d}}{\mathbf{dx}}[\phi(\mathbf{x})] = \mathbf{f}(\mathbf{x}),$$

then $\int \mathbf{f}(\mathbf{x})\, \mathbf{dx} = \phi(\mathbf{x}) + \mathbf{c}$, *where c is an arbitrary constant.*

Proof. $$\frac{d}{dx}[\phi(x) + c] = \frac{d}{dx}[\phi(x)] + 0 \qquad \left[\because \frac{d}{dx}(c) = 0\right]$$

$$= \frac{d}{dx}[\phi(x)] = f(x) \qquad \text{(Given) [Diff. coeff. form]}$$

$$\therefore \qquad \int f(x)\, dx = \phi(x) + c. \qquad \text{[Integral form (Note, Art. 2)]}$$

Note 1. *Method to prove the theorem of Art. 3.*

Differentiate the R.H.S. of the result to be proved, and write the resulting differential coefficient in the integral form. *(See Note, Art. 2)*

The same method will be used to prove two general theorems, viz., theorems II and III of Art. 7.

Note 2. Indefinite integral. *Since $\phi(x) + c$ contains an indefinite constant c, it is called the* ***indefinite integral*** *of $f(x)$.*

Note 3. Integration. Def. *The process of finding the integral is called* ***integration.***

"Find the integral of $f(x)$" is shortened into "integrate $f(x)$".

Integration with respect to x. If there is any likelihood of doubt regarding the variable with respect to which the function is to be integrated, then, for clearness, we say "integrate $f(x)$ **with respect to x.**

Note 4. Constant of integration. Def. *The arbitrary constant c in $\phi(x) + c$ is called the* **constant of integration.**

Customary omission of the arbitrary constant. *The arbitrary constant c is usually omitted in an indefinite integral but is always understood.*

4. Different methods to evaluate an indefinite integral generally answers in different forms. But it will be found that any two answers either identical, or they differ by a constant.

***How to write this step.**

$\int 0\, dx$ = the function whose diff. coeff. with respect to x is 0

$= c.$

i.e., any constant whatever.

Example. $\int \frac{1}{\sqrt{1-x^2}}\,dx = \sin^{-1} x$ ...(1) $\left[\because \frac{d}{dx}(\sin^{-1} x) = \frac{1}{\sqrt{1-x^2}}\right]$

Also $\int \frac{1}{\sqrt{1-x^2}}\,dx = -\cos^{-1} x$ ...(2)

$$\left[\because \frac{d}{dx}(-\cos^{-1} x) = -\left(-\frac{1}{\sqrt{1-x^2}}\right) = \frac{1}{\sqrt{1-x^2}}\right]$$

Caution. From (1) and (2), it does *not* follow that

$$\sin^{-1} x = -\cos^{-1} x.*$$

What follows is that $\sin^{-1} x$ and $-\cos^{-1} x$ *differ by a constant.*

In fact $\sin^{-1} x - (-\cos^{-1} x) = \sin^{-1} x + \cos^{-1} x = \frac{\pi}{2}$. [From Trigonometry]

Note. *Strictly speaking,*

from (1), $\int \frac{1}{\sqrt{1-x^2}}\,dx = \sin^{-1} x + c,$ [End of Note 4, Art. 3]

and from (2), $\int \frac{1}{\sqrt{1-x^2}}\,dx = -\cos^{-1} x + c',$

the arbitrary constants c, c′ being different.

SECTION II
STANDARD FORMS

[☞ *All the standard forms (marked I, II, III, ... in the both are of the greatest importance, and the student is strongly urged to come them to memory.*]

5. I. $\int \mathbf{x^n\,dx} = \frac{\mathbf{x^{n+1}}}{\mathbf{n+1}}$. **[n ≠ – 1†]** $\left[\because \frac{d}{dx}\left(\frac{x^{n+1}}{n+1}\right) = \frac{1}{n+1}(n+1)x^n = x^n\right]$

[Rule to integrate a power of x with a constant index :
Increase the index by 1, and divide by the new index.]

Cor. $\int \mathbf{1\,dx = x}$.

---

*For, if possible, suppose that $\sin^{-1} x = -\cos^{-1} x$.

Then putting $x = 0$ on both sides, we get

$$\sin^{-1} 0 = -\cos^{-1} 0, \quad \text{or} \quad 0 = -\frac{\pi}{2},$$

which is impossible.

† If $n = -1,$

then $\int x^n\,dx = \int x^{-1}\,dx = \int \frac{1}{x}\,dx = \log x.$ $\left[\because \frac{d}{dx}(\log x) = \frac{1}{x}\right]$

(See Art. 5, II.)

Proof. $\int 1\,dx = \int x^0\,dx$ [Form $\int x^n\,dx$ (*Note this step*)]

[Compare with $\int x^n\,dx$, here $n = 0$]

$$= \frac{x^{0+1}}{0+1} \qquad \left[\frac{x^{n+1}}{n+1} \text{ (Art. 5, I)}\right]$$

$$= \frac{x^1}{1} = x.$$

Example. *Write down the integrals of* $x^2, \frac{1}{x^2}, \frac{1}{\sqrt{x}}$.

(*i*) $\int x^2\,dx$ [Compare with $\int x^n\,dx$, here $n = 2$]

$$= \frac{x^{2+1}}{2+1} \qquad \left[\frac{x^{n+1}}{n+1} \text{ (Art. 5, I)}\right]$$

$$= \frac{x^3}{3}.$$

(*ii*) $\int \frac{1}{x^2}\,dx = \int x^{-2}\,dx$ [Form $\int x^n\,dx$ (*Note this step*)]

[Compare with $\int x^n\,dx$, here $n = -2$]

$$= \frac{x^{-2+1}}{-2+1} \qquad \left[\frac{x^{n+1}}{n+1} \text{ (Art. 5, I)}\right]$$

$$= \frac{x^{-1}}{-1} = -\frac{1}{x}.$$

(*iii*) $\int \frac{1}{\sqrt{x}}\,dx = \int \frac{1}{x^{\frac{1}{2}}}\,dx = \int x^{-\frac{1}{2}}\,dx$ [Form $\int x^n\,dx$ (*Note this step*)]

[Compare with $\int x^n\,dx$, here $n = -\frac{1}{2}$]

$$= \frac{x^{-\frac{1}{2}+1}}{-\frac{1}{2}+1} \qquad \left[\frac{x^{n+1}}{n+1} \text{ (Art. 5, I)}\right]$$

$$= \frac{x^{\frac{1}{2}}}{\frac{1}{2}} = 2\sqrt{x}.$$

EXAMPLES

Write down the integrals of (Exs. 1—4) :

1. x^a, x^9, x. **2.** $x^{-9}, \frac{1}{x^{11}}, \frac{1}{x^n}$.

3. $x^{\frac{2}{3}}, x^{\frac{1}{n}}, \sqrt{x}$. **4.** $x^{-\frac{2}{3}}, \frac{1}{x^{\frac{3}{4}}}$.

II. $\int \frac{1}{x}\,dx = \log x.$ $\left[\because \frac{d}{dx}(\log x) = \frac{1}{x}\right]$

Note. *Throughout this book we shall suppose that logarithms are the base e unless the contrary is expressly stated.*

Thus $\log x = \log_e x$; $\mathbf{\log e = \log_e e = 1}$.

****Example.** *From* $\int x^n\,dx = \dfrac{x^{n+1}}{n+1}$, *deduce that* $\int \dfrac{1}{x}\,dx = \log x$.

$$\int x^n\,dx = \frac{x^{n+1}}{n+1} \qquad \text{[Supply the arbitrary constant } c\text{]}$$

$$= \frac{x^{n+1}}{n+1} + c \qquad \left[\text{Subtract and add } \frac{1}{n+1}\right]$$

$$= \frac{x^{n+1}-1}{n+1} + \left(\frac{1}{n+1} + c\right)$$

or
$$\int x^n\,dx = \frac{x^{n+1}-1}{n+1} + c',$$

where $c'\left(=\dfrac{1}{n+1}+c\right)$ is an *arbitrary* constant.

Taking the limits when $n \to -1$, so that $n+1 \to 0$,

$$\int x^{-1}\,dx = \underset{n+1\to 0}{\text{Lt}} \frac{x^{n+1}-1}{n+1} + c'$$

$$\left[\text{Compare with } \underset{x\to 0}{\text{Lt}} \frac{a^x-1}{x}, \text{ here "}x\text{"} = n+1, \text{"}a\text{"} = x\right]$$

$$= \log x + c' \qquad [\log a = c']$$

[Omit the arbitrary constant c']

or
$$\int \frac{1}{x}\,dx = \log x.$$

III. $$\int \mathbf{a^x\,dx} = \frac{\mathbf{a^x}}{\mathbf{\log a}}. \qquad \left[\because \frac{d}{dx}\left(\frac{a^x}{\log a}\right) = \frac{1}{\log a}a^x \log a = a^x\right]$$

Example. *Write down the integral of* 2^x.

$$\int 2^x\,dx \qquad \text{[Compare with } \textstyle\int a^x\,dx, \text{ here } a = 2\text{]}$$

$$= \frac{2^x}{\log 2}. \qquad \left[\frac{a^x}{\log a} \text{ (Art. 5, III)}\right]$$

IV. $$\int \mathbf{e^x\,dx = e^x}. \qquad \left[\because \frac{d}{dx}(e^x) = e^x\right]$$

V. $$\int \mathbf{\sin x\,dx = -\cos x}. \qquad \left[\because \frac{d}{dx}(-\cos x) = -(-\sin x) = \sin x\right]$$

Note. *Throughout this book we shall suppose that angles are measured in circular measure unless the contrary is expressly stated.*

VI. $$\int \mathbf{\cos x\,dx = \sin x}. \qquad \left[\because \frac{d}{dx}(\sin x) = \cos x\right]$$

VII. $$\int \mathbf{\sec^2 x\,dx = \tan x}. \qquad \left[\because \frac{d}{dx}(\tan x) = \sec^2 x\right]$$

VIII. $$\int \mathbf{\text{cosec}^2 x\,dx = -\cot x}. \qquad \left[\because \frac{d}{dx}(-\cot x) = -(-\text{cosec}^2 x) = \text{cosec}^2 x\right]$$

IX. $\int$ **sec x tan x dx = sec x.** $\left[\because \dfrac{d}{dx}(\sec x) = \sec x \tan x\right]$

X. $\int$ **cosec x cot x dx = – cosec x.**

$$\left[\because \frac{d}{dx}(-\operatorname{cosec} x) = -(-\operatorname{cosec} x \cot x) = \operatorname{cosec} x \cot x\right]$$

EXAMPLE

Write down the integrals of $\dfrac{\sin x}{\cos^2 x}, \dfrac{\cos x}{\sin^2 x}$.

(*i*) $\displaystyle\int \frac{\sin x}{\cos^2 x}\, dx$ $\left[\text{Separate } \dfrac{1}{\cos x} \textit{ (Note this step)}\right]$

$$= \int \frac{1}{\cos x} \cdot \frac{\sin x}{\cos x}\, dx = \int \sec x \tan x\, dx = \sec x. \qquad \text{[Art. 5, IX]}$$

(*ii*) $\displaystyle\int \frac{\cos x}{\sin^2 x}\, dx$ $\left[\text{Separate } \dfrac{1}{\sin x} \textit{ (Note this step)}\right]$

$$= \int \frac{1}{\sin x} \cdot \frac{\cos x}{\sin x}\, dx = \int \operatorname{cosec} x \cot x\, dx = -\operatorname{cosec} x. \qquad \text{[Art. 5, X]}$$

[**Aid to memory.** (*i*) **For** $\int a^x\, dx = \dfrac{a^x}{\log a}$.

$$\frac{d}{dx}(a^x) = a^x \log a.$$

Thus for the differential coefficient of a^x, we multiply it by a (constant) $\log a$,

∴ for the integral of a^x which is the inverse (*i.e.*, opposite) of the differential coefficient, we must divide it by $\log a$.

Thus $$\int a^x\, dx = \frac{a^x}{\log a}.$$

Aid to memory. (*ii*) **Rule of "co".** All the results (circular functions) on the L.H.S. of the above list (I to X), which begin with the letters "co", *viz.*, $\cos x$, $\cot x$, $\operatorname{cosec} x$ have a negative sign before them.

Thus $\int \sin x\, dx = -\cos x$, $\int \operatorname{cosec}^2 x\, dx = -\cot x$, $\int \operatorname{cosec} x \cot x\, dx = -\operatorname{cosec} x$.

Reason. The circular functions which begin with the letters "co", *viz.*, $\cos x$, $\cot x$, $\operatorname{cosec} x$ decrease as x increases through the first quadrant,

∴ their differential coefficients are negative.]

6. Defs. In $\int f(x)\, dx$,

(1) the sign $\int$ is called the **integral sign** ;

(2) $f(x)$, the function to be integrated, is called the **integrand** ; and (3) the variable x in dx, with respect to which the function is to be integrated, is called the **independent variable.**

SECTION III

THREE GENERAL THEOREMS ON INDEFINITE INTEGRALS

7. Theorem I. $\frac{d}{dx}[\int f(x)\,dx] = f(x)$.

[In words : *The differential coefficient of the integral of a function is equal to the function itself.*]

****Proof.** Let $\int f(x)\,dx = \phi(x)$ ...(1) [*Integral form*]

Then $\frac{d}{dx}[\phi(x)] = f(x)$ ...(2) [*Diff. coeff. form* (Note, Art. 2)]

Substituting the value of $\phi(x)$ from (1) in (2),

$$\frac{d}{dx}[\int f(x)\,dx] = f(x).$$

Theorem II. If a is a constant and u is a function of x, then

$$\int a\,u\,dx = a\int u\,dx.$$

[In words : *The integral of the product of a constant and a function is equal to the product of the constant and the integral of the function.*

****Proof.** $\frac{d}{dx}\left[a\int u\,dx\right] = a\frac{d}{dx}\left[\int u\,dx\right]$ $\left[\because \frac{d}{dx}(c\,u) = c\frac{du}{dx}\right]$

$= a\,u$ $\left[\because \frac{d}{dx}[\int f(x)dx]\right] = f(x)$ (Theorem I)

[Diff. coeff. form]

$\therefore$ $\int a\,u\,dx = a\int u\,dx.$ [Integral form (Note, Art. 2)]

Note. Method to prove theorem II. See Note 1, Art. 3.

Example. *Write down the integral of* $\frac{x}{a}$.

$$\int \frac{x}{a}\,dx = \frac{1}{a}\int x\,dx \qquad \text{[Theorem II]}$$

$$= \frac{1}{a}\frac{x^2}{2} = \frac{x^2}{2a}.$$

EXAMPLE

Write down the integrals of ax, $\frac{a}{x}$.

Theorem III. If u, v, w,.... are functions of x, then

$$\int(\mathbf{u} + \mathbf{v} - \mathbf{w} + \ldots)\,\mathbf{dx} = \int \mathbf{u}\,\mathbf{dx} + \int \mathbf{v}\,\mathbf{dx} - \int \mathbf{w}\,\mathbf{dx} + \ldots .$$

[In words : *The integral of the algebraic sum of a finite number of functions is equal to the same algebraic sum of their integrals.*]

****Proof.** $\frac{d}{dx}[\int u\,dx + \int v\,dx - \int w\,dx + \ldots\ldots.]$

$$= \frac{d}{dx}(\int u\,dx) + \frac{d}{dx}(\int v\,dx) - \frac{d}{dx}(\int w\,dx) + \ldots$$

$$\left[\because \frac{d}{dx}(u+v-w+\ldots) = \frac{du}{dx} + \frac{dv}{dx} - \frac{dw}{dx} + \ldots\right]$$

$$= u + v - w + \ldots \qquad \left[\because \frac{d}{dx}[\int f(x)\,dx] = f(x) \text{(Theorem I)}\right]$$

[Diff. coeff. form]

$\therefore \int (u + v - w + \ldots.)\,dx = \int u\,dx + \int v\,dx - \int w\,dx + \ldots.$ [Integral form (Note, Art. 2)]

Note. Method to prove theorem III. See Note 1, Art. 3.

Example. *Write down the integral of* $\frac{a + bx + cx^2}{x^2}$.

$$\int \frac{a + bx + cx^2}{x^2}\,dx = \int \left[\frac{a}{x^2} + \frac{bx}{x^2} + \frac{cx^2}{x^2}\right] dx$$

$$= \int \left[\frac{a}{x^2} + \frac{b}{x} + c\right] dx$$

$$= \int \frac{a}{x^2}\,dx + \int \frac{b}{x}\,dx + \int c\,dx \qquad \text{[Theorem III]}$$

$$= a\int x^{-2}\,dx + b\int \frac{1}{x}\,dx + c\int 1\,dx \qquad \text{[Theorem II]}$$

$$= a\,\frac{x^{-1}}{-1} + b\log x + cx = -\frac{a}{x} + b\log x + cx.$$

[**Rule to integrate a fraction whose numerator contains an infinite number of terms and the denominator contains only one term or one product :**

Divide each term of the numerator by the denominator, and integrate term by term.]

EXAMPLES

Write down the integrals of (Exs. 1—3) :

1. (*i*) $\frac{a + x}{x}$. (*ii*) $a + bx - \frac{c}{x^3}$.

(*iii*) $\frac{a + bx^{-1} + cx^{-2}}{x^{-2}}$.

2. (*i*) $x^3 - 3^x$. (*ii*) $ax + a^x - x^a$.

3. $\cos^2 \frac{x}{2}$.

$$\int \cos^2 \frac{x}{2}\,dx = \int \frac{1 + \cos x}{2}\,dx$$

$$= \frac{1}{2}(x + \sin x).$$

$$\cos x = 2\cos^2 \frac{x}{2} - 1$$

$$\therefore \cos^2 \frac{x}{2} = \frac{1 + \cos x}{2}.$$

Note. Multiple angle method. *If the square or cube or any positive integral power of a sine or a cosine (or the product of such powers of a sine and a cosine) occurs in the integrand, it is generally useful to express the power (or the product of the powers) in a series of sines or cosines of multiple angles (by using the 2A or 3A formulae).*

4. Integrate (*i*) $\int \tan^2 x\, dx$. (*ii*) $\int \cot^2 x\, dx$.

(*i*) $\int \tan^2 x\, dx = \int (\sec^2 x - 1)\, dx$

$= \tan x - x.$

(*ii*) $\int \cot^2 x\, dx = \int (\text{cosec}^2 x - 1)\, dx$

$= -\cot x - x.$

$\sec^2 x = 1 + \tan^2 x$
$\therefore \tan^2 x = \sec^2 x - 1$
$\text{cosec}^2 x = 1 + \cot^2 x$
$\therefore \cot^2 x = \text{cosec}^2 x - 1$

Note. Important. *If* $\tan^2 x$ *occurs in the integrand, replace it by* $\sec^2 x - 1$, *if* $\cot^2 x$ *occurs, replace it by* $\text{cosec}^2 x - 1$.

Note on hyperbolic functions.

8. (*a*) Hyperbolic functions.

There are **six hyperbolic functions :**

(1) the *hyperbolic sine* of x which is written **sinh x** [read "shine x"]

(2) the *hyperbolic cosine* of x which is written **cosh x,**

(3) the *hyperbolic tangent* of x which is written **tanh x** [read "than x" (थैन x)],

(4) the *hyperbolic cotangent* of x which is written **coth x,**

(5) the *hyperbolic secant* of x which is written **sech x** [read "she-cant x"],

(6) the *hyperbolic cosecant* of x which is written **cosech x** [read "co-she-cant x"].

[Compare the six circular functions of x in Trigonometry, *viz.*, $\sin x$, $\cos x$, $\tan x$, $\cot x$, $\sec x$, $\text{cosec}\, x$.]

8. (*b*) Definitions of hyperbolic functions.

(1) $\sinh x = \dfrac{e^x - e^{-x}}{2}$.

(2) $\cosh x = \dfrac{e^x + e^{-x}}{2}$.

(3) $\tanh x = \dfrac{\sinh x}{\cosh x}$.

(4) $\coth x = \dfrac{\cosh x}{\sinh x}$.

(5) $\text{sech}\, x = \dfrac{1}{\cosh x}$.

(6) $\text{cosech}\, x = \dfrac{1}{\sinh x}$.

[**Aid to memory for defs. (3) to (6).**

$$\tan x = \frac{\sin x}{\cos x}\ ;\ \tanh x = \frac{\sinh x}{\cosh x}.$$

$$\cot x = \frac{\cos x}{\sin x}\ ;\ \coth x = \frac{\cosh x}{\sinh x}.$$

$$\sec x = \frac{1}{\cos x}\ ;\ \text{sech}\, x = \frac{1}{\cosh x}.$$

$$\text{cosec}\, x = \frac{1}{\sin x}\ ;\ \text{cosech}\, x = \frac{1}{\sinh x}\Bigg].$$

Cor. sinh 0 = 0 ; cosh 0 = 1.

Proofs. (*i*) $\sinh 0 = \dfrac{e^0 - e^{-0}}{2}$ $\left[\dfrac{e^x - e^{-x}}{2}\right]$

$= \dfrac{1-1}{2} = \dfrac{0}{2} = 0.$

(*ii*) $\cosh 0 = \dfrac{e^0 + e^{-0}}{2}$ $\left[\dfrac{e^x + e^{-x}}{2}\right]$

$= \dfrac{1+1}{2} = \dfrac{2}{2} = 1.$

[**Aid to memory.** $\sin 0 = 0$; $\sinh 0 = 0$.

$\cos 0 = 1$; $\cosh 0 = 1$.]

8. (*c*) **Fundamental square relation : $\cosh^2 x - \sinh^2 x = 1$.**

Proof. $\cosh^2 x - \sinh^2 x$

$$= \left(\frac{e^x + e^{-x}}{2}\right)^2 - \left(\frac{e^x - e^{-x}}{2}\right)^2 \qquad [\text{Form } a^2 - b^2]$$

$$= \left(\frac{e^x + e^{-x}}{2} + \frac{e^x - e^{-x}}{2}\right)\left(\frac{e^x + e^{-x}}{2} - \frac{e^x - e^{-x}}{2}\right) \qquad [(a+b)(a-b)]$$

$$= \left(\frac{e^x + e^{-x} + e^x - e^{-x}}{2}\right)\left(\frac{e^x + e^{-x} - e^x + e^{-x}}{2}\right)$$

$$= e^x \cdot e^{-x} = e^0 = 1.$$

Rule to write down a formula in hyperbolic functions corresponding to a formula in circular functions of general angles :

In the formula in circular functions, change sin x to i sinh x, cos x to cosh x,* $\therefore$ *tan x* $\left(= \dfrac{\sin x}{\cos x}\right)$ *to* $\left(\dfrac{i \sinh x}{\cosh x} = \right)$ *i tanh x, and similarly the other circular functions.*

[From Trigonometry]

**i* stands, as usual, for $\sqrt{-1}$.

Examples.

Formula in circular functions	Corresponding formula in hyperbolic functions
(1) $\cos^2 x + \sin^2 x = 1.$	(1) $\cosh^2 x + (i \sinh x)^2 = 1$ or $\cosh^2 x - \sinh^2 x = 1.$
(2) $\tan^2 x = \sec^2 x - 1.$	(2) $(i \tanh x)^2 = \text{sech}^2 x - 1$ or $-\tanh^2 x = \text{sech}^2 x - 1$ or $\tanh^2 x = 1 - \text{sech}^2 x.$
(3) $\sin x = 2 \sin \frac{x}{2} \cos \frac{x}{2}.$	(3) $i \sinh x = 2\left(i \sinh \frac{x}{2}\right) \cosh \frac{x}{2}$ or $\sinh x = 2 \sinh \frac{x}{2} \cosh \frac{x}{2}.$
(4) $\cos x = \cos^2 \frac{x}{2} - \sin^2 \frac{x}{2}.$	(4) $\cosh x = \cosh^2 \frac{x}{2} - \left(i \sinh \frac{x}{2}\right)$ $= \cosh^2 \frac{x}{2} - \left(-\sinh^2 \frac{x}{2}\right)$ $= \cosh^2 \frac{x}{2} + \sinh^2 \frac{x}{2}.$

8. (*d*) **Expressions for inverse hyperbolic functions logarithms.**

1. $\mathbf{\sinh^{-1} x = \log (x + \sqrt{1+x^2}).}$
2. $\mathbf{\cosh^{-1} x = \log (x + \sqrt{x^2 - 1}).}$
3. $\mathbf{\tanh^{-1} x = \frac{1}{2} \log \frac{1+x}{1-x}.}$ [From Trigonometry]

8. (*e*) **Differential coefficients of hyperbolic functions.**

1. $\mathbf{\frac{d}{dx} (\sinh x) = \cosh x.}$ 2. $\mathbf{\frac{d}{dx} (\cosh x) = \sinh x.}$
3. $\mathbf{\frac{d}{dx} (\tanh x) = sech^2 x.}$

Proofs. (1) $\frac{d}{dx} (\sinh x) = \frac{d}{dx}\left[\frac{e^x - e^{-x}}{2}\right] = \frac{1}{2}[e^x - e^{-x}(-1)]$

$$= \tfrac{1}{2}[e^x + e^{-x}] = \cosh x.$$

(2) $\frac{d}{dx}(\cosh x) = \frac{d}{dx}\left[\frac{e^x + e^{-x}}{2}\right] = \frac{1}{2}[e^x + e^{-x}(-1)]$

$$= \tfrac{1}{2}[e^x - e^{-x}] = \sinh x.$$

(3) $\frac{d}{dx}(\tanh x) = \frac{d}{dx}\left[\frac{\sinh x}{\cosh x}\right]$ $\left[\text{Form } \frac{d}{dx}\left(\frac{u}{v}\right)\right]$

$$= \frac{\cosh x \,.\, \cosh x - \sinh x \,.\, \sinh x}{\cosh^2 x}$$

$$= \frac{\cosh^2 x - \sinh^2 x}{\cosh^2 x} = \frac{1}{\cosh^2 x} = \text{sech}^2 x.$$

EXAMPLES

Establish the following results (Exs. 1—3) :

1. $\dfrac{d}{dx}(\coth x) = -\operatorname{cosech}^2 x.$ **2.** $\dfrac{d}{dx}(\operatorname{sech} x) = -\operatorname{sech} x \tanh x.$

3. $\dfrac{d}{dx}(\operatorname{cosech} x) = -\operatorname{cosech} x \coth x.$

Aid to memory for the differential coefficients of the hyperbolic functions. The differential coefficient of a hyperbolic function corresponds exactly to that of its corresponding circular function *except for cosh x and (its reciprocal) sinh x whose differential coefficients correspond to those of their corresponding circular functions*, **with the signs changed.**

Thus (1) $\dfrac{d}{dx}(\sin x) = \cos x$; $\dfrac{d}{dx}(\sinh x) = \cosh x.$

(2) $\dfrac{d}{dx}(\cos x) = -\sin x$; $\mathbf{\dfrac{d}{dx}(\cosh x) = -(-\sinh x) = \sinh x.}$

(3) $\dfrac{d}{dx}(\tan x) = \sec^2 x$; $\dfrac{d}{dx}(\tanh x) = \operatorname{sech}^2 x.$

(4) $\dfrac{d}{dx}(\cot x) = -\operatorname{cosec}^2 x$; $\dfrac{d}{dx}(\coth x) = -\operatorname{cosech}^2 x.$

(5) $\dfrac{d}{dx}(\sec x) = \sec x \tan x$; $\mathbf{\dfrac{d}{dx}(\operatorname{sech} x) = -\operatorname{sech} x \tanh x.}$

(6) $\dfrac{d}{dx}(\operatorname{cosec} x) = -\operatorname{cosec} x \cot x$;

$\dfrac{d}{dx}(\operatorname{cosech} x) = -\operatorname{cosech} x \coth x.$]

Two more standard forms.

9. XI. $\boldsymbol{\int \sinh x \, dx = \cosh x.}$ $\left[\because \dfrac{d}{dx}(\cosh x) = \sinh x \text{ (Art. 8, } (e), (2))\right]$

XII. $\boldsymbol{\int \cosh x \, dx = \sinh x.}$ $\left[\because \dfrac{d}{dx}(\sinh x) = \cosh x \text{ (Art. 8, } (e), (1))\right]$

Caution. The rule of "co" for the standard forms of integrals of circular functions (Aid to memory, *(ii)*, Art. 5) does *not* apply to standard forms of integrals in hyperbolic functions.

Thus $\int \sin x \, dx = -\cos x$, but $\int \sinh x \, dx = \cosh x$ (*not* $-\cosh x$).

EXAMPLES

Prove the following results (Exs. 1—4) :

1. $\int \operatorname{sech}^2 x \, dx = \tanh x.$

$\int \operatorname{sech}^2 x \, dx = \tanh x.$ $\left[\because \dfrac{d}{dx}(\tanh x) = \operatorname{sech}^2 x \text{ (Art. 8, } (e), (3))\right]$

2. $\int \text{cosech}^2 x\, dx = -\coth x.$

3. (*i*) $\int \text{sech}\, x \tanh x\, dx = -\text{sech}\, x.$

(*ii*) $\int \text{cosech}\, x \coth x\, dx = -\text{cosech}\, x.$

4. (*i*) $\int \frac{\sinh x}{\cosh^2 x}\, dx = -\text{sech}\, x.$ (*ii*) $\int \frac{\cosh x}{\sinh^2 x}\, dx = -\text{cosech}\, x.$

[**Aid to memory for the standard forms of integrals in hyperbolic functions.** Recollect the differential coefficients of the hyperbolic functions (see Aid to memory given after Ex. 3, Art. 8, (*e*)), and hence write down the integral.

Thus to write down $\int \sinh x\, dx$, we recollect $\frac{d}{dx}(\cosh x) = \sinh x,$

$\therefore \int \sinh x\, dx = \cosh x.$

Similarly for the other standard forms.]

SECTION IV

DEFINITE INTEGRAL

10. Definite integral. Definition and notation.

If $\int f(x)\, dx = \phi(x)$, then $\phi(b) - \phi(a)$ is called the **definite integral** of $f(x)$ between the limits a and b, and is written $\int_a^b f(x)\, dx$ [read "integral, from a to b, of $f(x)\, dx$"].

a is called the **lower limit** and b the **upper limit.**

Notation. $\phi(b) - \phi(a)$ is denoted by $[\phi(x)]_a^b$ [read "$\phi(x)$ from a to b"].

Thus $$\int_a^b \mathbf{f(x)\, dx} = \Big[\phi(\mathbf{x})\Big]_a^b = \phi(\mathbf{b}) - \phi(\mathbf{a}).$$

Note 1. Why the name "definite integral" ? *If instead of* $\int f(x)\, dx = \phi(x)$, *we take* $\int f(x)\, dx = \phi(x) + c$, *then*

$$\int_a^b f(x)\, dx = \Big[\phi(x) + c\Big]_a^b = [\phi(b) + c] - [\phi(a) + c] = \phi(b) - \phi(a),$$

so that the indefinite constant c disappears, and $\int_a^b f(x)\, dx$ *has a definite value* $[\phi(b) - \phi(a)]$. *Hence the name* **"definite integral"**.

[**Rule to evaluate** $\int_a^b \mathbf{f(x)\, dx}$ **:**

(*i*) *Integrate* $\int f(x)\, dx$, *omitting the arbitrary constant.*

(*ii*) *In this integral first substitute* x = *upper limit* (*b*), *and then substitute x = lower limit*

(a), and subtract the second result from the first.]

Note 2. Notation. *∫ dx is really one symbol denoting integration w.r.t.* x. It may, however, be regarded as a pair of brackets between which is inserted the function to be integrated, thus ∫ f(x) dx.*

But, for convenience

$$\int \frac{f(x)}{\phi(x)}\,dx \text{ is written } \int \frac{f(x)\,dx}{\phi(x)}, \text{ and } \int \frac{1}{\phi(x)}\,dx \text{ is written } \int \frac{dx}{\phi(x)}.$$

Thus $\int \frac{dx}{x} = \int \frac{1}{x}\,dx.$ [See Ex. 1, (*iii*) following.]

Example. *Write down the values of*

(*i*) $\int_a^b x^n\,dx.$ (*ii*) $\int_0^1 e^x\,dx.$ (*iii*) $\int_{-\frac{\pi}{2}}^{\frac{\pi}{2}} \cos x\,dx.$

(*i*) $\int_a^b x^n\,dx = \left[\frac{x^{n+1}}{n+1}\right]_a^b = \frac{1^*}{n+1}[b^{n+1} - a^{n+1}].$

***Note.** *A constant factor can be taken out from within the square brackets [ ].*

For $$\int_a^b x^n\,dx = \left[\frac{x^{n+1}}{n+1}\right]_a^b = \frac{b^{n+1}}{n+1} - \frac{a^{n+1}}{n+1} = \frac{1}{n+1}[b^{n+1} - a^{n+1}] \qquad ...(1)$$

and if we take out the constant factor $\frac{1}{n+1}$ from within the square brackets [ ], we get

$$\int_a^b x^n\,dx = \left[\frac{x^{n+1}}{n+1}\right]_a^b = \frac{1}{n+1}\left[x^{n+1}\right]_a^b = \frac{1}{n+1}[b^{n+1} - a^{n+1}],$$

which is the same as (1).

(*ii*) $\int_0^1 e^x\,dx = \left[e^x\right]_0^1 = e^1 - e^0 = e - 1.$

(*iii*) $$\int_{-\frac{\pi}{2}}^{\frac{\pi}{2}} \cos x\,dx = \left[\sin x\right]_{-\frac{\pi}{2}}^{\frac{\pi}{2}} = \sin\frac{\pi}{2} - \sin\left(-\frac{\pi}{2}\right) \qquad \left[\text{But } \sin\left(-\frac{\pi}{2}\right) = -\sin\frac{\pi}{2} = -1\right]$$

$$= 1 - (-1) = 2.$$

---

*The letters "w.r.t." stand, as usual, for "*with respect to*".

EXAMPLES

Evaluate (Exs. 1 - 3) :

1. (*i*) $\int_0^1 x^{99}\, dx.$ (*ii*) $\int_1^2 x^n\, dx.$ (*iii*) $\int_2^3 \frac{dx}{x}.$

2. (*i*) $\int_0^{\frac{\pi}{2}} \sin x\, dx.$ (*ii*) $\int_0^{\frac{\pi}{3}} \sec x \tan x\, dx.$ (*iii*) $\int_0^{\frac{\pi}{4}} \sec^2 x\, dx.$

3. $\int_a^b (x + \cos x)\, dx.$

SECTION V

DEFINITE INTEGRALS WITH INFINITE LIMITS

11. In the definition of $\int_a^b f(x)\, dx$ (Art. 10), it is assumed that a, b are finite. If, however, one of them is infinite*, we have the following definitions.

Def. 1. [*Upper limit infinite.*]

$$\int_a^\infty f(x)\, dx = \underset{b \to \infty}{\text{Lt}} \int_a^b f(x)\, dx,$$

provided the limit is a definite number.

Note 1. *Remember : There is no number 'infinity'.*

Note 2. Integral does not exist. Def. *If the limit is not a definite number, the integral is said* **not to exist**.

Def. 2. [*Lower limit infinite.*]

$$\int_{-\infty}^b f(x)\, dx = \underset{a \to -\infty}{\text{Lt}} \int_a^b f(x)\, dx,$$

provided the limit is a definite number.

Practical form. Def. 1′. [*Upper limit infinite.*]

If $\int f(x)\, dx = \phi(x),$

then
$$\int_a^\infty f(x)\, dx = \Big[\phi(x)\Big]_a^\infty = \underset{x \to \infty}{\text{Lt}}\ \phi(x) - \phi(a),$$

provided the limit is a definite number.

For, from def. 1,

$$\int_a^\infty f(x)\, dx = \underset{b \to \infty}{\text{Lt}} \int_a^b f(x)\, dx = \underset{b \to \infty}{\text{Lt}} \Big[\phi(x)\Big]_a^b = \underset{b \to \infty}{\text{Lt}}\ [\phi(b) - \phi(a)]$$

$$= \underset{b \to \infty}{\text{Lt}}\ \phi(b) - \phi(a) = \underset{x \to \infty}{\text{Lt}}\ \phi(x) - \phi(a).$$

---

*"infinite" means + ∞ or – ∞.

Similarly we have

Def. 2′. [*Lower limit infinite.*]

If $$\int f(x)\,dx = \phi(x),$$

then $$\int_{-\infty}^{b} f(x)\,dx = \Big[\phi(x)\Big]_{-\infty}^{b} = \phi(b) - \underset{x\to-\infty}{\text{Lt}}\ \phi(x),$$

provided the limit is a definite number.

EXAMPLES

1. Find $\int_1^\infty \frac{dx}{x^4}$.

$$\int_1^\infty \frac{1}{x^4}\,dx = \int_1^\infty x^{-4}\,dx = \left[\frac{x^{-3}}{-3}\right]_1^\infty = -\frac{1}{3}\left[\frac{1}{x^3}\right]_1^\infty = -\frac{1}{3}\left[\underset{x\to\infty}{\text{Lt}}\ \frac{1}{x^3} - 1\right]$$

$$= -\frac{1}{3}[0-1] \qquad \left[\because \frac{1}{\infty^3} = \frac{1}{\infty} = 0\right]$$

$$= \frac{1}{3}.$$

2. Find $\int_1^\infty \frac{dx}{x}$.

$$\int_1^\infty \frac{1}{x}\,dx = \Big[\log x\Big]_1^\infty = \underset{x\to\infty}{\text{Lt}}\ \log x - \log 1 = \log\infty - \log 1 = \infty - 0 = \infty,$$

which is not a definite number.

$\therefore$ the integral does not exist. [Def. (see Note 2, Art. 11)]

MISCELLANEOUS EXAMPLES ON CHAPTER I

Write down the integrals of (Exs. 1—4) :

1. (*i*) $\sqrt{x}\,(x^2 + 2x + 3)$. (*ii*) $\frac{4x^2 + 3x + 2}{x^3}$. (*iii*) $\left(x + \frac{1}{x}\right)^3$.

2. $\frac{(a^x + b^x)^2}{a^x b^x}$.

3. (*i*) $\sin x - \cos x$. (*ii*) $\sec x \tan x - 5\,\text{cosec}^2\, x$.

4. (*i*) $\sin x \sec^2 x$. (*ii*) $\frac{\sin^2 x - \cos^2 x}{\sin^2 x \cos^2 x}$.

5. Integrate $\int \frac{1}{1+\sin x}\,dx$.

$\int \frac{1}{1+\sin x}\, dx$ [Multiply the num. and denom. by $1 - \sin x$*]

$$= \int \frac{1-\sin x}{(1+\sin x)(1-\sin x)}\, dx = \int \frac{1-\sin x}{1-\sin^2 x}\, dx = \int \frac{1-\sin x}{\cos^2 x}\, dx$$

$$= \int \left[\frac{1}{\cos^2 x} - \frac{\sin x}{\cos^2 x}\right] dx$$

$$\left|\ \int \frac{\sin x}{\cos^2 x}\, dx = \int \frac{1}{\cos x} \cdot \frac{\sin x}{\cos x}\, dx = \int \sec x \tan x.\right.$$

$$= \int (\sec^2 x - \sec x \tan x)\, dx = \tan x - \sec x.$$

6. Show that, when $f(x)$ is of the form $a + bx + cx^2$,

$$\int_0^1 f(x)\, dx = \frac{1}{6}\{f(0) + 4f(\tfrac{1}{2}) + f(1)\}.$$

***Why this step.** In $\int \frac{1}{1+\sin x}\, dx$, the denominator is $1 + \sin x$, $\therefore$ in order to get $1 - \sin^2 x = \cos^2 x$ (one term) in the denominator, we multiply the numerator and denominator by $1 - \sin x$.

Had the integral been $\int \frac{1}{1-\sin x}\, dx$, we should have multiplied the numerator and denominator by $1 + \sin x$.

CHAPTER II

INTEGRATION BY SUBSTITUTION

SECTION I

CHANGE OF INDEPENDENT VARIABLE IN AN INDEFINITE INTEGRAL

12. To change the independent variable in $\int f(x)\, dx$ by the transformation $x = \phi(z)$.

Let $\qquad I = \int f(x)\, dx.$

Then $\qquad \dfrac{dI}{dx} = \dfrac{d}{dx}\left[\int f(x)\, dx\right] = f(x).$ [Theorem I, Art. 7]

Now $\qquad \dfrac{dI}{dz}^{*} = \dfrac{dI}{dx} \cdot \dfrac{dx}{dz} = f(x)\dfrac{dx}{dz}$ ...(1)

But $\qquad x = \phi(z)$ [Given],

$\therefore \qquad \dfrac{dx}{dz} = \phi'(z)$

$\therefore$ from (1), $\qquad \dfrac{dI}{dz} = f[\phi(z)]\, \phi'(z).$

Integrating both sides w.r.t. z,

$$I = \int f[\phi(z)]\, \phi'(z)\, dz$$

i.e., $$\int f(x)\, dx = \int f[\phi(z)]\, \phi'(z)\, dz.$$

[Rule to evaluate $\int f(x)\, dx$ by the substitution $x = \phi(z)$:

(*i*) *In the integrand, put $x = \phi(z)$ ...(1), and $dx = \phi'(z)\, dz$ [instead of $x = \phi'(z)$].*

(*ii*) *Evaluate the resulting integral in z.*

(*iii*) *Express the result in terms of x (from (1)).*]

Five standard forms.

13. To integrate

(*i*) $\displaystyle\int \frac{dx}{\sqrt{a^2 - x^2}}$. $\qquad$ (*ii*) $\displaystyle\int \frac{dx}{a^2 + x^2}$. $\qquad$ (*iii*) $\displaystyle\int \frac{dx}{x\sqrt{x^2 - a^2}}$.

(*i*) To integrate $\displaystyle\int \frac{1}{\sqrt{a^2 - x^2}}\, dx.$

Put $x = a \sin\theta$ $\qquad$ ...(1), then $dx = a \cos\theta\, d\theta$.

***Why this step.** Since we require an integral in which the independent variable is z, $\therefore$ we consider $\dfrac{dI}{dz}$.

$$\therefore \quad \int \frac{1}{\sqrt{a^2 - x^2}}\, dx = \int \frac{1}{\sqrt{a^2 - a^2 \sin^2 \theta}}\, a \cos \theta \, d\theta = \int \frac{1}{\sqrt{a^2 (1 - \sin^2 \theta)}}\, a \cos \theta \, d\theta$$

$$= \int \frac{1}{a \cos \theta}\, a \cos \theta \, d\theta = \int 1 \, d\theta = \theta = \sin^{-1} \frac{x}{a}.$$

$$\left[\because \text{ from (1), } \sin \theta = \frac{x}{a}, \therefore \ \theta = \sin^{-1} \frac{x}{a}\right]$$

(*ii*) To integrate $\int \frac{1}{a^2 + x^2}\, dx$.

Put $x = a \tan \theta$ ...(1), then $dx = a \sec^2 \theta \, d\theta$.

$$\therefore \quad \int \frac{1}{a^2 + x^2}\, dx = \int \frac{1}{a^2 + a^2 \tan^2 \theta}\, a \sec^2 \theta \, d\theta$$

$$= \int \frac{1}{a^2(1 + \tan^2 \theta)}\, a \sec^2 \theta \, d\theta$$

$$= \int \frac{1}{a \sec^2 \theta} \sec^2 \theta \, d\theta = \frac{1}{a} \int 1 \, d\theta = \frac{1}{a}\, \theta = \frac{1}{a} \tan^{-1} \frac{x}{a}$$

$$\left[\because \text{ from (1), } \tan \theta = \frac{x}{a}, \therefore \ \theta = \tan^{-1} \frac{x}{a}\right]$$

(*iii*) To integrate $\int \frac{1}{x\sqrt{x^2 - a^2}}\, dx$.

Put $x = a \sec \theta$ ...(1), then $dx = a \sec \theta \tan \theta \, d\theta$.

$$\therefore \quad \int \frac{1}{x\sqrt{x^2 - a^2}}\, dx = \int \frac{1}{a \sec \theta \sqrt{a^2 \sec^2 \theta - a^2}}\, a \sec \theta \tan \theta \, d\theta$$

$$= \int \frac{1}{\sqrt{a^2(\sec^2 \theta - 1)}} \tan \theta \, d\theta = \int \frac{1}{a \tan \theta} \tan \theta \, d\theta$$

$$= \frac{1}{a} \int 1 \, d\theta = \frac{1}{a}\, \theta = \frac{1}{a} \sec^{-1} \frac{x}{a}$$

$$\left[\because \text{ from (1), } \sec \theta = \frac{x}{a}, \therefore \ \theta = \sec^{-1} \frac{x}{a}\right]$$

Hence we have

Three more standard forms.

XIII. $\int \frac{1}{\sqrt{a^2 - x^2}}\, dx = \sin^{-1} \frac{x}{a}$.

XIV. $\int \frac{1}{a^2 + x^2}\, dx = \frac{1}{a} \tan^{-1} \frac{x}{a}$.

XV. $\int \frac{1}{x\sqrt{x^2 - a^2}}\, dx = \frac{1}{a} \sec^{-1} \frac{x}{a}$.

Example. *Integrate* $\dfrac{1}{\sqrt{9-4x^2}}$.

$$\int \frac{1}{\sqrt{9-4x^2}}\,dx$$

[Take the numerical value of coeff. of x^2 outside the quadratic (*Note this step*)]

$$= \int \frac{1}{\sqrt{4}\sqrt{\frac{9}{4}-x^2}}\,dx = \frac{1}{2}\int \frac{1}{\sqrt{(\frac{3}{2})^2-x^2}}\,dx$$

$$\left[\text{Compare with} \int \frac{1}{\sqrt{a^2-x^2}}\,dx, \text{ here } a = \frac{3}{2}\right]$$

$$= \frac{1}{2}\sin^{-1}\frac{x}{\frac{3}{2}} = \frac{1}{2}\sin^{-1}\frac{2x}{3}. \qquad \left[\sin^{-1}\frac{x}{a} \text{ (Art. 13, XIII)}\right]$$

EXAMPLES

Write down the indefinite integrals of (Exs. 1—2) :

1. (*i*) $\dfrac{1}{\sqrt{1-x^2}}$. (*ii*) $\dfrac{1}{9+x^2}$. (*iii*) $\dfrac{1}{x\sqrt{x^2-4}}$.

2. (*i*) $\dfrac{1}{4x^2+9}$. (*ii*) $\dfrac{1}{x\sqrt{3x^2-12}}$.

3. Integrate $\displaystyle\int \frac{x^2}{4+x^2}\,dx$.

$$\left[\int \frac{x^2}{4+x^2}\,dx = \int \frac{(4+x^2)-4^*}{4+x^2}\,dx = \int\left(1-\frac{4}{4+x^2}\right)dx.\right]$$

14. To integrate (*i*) $\displaystyle\int \frac{\mathbf{dx}}{\sqrt{\mathbf{a^2+x^2}}}$. (*ii*) $\displaystyle\int \frac{\mathbf{dx}}{\sqrt{\mathbf{x^2-a^2}}}$.

(*i*) To integrate $\displaystyle\int \frac{1}{\sqrt{a^2+x^2}}\,dx$.

Put $x = a\sinh\theta$ ...(1), then $dx = a\cosh\theta\, d\theta$.

$$\therefore \quad \int \frac{1}{\sqrt{a^2+x^2}}\,dx = \int \frac{1}{\sqrt{a^2+a^2\sinh^2\theta}}\,a\cosh\theta\,d\theta$$

$$= \int \frac{1}{\sqrt{a^2(1+\sinh^2\theta)}}\,a\cosh\theta\,d\theta$$

***How to write this step.** Write in the numerator the quadratic which occurs in the denominator, *viz.*, $4 + x^2$, and adjust the coefficient of x^2 ; now adjust the constant term [by subtracting 4], thus getting $x^2 = (4 + x^2) - 4$.

The same method will be used for $\displaystyle\int \frac{x^2}{\sqrt{1+x^2}}\,dx$. See Ex. 4, (*ii*), Art. 24.

$$= \int \frac{1}{a \cosh \theta} a \cosh \theta \, d\theta = \int 1 \, d\theta = \theta = \sinh^{-1} \frac{x}{a}$$

$$\left[\because \text{ from (1), } \sinh \theta = \frac{x}{a}, \therefore \ \theta = \sinh^{-1} \frac{x}{a}\right]$$

(*ii*) To integrate $\int \frac{1}{\sqrt{x^2 - a^2}} dx$.

Put $x = a \cosh \theta$ ...(1), then $dx = a \sinh \theta \, d\theta$.

$$\therefore \quad \int \frac{1}{\sqrt{x^2 - a^2}} dx = \int \frac{1}{\sqrt{a^2 \cosh^2 \theta - a^2}} a \sinh \theta \, d\theta$$

$$= \int \frac{1}{\sqrt{a^2 (\cosh^2 \theta - 1)}} a \sinh \theta \, d\theta$$

$$= \int \frac{1}{a \sinh \theta} a \sinh \theta \, d\theta = \int 1 \, d\theta = \theta = \cosh^{-1} \frac{x}{a}$$

$$\left[\because \text{ from (1), } \cosh \theta = \frac{x}{a}, \therefore \ \theta = \cosh^{-1} \frac{x}{a}\right]$$

Hence we have

Two more standard forms.

XVI. $\int \frac{1}{\sqrt{a^2 + x^2}} dx = \sinh^{-1} \frac{x}{a}$.

XVII. $\int \frac{1}{\sqrt{x^2 - a^2}} dx = \cosh^{-1} \frac{x}{a}$.

[**Aid to memory.**

$$\int \frac{1}{\sqrt{a^2 - x^2}} dx = \sin^{-1} \frac{x}{a} \ ; \ \int \frac{1}{\sqrt{a^2 + x^2}} dx = \sinh^{-1} \frac{x}{a}.\Big]$$

EXAMPLE

Write down the integrals of

(*i*) $\frac{1}{\sqrt{1 + x^2}}$. (*ii*) $\frac{1}{\sqrt{x^2 - 1}}$.

15. Rule to choose the right substitution.

If the integrand is the product of two factors, the first of which is a function of some quantity and the second is the differential coefficient of that quantity or becomes the differential coefficient of that quantity when multiplied by a constant put the quantity (whose function is the first factor) = z.

[Proof.** To integrate $\int f[\phi(x)] \, \phi'(x) \, dx$.

Put $\phi(x) = z$ ...(1), then $\phi'(x) \, dx = dz$. [Art. 12]

$$\therefore \quad \int f[\phi(x)] \, \phi'(x) \, dx = \int f(z) \, dz = \psi(z) \quad \text{(say)}$$

$$= \psi[\phi(x)]. \qquad \text{[From (1)]}$$

16. An important rule.

If the integral of a function of x is known, then if x is multiplied by a constant, and to the product is added another constant*, the integral is of the same form but is divided by the coefficient of x.*

[Proof.** Let $\int f(x)\,dx = \phi(x)$ ...(1) $\left[\text{To prove that } \int f(ax+b)\,dx = \dfrac{\phi(ax+b)}{a}\right]$.

Put $ax + b = z$† ...(2), then $a\,dx = dz$, or $dx = \dfrac{dz}{a}$.

$$\therefore \int f(ax+b)\,dx = \int f(z)\frac{dz}{a} = \frac{1}{a}\phi(z) \quad [\because \text{ from (1), changing } x \text{ to } z, \textstyle\int f(z)\,dz = \phi(z)]$$

$$= \frac{1}{a}\phi(ax+b). \qquad \text{[From (2)]}$$

Example 1. *Integrate* $\int (ax+b)^n\,dx$.

$\int (ax+b)^n\,dx$

[Compare with $\int x^n\,dx$, here x is multiplied by a, and to the product is added b]

$$= \frac{\dfrac{(ax+b)^{n+1}}{n+1}}{a} \qquad \left[\text{Of the same form } \frac{x^{n+1}}{n+1} \text{ but is divided by coeff. of } x\right]$$

$$= \frac{(ax+b)^{n+1}}{(n+1)a}.$$

Caution. *It is a* **common mistake** *not to divide by the coefficient of x but to give the answer as* $\dfrac{(ax+b)^{n+1}}{(n+1)}$.

Example 2. *Integrate* $\int \dfrac{dx}{ax+b}$.

$\int \dfrac{1}{ax+b}\,dx$

$\left[\text{Compare with } \int \dfrac{1}{x}\,dx, \text{ here } x \text{ is multiplied by } a, \text{ and to the product is added } b\right]$

$$= \frac{\log(ax+b)}{a}. \qquad \text{[Of the same form } \log x \text{ but is divided by coeff. of } x]$$

EXAMPLES

Integrate (Exs. 1—3) :

1. $\dfrac{1}{(a+x)^n}, \dfrac{1}{(a-x)^n}, \dfrac{1}{(a+x)^2} - \dfrac{1}{(a-x)^2}$.

---

*Positive or negative.

†**Why this step.** In $\int f(ax+b).\,1\,dx$, $f(ax+b)$ is a function of $ax+b$, and 1 becomes the differential coefficient of $ax+b$, when multiplied by a constant a,

$\therefore$ we put $ax + b = z$ [Rule (Art. 15)], so as to get $a\,dx = dz$, or $dx = \dfrac{dz}{a}$.

2. (*i*) $\dfrac{1}{x+1}, \dfrac{1}{a+x}$.

(*ii*) $\dfrac{1}{a-bx}, \dfrac{1}{x-a} - \dfrac{1}{x+a}, \dfrac{1}{a+x} - \dfrac{1}{a-x}$.

3. $\sqrt{2x+5}, \dfrac{1}{2x+1}$, $\sin(3x+2)$, $\text{cosec}^2 (a - bx)$.

4. Integrate $\displaystyle\int \frac{dx}{\sqrt{x-a}+\sqrt{x-b}}$.

[Rationalize the denom. (by multiplying the num. and denom. by $\sqrt{x-a} - \sqrt{x-b}$).]

Write down the indefinite integrals of (Exs. 5—7) :

5. $a + b^x - c^{2x}$.

6. $\dfrac{x}{1+x}, \dfrac{x}{x-a}$.

7. $\displaystyle\int \frac{2x^2+x}{x-1}$.

$$\left[\int \frac{2x^2+x}{x-1}\,dx = \int \left(2x+3+\frac{3}{x-1}\right) dx.\right.$$

$$\begin{array}{r} x-1)\;2x^2+x\;(2x+3 \\ 2x^2-2x \\ \hline 3x \\ 3x-3 \\ \hline 3 \end{array}$$

Note. Long division method. *If the degree of the numerator (of the integrand) is equal to, or greater than, that of the denominator, divide the numerator by the denominator until the degree of the remainder is less than that of the denominator.*]

8. Integrate $\displaystyle\int \sin mx \cos nx\, dx$.

[**Note. "A, B" formulae method.** *If the product of a sine and a cosine or of two (or more) sines or of two (or more) cosines occurs in the integrand, it is generally useful to apply the "A, B" formulae of Trigonometry and express the product in a series of sines or cosines of multiple angles.*]

9. Integrate $\displaystyle\int \sin mx \sin nx\, dx$ and deduce $\displaystyle\int \sin^2 mx\, dx$. Verify the result by independent integration.

10. Integrate $\displaystyle\int \cos mx \cos nx\, dx$ and deduce $\displaystyle\int \cos^2 mx\, dx$. Verify the result by independent integration.

11. Integrate (*i*) $\displaystyle\int \sin x \sin 2x \sin 3x\, dx$. (*ii*) $\displaystyle\int \sin mx \cos^2 x\, dx$.

Two important functional forms.

17. To integrate $\displaystyle\int [f(x)]^n f'(x)\, dx$.

Put $f(x) = z$* ...(1), then $f'(x)\,dx = dz$.

$$\therefore \quad \int [f(x)]^n f'(x)\,dx = \int z^n\,dz = \frac{z^{n+1}}{n+1} = \frac{[f(x)]^{n+1}}{n+1}. \qquad \text{[From (1)]}$$

Hence we have

An important form.

$$\int [\mathbf{f(x)}]^{\mathbf{n}}\, \mathbf{f'(x)\,dx} = \frac{[\mathbf{f(x)}]^{\mathbf{n+1}}}{\mathbf{n+1}}. \qquad [\mathbf{n \neq -1}]$$

Example 1. *Write down the integral of $(ax^2 + bx + c)^2\,(2ax + b)$.*

$$\int (ax^2 + bx + c)^2\,(2ax + b)dx \quad \left[\text{Compare with} \int [f(x)]^n f'(x)\,dx, \text{ here } f(x) = ax^2 + bx + c, \text{ so that } f'(x) = 2ax + b,\ n = 2\right]$$

$$= \frac{(ax^2 + bx + c)^{2+1}}{2+1} \qquad \left[\frac{[f(x)]^{n+1}}{n+1} \text{ (Art. 17)}\right]$$

$$= \frac{(ax^2 + bx + c)^3}{3}.$$

Example 2. *Integrate $\int (x - 1)\,(x^2 - 2x + 3)^{n/2}\,dx$.*

$$\int (x - 1)(x^2 - 2x + 3)^{n/2}\,dx = \frac{1}{2} \int (x^2 - 2x + 3)^{n/2}\,(2x - 2)\dagger\,dx$$

[*Note this step*]

$$\left[\text{Compare with} \int [f(x)]^n f'(x)\,dx, \text{ here } f(x) = x^2 - 2x + 3, \text{ so that } f'(x) = 2x - 2,\ \text{“}n\text{”} = \frac{n}{2}\right]$$

$$= \frac{1}{2}\,\frac{(x^2 - 2x + 3)^{\frac{n}{2}+1}}{\frac{n}{2} + 1} \qquad \left[\frac{[f(x)]^{n+1}}{n+1} \text{ (Art. 17)}\right]$$

$$= \frac{(x^2 - 2x + 3)^{\frac{n+2}{2}}}{n+2}.$$

EXAMPLES

1. Write down the integrals of

$$(ax^2 + bx + c)^n\,(2ax + b), \quad \frac{2ax + b}{(ax^2 + bx + c)^n}.$$

***Why this step.** In $\int [f(x)]^n f'(x)\,dx$, $[f(x)]^n$ is a function of $f(x)$, and $f'(x)$ is the differential coefficient of $f(x)$, $\therefore$ we put $f(x) = z$ [Rule (Art. 15)], so as to get $f'(x)\,dx = dz$.

2. (*i*) Integrate $x\sqrt{x^2+1}$, $x(a^2-x^2)^{\frac{n}{2}}$

(*ii*) Integrate $\int \frac{x+1}{\sqrt{x^2+2x+3}}\,dx$.

3. Integrate $\frac{x}{\sqrt{1-x^2}}, \frac{x}{\sqrt{x^2-1}}, \frac{x}{\sqrt{a^2+x^2}}$.

4. Integrate $\frac{x+b}{\sqrt{x^2+a^2}}, \frac{x-b}{\sqrt{x^2-a^2}}, \frac{2x+3}{\sqrt{1-x^3}}$.

Integrate (Exs. 5—6) :

5. (*i*) $\sin^5\theta\cos\theta$. (*ii*) $\cos^3 x\sin x$.

6. (*i*) $\sin 2x\cos^2 x$. (*ii*) $\tan^n x\sec^2 x$.

7. Integrate (*i*) $(e^x-a)^n e^x$. (*ii*) $\frac{1}{x(\log x)^2}$.

8. Integrate $\frac{\sin^{-1}x}{\sqrt{1-x^2}}, \frac{\tan^{-1}x}{1+x^2}, \frac{\sec^{-1}x}{x\sqrt{x^2-1}}$.

9. Integrate $\int\sqrt{\frac{x+1}{x-1}}\,dx$.

[Rationalize the num. (by multiplying the num. and denom. by $\sqrt{x+1}$).]

10. Integrate $\int\sqrt{\frac{1+x}{1-x}}\,dx$.

18. To integrate $\int \frac{f'(x)}{f(x)}\,dx$.

Put $f(x)=z$* ...(1), then $f'(x)\,dx=dz$,

$$\therefore \quad \int\frac{f'(x)}{f(x)}\,dx=\int\frac{dz}{z}=\int\frac{1}{z}\,dz=\log z=\log[f(x)].$$

Hence we have

An important form.

The integral of a fraction whose numerator is the differential coefficient of the denominator is **log (denominator).**

Example. *Integrate* $\frac{2ax+b}{ax^2+bx+c}, \frac{e^x+e^{-x}}{e^x-e^{-x}}, \frac{x}{x^2+a^2}$.

***Why this step.** In $\int\frac{f'(x)}{f(x)}\,dx$, $\frac{1}{f(x)}$ is a function of $f(x)$, and $f'(x)$ is the differential coefficient of $f(x)$, $\therefore$ we put $f(x)=z$ [Rule (Art. 15)], so as to get $f'(x)\,dx=dz$.

(i) $\int \frac{2ax+b}{ax^2+bx+c}\,dx$ [Num. is diff. coeff. of denom.]

$= \log(ax^2+bx+c).$ [log (denom.) (Art. 18)]

(ii) $\int \frac{e^x+e^{-x}}{e^x-e^{-x}}\,dx$ [Num. is diff. coeff. of denom.]

$= \log(e^x-e^{-x}).$ [log (denom.) (Art. 18)]

(iii) $\int \frac{x}{x^2+a^2}\,dx = \frac{1}{2}\int \frac{2x^*}{x^2+a^2}\,dx$ [*Note this step*]

[Num. is diff. coeff. of denom.]

$= \frac{1}{2}\log(x^2+a^2).$ [log (denom.) (Art. 18)]

EXAMPLES

1. Integrate $\frac{e^x}{e^x-a}, \frac{e^x-e^{-x}}{e^x+e^{-x}}, \frac{\sec^2 x}{\tan x}.$

2. Integrate $\frac{\frac{1}{\sqrt{1-x^2}}}{\sin^{-1}x}, \frac{1}{(1+x^2)\tan^{-1}x}, \frac{1}{x\sec^{-1}x\,.\sqrt{x^2-1}}.$

3. Integrate $\int \frac{dx}{x\log x}.$

4. Integrate $\frac{1}{x\log x\log(\log x)}, \frac{1}{x\log x\log(\log x)\log[\log(\log x)]},$

$\frac{1}{x\,l(x)\,l^2(x)\,l^3(x)\ldots l^n(x)},$

where $l^n(x)$ stands for log log log ... x, the log being repeated n times.

5. Integrate

(i) $\int \frac{\cot x}{\log\sin x}\,dx.$ (ii) $\frac{\sec x\,\mathrm{cosec}\,x}{\log\tan x}\,dx.$

6. Integrate $\frac{x^2}{x^3+a^3}, \frac{x^{n-1}}{x^n+a^n}.$ **7.** Integrate $\int \frac{x^{e-1}-e^{x-1}}{x^e-e^x}\,dx.$

19. Integrate (i) $\int$ **tan x dx.** (ii) $\int$ **cot x dx.**

(i) $\int \tan x\,dx = \int \frac{\sin x}{\cos x}\,dx = -\int \frac{-\sin x\,\dagger}{\cos x}\,dx$ [*Note this step*]

[Num. is diff. coeff. of denom.]

$= -\log\cos x.$ [log (denom.) (Art. 18)]

***Why this step.** In $\int \frac{x}{x^2+a^2}\,dx$, the differential coefficient of the denominator x^2+a^2 is $2x$. But the numerator is x. ∴ we take $2x$ (instead of x) in the numerator, and adjust by writing $\frac{1}{2}$ before the integral sign.

† **Why this step.** In $\int \frac{\sin x}{\cos x}\,dx$, the differential coefficient of the denominator $\cos x$ is $-\sin x$. But the numerator is $\sin x$, ∴ we take $-\sin x$ (instead of $\sin x$) in the numerator and adjust by writing (–) before the integral sign.

(*ii*) $$\int \cot x\, dx = \int \frac{\cos x}{\sin x} dx$$ [Num. is diff. coeff. of denom.]

$$= \log \sin x.$$ [log (denom.) (Art. 18)]

Hence we have

Two more standard forms.

XVIII. $\int \tan x\, dx = -\log \cos x.$ **XIX.** $\int \cot x\, dx = \log \sin x.$

[**Aid to memory.** In $-\log \cos x$, $\log \cos x$ involves $\cos x$ which begins with the letters "co", and $\log \cos x$, has a negative sign before it, thus

$$\int \tan x\, dx = -\log \cos x.$$

Compare the rule of "co" for the standard forms of integrals in circular functions (Aid to memory, (*ii*), Art. 5).]

EXAMPLES

1. Write down the indefinite integral of $\tan x + \cot x$.

2. Integrate $\int \cos x \left(\frac{1}{\sin x} + \frac{1}{\sin^2 x}\right) dx.$

20. To integrate (*i*) $\int$ **cosec x dx.** (*ii*) $\int$ **sec x dx.**

(*i*) $$\int \operatorname{cosec} x\, dx = \int \frac{1}{\sin x} dx = \int \frac{1}{2\sin\frac{x}{2}\cos\frac{x}{2}} dx$$

$\left[\text{Divide the num. and denom. by } \cos^2 \frac{x}{2}\right]$

$$= \int \frac{\sec^2 \frac{x}{2}}{2\tan\frac{x}{2}} dx = \int \frac{\frac{1}{2}\sec^2 \frac{x}{2}}{\tan\frac{x}{2}} dx$$ [Num. is diff. coeff. of denom.]

$$= \log \tan \frac{x}{2}.$$ [log (denom.) (Art. 18)]

(*ii*) $$\int \sec x\, dx = \int \frac{1}{\cos x} dx = \int \frac{1}{\sin\left(\frac{\pi}{2} + x\right)} dx = \int \operatorname{cosec}\left(\frac{\pi}{2} + x\right) dx$$

$$= \log \tan \frac{\left(\frac{\pi}{2} + x\right)}{2}$$ $\left[\because \int \operatorname{cosec} x\, dx = \log \tan \frac{x}{2}\right]$

$$= \log \tan \left(\frac{\pi}{4} + \frac{x}{2}\right).$$

Hence we have

Two more standard forms.

XX. $\int \sec x\, dx = \log \tan\left(\frac{\pi}{4} + \frac{x}{2}\right)$.

XXI. $\int \operatorname{cosec} x\, dx = \log \tan \frac{x}{2}$.

Example. *Integrate* $\int \frac{dx}{a \sin x + b \cos x}$.

Put $a = r \cos \alpha$ ...(1), $b = r \sin \alpha$ ...(2)

[so that squaring (1) and (2), and adding,

$$a^2 + b^2 = r^2 . 1, \quad \therefore \quad r = \sqrt{a^2 + b^2} \quad ...(3)$$

and dividing (2) by (1),

$$\frac{b}{a} = \frac{\sin \alpha}{\cos \alpha} = \tan \alpha, \quad \therefore \quad \alpha = \tan^{-1} \frac{b}{a} \quad ...(4)$$

Then $\int \frac{1}{a \sin x + b \cos x} dx = \int \frac{1}{r \cos \alpha \sin x + r \sin \alpha \cos x}$

$$= \int \frac{1}{r(\sin x \cos \alpha + \cos x \sin \alpha)} dx = \int \frac{1}{r \sin (x + \alpha)} dx$$

$$= \frac{1}{r} \int \operatorname{cosec} (x + \alpha)\, dx = \frac{1}{r} \log \tan \frac{x + \alpha}{2}$$

$$\left[\because \int \operatorname{cosec} x\, dx = \log \tan \frac{x}{2} \text{ (Art. 20, XXI)}\right]$$

[Substitute the values of r and α from (3) and (4)]

$$= \frac{1}{\sqrt{a^2 + b^2}} \log \tan\left(\frac{x}{2} + \frac{1}{2} \tan^{-1} \frac{b}{a}\right).$$

EXAMPLES

1. Integrate $\int \frac{\sin x}{\sin 2x} dx$.

Write down the integrals of (Exs. 2—3) :

2. (*i*) $\operatorname{cosec} (ax + b)$. (*ii*) $\operatorname{cosec} 2x$.

3. $\frac{1}{\cos^2 x - \sin^2 x}, \frac{1 + \tan^2 x}{1 - \tan^2 x}, \frac{1}{3 \sin x - 4 \sin^3 x}$.

4. Integrate

(*i*) $\int \frac{dx}{\sin x + \cos x}$. (*ii*) $\int \frac{\sec x\, dx}{a + b \tan x}$.

5. Integrate $\int \frac{dx}{(a \sin x + b \cos x)^2}$.

6. Integrate

(*i*) $\int \sqrt{1+\sin x}\, dx.$ (*ii*) $\int \frac{\sin x}{\sqrt{1+\sin x}}\, dx.$

7. Integrate $\int \operatorname{cosech} x\, dx.$

21. Two typical examples.

Example 1. *Integrate* $\int \frac{e^{\tan^{-1} x}}{1+x^2}\, dx.$

Put $\tan^{-1} x = z$* ...(1), then $\frac{1}{1+x^2}\, dx = dz.$

$$\int \frac{e^{\tan^{-1} x}}{1+x^2}\, dx = \int e^z\, dz = e^z = e^{\tan^{-1} x}.$$

Note. Important. *It is generally useful to put the function of x in the index of the power of e (in solved Ex. 1,* $\tan^{-1} x) = z.$

Example 2. *Evaluate* $\int \frac{1}{\sqrt{x}} \cos \sqrt{x}\, dx.$

Put $\sqrt{x} = z$† ...(1), then $\frac{1}{2\sqrt{x}}\, dx = dz$, or $\frac{1}{\sqrt{x}}\, dx = 2\, dz.$

$$\therefore \quad \int \frac{1}{\sqrt{x}} \cos \sqrt{x}\, dx = \int 2 \cos z\, dz = 2 \sin z = 2 \sin \sqrt{x}. \qquad \text{[From (1)]}$$

Note. Important. *It is generally useful to put the function of x the angle (in solved Ex. 2,* $\sqrt{x}) = z.$

EXAMPLES

1. Integrate $\int \frac{e^{m \tan^{-1} x}}{1+x^2}\, dx.$

2. Integrate (*i*) $nx^{n-1} \sin x^n.$ (*ii*) $x^2 \sec^2 x^3.$

3. Integrate (*i*) $e^x \sin e^x.$ (*ii*) $ae^x \cos e^x + b \tanh x.$

***Why this step.** In $\int e^{\tan^{-1} x} \frac{1}{1+x^2}\, dx$, $e^{\tan^{-1} x}$ is a function of $\tan^{-1} x$, $\frac{1}{1+x^2}$ is the differential coefficient of $\tan^{-1} x$, $\therefore$ we put $\tan^{-1} x = z$ [Rule (Art. 13)], so as to get $\frac{1}{1+x^2}\, dx = dz.$

† **Why this step.** In $\int \cos \sqrt{x} \cdot \frac{1}{\sqrt{x}}\, dx$, $\cos \sqrt{x}$ is a function of $\sqrt{x}$, and $\frac{1}{\sqrt{x}}$ becomes the differential coefficient of $\sqrt{x}$ when multiplied by $\frac{1}{2}$, $\therefore$ we put $\sqrt{x} = z$ [Rule (Art. 15)], so as to get $\frac{1}{2\sqrt{x}}\, dx = dz$, or $\frac{1}{\sqrt{x}}\, dx = 2dz.$

4. Integrate $\int \frac{1}{x} \cos(\log x)\, dx$.

5. Evaluate (*i*) $\int \frac{2x}{1+x^4}\, dx$. (*ii*) $\int \frac{x^2}{1+x^6}\, dx$.

6. Evaluate (*i*) $\int \frac{x\, dx}{\sqrt{a^4 - x^4}}$. (*ii*) $\int \frac{x^3\, dx}{\sqrt{x^8+1}}$.

7. Integrate (*i*) $\int \frac{d\theta}{a^2 \sin^2\theta + b^2 \cos^2\theta}$. (*ii*) $\int \frac{dx}{a^2 - b^2 \cos^2 x}$ $a > b$.

[**Rule to integrate a fraction whose numerator is 1 and denominator is a homogeneous function of the second degree in cos x and sin x :**

Divide the numerator and denominator by $\cos^2 x$, *and put* $\tan x = z$.]

8. Integrate $\int \frac{\tan x\, dx}{\sqrt{a + b\tan^2 x}}$. **9.** Integrate $\int \frac{\sin 2x\, dx}{(a + b\cos x)^2}$.

10. Integrate $\int \frac{\sin\theta - \cos\theta}{\sqrt{\sin 2\theta}}\, d\theta$.

Put $\sin\theta + \cos\theta = t$, $\sin^2\theta + \cos^2\theta + 2\sin\theta\cos\theta = t^2$, $2\sin\theta\cos\theta = t^2 - 1$ $\sqrt{\sin 2\theta} = \sqrt{t^2 - 1}$.

SECETION II

CHANGE OF INDEPENDENT VARIABLE IN A DEFINITE INTEGRAL

****22. To change the independent variable in $\int_a^b f(x)\, dx$ by the tranformation $x = \phi(z)$.**

Let $\int f(x)\, dx = \psi(x)$ ...(1)

In this relation, put $x = \phi(z)$ ...(2), then $dx = \phi'(z)\, dz$.

$\therefore$ $\int f[\phi(z)]\, \phi'(z)\, dz = \psi[\phi(z)]$ ...(3) [Art. 12]

From (2), when $x = a$, let $z = \alpha$, so that $a = \phi(\alpha)$
when $x = b$, let $z = \beta$, so that $b = \phi(\beta)$ ...(4)

Now from (1), $\int_a^b f(x)\, dx = \Big[\psi(x)\Big]_a^b = \psi(b) - \psi(a)$ ...(5)

and from (3), $\int_\alpha^\beta f[\phi(z)]\, \phi'(z)\, dz = \Big[\psi[\phi(z)]\Big]_\alpha^\beta = \psi[\phi(\beta)] - \psi[\phi(\alpha)]$, which from (4),

$= \psi(b) - \psi(a)$ ...(6)

$\therefore$ from (5) and (6), $\int_a^b f(x)\, dx = \int_\alpha^\beta f[\phi(z)]\, \phi'(z)\, dz$.

[Rule to evaluate $\int_a^b$ f(x) dx by the substitution of x = $\phi(z)$:

(i) In the integrand, put $x = \phi(z)$ *...(1), and* $dx = \phi'(z)\,dz$ $\left[\text{instead of } \frac{dx}{dz} = \phi'(z)\right]$.

(ii) From (1), when $x = a$, *find the value of z* [= α *(say)*],

when $x = b$, *find the value of z* [= β *(say)*].

(iii) Evaluate the resulting integral in z between the new limits α *and* β.

Caution 1. While evaluating the definite integral $\int_a^b f(x)\,dx$ by the substitution $x = \phi(z)$, the limits of integration must also be changed,*the new limits being the values of (the new independent variable) z corresponding to the values a and b of x.

It is a **common mistake** not to change the limits of integration but to evaluate the resulting integral in z between the old limits a and b of which are the values of x and not of z.

****Caution 2.** While evaluating the definite integral $\int_a^b f(x)\,dx$ by the substitution $x = \phi(z)$, the new limits of integration α and β should be such that, as x increases continuously from a to b, z *increases continuously* from α to β, or z *decreases continuously* from α to β.

Example 1. *Evaluate* $\int_0^1 \frac{dx}{e^x + e^{-x}}$.

$$\int_0^1 \frac{1}{e^x + e^{-x}}\,dx = \int_0^1 \frac{1}{e^x + \frac{1}{e^x}}\,dx = \int_0^1 \frac{e^x}{e^{2x} + 1}\,dx$$

[Put $e^x = z$ ...(1), then $e^x\,dx = dz$

and from (1), when $x = 0, z = 1$; when $x = 1, z = e$

$$= \int_1^e \frac{dz}{z^2 + 1} = \int_1^e \frac{1}{z^2 + (1)^2}\,dz$$

$$= \left[\frac{1}{1}\tan^{-1}\frac{z}{1}\right]_1^e$$

$$= \tan^{-1} e - \tan^{-1} 1$$

$$= \tan^{-1}\frac{e - 1\dagger}{1 + e(1)}$$

$$= \tan^{-1}\frac{e-1}{e+1}.$$

***Why this step.** In order to avoid the necessity of expressing the result in terms of the old independent variable x, the limits of integration must also be changed.

Note. Remember : $\mathbf{\tan^{-1} a - \tan^{-1} b = \tan^{-1} \frac{a - b}{1 + ab}}$. [From Trigonometry]

$$\therefore \qquad \tan^{-1} e - \tan^{-1} 1 = \tan^{-1} \frac{e-1}{1+e(1)} = \tan^{-1} \frac{e-1}{e+1}.$$

While evaluating definite integrals, it is customary to give the answer in this single inverse-tangent form.

Example 2. *Evaluate* $\int_0^1 \frac{x}{\sqrt{1+x^2}}\, dx.$

$$\int_0^1 \frac{x}{\sqrt{1+x^2}}\, dx = \frac{1}{2}\int_0^1 (1+x^2)^{-\frac{1}{2}} (2x)\, dx$$

$$= \frac{1}{2}\left[\frac{(1+x^2)^{\frac{1}{2}}}{\frac{1}{2}}\right]_0^1 \qquad \text{[Art. 17]}$$

$$= \left[\sqrt{1+x^2}\right]_0^1$$

$$= \sqrt{2} - 1.$$

Example 3. *Write down the value of* $\int_0^\infty e^{-x}\, dx$.

$$\int_0^\infty e^{-x}\, dx = \left[\frac{e^{-x}}{-1}\right]_0^\infty$$

$$= -\left[\underset{x \to \infty}{\text{Lt}}\ e^{-x} - e^0\right] \qquad \text{[Art. 11, Def. 1']}$$

$$= -[0 - 1] \qquad \left[\because\ e^{-\infty} = \frac{1}{e^\infty} = \frac{1}{\infty} (\because\ e > 1) = 0\right]$$

$$= 1.$$

EXAMPLES

1. Evaluate (*i*) $\int_0^1 \frac{1}{\sqrt{1-x^2}}\, dx$*. (*ii*) $\int_0^1 \frac{1}{1+x^2}\, dx$.

Note on the principal values of inverse circular functions. Important. *While evaluating a definite integral, take the* ***principal values*** *of the inverse circular functions as follows :*

(1) $\sin^{-1} x$ (or $\text{cosec}^{-1} x$) = the angle between $-\frac{\pi}{2}$ and $\frac{\pi}{2}$ (both values inclusive), whose sine (or cosecant) is x.

---

*For a rigorous treatment, see solved Ex. 2, Art. 86.

Examples. $\sin^{-1} 1 = \frac{\pi}{2}$, $\sin^{-1}(-1) = -\frac{\pi}{2}$ $\left(not\ \frac{3\pi}{2}\right)$.

(2) $\tan^{-1} x$ (or $\cot^{-1} x$) = the angle between $-\frac{\pi}{2}$ and $\frac{\pi}{2}$ (both values inclusive), whose tangent (or cotangent) is x.

Examples. $\tan^{-1} 1 = \frac{\pi}{4}$, $\tan^{-1}(-1) = -\frac{\pi}{4}$ $\left(not\ \frac{3\pi}{4}\right)$.

(3) $\cos^{-1} x$ (or $\sec^{-1} x$) = the angle between 0 and π (both values inclusive), whose cosine (or secant) is x.

Examples. $\cos^{-1} \frac{1}{2} = \frac{\pi}{3}$, $\cos^{-1}\left(-\frac{1}{2}\right) = \frac{2\pi}{3}$ $\left(not\ \frac{4\pi}{3}\right)$.

Note that the angles are always taken in *circular measure.*

[**Aid to memory.** In the case of the inverse sine (or inverse cosecant), inverse tangent (or inverse cotangent) take the positive acute or negative acute angle having the given sine (or cosecant), or tangent (or cotangent).

But in the case of the inverse cosine (or inverse secant) take the positive acute or positive obtuse angle having the given cosine (or secant).]

2. Evaluate (*i*) $\int \frac{dx}{e^x + e^{-x}}$. (*ii*) $\int_0^1 \operatorname{sech} x \, dx$.

3. Evaluate $\int_0^{\pi/2} \frac{dx}{a^2 \sin^2 x + b^2 \cos^2 x}$.

4. Evaluate $\int_0^{\pi/4} \frac{\sin 2\theta}{\sin^4 \theta + \cos^4 \theta} d\theta$.

[Divide the numerator and denominator by $\cos^4 \theta$, and put $\tan^2 \theta = z$.]

5. Prove $\int_a^b \frac{\log x}{x} dx = \frac{1}{2} \log\left(\frac{b}{a}\right) \log(ab)$.

MISCELLANEOUS EXAMPLES ON CHAPTER II

1. Evaluate (*i*) $\int_0^1 \frac{1-x}{1+x} dx$. (*ii*) $\int \frac{dx}{\sqrt{1+x} - \sqrt{x}}$.

2. Prove that, if $0 \le \alpha \le 1$ and the positive value of the square root is taken,

$$\int_{-1}^{1} \frac{dx}{\sqrt{1-2\alpha x+\alpha^2}} = 2.$$

What is the value of the integral if $\alpha > 1$?

3. Show that

(*i*) $\int_0^{\pi} \sin mx \sin nx \, dx$ and (*ii*) $\int_0^{\pi} \cos mx \cos nx \, dx$ are zero, m and n are unequal integers, and $= \frac{\pi}{2}$ if m and n are equal integers.

****4.** If m and n are integers, prove that

$$\int_0^{\pi} \cos mx \sin nx \, dx = \frac{2n}{n^2 - m^2} \text{ or } 0$$

according as $n - m$ is odd or even.

5. Evaluate $\int_0^{\frac{\pi}{2}} \frac{\sin^2 \theta}{(1+\cos\theta)^2} d\theta.$

6. Integrate (*i*) $\int \frac{1}{x(1+\log x)} dx.$ (*ii*) $\int_0^{\frac{\pi}{3}} \frac{\cos x \, dx}{3+4\sin x}.$

(*iii*) $\int \frac{\sin x \cos x \, dx}{a^2 \sin^2 x + b^2 \cos^2 x}.$

7. Integrate

(*i*) $\int \frac{\cos 2x}{\cos x} dx.$ (*ii*) $\int \frac{dx}{(2\sin x + \cos x)^2}.$

8. Integrate

(*i*) $\int \sec x \log(\sec x + \tan x) \, dx.$ (*ii*) $\int \frac{\sqrt{x} \, dx}{\sqrt{a^3 - x^3}}.$

9. Integrate $\int \frac{dx}{1+3\sin^2 x}.$

10. Show that $\int \sec x \, dx = \cosh^{-1}(\sec x).$

[Put $\sec x = z$].*

11. Integrate (*i*) $\int \frac{\sin x}{\sin(x-a)} dx.$ (*ii*) $\int \frac{dx}{\sin(x-a)\sin(x-b)}.$

***Why this step.** As suggested by $\sec x$ in $\cosh^{-1}(\sec x)$ on the R.H.S. of result to be proved, put $\sec x = z$.

$$\left[(i) \int \frac{\sin x}{\sin(x-a)} dx = \int \frac{\sin[(x-a)+a]}{\sin(x-a)} dx = \int \frac{\sin(x-a)\cos a + \cos(x-a)\sin a}{\sin(x-a)} dx\right.$$

$$= \int [\cos a + \sin a \cot(x-a)]\, dx.$$

$$(ii) \int \frac{1}{\sin(x-a)\sin(x-b)} dx = \frac{1}{\sin(a-b)} \int \frac{\sin(a-b)^*}{\sin(x-a)\sin(x-b)} dx$$

$$= \frac{1}{\sin(a-b)} \int \frac{\sin[(x-b)-(x-a)]}{\sin(x-a)\sin(x-b)} dx$$

$$= \frac{1}{\sin(a-b)} \int \frac{\sin(x-b)\cos(x-a) - \cos(x-b)\sin(x-a)}{\sin(x-a)\sin(x-b)} dx$$

$$\left.= \frac{1}{\sin(a-b)} \int [\cot(x-a) - \cot(x-b)]\, dx.\right]$$

**12. Integrate $\int \sqrt{\frac{\sin(x-\alpha)}{\sin(x+\alpha)}}\, dx$.

[Rationalize the num. (by multiplying the num. and denom. by $\sqrt{\sin(x-\alpha)}$. Then

$$\int \sqrt{\frac{\sin(x-\alpha)}{\sin(x+\alpha)}}\, dx = \int \frac{\sin(x-\alpha)}{\sqrt{\sin^2 x - \sin^2 \alpha}} dx = \int \frac{\sin x \cos\alpha - \cos x \sin\alpha}{\sqrt{\sin^2 x - \sin^2\alpha}} dx$$

$$= \cos\alpha \int \frac{\sin x\, dx}{\sqrt{\sin^2 x - \sin^2\alpha}} - \sin\alpha \int \frac{\cos x\, dx}{\sqrt{\sin^2 x - \sin^2\alpha}}$$

$$= \cos\alpha \int \frac{\sin x\, dx}{\sqrt{1-\cos^2 x - \sin^2\alpha}} - \sin\alpha \int \frac{\cos x\, dx}{\sqrt{\sin^2 x - \sin^2\alpha}}$$

$$\left.= \cos\alpha \int \frac{\sin x\, dx}{\sqrt{\cos^2\alpha - \cos^2 x}} - \sin\alpha \int \frac{\cos x\, dx}{\sqrt{\sin^2 x - \sin^2\alpha}}.\right]$$

**13. Integrate $\int \frac{1}{\sqrt{\sin^3 x \sin(x+\alpha)}}\, dx$.

[(The denom. is a homogeneous function of the second degree $\cos x$ and $\sin x$.) Divide the num. and denom. by $\sin^2 x$†. It will be found that integral $= \int \frac{\operatorname{cosec}^2 x\, dx}{\sqrt{\cos\alpha + \cot x \sin\alpha}}$.]

---

*Why this step. In $\int \frac{1}{\sin(x-a)\sin(x-b)} dx$, the *difference* of the two angles $x-a$ and $x-b$ in the denominator $= (x-b)-(x-a) = a-b$ (constant),

$\therefore$ we take $\sin(a-b)$ in the numerator, and adjust by writing $\frac{1}{\sin(a-b)}$ before the integral sign.

†As suggested by the power of $\sin x$ in the denominator, we divide the numerator and denominator by $\sin^2 x$. Compare the Rule given in Ex. 7, Art. 21.

CHAPTER III

INTEGRATION BY PARTS

SECTION I

RULE OF INTEGRATION BY PARTS

23. To prove the rule of integration by parts.

Let u, v be two functions of x. Then

$$\frac{d}{dx}(uv) = u\frac{dv}{dx} + v\frac{du}{dx}.$$

Integrating both sides w.r.t. x,

$$uv = \int u\frac{dv}{dx}\,dx + \int v\frac{du}{dx}\,dx.$$

Transposing, $\int u\frac{dv}{dx}\,dx = uv - \int v\frac{du}{dx}\,dx$ ...(1)

Put $\frac{dv}{dx} = w$, then, integrating both sides w.r.t. x,

$$v = \int w\,dx.$$

$\therefore$ from (1), $\int uw\,dx = u\int w\,dx - \int \frac{du}{dx}\left(\int w\,dx\right)dx.$

In words :

Integral of product of two functions

= 1st function × integral of 2nd

– integral of [diff. coeff. of 1st × integral of 2nd].

Note. Rule of integration by parts. *The above rule is called* **rule of integration by parts** *and is extremely important. In applying it, the first function should be chosen such that the integral of the R.H.S. is more easily integrable than the original integral.*

Example 1. *Integrate* $\int x\cos nx\,dx.$

$\int x\cos nx\,dx$ [Integrate by parts]

$$= x\,.\,\frac{\sin nx}{n} - \int 1.\frac{\sin nx}{n}\,dx \quad ...(1)$$

[(1st fun. × int. of 2nd) – int. of (diff. coeff. of 1st × int. of 2nd) (Art. 23)]

$$= \frac{x\sin nx}{n} - \frac{1}{n}\int \sin nx\,dx$$

$$= \frac{x\sin nx}{n} - \frac{1}{n}\left(-\frac{\cos nx}{n}\right) = \frac{x\sin nx}{n} + \frac{\cos nx}{n^2}.$$

Note. *In integrating* $\int x \cos nx \, dx$ *by parts, we have taken x as the first function because by doing so the integral* $\int 1 . \frac{\sin nx}{n} dx$ *on the R.H.S. of (1) is more easily integrable than the original integral* $\int x \cos nx \, dx$.

In general we have the

Rule to choose the first function rightly in the rule of integration by parts :

If one factor of the product is a power of x, take it as the first function except when the other factor is an inverse circular (or hyperbolic) function or a logarithm in which case take that factor as the first function.

Example 2. *Integrate* $\int x^n \log x \, dx$.

$\int x^n \log x \, dx$ [Take log x as first function (Rule)]

$$= \int \log x . x^n \, dx \quad \text{[Integrate by parts]}$$

$$= \log x . \frac{x^{n+1}}{n+1} - \int \frac{1}{x} . \frac{x^{n+1}}{n+1} dx$$

[1st fun. × int. of 2nd – int. of (diff. coeff. of 1st × int. of 2nd) (Art. 23)]

$$= \frac{x^{n+1}}{n+1} \log x - \frac{1}{n+1} \int x^n \, dx = \frac{x^{n+1}}{n+1} \log x - \frac{1}{n+1} \frac{x^{n+1}}{n+1}$$

$$= \frac{x^{n+1}}{n+1} \log x - \frac{x^{n+1}}{(n+1)^2} .$$

Example 3. *Integrate* $\int \log x \, dx$.

$$\int \log x \, dx = \int \log x . 1 \, dx$$

[Take log x as first function and 1 as second function]

[Integrate by parts]

$$= \log x . x - \int \frac{1}{x} . x \, dx$$

[1st fun. × int. of 2nd – int. of (diff. coeff. of 1st × int. of 2nd) (Art. 23)]

$$= x \log x - \int 1 \, dx = x \log x - x.$$

Note. Rule to integrate by parts a single inverse circular (or hyperbolic) function or a logarithm :

Take the inverse circular (or hyperbolic) function or the logarithm as the first function and 1 as the second function.

Example 4. *Calculate* $\int_0^{\pi} x \sin x \, dx$.

$\int_0^{\pi} x \sin x \, dx$ [Integrate by parts]

$$= \Big[x(-\cos x) \Big]_0^{\pi} - \int_0^{\pi} 1(-\cos x)\, dx$$

$$= \pi(1) - 0 + \int_0^{\pi} \cos x\, dx = \pi + \Big[\sin x \Big]_0^{\pi} = \pi + 0 - 0 = \pi.$$

EXAMPLES

Integrate (Exs. 1—2) :

1. xe^x, $x \cos x$, $x \sec^2 x$.

2. $x \sinh x$, $x \cos 2x$.

3. Integrate (*i*) $\int x \sin x \cos x\, dx$.

(*ii*) $\int x \cos^2 x\, dx$. (*iii*) $\int x \cos^3 x\, dx$.

4. Integrate $x \sin x \sin 2x \sin 3x$.

5. Integrate (*i*) $\int x^2 \log x\, dx$.

(*ii*) $\int x^3 \tan^{-1} x\, dx$. (*iii*) $\int x^2 \sin^{-1} x\, dx$.

6. Integrate (*i*) $\int \sin^{-1} x\, dx$.

(*ii*) $\int \tan^{-1} x\, dx$. (*iii*) $\int \cos^{-1} \dfrac{1}{x}\, dx$.

7. Integrate $\int \dfrac{x \sin^{-1} x}{(1-x^2)^{\frac{1}{2}}}\, dx$.

[**Note. Important.** *If an inverse function such as* $\sin^{-1} x$ *(or* $\tan^{-1} x$*) occurs in the integrand, it is generally useful to put* $x = \sin\theta$ *(or* $x = \tan\theta$*), and then work with the direct function so that* $dx = \cos\theta\, d\theta$ *(or* $dx = \sec^2\theta\, d\theta$*).]*

8. Integrate $\int \sin^{-1} \sqrt{\dfrac{x}{a+x}}\, dx$.

$\left[\text{Put } \sqrt{\dfrac{x}{a+x}} = \sin\theta \text{ or, squaring, } x = (a+x)\sin^2\theta\right.$

or $x(1 - \sin^2\theta) = a \sin^2\theta$, or $x = a \tan^2\theta$, then $dx = a.\, 2 \tan\theta \sec^2\theta\, d\theta$.

$$\therefore \quad \int \sin^{-1} \sqrt{\frac{x}{a+x}}\, dx = \int \theta \,.\, 2a \tan\theta \sec^2\theta\, d\theta$$

$$= a \int \theta\, (2 \tan\theta \sec^2\theta)\, d\theta.$$ [Integrate by parts.]

9. Evaluate $\int_0^{\pi} x \sin^2 x\, dx$.

10. Integrate $\int \sin^{-1} \dfrac{2x}{1+x^2}\, dx$.

[**Note.** *To integrate* $\sin^{-1}$ *(algebraic expression in x), we put x = such a circular function of θ that the algebraic expression in x may become the* **sine** *of some angle.*

Thus to integrate $\int \sin^{-1} \frac{2x}{1+x^2}\,dx$, *we put* $x = \tan\theta$.

$\because$ *then* $$\frac{2x}{1+x^2} = \frac{2\tan\theta}{1+\tan^2\theta} = \mathbf{\sin}\, 2\theta.$$

Similarly to integrate $\cos^{-1}$ *(algebraic expression in x), we put x = such a circular function of θ that the algebraic expression in x may become the* **cosine** *of some angle, and so on.]*

11. Integrate $\int \cos 2\theta \log \frac{\cos\theta + \sin\theta}{\cos\theta - \sin\theta}\, d\theta.$

12. Integrate $\int \frac{x^2\,dx}{(x\sin x + \cos x)^2}.$

$$\left[\frac{d}{dx}(x\sin x + \cos x) = 1 . \sin x + x\cos x - \sin x = x\cos x.\right.$$

$$\therefore \int \frac{x^2}{(x\sin x + \cos x)^2}\,dx = \int \left(\frac{x}{\cos x}\right) . \frac{x\cos x}{(x\sin x + \cos x)^2}\,dx. \qquad \text{Integrate by parts.]}$$

13. Integrate $\int \frac{\cosh x + \sinh x \sin x}{1 + \cos x}\,dx.$

Three standard forms.

24. To integrate

(*i*) $\int \sqrt{\mathbf{a^2 - x^2}}\,\mathbf{dx}$. (*ii*) $\int \sqrt{\mathbf{a^2 + x^2}}\,\mathbf{dx}$. (*iii*) $\int \sqrt{\mathbf{x^2 - a^2}}\,\mathbf{dx}$.

(*i*) $\int \sqrt{a^2 - x^2}\,dx = \int \sqrt{a^2 - x^2}\, .\, 1\,dx$ [*Note this step*]

[Integrate by parts]

$$= \sqrt{a^2 - x^2}\, .\, x - \int \frac{1}{2\sqrt{a^2 - x^2}}(-2x)\, .\, x\,dx$$

$$= x\sqrt{a^2 - x^2} + \int \frac{x^2}{\sqrt{a^2 - x^2}}\,dx.$$

$$= x\sqrt{a^2 - x^2} + \int \frac{-(a^2 - x^2) + a^{2*}}{\sqrt{a^2 - x^2}}\,dx$$ [*Note this step*]

***How to write this step.** Write in the numerator the quadratic whose square root occurs in the denominator, *viz.*, $a^2 - x^2$, and adjust the coefficient of x^2 by multiplying $a^2 - x^2$ by (– 1) [$\because$ we have got (– 1) as the coefficient of x^2 in $a^2 - x^2$ but require 1 as the coefficient of x^2 in x^2] thus getting $-(a^2 - x^2)$;

now adjust the constant term by adding a^2 [$\because$ we have got $(-a^2)$ as the constant term in $-(a^2 - x^2)$ but require 0 as the constant term in x^2] thus getting $x^2 = -(a^2 - x^2) + a^2$.

The same step will be used in Art. 24, (*ii*) and (*iii*).

$$= x\sqrt{a^2 - x^2} - \int \sqrt{a^2 - x^2}\, dx + a^2 \int \frac{1}{\sqrt{a^2 - x^2}}\, dx$$

$$= x\sqrt{a^2 - x^2} - \int \sqrt{a^2 - x^2}\, dx + a^2 \sin^{-1} \frac{x}{a}.$$

Transposing*,

$$2\int \sqrt{a^2 - x^2}\, dx = x\sqrt{a^2 - x^2} + a^2 \sin^{-1} \frac{x}{a}$$

$$\therefore \quad \int \sqrt{a^2 - x^2}\, dx = \frac{x\sqrt{a^2 - x^2}}{2} + \frac{a^2}{2} \sin^{-1} \frac{x}{a}.$$

***Note.** *If the original integral re-appears on the R.H.S. (as here), we transpose it to the L.H.S. and thus evaluate the original integral. The same step will be used in Art. 24, (ii) and (iii), and in Art. 27, (i) and (ii).*

(ii) $\int \sqrt{a^2 + x^2}\, dx = \int \sqrt{a^2 + x^2} \cdot 1\, dx$ [*Note this step*]

[Integrate by parts]

$$= \sqrt{a^2 + x^2} \cdot x - \int \frac{1}{2\sqrt{a^2 + x^2}} (2x) \cdot x\, dx$$

$$= x\sqrt{a^2 + x^2} - \int \frac{x^2}{\sqrt{a^2 + x^2}}\, dx$$

$$= x\sqrt{a^2 + x^2} - \int \frac{(a^2 + x^2) - a^2 \,\dagger}{\sqrt{a^2 + x^2}}\, dx \qquad \text{[Note this step]}$$

$$= x\sqrt{a^2 + x^2} - \int \sqrt{a^2 + x^2}\, dx + a^2 \int \frac{1}{\sqrt{a^2 + x^2}}\, dx$$

$$= x\sqrt{a^2 + x^2} - \int \sqrt{a^2 + x^2}\, dx + a^2 \sin^{-1} \frac{x}{a}.$$

Transposing,

$$2\int \sqrt{a^2 + x^2}\, dx = x\sqrt{a^2 + x^2} + a^2 \sinh^{-1} \frac{x}{a}$$

$$\therefore \quad \int \sqrt{a^2 + x^2}\, dx = \frac{x\sqrt{a^2 + x^2}}{2} + \frac{a^2}{2} \sinh^{-1} \frac{x}{a}.$$

(iii) $\int \sqrt{x^2 - a^2}\, dx = \int \sqrt{x^2 - a^2} \cdot 1dx$ [*Note this step*]

[Integrate by parts]

$$= \sqrt{x^2 - a^2} \cdot x - \int \frac{1}{2\sqrt{x^2 - a^2}} (2x) \cdot x\, dx$$

$$= x\sqrt{x^2 - a^2} - \int \frac{x^2}{\sqrt{x^2 - a^2}}\, dx$$

† **How to write this step.** Proceed as in the footnote* on p. 39.

$$= x\sqrt{x^2 - a^2} - \int \frac{(x^2 - a^2) + a^2}{\sqrt{x^2 - a^2}}\,dx \quad \text{* [Note this step]}$$

$$= x\sqrt{x^2 - a^2} - \int \sqrt{x^2 - a^2}\,dx - a^2 \int \frac{1}{\sqrt{x^2 - a^2}}\,dx$$

$$= x\sqrt{x^2 - a^2} - \int \sqrt{x^2 - a^2}\,dx - a^2 \cosh^{-1}\frac{x}{a}.$$

Transposing,

$$2\int \sqrt{x^2 - a^2}\,dx = x\sqrt{x^2 - a^2} - a^2\cosh^{-1}\frac{x}{a}$$

$$\therefore \quad \int \sqrt{x^2 - a^2}\,dx = \frac{x\sqrt{x^2 - a^2}}{2} - \frac{a^2}{2}\cosh^{-1}\frac{x}{a}.$$

Hence we have

Three more standard forms.

XXII. $\int \sqrt{a^2 - x^2}\,dx = \frac{x\sqrt{a^2 - x^2}}{2} + \frac{a^2}{2}\sin^{-1}\frac{x}{a}.$

XXIII. $\int \sqrt{a^2 + x^2}\,dx = \frac{x\sqrt{a^2 + x^2}}{2} + \frac{a^2}{2}\sinh^{-1}\frac{x}{a}.$

XXIV. $\int \sqrt{x^2 - a^2}\,dx = \frac{x\sqrt{x^2 - a^2}}{2} - \frac{a^2}{2}\cosh^{-1}\frac{x}{a}.$

[**Aid to memory** for $\int \sqrt{a^2 - x^2}\,dx$, $\int \sqrt{a^2 + x^2}\,dx$, $\int \sqrt{x^2 - a^2}\,dx$. Integral $= \frac{x \times \text{integrand}}{2} + \frac{a^2 \text{ or } -a^2 \text{ as in integrand}}{2} \times$ integral of (reciprocal of integrand).]

Example 1. *Integrate* $\sqrt{2 - 3x^2}\,dx.$

$\int \sqrt{2 - 3x^2}\,dx$ [Take the numerical value of coeff. of x^2 outside the quadratic]

$$= \int \sqrt{3\left(\tfrac{2}{3} - x^2\right)}\,dx = \sqrt{3}\int \sqrt{\left(\sqrt{\tfrac{2}{3}}\right)^2 - x^2}\,dx$$

$$\left[\text{Compare with} \int \sqrt{a^2 - x^2}\,dx, \text{ here } a = \sqrt{\tfrac{2}{3}}\right]$$

$$= \sqrt{3}\left[\frac{x\sqrt{(\sqrt{\frac{2}{3}})^2 - x^2}}{2} + \frac{(\sqrt{\frac{2}{3}})^2}{2}\sin^{-1}\frac{x}{\sqrt{\frac{2}{3}}}\right] \left[\frac{x\sqrt{a^2 - x^2}}{2} + \frac{a^2}{2}\sin^{-1}\frac{x}{a} \text{ (Art. 24, XXII)}\right]$$

$$= \sqrt{3}\left[\frac{x\sqrt{\frac{2}{3} - x^2}}{2} + \frac{\frac{2}{3}}{2}\sin^{-1} x\sqrt{\tfrac{3}{2}}\right] = \sqrt{3}\left[\frac{x\sqrt{2 - 3x^2}}{2\sqrt{3}} + \frac{1}{3}\sin^{-1} x\sqrt{\tfrac{3}{2}}\right]$$

$$= \frac{x\sqrt{2 - 3x^2}}{2} + \frac{1}{\sqrt{3}}\sin^{-1} x\sqrt{\tfrac{3}{2}}.$$

***How to write this step.** Proceed as in the footnote *on p. 39.

Example 2. *Integrate* $\int \sec^3 x\, dx$.

Put $\tan x = z$ ...(1), then $\sec^2 x\, dx = dz$, or $dx = \dfrac{dz}{\sec^2 x}$.

$$\therefore \quad \int \sec^3 x\, dx = \int \sec^3 x \cdot \frac{dz}{\sec^2 x} = \int \sec x\, dz = \sqrt{1+\tan^2 x}\, dz$$

$$= \int \sqrt{1+z^2}\, dz = \int \sqrt{(1)^2+z^2}\, dz \qquad \left[\text{Compare with } \int \sqrt{a^2+x^2}\, dx, \text{ here } a = 1\right]$$

$$= \frac{z\sqrt{(1)^2+z^2}}{2} + \frac{(1)^2}{2}\sinh^{-1} z \qquad \left[\frac{x\sqrt{a^2+x^2}}{2} + \frac{a^2}{2}\sinh^{-1}\frac{x}{a} \text{ (Art. 24, XXIII)}\right]$$

$$= \frac{\tan x\sqrt{1+\tan^2 x}}{2} + \tfrac{1}{2}\sinh^{-1}(\tan x) \qquad \text{[From (1)]}$$

$$= \frac{\tan x \sec x}{2} + \tfrac{1}{2}\sinh^{-1}(\tan x).$$

EXAMPLES

1. Write down the integrals of $\sqrt{1-x^2}$, $\sqrt{x^2-1}$, $\sqrt{1+x^2}$.

2. Show that (*i*) $\displaystyle\int_0^a \sqrt{a^2-x^2}\, dx = \frac{\pi a^2}{4}$.

(*ii*) $\displaystyle\int_2^{\frac{5}{2}} \sqrt{x^2-4}\, dx = \frac{15}{8} - 2\log 2$.

[(*ii*) It will be found that

$$\text{integral} = \frac{15}{8} - 2\cosh^{-1}\frac{5}{4} = \frac{15}{8} - 2\log\left[\frac{5}{4} + \sqrt{(\tfrac{5}{4})^2 - 1}\right]$$

[$\because \cosh^{-1} x = \log(x + \sqrt{x^2-1})$ (Art. 8, (*d*), (2))]

3. Integrate $\int \operatorname{cosec}^3 x\, dx$.

4. Integrate (*i*) $\displaystyle\int \frac{x^2}{\sqrt{1-x^2}}\, dx$. (*ii*) $\displaystyle\int \frac{x^2}{\sqrt{1+x^2}}\, dx$.

5. Integrate (*i*) $\displaystyle\int \frac{x^2+1}{\sqrt{x^2+4}}\, dx$. (*ii*) $\displaystyle\int (x+1)\sqrt{x^2+1}\, dx$.

6. Integrate (*i*) $\displaystyle\int \frac{x^2-2x+3}{\sqrt{1-x^2}}\, dx$. (*ii*) $\displaystyle\int \frac{x^2+2x+3}{\sqrt{x^2+1}}\, dx$.

7. Integrate (*i*) $\displaystyle\int x\sqrt{\frac{1-x}{1+x}}\, dx$. (*ii*) $\displaystyle\int \frac{(x+1)\sqrt{x+2}}{\sqrt{x-2}}\, dx$.

8. Integrate $\int x\sin^{-1} x\, dx$.

25. To integrate $\int [f(x) + f'(x)]\, e^x\, dx$.

$$\int [f(x) + f'(x)]\, e^x\, dx = \int f(x)\, e^x\, dx + \int f'(x)\, e^x\, dx$$

[Integrate only 1st integral on R.H.S. by parts]

$$= f(x)\, .\, e^x - \int f'(x)\, .\, e^x\, dx + \int f'(x)\, e^x\, dx = f(x)\, e^x.$$

[$\because$ last two integrals on R.H.S. cancel]

[Rule to integrate $\int [f(x) + f'(x)]\, e^x\, dx$:

Split the integral into two, and integrate only the first integral by parts.]

Example. *Integrate* $\int \dfrac{xe^x}{(x+1)^2}\, dx.$

$$\int \frac{x\,e^x}{(x+1)^2}\, dx = \int \frac{(x+1)-1}{(x+1)^2}\, e^x\, dx$$

$$= \int \left[\frac{1}{x+1} - \frac{1}{(x+1)^2}\right] e^x\, dx \qquad [\text{Form } \textstyle\int [f(x) + f'(x)]\, e^x\, dx]$$

$$= \int \frac{1}{x+1}\, e^x\, dx - \int \frac{1}{(x+1)^2}\, e^x\, dx$$

[Integrate only 1st integral on R.H.S. by parts]

$$= \frac{1}{x+1}\, .\, e^x - \int -\frac{1}{(x+1)^2}\, .\, e^x\, dx - \int \frac{1}{(x+1)^2}\, e^x\, dx$$

$$= \frac{e^x}{x+1}.$$

[$\because$ last two integrals on R.H.S. cancel]

EXAMPLES

1. Integrate $\int e^x (\sin x + \cos x)\, dx.$

2. Integrate

(*i*) $\int e^x \dfrac{1+\sin x}{1+\cos x}\, dx.$

(*ii*) $\int e^x \dfrac{1-\sin x}{1-\cos x}\, dx.$

26. Repeated application of the rule of integration by parts.

Sometimes it is necessary to apply the rule of integration by parts repeatedly to evaluate an integral completely.

Example. *Integrate* $\int x^2 \cos nx\, dx.$

$$\int x^2 \cos nx\, dx \qquad [\text{Integrate by parts}]$$

$$= x^2\, \frac{\sin nx}{n} - \int 2x\, \frac{\sin nx}{n}\, dx$$

$$= \frac{x^2 \sin nx}{n} - \frac{2}{n}\int x \sin nx\, dx \qquad \text{[Integrate again by parts]}$$

$$= \frac{x^2 \sin nx}{n} - \frac{2}{n}\left[x\left(-\frac{\cos nx}{n}\right) - \int 1.\left(-\frac{\cos nx}{n}\right)dx\right]$$

$$= \frac{x^2 \sin nx}{n} - \frac{2}{n}\left[-\frac{x\cos nx}{n} + \frac{1}{n}\int \cos nx\, dx\right]$$

$$= \frac{x^2 \sin nx}{n} - \frac{2}{n}\left[-\frac{x\cos nx}{n} + \frac{1}{n}\frac{\sin nx}{n}\right]$$

$$= \frac{x^2 \sin nx}{n} - \frac{2}{n}\left[-\frac{x\cos nx}{n} + \frac{\sin nx}{n^2}\right]$$

$$= \frac{x^2 \sin nx}{n} + \frac{2x\cos nx}{n^2} - \frac{2\sin nx}{n^3}.$$

EXAMPLES

1. Integrate $\int x^2 e^x\, dx$.

2. Integrate (*i*) $\int x^2 \cos x\, dx$.

(*ii*) $\int x^2 \sin x\, dx$. (*iii*) $\int x^2 \cosh x\, dx$.

3. Calculate $\int_0^{\pi} x^2 \sin x\, dx$.

4. Integrate $\int x^n (\log x)^2\, dx$.

5. Prove that

$$\int u \frac{d^2v}{dx^2}\, dx = u\frac{dv}{dx} - v\frac{du}{dx} + \int v \frac{d^2u}{dx^2}\, dx.$$

Two standard integrals.

27. To integrate (*i*) $\int e^{ax} \sin bx\, dx$. (*ii*) $\int e^{ax} \cos bx\, dx$.

(*i*) $\int e^{ax} \sin bx\, dx$ [Integrate by parts]

$$= e^{ax}\left(-\frac{\cos bx}{b}\right) - \int e^{ax}\, a\left(-\frac{\cos bx}{b}\right)dx$$

$$= -\frac{e^{ax}\cos bx}{b} + \frac{a}{b}\int e^{ax} \cos bx\, dx \qquad \text{[Integrate again by parts]}$$

$$= -\frac{e^{ax}\cos bx}{b} + \frac{a}{b}\left[e^{ax}.\frac{\sin bx}{b} - \int e^{ax}\, a.\frac{\sin bx}{b}\, dx\right]$$

$$= -\frac{e^{ax}\cos bx}{b} + \frac{a}{b}\left[\frac{e^{ax}\sin bx}{b} - \frac{a}{b}\int e^{ax} \sin bx\, dx\right]$$

$$= -\frac{e^{ax}\cos bx}{b} + \frac{a}{b^2} e^{ax} \sin bx - \frac{a^2}{b^2}\int e^{ax} \sin bx\, dx.$$

Transposing,

$$\left(1+\frac{a^2}{b^2}\right)\int e^{ax}\sin bx\,dx = \frac{ae^{ax}\sin bx - be^{ax}\cos bx}{b^2}$$

$$\therefore \quad \int e^{ax}\sin bx\,dx = \frac{e^{ax}(a\sin bx - b\cos bx)}{b^2\left(1+\frac{a^2}{b^2}\right)} = \frac{e^{ax}(a\sin bx - b\cos bx)}{a^2+b^2} \qquad ...(1)$$

Put $a = r\cos\alpha$ ...(2), $b = r\sin\alpha$ ...(3)

[so that squaring (2) and (3), and adding,

$$a^2 + b^2 = r^2 \,.\, 1, \qquad \therefore \quad r = \sqrt{a^2+b^2} \qquad ...(4)$$

and dividing (3) by (2),

$$\frac{b}{a} = \frac{\sin\alpha}{\cos\alpha} = \tan\alpha, \quad \therefore \quad \alpha = \tan^{-1}\frac{b}{a} \qquad ...(5)]$$

Then from (1),

$$\int e^{ax}\sin bx\,dx = \frac{e^{ax}(r\cos\alpha\sin bx - r\sin\alpha\cos bx)}{a^2+b^2}$$

$$= \frac{e^{ax}\,r(\sin bx\cos\alpha - \cos bx\sin\alpha)}{a^2+b^2} = \frac{e^{ax}\,r\sin(bx-\alpha)}{a^2+b^2}$$

[Substitute the values of r and α from (4) and (5)]

$$= \frac{e^{ax}\sqrt{a^2+b^2}\sin\left(bx - \tan^{-1}\frac{b}{a}\right)}{a^2+b^2} = \frac{e^{ax}\sin\left(bx-\tan^{-1}\frac{b}{a}\right)}{\sqrt{a^2+b^2}}$$

(ii) $\int e^{ax}\cos bx\,dx$ [Integrate by parts]

$$= e^{ax}.\frac{\sin bx}{b} - \int e^{ax}\,a\,.\frac{\sin bx}{b}\,dx$$

$$= \frac{e^{ax}\sin bx}{b} - \frac{a}{b}\int e^{ax}\sin bx\,dx \qquad \text{[Integrate again by parts]}$$

$$= \frac{e^{ax}\sin bx}{b} - \frac{a}{b}\left[e^{ax}\left(-\frac{\cos bx}{b}\right) - \int e^{ax}\,a\left(-\frac{\cos bx}{b}\right)dx\right]$$

$$= \frac{e^{ax}\sin bx}{b} - \frac{a}{b}\left[-\frac{e^{ax}\cos bx}{b} + \frac{a}{b}\int e^{ax}\cos bx\,dx\right]$$

$$= \frac{e^{ax}\sin bx}{b} + \frac{a}{b^2}e^{ax}\cos bx - \frac{a^2}{b^2}\int e^{ax}\cos bx\,dx$$

Transposing,

$$\left(1+\frac{a^2}{b^2}\right)\int e^{ax}\cos bx\,dx = \frac{a\,e^{ax}\cos bx + b\,e^{ax}\sin bx}{b^2}$$

$$\therefore \quad \int e^{ax} \cos bx\, dx = \frac{e^{ax}(a \cos bx + b \sin bx)}{b^2\left(1+\dfrac{a^2}{b^2}\right)} = \frac{e^{ax}(a \cos bx + b \sin bx)}{a^2+b^2} \quad ...(1)$$

Put $a = r \cos \alpha$ ...(2), $b = r \sin \alpha$ ...(3)

[so that squaring (2) and (3), and adding,

$$a^2 + b^2 = r^2 . 1, \qquad \therefore \quad r = \sqrt{a^2+b^2} \quad ...(4)$$

and dividing (3) by (2),

$$\frac{b}{a} = \frac{\sin \alpha}{\cos \alpha} = \tan \alpha, \quad \therefore \quad \alpha = \tan^{-1}\frac{b}{a} \quad ...(5)]$$

Then from (1),

$$\int e^{ax} \cos bx\, dx = \frac{e^{ax}(r \cos \alpha \cos bx + r \sin \alpha \sin bx)}{a^2+b^2}$$

$$= \frac{e^{ax} r(\cos bx \cos \alpha + \sin bx \sin \alpha)}{a^2+b^2}$$

$$= \frac{e^{ax} r \cos(bx - \alpha)}{a^2+b^2}$$ [Substitute the values of r and α from (4) and (5)]

$$= \frac{e^{ax}\sqrt{a^2+b^2}\cos\left(bx - \tan^{-1}\dfrac{b}{a}\right)}{a^2+b^2} = \frac{e^{ax}\cos\left(bx - \tan^{-1}\dfrac{b}{a}\right)}{\sqrt{a^2+b^2}}.$$

Hence we have

Two standard integrals.

XXV. $$\int \mathbf{e^{ax} \sin bx\, dx} = \frac{1}{a^2+b^2} e^{ax}(a \sin bx - b \cos bx)$$

$$= \frac{1}{\sqrt{a^2+b^2}} e^{ax} \sin\left(bx - \tan^{-1}\frac{b}{a}\right).$$

XXVI. $$\int \mathbf{e^{ax} \cos bx\, dx} = \frac{1}{a^2+b^2} e^{ax}(a \cos bx + b \sin bx)$$

$$= \frac{1}{\sqrt{a^2+b^2}} e^{ax} \cos\left(bx - \tan^{-1}\frac{b}{a}\right).$$

Note. Which form of the answer to use, and where ?

For evaluating indefinite integrals, when a is positive, either form may be used, and *generally* the shorter, *viz.*,

$$\int e^{ax} \sin bx\, dx = \frac{1}{\sqrt{a^2+b^2}} e^{ax} \sin\left(bx - \tan^{-1}\frac{b}{a}\right) \text{ is used.}$$

But (1) for evaluating indefinite integrals *when a is negative,*

and (2) for evaluating definite integrals,

$$\int e^{ax} \sin bx \, dx = \frac{1}{a^2 + b^2} e^{ax} (a \sin bx - b \cos bx) \text{ is used.}$$

For, in (1), it will be found (see Ex. 7, (*i*), following) that

$$\int e^{-ax} \sin bx \, dx = -\frac{1}{\sqrt{a^2 + b^2}} e^{-ax} \sin\left(bx + \tan^{-1}\frac{b}{a}\right),$$

and in (2), the trigonometric calculations in the answer are easier.

Similarly for the two forms of the answer for $\int e^{ax} \cos bx \, dx$.

Example 1. *Integrate* $\int e^{3x} \sin 4x \, dx$.

$\int e^{3x} \sin 4x \, dx$ [Compare with $\int e^{ax} \sin bx \, dx$, here $a = 3$, $b = 4$]

$$= \frac{1}{\sqrt{(3)^2 + (4)^2}} e^{3x} \sin (4x - \tan^{-1} \tfrac{4}{3})$$

$$\left[\frac{1}{\sqrt{a^2 + b^2}} e^{ax} \sin\left(bx - \tan^{-1}\frac{b}{a}\right) \text{(Art. 27, XXV)}\right]$$

$$= \tfrac{1}{5} e^{3x} \sin (4x - \tan^{-1} \tfrac{4}{3}).$$

Example 2. *Integrate* $\int e^{x} \sin^2 x \cos^3 x \, dx$.

$\int e^{x} \sin^2 x \cos^3 x \, dx$

$$= \int e^{x} (\sin^2 x \cos^2 x) \cos x \, dx$$

$$\left| \begin{array}{l} \sin 2x = 2 \sin x \cos x \\ \therefore \sin x \cos x = \dfrac{\sin 2x}{2} \end{array} \right.$$

$$= \int e^{x} \frac{\sin^2 2x}{4} \cos x \, dx = \int e^{x} \frac{1}{4}\left(\frac{1 - \cos 4x}{2}\right) \cos x \, dx$$

$$= \int e^{x} \frac{1}{8} (\cos x - \cos 4x \cos x) \, dx = \int e^{x} \frac{1}{16} (2 \cos x - 2 \cos 4x \cos x) \, dx$$

$$= \int e^{x} \frac{1}{16} [2 \cos x - (\cos 5x + \cos 3x)] \, dx$$

$$= \frac{1}{16} [2 \int e^{x} \cos x \, dx - \int e^{x} \cos 3x \, dx - \int e^{x} \cos 5x \, dx]$$

$$= \frac{1}{16} \left[2 \frac{1}{\sqrt{(1)^2 + (1)^2}} e^{x} \cos (x - \tan^{-1} \tfrac{1}{1}) - \frac{1}{\sqrt{(1)^2 + (3)^2}} \right.$$

$$\left. \times e^{x} \cos (3x - \tan^{-1} \tfrac{3}{1}) - \frac{1}{\sqrt{(1)^2 + (5)^2}} e^{x} \cos (5x - \tan^{-1} \tfrac{5}{1}) \right]$$

$$= \frac{1}{16} \left[\sqrt{2} \, e^{x} \cos\left(x - \frac{\pi}{4}\right) - \frac{1}{\sqrt{10}} e^{x} \cos (3x - \tan^{-1} 3) - \frac{1}{\sqrt{26}} e^{x} \cos (5x - \tan^{-1} 5) \right].$$

EXAMPLES

1. Integrate (*i*) $\int e^{mx} \sin x \cos x \, dx$.

(*ii*) $\int (a \sin^2 x + 2b \sin x \cos x + c \cos^2 x) e^{x} \, dx$.

2. Integrate $\int e^{mx} \sin x \cos x \cos 2x \, dx$.

3. Integrate $\int e^{m \sin^{-1} x}\, dx$.

4. Integrate (*i*) $\dfrac{e^{m \tan^{-1} x}}{(1+x^2)^{\frac{3}{2}}}\, dx$.

(*ii*) $\displaystyle\int \frac{e^{m \tan^{-1} x}}{(1+x^2)^2}\, dx$. (*iii*) $\displaystyle\int \frac{e^{m \tan^{-1} x}}{(1+x^2)^{\frac{5}{2}}}\, dx$.

5. Integrate $\displaystyle\int \cos\left\{b \log \frac{x}{a}\right\} dx$.

6. If $u = \int e^{ax} \cos bx\, dx$, $v = \int e^{ax} \sin bx\, dx$, prove that

$$\tan^{-1} \frac{v}{u} + \tan^{-1} \frac{b}{a} = bx, \text{ and that } (a^2 + b^2)(u^2 + v^2) = e^{2ax}.$$

****7.** Prove that

(*i*) $\displaystyle\int e^{-ax} \sin bx\, dx = -\frac{1}{\sqrt{a^2+b^2}}\, e^{-ax} \sin\left(bx + \tan^{-1} \frac{b}{a}\right)$.

(*ii*) $\displaystyle\int e^{-ax} \cos bx\, dx = -\frac{1}{\sqrt{a^2+b^2}}\, e^{-ax} \cos\left(bx + \tan^{-1} \frac{b}{a}\right)$.

[(*i*) $\displaystyle\int e^{-ax} \sin bx\, dx = \frac{1}{(-a)^2+b^2}\, e^{-ax} (-a \sin bx - b \cos bx)$

$\left[\dfrac{1}{a^2+b^2}\, e^{ax} (a \sin bx - b \cos bx\right.$ (Art. 27, XXV), Note on p. 46)$\Big]$

$$= -\frac{1}{a^2+b^2}\, e^{-ax} (a \sin bx + b \cos bx) \qquad \text{...(1)}$$

Put $a = r \cos \alpha$ ...(2), $b = r \sin \alpha$ ...(3), and proceed as in Art. 27, (*i*).]

8. Evaluate $\displaystyle\int_0^\infty e^{-ax} \sin bx\, dx$.

$$\int_0^\infty e^{-ax} \sin bx\, dx = \left[\frac{1}{(-a)^2+b^2}\, e^{-ax} (-a \sin bx - b \cos bx)\right]_0^\infty$$

$\left[\dfrac{1}{a^2+b^2}\, e^{ax} (a \sin bx - b \cos bx\right.$ (Art. 27, XXV), Note on p. 46$\Big]$

$$= -\frac{1}{a^2+b^2}\left[e^{-ax} (a \sin bx + b \cos bx)\right]_0^\infty$$

$$= -\frac{1}{a^2+b^2}\left[\underset{x \to \infty}{\text{Lt}}\, e^{-ax} (a \sin bx + b \cos bx) - b\right]$$

$$\left[\text{But } \underset{x \to \infty}{\text{Lt}}\, e^{-ax} = 0 \left(\because\ e^{-\infty} = \frac{1}{e^{\infty}} = \frac{1}{\infty} = 0\right) \text{ and } \underset{x \to \infty}{\text{Lt}} \sin bx, \quad \underset{x \to \infty}{\text{Lt}} \cos bx \text{ cannot be numerically} > 1\right]$$

$$= -\frac{1}{a^2 + b^2}[0 - b] = \frac{b}{a^2 + b^2}.$$

9. Integrate $\int \cosh ax \sin bx\, dx$.

SECTION II

GENERAL RULE OF INTEGRATION BY PARTS

28. To prove the general rule of integration by parts.

Let u, v be two functions of x, and dashes denote differentiations and suffixes integrations w.r.t. x.

Then $\int u\, v\, dx$ [Integrate by parts]

$$= uv_1 - \int u'v_1\, dx \qquad ...(1) \qquad \text{[Integrate again by parts]}$$

$$= uv_1 - \left[u'v_2 - \int u''v_2\, dx\right]$$

$$= uv_1 - u'v_2 + \int u''v_2\, dx \qquad ...(2) \qquad \text{[Integrate again by parts]}$$

$$= uv_1 - u'v_2 + u''v_3 - \int u'''v_3 dx \qquad ...(3)$$

Generalising from (1), (2), (3),

$$\int uv\, dx = uv_1 - u'v_2 + u''v_3 - \dots \text{ to } n \text{ terms} + (-1)^n \int u^{(n)} v_n\, dx^*,$$

where $u^{(n)}$ stands for u with n dashes.

In words : *Integral of product of two functions*

= *1st function* × *integral of 2nd – go on differentiating 1st, integrating 2nd, signs alternately + ve and – ve, ...*

Note. General rule of integration by parts. *The above rule is called the* **general rule of integration by parts.** *It enables us to write down most quickly the integral of the product of two functions, if the first is a power of x and the second can be integrated at once any number of times (e.g., e^{ax}, cos ax, sin ax, cosh ax, sinh ax, etc.).*

Example. *Integrate* $\int x^4 \sin x \cos x\, dx$.

$$\int x^4 \sin x \cos x\, dx = \int x^4 \frac{\sin 2x}{2}\, dx \qquad \left|\ \begin{array}{l} \sin 2x = 2 \sin x \cos x \\ \therefore\ \sin x \cos x = \dfrac{\sin 2x}{2}. \end{array}\right.$$

***How to write this step.**

In (1), the last term $= -\int u'v_1\, dx = +(-1)^1 \int u'v_1\, dx$;

in (2), the last term $= +\int u''v_2\, dx = +(-1)^2 \int u''v_2 dx$;

in (3), the last term $= -\int u'''v_3 dx = +(-1)^3 \int u'''\, v_3\, dx$; ... ;

$\therefore$ in general, the last term $= +(-1)^n \int u^{(n)}\, v_n\, dx$.

$$= \frac{1}{2}\int x^4 \sin 2x\, dx \qquad \text{[Integrate by general rule of integration by parts]}$$

$$= \frac{1}{2}\left[x^4\left(-\frac{\cos 2x}{2}\right) - 4x^3\left(-\frac{\sin 2x}{4}\right) + 12x^2\left(\frac{\cos 2x}{8}\right) - 24x\left(\frac{\sin 2x}{16}\right) + 24\left(-\frac{\cos 2x}{32}\right) - 0\right] \quad ...(1)$$

[1st fun. × int. of 2nd – go on differentiating 1st, integrating 2nd, signs alternately + ve and – ve (Art. 28)]

$$= \frac{1}{2}\left[-\frac{x^4\cos 2x}{2} + x^3\sin 2x + \frac{3x^2\cos 2x}{2} - \frac{3x\sin 2x}{2} - \frac{3\cos 2x}{4}\right] \quad ...(2)$$

Caution. *Do not effect any arithmetical simplification until the integration is complete. (See (1) above.) The whole process is much less liable to error if simplification is postponed to the end. (See (2) above.)*

EXAMPLES

1. Integrate (*i*) $\int x^3 e^x\, dx$. (*ii*) $\int x^4 e^x\, dx$.

2. Evaluate $\int_0^1 x^6 e^x\, dx$.

3. Integrate (*i*) $\int x^3 \sin x\, dx$. (*ii*) $\int x^3 \sin^2 x\, dx$.

4. Evaluate (*i*) $\int_0^\pi x^5 \sin x\, dx$. (*ii*) $\int_0^\pi x^5 \cos x\, dx$.

5. Integrate (*i*) $\int x^3 \sinh x\, dx$. (*ii*) $\int x^5 \cosh x\, dx$.

6. Show that if u is a rational integral function* of x,

$$\int e^{x/a} u\, dx = ae^{x/a}\left\{u - a\frac{du}{dx} + a^2\frac{d^2u}{dx^2} - a^3\frac{d^3u}{dx^3} + \ldots\right\},$$

where the series within the brackets is necessarily finite.

29. To evaluate integrals of the form

$$\int x^n e^{ax} \sin bx\, dx \quad \text{or} \quad \int x^n e^{ax} \cos bx\, dx.$$

$$\int x^n e^{ax} \sin bx\, dx = \int x^n (e^{ax} \sin bx)\, dx,$$

which can be integrated by the general rule of integration by parts.

Similarly for $\int x^n e^{ax} \cos bx\, dx$.

Example. *Integrate* $\int x^3 e^x \cos x\, dx$.

---

*__Rational integral function or polynomial. Def. A rational integral function of x or a polynomial in x__ is a function of the form

$$a_0x^n + a_1x^{n-1} + a_2x^{n-2} + \ldots + a_n,$$

where $a_0, a_1, a_2, \ldots, a_n$ are constants, and n is a positive integer.

$$\int e^x \cos x\, dx = \frac{1}{\sqrt{(1)^2 + (1)^2}} e^x \cos\left(x - \tan^{-1}\frac{1}{1}\right) = \frac{1}{\sqrt{2}} e^x \cos\left(x - \frac{\pi}{4}\right).$$

[Thus for the integral of $e^x \cos x$, we divide it by $\sqrt{2}$, and subtract $\frac{\pi}{4}$ from the angle]

$\therefore \int x^3 (e^x \cos x)\, dx$ [Integrate by general rule of integration by parts]

$$= x^3 . \frac{1}{2^{\frac{1}{2}}} e^x \cos\left(x - \frac{\pi}{4}\right) - 3x^2 . \frac{1}{2^{\frac{2}{2}}} e^x \cos\left(x - \frac{2\pi}{4}\right)$$

$$+ 6x . \frac{1}{2^{\frac{3}{2}}} e^x \cos\left(x - \frac{3\pi}{4}\right) - 6 . \frac{1}{2^{\frac{4}{2}}} e^x \cos\left(x - \frac{4\pi}{4}\right) + 0$$

$$= e^x \left[\frac{1}{\sqrt{2}} x^3 \cos\left(x - \frac{\pi}{4}\right) - \frac{3}{2} x^2 \sin x + \frac{3}{\sqrt{2}} x \cos\left(x - \frac{3\pi}{4}\right) + \frac{3}{2} \cos x\right]$$

EXAMPLES

1. Integrate $\int x\, e^x \sin^2 x\, dx$.

2. Integrate $\int x^2\, 2^x \sin 2x\, dx$.

MISCELLANEOUS EXAMPLES ON CHAPTER III

1. Integrate (*i*) $\int x\, e^{ax}\, dx$.

(*ii*) $\int x \tan^2 x\, dx$. (*iii*) $\int \frac{x + \sin x}{1 + \cos x}\, dx$.

2. Integrate (*i*) $\int x \log x\, dx$. (*ii*) $\int x \tan^{-1} x\, dx$.

3. Integrate (*i*) $\int x^2 \tan^{-1} x\, dx$. (*ii*) $\int \frac{x^2 \tan^{-1} x}{1 + x}\, dx$.

4. Integrate (*i*) $\int \log(x + \sqrt{a^2 + x^2})\, dx$.

(*ii*) $\int \log(1 + x^2)\, dx$.

5. Integrate

(*i*) $\int \frac{x \tan^{-1} x}{(1 + x^2)^{\frac{3}{2}}}\, dx$. (*ii*) $\int \frac{\sin^{-1} x}{(1 - x^2)^{\frac{3}{2}}}\, dx$.

6. (*a*) Prove that $\int_0^1 \frac{x^3 \sin^{-1} x}{\sqrt{1 - x^2}}\, dx = \frac{7}{9}$.

**(*b*) Integrate $\int x \sin^{-1} \frac{1}{2} \sqrt{\frac{2a - x}{a}}\, dx$.

7. Integrate (*i*) $\int e^x \frac{1 + x \log x}{x}\, dx.$ (*ii*) $\int e^x \frac{1 + x^2}{(1 + x)^2}\, dx.$

8. Integrate (*i*) $\int e^{ax} \cos (bx + c)\, dx.$

(*ii*) $\int e^{2x} \cos^3 x\, dx.$ (*iii*) $\int e^{ax} \sin^3 x\, dx.$

****9.** Evaluate $\int_0^\infty e^{-\alpha x} \cos \beta x \cos \gamma x\, dx$, where $\alpha > 0$.

10. Integrate (*i*) $\int x^2 e^{mx}\, dx.$

(*ii*) $\int x^2 (\log x)^2\, dx.$ (*iii*) $\int (\log x)^2\, dx.$

****11.** Evaluate $\int_0^\infty x e^{-2x} \cos x\, dx.$

[$\int_0^\infty x(e^{-2x} \cos x)dx$ [Integrate by parts]

$$= \left[x \cdot \frac{1}{5} e^{-2x} (-2 \cos x + \sin x)\right]_0^\infty - \int_0^\infty 1 \cdot \frac{1}{5} e^{-2x} (-2 \cos x + \sin x)\, dx$$

$$= 0 - 0 + \frac{2}{5} \int_0^\infty e^{-2x} \cos x\, dx - \frac{1}{5} \int_0^\infty e^{-2x} \sin x\, dx.]$$

12. (*a*) Evaluate $\int_0^{\frac{\pi}{2}} x^3 \sin 3x\, dx.$

(*b*) Integrate $\int e^x (x \cos x + \sin x)\, dx.$

CHAPTER IV

INTEGRATION OF RATIONAL ALGEBRAIC FUNCTIONS

30. Polynomial. Def. A *polynomial in x* is a function of the form

$$a_0x^n + a_1x^{n-1} + a_2x^{n-2} + \dots + a_n,$$

where $a_0, a_1, a_2, \dots, a_n$ are constants, and n is a positive integer.

Rational algebraic fraction. Def. A *rational algebraic fraction* is a fraction of the form

$$\frac{a_0x^n + a_1x^{n-1} + a_2x^{n-2} + \dots + a_n}{b_0x^p + b_1x^{p-1} + b_2x^{p-2} + \dots + b_p},$$

where $a_0, a_1, a_2, \dots, a_n$; $b_0, b_1, b_2, \dots, b_p$ are constants, and n and p are positive integers.

Proper rational algebraic fraction. Def. A *proper rational algebraic fraction* is a fraction in which the degree of the numerator is less than that of the denominator.

Note on partial fractions.

31. (a) If the degree of the numerator is equal to, or greater than, that of the denominator, we can divide the numerator by the denominator until the degree of the remainder is less than that of the denominator. Then the given fraction = a polynomial + a proper rational algebraic fraction.

Hence we have only to resolve a proper rational algebraic fraction into partial fractions.

Example.

$$\frac{x^3}{(x-1)(x-2)} = x + 3 + \frac{7x-6}{(x-1)(x-2)}.$$

$$\begin{array}{r|l} x^2 - 3x + 2)\ x^3 & (x+3 \\ x^3 - 3x^2 + 2x & \\ \hline 3x^2 - 2x & \\ 3x^2 - 9x + 6 & \\ \hline 7x - 6 & \end{array}$$

31. (b) Partial fractions of the correct form.

First resolve the denominator into its real prime factors. We know from Algebra that these factors will be linear or quadratic, non-repeated or repeated, and that for a proper rational algebraic fraction—

(1) corresponding to a non-repeated linear factor $(x - a)$ in the denominator, there is a partial fraction of the form $\dfrac{A}{x-a}$;

(2) corresponding to a repeated linear factor $(x - a)^r$ in the denominator,

there are r partial fractions of the form

$$\frac{A_1}{x-a}+\frac{A_2}{(x-a)^2}+\dots+\frac{A_r}{(x-a)^r};$$

(3) corresponding to a non-repeated quadratic factor (x^2+ax+b) in the denominator, there is a partial fraction of the form $\dfrac{Ax+B}{x^2+ax+b}$;

(4) corresponding to a repeated quadratic factor $(x^2+ax+b)^r$ in the denominator, there are r partial fractions of the form

$$\frac{A_1x+B_1}{x^2+ax+b}+\frac{A_2x+B_2}{(x^2+ax+b)^2}+\dots+\frac{A_rx+B_r}{(x^2+ax+b)^r}.$$

The values of the unknown constants A, A_1, ..., B, B_1, ... can be found as shown in the following Arts.

32. Case I. Rule to find the partial fraction corresponding to a non-repeated linear factor (x – a) in the denominator of a proper rational algebraic fraction :

(i) Put the factor $(x-a)=0$, and find the value of $x(=a)$.

(ii) Substitute this value of x $(=a)$ everywhere in the given fraction except in the factor $(x-a)$ itself.

Note. *The above rule enables us also to find the partial fraction corresponding to a non-repeated linear factor $(ax+b)$ in the denominator of a proper rational algebraic fraction. We, however, give below the proof for $(x-a)$. A similar proof holds for $(ax+b)$.*

** [**Proof.** Let the given fraction be $\dfrac{f(x)}{(x-a)\phi(x)}$.

Let $$\frac{f(x)}{(x-a)\phi(x)}=\frac{A}{x-a}+\frac{\psi(x)}{\phi(x)} \quad ...(1) \quad | \ (x-a)\,\phi(x)$$

Multiplying both sides by $(x-a)\,\phi(x)$ [Denom. of L.H.S.]

$$f(x)=A\phi(x)+\psi(x)(x-a).$$

Putting $x=a$ on both sides,

$$f(a)=A\phi(a)+\psi(a)(0),$$

$$\therefore \quad A=\frac{f(a)}{\phi(a)}.$$

$\therefore$ from (1), the partial fraction corresponding to $(x-a)$ is

$$\frac{A}{x-a}=\frac{f(a)}{(x-a)\phi(a)},$$

which can be written down by putting $x=a$ everywhere in the given fraction $\dfrac{f(x)}{(x-a)\phi(x)}$ except in the factor $(x-a)$ itself.]

Two standard forms.

33. To integrate (*i*) $\int \frac{dx}{x^2 - a^2}$ **(x > a).** (*ii*) $\int \frac{dx}{a^2 - x^2}$ **(x < a).**

(*i*) $$\frac{1}{x^2 - a^2} = \frac{1}{(x+a)(x-a)} = \frac{1}{(x+a)(-2a)} + \frac{1}{2a(x-a)} \quad \text{[Rule* (Art. 32)]}$$

$$= \frac{1}{2a}\left[\frac{1}{x-a} - \frac{1}{x+a}\right].$$

$$\therefore \int \frac{1}{x^2 - a^2}\, dx = \frac{1}{2a}\int \left[\frac{1}{x-a} - \frac{1}{x+a}\right] dx$$

$$= \frac{1}{2a}[\log(x-a) - \log(x+a)] = \frac{1}{2a}\log\frac{x-a}{x+a} \quad (x > a).$$

(*ii*) $$\frac{1}{a^2 - x^2} = \frac{1}{(a+x)(a-x)} = \frac{1}{(a+x)(2a)} + \frac{1}{(2a)(a-x)} \quad \text{[Rule (Art. 32)]}$$

$$= \frac{1}{2a}\left[\frac{1}{a+x} + \frac{1}{a-x}\right].$$

$$\therefore \int \frac{1}{a^2 - x^2}\, dx = \frac{1}{2a}\int \left[\frac{1}{a+x} + \frac{1}{a-x}\right] dx = \frac{1}{2a}\left[\log(a+x) + \frac{\log(a-x)}{-1}\right]$$

$$= \frac{1}{2a}[\log(a+x) - \log(a-x)] = \frac{1}{2a}\log\frac{a+x}{a-x} \quad (x < a).$$

Hence we have

Two more standard forms.

XXVII. $$\int \frac{1}{x^2 - a^2}\, dx = \frac{1}{2a}\log\frac{x-a}{x+a} \quad (x > a).$$

XXVIII. $$\int \frac{1}{a^2 - x^2}\, dx = \frac{1}{2a}\log\frac{a+x}{a-x} \quad (x < a).$$

***Explanation.** To find the partial fraction corresponding to the non-repeated linear factor $(x + a)$ in the denominator :

(*i*) Put $x + a = 0$, then $x = -a$.

(*ii*) Substitute this value of $x(= -a)$ everywhere in the given fraction $\frac{1}{(x+a)(x-a)}$ except in the factor $(x + a)$ itself (from which we found the value of x), thus getting

$$\frac{1}{(x+a)(-a-a)} = \frac{1}{(x+a)(-2a)}.$$

Similarly for the partial fraction corresponding to the factor $(x - a)$ in the denominator.

Example. *Integrate* $\int \frac{dx}{20-5x^2}$.

$$\int \frac{1}{20-5x^2}\,dx \quad \text{[Take the numerical value of coeff. of } x^2 \text{ outside the quadratic]}$$

$$= \int \frac{1}{5(4-x^2)}\,dx$$

$$= \frac{1}{5}\int \frac{1}{(2)^2-x^2}\,dx \qquad \left[\text{Compare with } \int \frac{1}{a^2-x^2}\,dx, \text{ here } a=2\right]$$

$$= \frac{1}{5}\cdot\frac{1}{2(2)}\log\frac{2+x}{2-x} \qquad \left[\frac{1}{2a}\log\frac{a+x}{a-x} \text{ (Art. 33, XXVIII)}\right]$$

$$= \frac{1}{20}\log\frac{2+x}{2-x}.$$

EXAMPLES

Integrate (Exs. 1—2) :

1. (*i*) $\int \frac{dx}{x^2-4}$. (*ii*) $\int \frac{x^2\,dx}{x^2-4}$. **2.** $\int \frac{\sin x}{\sin 3x}\,dx$.

34. To integrate $\int \frac{\mathbf{dx}}{\mathbf{ax^2+bx+c}}$.

$$\int \frac{dx}{ax^2+bx+c}\,dx = \int \frac{1}{a\left(x^2+\frac{b}{a}x+\frac{c}{a}\right)}\,dx$$

$$= \frac{1}{a}\int \frac{1}{\left(x^2+\frac{b}{a}x+\frac{b^2}{4a^2}\right)-\frac{b^2}{4a^2}+\frac{c}{a}}\,dx$$

$$= \frac{1}{a}\int \frac{1}{\left(x+\frac{b}{2a}\right)^2+\frac{4ac-b^2}{4a^2}}\,dx \qquad \text{...(1)}$$

Case I. If $b^2 > 4ac$, then from (1),

$$\int \frac{1}{ax^2+bx+c}\,dx = \frac{1}{a}\int \frac{1}{\left(x+\frac{b}{2a}\right)^2-\frac{b^2-4ac}{4a^2}}\,dx$$

$$= \frac{1}{a}\int \frac{1}{\left(x+\frac{b}{2a}\right)^2-\left(\frac{\sqrt{b^2-4ac}}{2a}\right)^2}\,dx$$

$$= \frac{1}{a}\cdot\frac{1}{2\,\frac{\sqrt{b^2-4ac}}{2a}}\log\frac{x+\frac{b}{2a}-\frac{\sqrt{b^2-4ac}}{2a}}{x+\frac{b}{2a}+\frac{\sqrt{b^2-4ac}}{2a}}$$

$$= \frac{1}{\sqrt{b^2 - 4ac}} \log \frac{2ax + b - \sqrt{b^2 - 4ac}}{2ax + b + \sqrt{b^2 - 4ac}}.$$

Case II. If $b^2 < 4ac$, then from (1),

$$\int \frac{1}{ax^2 + bx + c}\, dx = \frac{1}{a} \int \frac{1}{\left(x + \frac{b}{2a}\right)^2 + \left(\frac{\sqrt{4ac - b^2}}{2a}\right)^2}\, dx$$

$$= \frac{1}{a} \cdot \frac{1}{\frac{\sqrt{4ac - b^2}}{2a}} \tan^{-1} \frac{x + \frac{b}{2a}}{\frac{\sqrt{4ac - b^2}}{2a}} = \frac{2}{\sqrt{4ac - b^2}} \tan^{-1} \frac{2ax + b}{\sqrt{4ac - b^2}}.$$

Note. Rule to integrate $\int \frac{1}{\textbf{quadratic}}\,$ **dx.**

Take the numerical value of the coefficient of x^2 outside the quadratic. Complete the square in the terms containing x, and use the proper standard form.

The same rule will be used to integrate $\int \frac{1}{\sqrt{quadratic}}\, dx$ (Art. 41) *and* $\int \sqrt{quadratic}\, dx$. (Art. 41, Ex. 4).

EXAMPLES

1. Integrate (*i*) $\int \frac{Ax + B}{ax^2 + bx + c}$. (*ii*) $\int \frac{dx}{2x^2 - 2x + 1}$.

2. Evaluate $\int_0^1 \frac{dx}{3 + 2x + x^2}$.

35. To integrate $\int \frac{\mathbf{Ax + B}}{\mathbf{ax^2 + bx + c}}$ **dx.**

$$\int \frac{Ax + B}{ax^2 + bx + c}\, dx = \int \frac{\frac{A}{2a}(2ax + b) - \frac{Ab}{2a} + B\,*}{ax^2 + bx + c}\, dx$$

***How to write this step.** *Write in the numerator the differential coefficient of the quadratic $(ax^2 + bx + c)$ in the denominator, viz., $2ax + b$, and adjust the coefficient of x by multiplying $2ax + b$ by* $\frac{A}{2a}$ ($\because$ we have got $2a$ as the coefficient of x in $2ax + b$ but we require A as the coefficient of x in $Ax + B$), thus getting $\frac{A}{2a}(2ax + b)$; *now adjust the constant term by subtracting* $\frac{Ab}{2a}$ *and adding B* ($\because$ we have got $\frac{Ab}{2a}$ as the constant term in $\frac{A}{2a}(2ax + b)$, but we require B as the constant term in $Ax + B$), thus getting $Ax + B = \frac{A}{2a}(2ax + b) - \frac{Ab}{2a} + B$.

$$= \frac{A}{2a} \int \frac{2ax + b}{ax^2 + bx + c}\, dx + \left(B - \frac{Ab}{2a}\right) \int \frac{1}{ax^2 + bx + c}\, dx$$

$$= \frac{A}{2a} \log (ax^2 + bx + c) + \left(B - \frac{Ab}{2a}\right) \int \frac{1}{ax^2 + bx + c}\, dx \qquad ...(1)$$

Now $\int \frac{1}{ax^2 + bx + c}\, dx$ can be integrated (Art. 34),

$\therefore$ from (1), $\int \frac{Ax + B}{ax^2 + bx + c}\, dx$ can be integrated.

Example. *Integrate* $\int \frac{x}{x^2 + 2x + 2}\, dx.$

$$\int \frac{x}{x^2 + 2x + 2}\, dx = \int \frac{\frac{1}{2}(2x + 2) - 1^*}{x^2 + 2x + 2}\, dx$$

$$= \frac{1}{2} \int \frac{2x + 2}{x^2 + 2x + 2}\, dx - \int \frac{1}{x^2 + 2x + 2}\, dx$$

$$= \frac{1}{2} \log (x^2 + 2x + 2) - \int \frac{1}{x^2 + 2x + 2}\, dx \qquad ...(1)$$

Now $\int \frac{1}{(x^2 + 2x) + 2}\, dx = \int \frac{1}{(x^2 + 2x + 1) + 1}\, dx = \int \frac{1}{(x + 1)^2 + (1)^2}\, dx$

$$= \frac{1}{1} \tan^{-1} \frac{x + 1}{1} = \tan^{-1} (x + 1).$$

$\therefore$ from (1), $\int \frac{x}{x^2 + 2x + 2}\, dx = \frac{1}{2} \log (x^2 + 2x + 2) - \tan^{-1} (x + 1).$

EXAMPLES

Integrate (Exs. 1—5) :

1. $\int \frac{x + 1}{x^2 + 4x + 5}\, dx$

2. $\int \frac{x\, dx}{x^2 + 2x + 1}.$

3. $\int \frac{2x - 1}{2x^2 + 2x + 1}\, dx.$

4. $\int \frac{(x + 1)dx}{3 + 2x - x^2}.$ $\left[\int \frac{x + 1}{3 + 2x - x^2}\, dx = \int \frac{x + 1}{(1 + x)(3 - x)}\, dx = \int \frac{1}{3 - x}\, dx.\right]$

5. (*i*) $\int \frac{2x^2 + 3x + 4}{x^2 + 4x + 5}\, dx.$ (*ii*) $\int \frac{(x - 1)^2\, dx}{x^2 + 2x + 10}.$

Integration of a rational algebraic fraction.

36. To integrate a rational algebraic fraction whose denominator can be resolved into linear factors.

__How to write this step.__ Proceed as in the footnote on p. 57.

Example. *Integrate* $\int \frac{x^2+1}{(2x+1)(x-1)(x+1)}\,dx.$

$$\frac{x^2+1}{(2x+1)(x-1)(x+1)}$$

$$= \frac{\frac{5}{4}}{(2x+1)(-\frac{3}{2})(\frac{1}{2})} + \frac{2}{(3)(x-1)(2)} + \frac{2}{(-1)(-2)(x+1)} \qquad \text{[Rule (Art. 32)]}$$

$$= -\frac{5}{3(2x+1)} + \frac{1}{3(x-1)} + \frac{1^*}{x+1} \qquad ...(1)$$

$$\therefore \quad \int \frac{x^2+1}{(2x+1)(x-1)(x+1)}\,dx = -\frac{5}{3}\int \frac{1}{2x+1}\,dx + \frac{1}{3}\int \frac{1}{x-1}\,dx + \int \frac{1}{x+1}\,dx$$

$$= -\frac{5}{3}\,\frac{\log(2x+1)}{2} + \frac{1}{3}\log(x-1) + \log(x+1)$$

$$= -\tfrac{5}{6}\log(2x+1) + \tfrac{1}{3}\log(x-1) + \log(x+1).$$

EXAMPLES

Integrate (Exs. 1—6) :

1. (*i*) $\int \frac{x\,dx}{(x-a)(x-b)}.$ (*ii*) $\int \frac{x^2}{(x+a)(x+b)(x+c)}\,dx.$

2. (*i*) $\int \frac{x^2\,dx}{(x+1)(x-2)(x+3)}.$ (*ii*) $\int \frac{x^2+lx+m}{(x-a)(x-b)(x-c)}\,dx.$

3. (*i*) $\int \frac{1}{x(x^2-4)}\,dx.$ (*ii*) $\int \frac{x^2+1}{x(x^2-1)}\,dx.$

(*iii*) $\int \frac{1-2x^2}{2x-x^3}\,dx.$

4. $\int x^2(x-a)^{-1}(x-b)^{-1}\,dx.$

$\left[x^2(x-a)^{-1}(x-b)^{-1} = \frac{x^2}{(x-a)(x-b)}\right.$.

Here the degree of the num. = that of the denom.

$\therefore$ we divide the num. by the denom.,

the quotient is 1.

$$x^2 \ldots)\, x^2\, (1$$
$$\underline{x^2 \ldots}$$

---

*__Check.__ On both sides of (1), put x = a number for which the denominator of the L.H.S. is not zero, say, $x = 2$, we get

$$\frac{5}{(5)(1)(3)} = -\frac{5}{3(5)} + \frac{1}{3(1)} + \frac{1}{3}, \text{ or } \frac{1}{3} = -\frac{1}{3} + \frac{1}{3} + \frac{1}{3}, \text{ or } \frac{1}{3} = \frac{1}{3},$$

which is true.

Similarly the partial fractions of any rational algebraic fraction should be checked.

$$\therefore \qquad \frac{x^2}{(x-a)(x-b)} = 1 + \frac{a^2}{(x-a)(a-b)} + \frac{b^2}{(b-a)(x-b)}. \qquad \text{[Rule (Art. 32)]}$$

Note. Important. *After having found the quotient, 1, we have found the partial fractions corresponding to $(x-a)$, $(x-b)$ in the denominator of the given fraction by applying the usual rule (Art. 32) to the given fraction $\frac{x^2}{(x-a)(x-b)}$* instead of the remaining fraction left after division.*

This method is general, and usually saves much labour.

5. (*i*) $\int \frac{(x-a)(x-b)}{(x-c)(x-d)}\, dx.$ (*ii*) $\int \frac{(x-a)(x-b)(x-c)}{(x-a')(x-b')(x-c')}\, dx.$

****6.** $\int \frac{x^5}{x^3 - 2x^2 - 5x + 6}\, dx.$

7. (*i*) Investigate $\int_0^{\frac{\pi}{2}} \frac{\cos x\, dx}{(1+\sin x)(2+\sin x)}.$

(*ii*) Integrate $\int \frac{dx}{\sin x + \sin 2x}.$

[**Note. Important.** *If* $\cos x$ *is a factor of the numerator or denominator, and the remaining fraction is a rational function of* $\sin x$, *put* $\sin x = z$.

---

*[**Proof.** $\frac{x^2}{(x-a)(x-b)} = 1 + \frac{\text{an expression of first degree}}{(x-a)(x-b)}$ $x^2 \ldots)\ x^2\ (1$

$x^2 \ldots$

$$= 1 + \frac{A}{x-a} + \frac{B}{x-b} \qquad \ldots(1) \quad \mid (x-a)(x-b)$$

Multiplying both sides by $(x-a)(x-b)$, [Denom. of L.H.S.]

$$x^2 = (x-a)(x-b) + A(x-b) + B(x-a).$$

Putting $x = a$ on both sides,

$$a^2 = 0 + A(a-b) + B(0),$$

$$\therefore \qquad A = \frac{a^2}{a-b}.$$

$\therefore$ from (1), the partial fraction corresponding to $(x-a)$ is

$$\frac{A}{x-a} = \frac{a^2}{(x-a)(a-b)},$$

which can be found by applying the rule (Art. 32) to the *given fraction*

$$\frac{x^2}{(x-a)(x-b)}.$$

Similarly for the partial fraction corresponding to $(x-b)$.]

Similarly, if sin x is a factor of the numerator or denominator, and the remaining fraction is a rational function of cos x, put cos x = z.]

8. Integrate $\int \dfrac{dx}{x(1+4x^3+3x^6)}$.

[$\displaystyle\int \frac{1}{x(1+4x^3+3x^6)}\,dx$ (Multiply the num. and denom. by x^2)

$$= \int \frac{x^2\,dx}{x^3(1+4x^3+3x^6)}. \text{ Put } x^3 = z\dagger.$$

†**Note.** *If x^{n-1} is a factor of the numerator, and the remaining fraction is a rational function of x^n, put $x^n = z$.*]

9. Integrate $\int \dfrac{1}{x(x^5+1)}\,dx$.

10. Integrate (*i*) $\int \dfrac{dx}{e^x-1}$.

(*ii*) $\int \dfrac{dx}{1+3e^x+2e^{2x}}$.

[**Rule to integrate a rational function of e^x :** *Put $e^x = z$.*]

11. Evaluate $\int x^2 \log(1-x^2)\,dx$, and deduce that

$$\frac{1}{1.5}+\frac{1}{2.7}+\frac{1}{3.9}+\ldots = \frac{8}{9}-\frac{2}{3}\log_e 2.$$

[$\int x^2 \log(1-x^2)\,dx = \int \log(1-x^2)\,.\,x^2\,dx$. Integrate by parts.

It will be found that

$$\int x^2\log(1-x^2)\,dx = \frac{x^3}{3}\log(1-x^2)+\frac{2}{3}\left[-\frac{x^3}{3}-x+\frac{1}{2}\log\frac{1+x}{1-x}\right]$$

$$= \frac{x^3}{3}[\log(1+x)+\log(1-x)] - \frac{2}{3}\left(\frac{x^3}{3}+x\right) + \frac{1}{3}[\log(1+x)-\log(1-x)]$$

$$= \frac{1+x^3}{3}\log(1+x) - \frac{1-x^3}{3}\log(1-x) - \frac{2}{3}\left(\frac{x^3}{3}+x\right).$$

$$\therefore \int_0^1 x^2\log(1-x^2)\,dx$$

$$= \left[\frac{1+x^3}{3}\log(1+x) - \frac{1-x^3}{3}\log(1-x) - \frac{2}{3}\left(\frac{x^3}{3}+x\right)\right]_0^1$$

Evaluate ...(1)

***Why this step.** In order to make x^3 of x in the denominator, the remaining fraction $\dfrac{1}{1+4x^3+3x^6}$ being a function of x^3, we multiply the numerator and denominator by x^2.

Also $\int_0^1 x^2 \log(1-x^2)\,dx$ [Expand $\log(1-x^2)$ in powers of x, multiply each term by x^2, and integrate term by term]

Evaluate ...(2)

Equate R.H.S.'s of (1) and (2).]

37. Case II. Rule to find the partial fractions corresponding to a repeated linear factor $(x-a)^r$ in the denominator of a proper rational algebraic fraction :

(i) *Put the first power of the repeated factor, i.e., $x-a=y$, and find the value of x $(=a+y)$.*

(ii) *Substitute this value of $x(=a+y)$ everywhere in the given fraction, and expand the numerator and denominator* **in ascending powers of y.**

(iii) *Omit the factor $\frac{1}{y^r}$, and divide the numerator of the remaining fraction by its denominator until y^r is a factor of the remainder.*

(iv) *Divide by y^r, and put $y=x-a$.*

This gives all the partial fractions corresponding to the factor $(x-a)^r$ and a proper rational algebraic fraction which may again be resolved into partial fractions.

** **Proof.** Let the given fraction be $\frac{f(x)}{(x-a)^r\,\phi(x)}$.

Put $x-a=y$, then $x=a+y$.

$$\therefore \quad \frac{f(x)}{(x-a)^r\,\phi(x)} = \frac{f(a+y)}{y^r\,\phi(a+y)}$$

[Expand the num. and denom. **in ascending powers of y**]

$$= \frac{1}{y^r}\cdot\frac{A_0+A_1y+A_2y^2+\ldots}{B_0+B_1y+B_2y^2+\ldots} \qquad \ldots(1)$$

Omit the factor $\frac{1}{y^r}$, and

divide $A_0+A_1y+A_2y^2+\ldots$ by $B_0+B_1y+B_2y^2+\ldots$ until y^r is a factor of the remainder thus :

$B_0+B_1y+B_2y^2+\ldots)A_0+A_1y+A_2y^2+\ldots\,(C_0+C_1y+C_2y^2+\ldots+C_{r-1}y^{r-1})$

....

$D_0y^r+D_1y^{r+1}+D_2y^{r+2}+\ldots.$

[Note that y^r is a factor of the remainder. For,

the remainder $= D_0y^r+D_1y^{r+1}+D_2y^{r+2}+\ldots = y^r(D_0+D_1y+D_2y^2+\ldots)$]

$\therefore$ from (1),

$$\frac{f(x)}{(x-a)^r\,\phi(x)} = \frac{1}{y^r}\left[C_0 + C_1 y + C_2 y^2 + \ldots + C_{r-1}y^{r-1} + \frac{D_0 y^r + D_1 y^{r+1} + D_2 y^{r+2} + \ldots}{B_0 + B_1 y + B_2 y^2 + \ldots}\right]$$

$$= \frac{C_0}{y^r} + \frac{C_1}{y^{r-1}} + \frac{C_2}{y^{r-2}} + \ldots + \frac{C_{r-1}}{y} + \frac{D_0 + D_1 y + D_2 y^2 + \ldots}{\phi(a+y)} \qquad [\text{ Put } y = x - a]$$

$$= \frac{C_0}{(x-a)^r} + \frac{C_1}{(x-a)^{r-1}} + \frac{C_2}{(x-a)^{r-2}} + \ldots + \frac{C_{r-1}}{x-a} + \frac{D_0 + D_1(x-a) + D_2(x-a)^2 + \ldots}{\phi(x)}.$$

$\therefore$ the partial fractions corresponding to $(x-a)^r$ are

$$\left.\frac{C_0}{(x-a)^r} + \frac{C_1}{(x-a)^{r-1}} + \frac{C_2}{(x-a)^{r-2}} + \ldots + \frac{C_{r-1}}{x-a}.\right]$$

Example. *Integrate* $\displaystyle\int \frac{x^2}{(x-1)^3(x+1)}\,dx.$

Put $x - 1^* = y$, then $x = 1 + y$.

$$\therefore \qquad \frac{x^2}{(x-1)^3(x+1)} = \frac{(1+y)^2}{y^3(2+y)}$$

[Expand the num. and denom. **in ascending powers of y** (*Note this step*)]

$$= \frac{1}{y^3}\cdot\frac{1+2y+y^2}{2+y} \qquad \ldots(1) \qquad \left[\text{Omit the factor } \frac{1}{y^3}\right]$$

Divide $1 + 2y + y^2$ by $2 + y$ until y^3 is a factor of the remainder thus :

$$2+y\,\big)\,1+2y+y^2\,\left(\frac{1}{2} + \frac{3}{4}y + \frac{1}{8}y^2\right.$$

$$\underline{1 + \frac{1}{2}y}$$

$$\frac{3}{2}y + y^2$$

$$\underline{\frac{3}{2}y + \frac{3}{4}y^2}$$

$$\frac{1}{4}y^2$$

$$\underline{\frac{1}{4}y^2 + \frac{1}{8}y^3}$$

$$-\frac{1}{8}y^3$$

$\therefore$ from (1),

$$\frac{x^2}{(x-1)^3(x+1)} = \frac{1}{y^3}\left[\frac{1}{2} + \frac{3}{4}y + \frac{1}{8}y^2 + \frac{-\frac{1}{8}y^3}{2+y}\right]$$

$$= \frac{1}{2}\frac{1}{y^3} + \frac{3}{4}\frac{1}{y^2} + \frac{1}{8}\frac{1}{y} - \frac{1}{8}\frac{1}{2+y} \qquad [\text{Put } y = x - 1]$$

$$= \frac{1}{2}\frac{1}{(x-1)^3} + \frac{3}{4}\frac{1}{(x-1)^2} + \frac{1}{8}\frac{1}{x-1} - \frac{1}{8}\frac{1}{x+1}.$$

**i.e.*, first power of the repeated factor $(x-1)^3$.

$$\therefore \int \frac{x^2}{(x-1)^3(x+1)}\,dx$$

$$= \frac{1}{2}\int (x-1)^{-3}\,dx + \frac{3}{4}\int (x-1)^{-2}\,dx + \frac{1}{8}\int \frac{1}{x-1}\,dx - \frac{1}{8}\int \frac{1}{x+1}\,dx$$

$$= \frac{1}{2}\frac{(x-1)^{-2}}{-2} + \frac{3}{4}\frac{(x-1)^{-1}}{-1} + \frac{1}{8}\log(x-1) - \frac{1}{8}\log(x+1)$$

$$= -\frac{1}{4}\frac{1}{(x-1)^2} - \frac{3}{4}\frac{1}{x-1} + \frac{1}{8}\log\frac{x-1}{x+1}.$$

Note. *If there is only one other non-repeated linear factor in the denominator (as here) or only one other non-repeated quadratic factor in the denominator (as in solved Ex. 2, Art. 38), the rule of Art. 37 gives all the partial fractions.*

EXAMPLES

Evaluate (Exs. 1—5) :

1. (*i*) $\int \frac{dx}{(x-1)^2(x+1)}$. (*ii*) $\int \frac{dx}{(x+1)^4(x-1)}$.

2. (*i*) $\int (ax^2 + bx^3)^{-1}\,dx$. **(*ii*) $\int \frac{x\,dx}{(x+1)^2(x+2)(x-1)}$.

3. (*i*) $\int \frac{(x+1)\,dx}{(x-1)^2(x+2)^2}$. (*ii*) $\int_0^{\frac{1}{2}} (1-x^2)^{-2}\,dx$.

[(*i*) Put $x - 1 = y$ etc. It will be found that

$$\frac{x+1}{(x-1)^2(x+2)^2} = \frac{2}{9}\frac{1}{(x-1)^2} - \frac{1}{27}\frac{1}{x-1} + \frac{1}{27}\frac{x-1}{(x+2)^2}.$$

Now to resolve the last term $\frac{x-1}{(x+2)^2}$ again into partial fractions, put $x + 2 = z$, then $x = -2 + z$

$$\therefore \quad \frac{x-1}{(x+2)^2} = \frac{-3+z}{z^2} = -\frac{3}{z^2} + \frac{1}{z} = -\frac{3}{(x+2)^2} + \frac{1}{x+2}.]$$

4. $\int \frac{\sec x\,dx}{1+\operatorname{cosec} x}$. **5.** $\int \frac{dx}{(e^x - 1)^2}$.

38. Case III. Rule to find the partial fractions if there is a non-repeated quadratic factor ($x^2 + ax + b$), or a repeated quadratic factor ($x^2 + ax + b$)^r in the denominator of a proper rational algebraic fraction :

(*i*) *Let the given fraction = the sum of the partial fractions of the correct form [see Art. 31, (b)], the partial fractions corresponding to the non-repeated and repeated linear factors having been found by the rules (Arts. 32 and 37).*

(ii) Multiply both sides by the denominator of the L.H.S.

(iii) Equate coefficients of like powers of x on both sides, and find the values of the unknown constants.

Example 1. *Integrate* $\int \frac{x}{(x-1)(x^2+4)} dx.$

Let $$\frac{x}{(x-1)(x^2+4)} = \frac{1}{(x-1)5} + \frac{Ax+B}{x^2+4} \qquad \text{...(1)} \quad | \ (x-1)(x^2+4)$$

[Finding the partial fraction corresponding to $(x-1)$ by the rule (Art. 32)]

Multiplying both sides by $(x-1)(x^2+4)$, [Denom. of L.H.S.]

$$x = \frac{1}{5}(x^2+4) + (Ax+B)(x-1).$$

Equating coeffs. of x^2, $0 = \frac{1}{5} + A$, $\therefore \ A = -\frac{1}{5}$.

Equating coeffs. of x, $1 = -A + B$ (Substitute the value of A found above)

or $$1 = \frac{1}{5} + B, \ \therefore \ B = \frac{4}{5}.$$

Substituting these values of A, B in (1),

$$\frac{x}{(x-1)(x^2+4)} = \frac{1}{5(x-1)} + \frac{-\frac{1}{5}x + \frac{4}{5}}{x^2+4}.$$

$$\therefore \quad \int \frac{x}{(x-1)(x^2+4)} dx = \frac{1}{5}\int \frac{1}{x-1} dx - \frac{1}{5}\int \frac{x}{x^2+4} dx + \frac{4}{5}\int \frac{1}{x^2+4} dx \qquad \text{...(2)}$$

Now $$\int \frac{1}{x-1} dx = \log(x-1),$$

$$\int \frac{x}{x^2+4} dx = \frac{1}{2}\int \frac{2x}{x^2+4} dx = \frac{1}{2}\log(x^2+4),$$

$$\int \frac{1}{x^2+4} dx = \int \frac{1}{x^2+(2)^2} dx = \frac{1}{2}\tan^{-1}\frac{x}{2}.$$

$\therefore$ from (2),

$$\int \frac{x}{(x-1)(x^2+4)} dx = \frac{1}{5}\log(x-1) - \frac{1}{5}\cdot\frac{1}{2}\log(x^2+4) + \frac{4}{5}\cdot\frac{1}{2}\tan^{-1}\frac{x}{2}$$

$$= \frac{1}{5}\log(x-1) - \frac{1}{10}\log(x^2+4) + \frac{2}{5}\tan^{-1}\frac{x}{2}.$$

Example 2. *Integrate* $\int \frac{dx}{(x+1)^2(x^2+1)}$.

[To find the partial fractions corresponding to $(x+1)^2$ by the rule (Art. 37).]

Put $x + 1 = y$, then $x = -1 + y$.

$$\therefore \quad \frac{1}{(x+1)^2(x^2+1)} = \frac{1}{y^2[(-1+y)^2+1]}$$

[Expand the (num. and) denom. **in ascending powers of y** (*Note this step*)]

$$= \frac{1}{y^2(2-2y+y^2)}. \qquad \left[\text{Omit } \frac{1}{y^2}\right]$$

Divide 1 by $2 - 2y + y^2$ until y^2 is a factor of the remainder thus :

$$2-2y+y^2 \overline{)\,1\,}\left(\frac{1}{2}+\frac{1}{2}y\right.$$

$$\frac{1-y+\frac{1}{2}y^2}{y-\frac{1}{2}y^2}$$

$$\frac{y-y^2+\frac{1}{2}y^3}{\frac{1}{2}y^2-\frac{1}{2}y^3}$$

$$\therefore \quad \frac{1}{(x+1)^2(x^2+1)} = \frac{1}{y^2}\left[\frac{1}{2}+\frac{1}{2}y+\frac{\frac{1}{2}y^2-\frac{1}{2}y^3}{2-2y+y^2}\right] = \frac{1}{2}\frac{1}{y^2}+\frac{1}{2}\frac{1}{y}+\frac{\frac{1}{2}-\frac{1}{2}y}{2-2y+y^2}$$

$$= \frac{1}{2}\frac{1}{(x+1)^2}+\frac{1}{2}\frac{1}{x+1}+\frac{1}{2}\left(\frac{-x}{x^2+1}\right). \qquad \left|\begin{array}{l} 1-y \\ =1-(x+1)\end{array}\right.$$

$$\therefore \int \frac{1}{(x+1)^2(x^2+1)}\,dx = \frac{1}{2}\int (x+1)^{-2}\,dx + \frac{1}{2}\int \frac{1}{x+1}\,dx - \frac{1}{4}\int \frac{2x}{x^2+1}\,dx$$

$$= \frac{1}{2}\frac{(x+1)^{-1}}{-1} + \frac{1}{2}\log(x+1) - \frac{1}{4}\log(x^2+1)$$

$$= -\frac{1}{2}\frac{1}{x+1} + \frac{1}{2}\log(x+1) - \frac{1}{4}\log(x^2+1).$$

Note. *See Note in solved Ex., Art. 37.*

EXAMPLES

Integrate (Exs. 1—3) :

1. (*i*) $\int \frac{dx}{(x-3)(x^2+9)}$. (*ii*) $\int \frac{(x-2)dx}{(x+2)(x^2+4)}$.

2. (*i*) $\int \frac{x\,dx}{x^3-1}$. (*ii*) $\int \frac{dx}{1+x+x^2+x^3}$.

3. $\int \frac{dx}{(x+1)(x^3-1)}$.

4. Evaluate (*i*) $\int_0^1 \frac{dx}{1+2x+2x^2+2x^3+x^4}$. (*ii*) $\int \frac{dx}{(1-x)^2(1+x^2)}$.

Integrate (Exs. 5—6) :

5. (*i*) $\int \frac{dx}{(2+x)^2(1+x+x^2)}$. (*ii*) $\int \frac{x^2\,dx}{(x-1)^2(x^2-x+1)}$.

**6. (i) $\int \frac{(x+a)\,dx}{x^2(x-a)(x^2+a^2)}$. (ii) $\int \frac{dx}{x(x+1)^2(x^2+1)}$.

[(i) First find the partial fractions corresponding to the repeated linear factor x^2 by dividing $a + x$ by $[(x-a)(x^2+a^2) = x^3 + a^2x - ax^2 - a^3$, *i.e.*, by$] - a^3 + a^2x - ax^2 + x^3$ (*ascending powers of* x) until x^2 is a factor of the remainder. It will be found that

$$\frac{x+a}{x^2(x-a)(x^2+a^2)} = \frac{1}{x^2}\left[-\frac{1}{a^2} - \frac{2}{a^3}x + \frac{\frac{1}{a}x^2 - \frac{1}{a^2}x^3 + \frac{2}{a^3}x^4}{(x-a)(x^2+a^2)}\right]$$

$$= -\frac{1}{a^2}\frac{1}{x^2} - \frac{2}{a^3}\frac{1}{x} + \frac{\frac{1}{a} - \frac{1}{a^2}x + \frac{2}{a^3}x^2}{(x-a)(x^2+a^2)}.$$

Now resolve the last term $\frac{\frac{1}{a} - \frac{1}{a^2}x + \frac{2}{a^3}x^2}{(x-a)(x^2+a^2)}$ into partial fractions. (Proceed as in solved Ex. 1, Art. 38.)]

**7. Integrate $\int \frac{dx}{(x^2+1)(x^2-x+1)}$.

[Let $\frac{1}{(x^2+1)(x^2-x+1)} = \frac{Ax+B}{x^2+1} + \frac{Cx+D}{x^2-x+1}$ ...(1) | $(x^2+1)(x^2-x+1)$

Multiplying both sides by $(x^2+1)(x^2-x+1)$, [Denom. of L.H.S.]

$$1 = (Ax+B)(x^2-x+1) + (Cx+D)(x^2+1) \qquad ...(2)$$

Putting (one quadratic factor in the denom. of the given fraction, *i.e.*,) $x^2 + 1 = 0, x^2 = -1$.

Substituting this value of x^2 $(= -1)$ on both sides of (2),

$$1 = (Ax+B)(-x) + (Cx+D)\,0 = -Ax^2 - Bx \qquad (\text{Put } x^2 = -1)$$

$$= A - Bx.$$

Equating coeffs. of x, $0 = -B$
Equating constant terms, $1 = A$
} Find the values of A, B.

Again putting (second quadratic factor in the denom. of the given fraction, *i.e.*,) $x^2 - x + 1 = 0, x^2 = x - 1$.

Substituting this value of x^2 $(= x - 1)$ on both sides of (2),

$$1 = (Ax+B)(0) + (Cx+D)(x) = Cx^2 + Dx \qquad (\text{Put } x^2 = x-1)$$

$$= C(x-1) + Dx.$$

Equating coeffs. of x, $0 = C + D$
Equating constant terms, $1 = -C$
} Find the values of C,D.]

8. Integrate $\int \frac{dx}{(a^2+x^2)^2}$.

Put $x = a\tan\theta$ ...(1), [Cf. Art. 13, (ii)]

then $$dx = a \sec^2 \theta \, d\theta.$$

$$\therefore \int \frac{1}{(a^2+x^2)^2} dx = \int \frac{1}{(a^2+a^2\tan^2\theta)^2} a \sec^2\theta \, d\theta$$

$$= \int \frac{1}{a^4(1+\tan^2\theta)^2} a\sec^2\theta \, d\theta = \int \frac{1}{a^3 \sec^4\theta} \sec^2\theta \, d\theta$$

$$= \frac{1}{a^3}\int \frac{1}{\sec^2\theta} d\theta = \frac{1}{a^3}\int \cos^2\theta \, d\theta = \frac{1}{a^3}\int \frac{1+\cos 2\theta}{2} d\theta$$

$$= \frac{1}{2a^3}\left[\theta + \frac{\sin 2\theta}{2}\right] = \frac{1}{2a^3}\left[\theta + \frac{2\sin\theta\cos\theta}{2}\right] = \frac{1}{2a^3}[\theta + \sin\theta\cos\theta]$$

$$\left[\text{Now from (1), } \tan\theta = \frac{x}{a}\right.$$

$$\left.\therefore \sin\theta = \frac{x}{\sqrt{a^2+x^2}}, \cos\theta = \frac{a}{\sqrt{a^2+x^2}}\right]$$

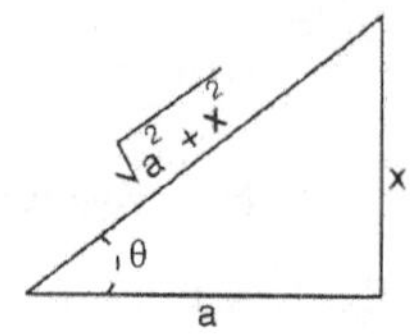

$$= \frac{1}{2a^3}\left[\tan^{-1}\frac{x}{a} + \frac{x}{\sqrt{a^2+x^2}} \cdot \frac{a}{\sqrt{a^2+x^2}}\right] = \frac{1}{2a^3}\left[\tan^{-1}\frac{x}{a} + \frac{ax}{a^2+x^2}\right].$$

[**Note.** *If a power of* (a^2+x^2) *occurs in the denominator of the integrand, it is generally useful to put* $x = a \tan\theta$.]

9. Evaluate $\displaystyle\int_0^a \frac{a^2-x^2}{(a^2+x^2)^2} dx.$

10. Integrate $\displaystyle\int \frac{2x \, dx}{(1-x)(1+x^2)^2}.$

$\left[\displaystyle\int \frac{2x}{(1-x)(1+x^2)^2} dx = \int \frac{1}{1-x} \cdot \frac{2x}{(1+x^2)^2} dx. \text{ Integrate by parts.}\right]$

11. Show that the sum of the infinite series

$$\frac{1}{a} - \frac{1}{a+b} + \frac{1}{a+2b} - \frac{1}{a+3b} + \ldots \; (a > 0, b > 0)$$

can be expressed in the form $\displaystyle\int_0^1 \frac{t^{a-1}}{1+t^b} dt.$

Hence prove that

$$\frac{1}{1} - \frac{1}{4} + \frac{1}{7} - \frac{1}{10} + \frac{1}{13} - \frac{1}{16} + \ldots = \frac{1}{3}(\pi \, 3^{-1/2} + \log_e 2).$$

$\left[\text{Evaluate } \displaystyle\int_0^1 \frac{t^{a-1}}{1+t^b} dt. \right.$ (Expand $(1+t^b)^{-1}$ by the Binomial theorem, multiply each term by t^{a-1}, and integrate term by term.)

It will be found that $\int_0^1 \frac{t^{a-1}}{1+t^b}\,dt = \frac{1}{a} - \frac{1}{a+b} + \frac{1}{a+2b} - \frac{1}{a+3b} + \ldots$

or $$\frac{1}{a} - \frac{1}{a+b} + \frac{1}{a+2b} - \frac{1}{a+3b} + \ldots = \int_0^1 \frac{t^{a-1}}{1+t^b}\,dt \quad \ldots(1)$$

Comparing $\frac{1}{a} - \frac{1}{a+b} + \frac{1}{a+2b} - \frac{1}{a+3b} + \ldots$ with the given series

$$\frac{1}{1} - \frac{1}{4} + \frac{1}{7} - \frac{1}{10} + \ldots,$$

we get $$a = 1,\ a + b = 4, \quad \therefore \quad b = 3.$$

Substituting these values of a, b in (1),

$$\frac{1}{1} - \frac{1}{4} + \frac{1}{7} - \frac{1}{10} + \ldots = \int_0^1 \frac{1}{1+t^3}\,dt.$$ Evaluate this integral.]

39. Only even powers of x occur. Important type. Shorter rule to find the partial fractions of a rational algebraic fraction in which only even powers of x occur in both the numerator and denominator :

(i) Put $x^2 \equiv y$, and resolve the resulting fraction in y into partial fractions by the usual rule (s).

(ii) In the resulting partial fractions, put $y = x^2$.

Example. *Integrate* $\int \frac{x^2\,dx}{(x^2+1)(3x^2+1)}$.

Put $x^2 = y$, then

$$\frac{x^2}{(x^2+1)(3x^2+1)} = \frac{y}{(y+1)(3y+1)} = \frac{-1}{(y+1)(-2)} + \frac{-\frac{1}{3}}{(\frac{2}{3})(3y+1)} \quad \text{[Rule (Art. 32)]}$$

$$= \frac{1}{2(y+1)} - \frac{1}{2(3y+1)} \quad [\text{Put } y = x^2]$$

$$= \frac{1}{2(x^2+1)} - \frac{1}{2(3x^2+1)}.$$

$$\therefore \int \frac{x^2}{(x^2+1)(3x^2+1)}\,dx = \frac{1}{2}\int \frac{1}{x^2+1}\,dx - \frac{1}{2}\int \frac{1}{3x^2+1}\,dx \quad \ldots(1)$$

Now $$\int \frac{1}{x^2+1}\,dx = \int \frac{1}{x^2+(1)^2}\,dx$$

$$= \frac{1}{1}\tan^{-1}\frac{x}{1} = \tan^{-1} x,$$

$$\int \frac{1}{3x^2+1}\,dx = \int \frac{1}{3(x^2+\frac{1}{3})}\,dx = \frac{1}{3}\int \frac{1}{x^2+\left(\frac{1}{\sqrt{3}}\right)^2}\,dx$$

$$= \frac{1}{3}\cdot\frac{1}{\frac{1}{\sqrt{3}}}\tan^{-1}\frac{x}{\frac{1}{\sqrt{3}}} = \frac{1}{\sqrt{3}}\tan^{-1} x\sqrt{3}.$$

$\therefore$ from (1), $\displaystyle\int \frac{x^2}{(x^2+1)(3x^2+1)}\,dx = \frac{1}{2}\tan^{-1} x - \frac{1}{2\sqrt{3}}\tan^{-1} x\sqrt{3}.$

EXAMPLES

Integrate (Exs. 1—3) :

1. (*i*) $\displaystyle\int \frac{(x^2+1)\,dx}{(x^2+2)(2x^2+1)}.$

(*ii*) $\displaystyle\int \frac{x^2\,dx}{(x^2-1)(x^2+2)}.$ (*iii*) $\displaystyle\int \frac{dx}{x^4+3x^2-4}.$

2. $\displaystyle\int \frac{dx}{(x^2+a^2)(x^2+b^2)}.$ **3.** $\displaystyle\int \frac{(x^2+a^2)(x^2+b^2)}{(x^2+c^2)(x^2+d^2)}\,dx.$

4. Show that

$$\int_0^\infty \frac{x^2\,dx}{(x^2+a^2)(x^2+b^2)(x^2+c^2)} = \frac{\pi}{2(a+b)(b+c)(c+a)}.$$

5. Integrate $\displaystyle\int \frac{\sin x}{\sin 4x}\,dx.$ **6.** Integrate $\displaystyle\int \frac{\cot\theta - 3\cot 3\theta}{3\tan 3\theta - \tan\theta}\,d\theta.$

Shorter methods for two important types.

40-I. Type I. Rule to integrate a rational algebraic fraction in which only odd powers of x occur in the numerator, and only even in the denominator, or vice versa :

Put $x^2 = z$, *then* $2x\,dx = dz$, *or* $dx = \dfrac{dz}{2x}$.

Now resolve the resulting integrand in z into partial fractions, if necessary.

Example. *Integrate* $\displaystyle\int \frac{x\,dx}{x^4+x^2+1}.$

[Here only odd powers of x occur in the num., and only even in the denom.]

Put $x^2 = z$ ...(1), then $2x\,dx = dz$, or $x\,dx = \dfrac{dz}{2}$.

$$\therefore \quad \int \frac{x\,dx}{x^4+x^2+1} = \int \frac{dz}{2(z^2+z+1)} = \frac{1}{2}\int \frac{1}{(z^2+z+\frac{1}{4})+\frac{3}{4}}\,dz$$

$$= \frac{1}{2}\int \frac{1}{(z+\frac{1}{2})^2 + \left(\frac{\sqrt{3}}{2}\right)^2} dz = \frac{1}{2} \cdot \frac{1}{\frac{\sqrt{3}}{2}} \tan^{-1} \frac{z+\frac{1}{2}}{\frac{\sqrt{3}}{2}} = \frac{1}{\sqrt{3}} \tan^{-1} \frac{2z+1}{\sqrt{3}}$$

$$= \frac{1}{\sqrt{3}} \tan^{-1} \frac{2x^2+1}{\sqrt{3}}. \qquad \text{[From (1)]}$$

EXAMPLES

1. Integrate

(*i*) $\int \frac{5x^2+1}{x(x^4-1)} dx.$ (*ii*) $\int \frac{1}{x(x^2+1)^3} dx.$

2. Integrate $\int \frac{x^5-2}{x^5-x} dx.$

40-II. Type II. Rule to integrate $\int \frac{x^2+1}{x^4+kx^2+1} dx$, or $\int \frac{x^2-1}{x^4+kx^2+1} dx$, where k is a constant positive, negative, or zero :

Divide the numerator and denominator by x^2, and put $x - \frac{1}{x} = z$, or $x + \frac{1}{x} = z$, whichever substitution gives on differentiation the numerator of the resulting integrand.

Cor. 1. To integrate $\int \frac{x^2}{x^4+kx^2+1} dx$.

$$\int \frac{x^2}{x^4+kx^2+1} dx = \frac{1}{2}\int \frac{2x^2}{x^4+kx^2+1} dx = \frac{1}{2}\int \frac{(x^2+1)+(x^2-1)}{x^4+kx^2+1} dx$$

$$= \frac{1}{2}\int \frac{x^2+1}{x^4+kx^2+1} dx + \frac{1}{2}\int \frac{x^2-1}{x^4+kx^2+1} dx,$$

each of which can be integrated (Art. 40-II).

Cor. 2. To integrate $\int \frac{1}{x^4+kx^2+1} dx$.

$$\int \frac{1}{x^4+kx^2+1} dx = \frac{1}{2}\int \frac{2}{x^4+kx^2+1} dx = \frac{1}{2}\int \frac{(x^2+1)-(x^2-1)}{x^4+kx^2+1} dx$$

$$= \frac{1}{2}\int \frac{x^2+1}{x^4+kx^2+1} dx - \frac{1}{2}\int \frac{x^2-1}{x^4+kx^2+1} dx,$$

each of which can be integrated (Art. 40-II).

Example. *Integrate* $\int \frac{x^2+1}{x^4-x^2+1} dx.$

$$\int \frac{x^2+1}{x^4-x^2+1} dx \qquad \text{[Divide the num. and denom. by } x^2\text{]}$$

$$= \int \frac{1+\frac{1}{x^2}}{x^2 - 1 + \frac{1}{x^2}}\, dx$$

$\left[\text{Put } x - \frac{1}{x} = z^* \quad ...(1), \text{ then } \left(1+\frac{1}{x^2}\right) dx = dz.\right.$

From (1), squaring, $x^2 + \frac{1}{x^2} - 2 = z^2$, or $x^2 + \frac{1}{x^2} = z^2 + 2]$

$$= \int \frac{dz}{z^2+2-1} = \int \frac{dz}{z^2+1} = \int \frac{1}{z^2+(1)^2}\, dz \qquad ...(2)$$

$$= \frac{1}{1}\tan^{-1}\frac{z}{1} = \tan^{-1} z = \tan^{-1}\left(x - \frac{1}{x}\right) \qquad \text{[From (1)]}$$

$$= \tan^{-1}\frac{x^2-1}{x}.$$

EXAMPLES

Integrate (Exs. 1—4) :

1. (*i*) $\int \frac{x^2-1}{x^4+x^2+1}\, dx.$ (*ii*) $\int \frac{x^2+1}{x^4+1}\, dx.$

2. (*i*) $\int \frac{x^2-3x+1}{x^4-x^2+1}\, dx.$ (*ii*) $\int \frac{(x-1)^2}{x^4+x^2+1}\, dx.$

3. (*i*) $\int \frac{dx}{x^4+1}.$ (*ii*) $\int \frac{x^2\, dx}{x^4+1}.$

$\left[(i) \int \frac{1}{x^4+1}\, dx = \frac{1}{2}\int \frac{2}{x^4+1}\, dx = \frac{1}{2}\int \frac{(x^2+1)-(x^2-1)}{x^4+1}\, dx\right.$

$$= \frac{1}{2}\int \frac{x^2+1}{x^4+1} dx - \frac{1}{2}\int \frac{x^2-1}{x^4+1}\, dx.$$

To integrate $\int \frac{x^2+1}{x^4+1}\, dx$, divide the num. and denom. by x^2, and put $x - \frac{1}{x} = z$.

To integrate $\int \frac{x^2-1}{x^4+1}\, dx$, divide the num. and denom. by x^2, and put $x + \frac{1}{x} = u.]$

4. (*i*) $\int \frac{x^2+a^2}{x^4-a^2x^2+a^4}\, dx.$ (*ii*) $\int (x^2 - a^2)(x^4 + a^2x^2 + a^4)^{-1}\, dx.$

---

*__Why this step.__ $\because$ on differentiation we get $\left(1+\frac{1}{x^2}\right) dx = dz$ as required in $\int \frac{1+\frac{1}{x^2}}{x^2-1+\frac{1}{x^2}}\, dx.$

$\left[(i) \int \frac{x^2 + a^2}{x^4 - a^2x^2 + a^4} dx \right.$ [Divide the num. and denom. by x^2]

$$= \int \frac{1 + \frac{a^2}{x^2}}{x^2 - a^2 + \frac{a^4}{x^2}} dx. \qquad \text{Put } x - \frac{a^2}{x} = z^* \ldots(1), \text{ then } \left(1 + \frac{a^2}{x^2}\right) dx = dz. \Bigg]$$

****5.** Evaluate

(i) $\int_0^{\frac{\pi}{4}} \sqrt{\tan\theta}\, d\theta.$ (ii) $\int_0^{\frac{\pi}{4}} \sqrt{\cot\theta}\, d\theta.$

[(i) Put $\tan\theta = z^2$.]

****6.** Evaluate $\int_0^{\frac{\pi}{4}} \frac{dx}{\cos^4 x - \cos^2 x \sin^2 x + \sin^4 x}$.

[Rule to integrate a fraction whose numerator is 1 and denominator is a homogeneous function of the fourth degree in cos x and sin x :

Divide the numerator and denominator by $\cos^4 x$, and put $\tan x = z$. (Compare the Rule in Ex. 7, Art. 21.)]

MISCELLANEOUS EXAMPLES ON CHAPTER IV

1. (i) Integrate $\int \frac{dx}{x^2 - x - 6}$.

**(ii) Evaluate $\int_0^1 \frac{dx}{1 + 2x\cosh\alpha + x^2}$.

$\left[(ii) \text{ Use } \int \frac{1}{x^2 - a^2} dx = -\frac{1}{a}\coth^{-1}\frac{x}{a} \dagger. \right]$

[**Proof.** Put $x = a\coth\theta$ ...(1), then $dx = -a\,\text{cosech}^2\theta\, d\theta$.

$$\therefore \int \frac{1}{x^2 - a^2} dx = \int \frac{1}{a^2\coth^2\theta - a^2}(-a\,\text{cosech}^2\theta\, d\theta)$$

$$= \int \frac{1}{a^2(\coth^2\theta - 1)}(-a\,\text{cosech}^2\theta\, d\theta)$$

$$= \int -\frac{1}{a\,\text{cosech}^2\theta}\,\text{cosech}^2\theta\, d\theta$$

$$= -\frac{1}{a}\int 1\, d\theta = -\frac{1}{a}\theta$$

***Why this step.** $\because$ on differentiation we get $\left(1 + \frac{a^2}{x^2}\right) dx = dz$ as required in $\int \frac{1 + \frac{a^2}{x^2}}{x^2 - a^2 + \frac{a^4}{x^2}} dx.$

$$= -\frac{1}{a}\coth^{-1}\frac{x}{a}\Bigg] \qquad \left(\because \text{ from (1), } \coth\theta = \frac{x}{a}, \therefore \theta = \coth^{-1}\frac{x}{a}\right)$$

2. Integrate.

(*i*) $\int \frac{3x-5}{x^2+8x-20}\,dx.$ (*ii*) $\int \frac{dx}{x(x^n+1)}.$

[(*i*) Short cut. $\int \frac{3x-5}{x^2+8x-20}\,dx = \int \frac{3x-5}{(x+10)(x-2)}\,dx.$

Resolve the integrand into partial fractions.]

3. (*i*) Evaluate $\int_1^2 \frac{dx}{x(1+2x)^2}.$

(*ii*) Integrate $\int \frac{x}{(x-1)^2(x+2)}\,dx.$

4. Integrate

(*i*) $\int \frac{x^3+2}{(x-1)(x-2)^3}\,dx.$

(*ii*) $\int \frac{(x^2+1)}{(x^2-1)^2}\,dx.$

$$\left[(ii)\ \int \frac{x^2+1}{(x^2-1)^2}\,dx = \int \frac{x^2+1}{(x+1)^2(x-1)^2}\,dx.\right]$$

5. (*i*) Evaluate $\int_1^\infty \frac{dx}{x(1+x^2)}.$

(*ii*) Integrate $\int \frac{dx}{x^3-1}.$

Integrate (Exs. 6—7) :

****6.** $\int \frac{x+2}{(2x^2+4x+3)^2}\,dx.$

$$\left[(i)\ \int \frac{x+2}{(2x^2+4x+3)^2}\,dx = \int \frac{\frac{1}{4}(4x+4)+1}{(2x^2+4x+3)^2}\,dx\right.$$

$$= \frac{1}{4}\int (2x^2+4x+3)^{-2}(4x+4)\,dx + \int \frac{1}{(2x^2+4x+3)^2}\,dx \qquad ...(1)$$

Now $\int \frac{1}{(2x^2+4x+3)^2}\,dx = \int \frac{1}{4(x^2+2x+\frac{3}{2})^2}\,dx = \frac{1}{4}\int \frac{1}{[(x^2+2x+1)+\frac{1}{2}]^2}\,dx$

$$\left.= \frac{1}{4}\int \frac{1}{\left[(x+1)^2+\left(\frac{1}{\sqrt{2}}\right)^2\right]^2}\,dx. \text{ Put } x+1 = \frac{1}{\sqrt{2}}\tan\theta.\right]$$

7. (*i*) $\int \frac{dx}{(2+x^2)(1-x^2)}.$ (*ii*) $\int \frac{1}{x^4-1}\,dx.$

8. Prove that $\int_1^{\infty} \frac{(x^2+3)\,dx}{x^6(x^2+1)} = \frac{1}{30}(58-15\pi)$.

****9.** (*i*) Integrate $\int \frac{x^4}{(x^2+1)^2}\,dx$.

(*ii*) Evaluate $\int_1^{\infty} \frac{x^4+1}{x^2(x^2+1)^2}\,dx$.

10. Integrate $\frac{x^3}{(x^2+1)^3}$, (*a*) by the substitution $x = \tan\theta$, (*b*) by the substitution $u = x^2 + 1$, and show that the results you obtain by the two methods are in accordance.

[(*ii*) {Ans. obtained by (*a*)} – {Ans. obtained by (*b*)} = $\frac{1}{4}$, constant. See Art. 4.]

11. Integrate $\int \frac{2x\,dx}{(x^2+1)(x^2+3)}$.

12. (*i*) Integrate $\int \frac{x^2 dx}{x^4+x^2+1}$.

(*ii*) Prove that $\int_x^{\frac{1}{x}} \frac{dx}{1+x^4} = \frac{1}{\sqrt{2}} \cot^{-1} \frac{\sqrt{2}}{\frac{1}{x}-x}$.

13. Integrate

(*i*) $\int \frac{x^2 dx}{x^4+a^2}$ (*ii*) $\int \frac{dx}{x^4+8x^2+9}$.

[(*ii*) Compare the denom. of the integrand, *viz.*, $x^4 + 8x^2 + 9$ with $x^4 + kx^2 + c^2$, here $c^2 = 9$, $\therefore c = 3$. $\therefore$ we write the num. of the integrand, *viz.*, 1 in terms of $x^2 + c$ and $x^2 - c$, *i.e.*, in terms of $x^2 + 3$ and $x^2 - 3$ thus :

$$\int \frac{1}{x^4+8x^2+9}\,dx = \frac{1}{6}\int \frac{(x^2+3)-(x^2-3)}{x^4+8x^2+9}\,dx$$

$$= \frac{1}{6}\int \frac{x^2+3}{x^4+8x^2+9}\,dx - \frac{1}{6}\int \frac{x^2-3}{x^4+8x^2+9}\,dx.$$

To integrate $\int \frac{x^2+3}{x^4+8x^2+9}\,dx$, divide the num. and denom. by x^2 and put $x - \frac{3}{x} = z$.

To integrate $\int \frac{x^2-3}{x^4+8x^2+9}\,dx$, divide the num. and denom. by x^2 and put $x + \frac{3}{x} = u$.]

CHAPTER V

INTEGRATION OF IRRATIONAL ALGEBRAIC FUNCTIONS

41. To integrate $\displaystyle\int \frac{dx}{\sqrt{ax^2+bx+c}}$.

Case I. If a is $+ve$, then

$$\int \frac{1}{\sqrt{ax^2+bx+c}}\,dx = \int \frac{1}{\sqrt{a}\sqrt{\left(x^2+\frac{b}{a}x\right)+\frac{c}{a}}}\,dx$$

$$= \frac{1}{\sqrt{a}}\int \frac{1}{\sqrt{\left(x^2+\frac{b}{a}x+\frac{b^2}{4a^2}\right)-\frac{b^2}{4a^2}+\frac{c}{a}}}\,dx$$

$$= \frac{1}{\sqrt{a}}\int \frac{1}{\sqrt{\left(x+\frac{b}{2a}\right)^2+\frac{4ac-b^2}{4a^2}}}\,dx \qquad \ldots(1)$$

Sub-case (i). If $b^2 > 4ac$, then from (1),

$$\int \frac{1}{\sqrt{ax^2+bx+c}}\,dx = \frac{1}{\sqrt{a}}\int \frac{1}{\sqrt{\left(x+\frac{b}{2a}\right)^2-\frac{b^2-4ac}{4a^2}}}\,dx$$

$$= \frac{1}{\sqrt{a}}\int \frac{1}{\sqrt{\left(x+\frac{b}{2a}\right)^2-\left(\frac{\sqrt{b^2-4ac}}{2a}\right)^2}}\,dx$$

$$= \frac{1}{\sqrt{a}}\cosh^{-1}\frac{x+\frac{b}{2a}}{\frac{\sqrt{b^2-4ac}}{2a}} = \frac{1}{\sqrt{a}}\cosh^{-1}\frac{2ax+b}{\sqrt{b^2-4ac}}.$$

Sub-case (ii). If $b^2 < 4ac$, then from (1),

$$\int \frac{1}{\sqrt{ax^2+bx+c}}\,dx = \frac{1}{\sqrt{a}}\int \frac{1}{\sqrt{\left(x+\frac{b}{2a}\right)^2+\left(\frac{\sqrt{4ac-b^2}}{2a}\right)^2}}\,dx$$

$$= \frac{1}{\sqrt{a}}\sinh^{-1}\frac{x+\frac{b}{2a}}{\frac{\sqrt{4ac-b^2}}{2a}} = \frac{1}{\sqrt{a}}\sinh^{-1}\frac{2ax+b}{\sqrt{4ac-b^2}}.$$

Case II. If a is $-ve = -A$ (say, where A is +ve), then

$$\int \frac{1}{\sqrt{ax^2+bx+c}}\,dx = \int \frac{1}{\sqrt{-Ax^2+bx+c}}\,dx$$

$$= \int \frac{1}{\sqrt{A}\sqrt{-x^2+\frac{b}{A}x+\frac{c}{A}}}\,dx = \frac{1}{\sqrt{A}}\int \frac{1}{\sqrt{-\left(x^2-\frac{b}{A}x\right)+\frac{c}{A}}}\,dx$$

$$= \frac{1}{\sqrt{A}}\int \frac{1}{\sqrt{-\left(x^2-\frac{b}{A}x+\frac{b^2}{4A^2}\right)+\frac{b^2}{4A^2}+\frac{c}{A}}}\,dx$$

$$= \frac{1}{\sqrt{A}}\int \frac{1}{\sqrt{\frac{b^2+4Ac}{4A^2}-\left(x-\frac{b}{2A}\right)^2}}\,dx$$

$$= \frac{1}{\sqrt{A}}\sin^{-1}\frac{x-\frac{b}{2A}}{\frac{\sqrt{b^2+4Ac}}{2A}} = \frac{1}{\sqrt{A}}\sin^{-1}\frac{2Ax-b}{\sqrt{b^2+4Ac}} \qquad [\text{Put } A=-a]$$

$$= \frac{1}{\sqrt{-a}}\sin^{-1}\frac{-2ax-b}{\sqrt{b^2-4ac}}.$$

[**Rule to integrate** $\int \frac{1}{\sqrt{\textbf{quadratic}}}\,\mathbf{dx}$ **:**

See the rule of Note, Art. 34.]

Example 1. *Integrate* $\int \frac{dx}{\sqrt{2x^2+3x+4}}$.

$$\int \frac{1}{\sqrt{2x^2+3x+4}}\,dx = \int \frac{1}{\sqrt{2}\sqrt{(x^2+\frac{3}{2}x)+2}}\,dx$$

$$= \frac{1}{\sqrt{2}}\int \frac{1}{\sqrt{(x^2+\frac{3}{2}x+\frac{9}{16})-\frac{9}{16}+2}}\,dx$$

$$= \frac{1}{\sqrt{2}}\int \frac{1}{\sqrt{\left(x+\frac{3}{4}\right)^2+\frac{23}{16}}}\,dx = \frac{1}{\sqrt{2}}\int \frac{1}{\sqrt{\left(x+\frac{3}{4}\right)^2+\left(\frac{\sqrt{23}}{4}\right)^2}}\,dx$$

$$= \frac{1}{\sqrt{2}}\sinh^{-1}\frac{x+\frac{3}{4}}{\frac{\sqrt{23}}{4}} = \frac{1}{\sqrt{2}}\sinh^{-1}\frac{4x+3}{\sqrt{23}}.$$

Example 2. *Integrate* $\int \frac{dx}{\sqrt{4+3x-2x^2}}$.

$$\int \frac{1}{\sqrt{4+3x-2x^2}}\,dx = \int \frac{1}{\sqrt{2+\frac{3}{2}x-x^2}}\,dx = \frac{1}{\sqrt{2}}\int \frac{1}{\sqrt{2-(x^2-\frac{3}{2}x)}}\,dx$$

$$= \frac{1}{\sqrt{2}}\int \frac{1}{\sqrt{2-(x^2-\frac{3}{2}x+\frac{9}{16})+\frac{9}{16}}}\,dx = \frac{1}{\sqrt{2}}\int \frac{1}{\sqrt{\frac{41}{16}-(x-\frac{3}{4})^2}}\,dx$$

$$= \frac{1}{\sqrt{2}}\int \frac{1}{\sqrt{\left(\frac{\sqrt{41}}{4}\right)^2-(x-\frac{3}{4})^2}}\,dx = \frac{1}{\sqrt{2}}\sin^{-1}\frac{x-\frac{3}{4}}{\frac{\sqrt{41}}{4}}$$

$$= \frac{1}{\sqrt{2}}\sin^{-1}\frac{4x-3}{\sqrt{41}}.$$

EXAMPLES

1. Integrate (*i*) $\int \frac{dx}{\sqrt{x^2+2x+4}}$.

(*ii*) $\int \frac{dx}{\sqrt{x^2+x-2}}$. (*iii*) $\int \frac{dx}{\sqrt{2x^2-2x+3}}$.

2. Evaluate (*i*) $\int_0^1 \frac{dx}{\sqrt{2-2x-x^2}}$, (*ii*) $\int \frac{dx}{\sqrt{3+8x-3x^2}}$.

3. Integrate

(*a*) (*i*) $\int \frac{dx}{\sqrt{2ax-x^2}}$. (*ii*) $\int \frac{dx}{\sqrt{x-x^2}}$. (*iii*) $\int \frac{1}{2\sqrt{x}\,\sqrt{1-x}}\,dx$.

(*b*) Prove that, with certain limitations on the values of the constants involved,

$$\int_\beta^x \frac{dx}{(\alpha-x)(x-\beta)} = 2 \text{ arc cos } \sqrt{\frac{\alpha-x}{\alpha-\beta}}.$$

[**Notation.** *arc cos x denotes* $\cos^{-1} x$.]

[(*b*) **Shortcut.** Put $\alpha - x = z^2$.]

4. Integrate (*i*) $\int \sqrt{x^2-2x}\,dx$.

(*ii*) $\int \sqrt{a+bx+cx^2}\,dx$ (*c* positive).

(*iii*) $\int \sqrt{a+bx-cx^2}\,dx$ (*c* positive).

[**Rule to integrate** $\int \sqrt{\textbf{quadratic}}\,\textbf{dx}$ **:** *See the rule of Note, Art. 34.*]

5. Integrate $\int_\alpha^\beta \sqrt{(x-\alpha)(\beta-x)}\,dx$.

42. To integrate $\int \frac{\mathbf{Ax+B}}{\sqrt{\mathbf{ax^2+bx+c}}}\,\mathbf{dx}$.

$$\int \frac{Ax + B}{\sqrt{ax^2 + bx + c}}\,dx = \int \frac{\frac{A}{2a}(2ax + b) - \frac{Ab}{2a} + B^*}{\sqrt{ax^2 + bx + c}}\,dx$$

$$= \frac{A}{2a}\int (ax^2 + bx + c)^{-1/2}\,(2ax + b)dx + \left(B - \frac{Ab}{2a}\right)\int \frac{1}{\sqrt{ax^2 + bx + c}}\,dx$$

$$= \frac{A}{2a}\,\frac{(ax^2 + bx + c)^{\frac{1}{2}}}{\frac{1}{2}} + \left(B - \frac{Ab}{2a}\right)\int \frac{1}{\sqrt{ax^2 + bx + c}}\,dx$$

$$= \frac{A}{a}\sqrt{ax^2 + bx + c} + \left(B - \frac{Ab}{2a}\right)\int \frac{1}{\sqrt{ax^2 + bx + c}}\,dx \qquad \text{...(1)}$$

Now $\int \frac{1}{\sqrt{ax^2 + bx + c}}\,dx$ can be integrated (Art. 41),

$\therefore$ from (1), $\int \frac{Ax + B}{\sqrt{ax^2 + bx + c}}\,dx$ can be integrated.

Note. $\int (Ax + B)\sqrt{ax^2 + bx + c}\;dx$ *can also be integrated in a similar manner.*

EXAMPLES

Integrate (Exs. 1 – 4) :

1. (*i*) $\int \frac{2x + 3}{\sqrt{x^2 + x + 1}}\,dx.$ (*ii*) $\int \frac{x\,dx}{\sqrt{9 + 8x - x^2}}.$

2. $\int \frac{x^2 + x + 1}{\sqrt{x^2 + 2x + 3}}\,dx.$

3. $\int (x + 1)\sqrt{x^2 - x + 1}\,dx.$

****4.** $\int \frac{x^3 + x^2 + x + 1}{\sqrt{x^2 + 2x + 3}}\,dx.$

***How to write this step.** *Write in the numerator the differential coefficient of the quadratic* $(ax^2 + bx + c)$ *whose square root occurs in the denominator, viz.,* $2ax + b$, *and adjust the coefficient of* x *by multiplying* $2ax + b$ *by* $\frac{A}{2a}$. [$\because$ we have got $2a$ as the coefficient of x in $2ax + b$ but we require A as the coefficient of x in $Ax + B$], thus getting $\frac{A}{2a}(2ax + b)$; *now adjust the constant term by subtracting* $\frac{Ab}{2a}$ *and adding B* [$\because$ we have got $\frac{Ab}{2a}$ as the constant term in $\frac{A}{2a}(2ax + b)$, but we require B as the constant term in $Ax + B$], thus getting $Ax + B = \frac{A}{2a}(2ax + b) - \frac{Ab}{2a} + B$. Compare the footnote on p. 57.

$$\left[\int \frac{x^3+x^2+x+1}{\sqrt{x^2+2x+3}}\,dx = \int \frac{x(x^2+2x+3)-x^2-2x+1^*}{\sqrt{x^2+2x+3}}\,dx\right.$$

$$\left. = \int x\sqrt{x^2+2x+3}\,dx - \int \frac{x^2+2x-1}{\sqrt{x^2+2x+3}}\,dx.\right]$$

Four standard cases of $\int \frac{1}{X\sqrt{Y}}$ dx,

where X, Y are linear or quadratic functions of x.

43. Case I. Rule to integrate $\int \frac{1}{\text{linear}\ \sqrt{\text{linear}}}$ dx.

Put $\sqrt{linear} = z.$

[Proof.** To integrate $\int \frac{1}{(ax+b)\sqrt{cx+e}}\,dx$.†

Put $\sqrt{cx+e} = z$ ...(1), then $\frac{1}{2\sqrt{cx+e}}\,c\,dx = dz$, or $\frac{1}{\sqrt{cx+e}}\,dx = \frac{2dz}{c}$.

$\therefore \int \frac{1}{(ax+b)\sqrt{cx+e}}\,dx$ | From (1), squaring, $cx+e=z^2$, or $x = \frac{z^2-e}{c}$

$$= \int \frac{1}{\left[a\left(\frac{z^2-e}{c}\right)+b\right]}\frac{2dz}{c} = 2\int \frac{1}{az^2-ae+bc}\,dz,$$

which can be integrated (Arts. 13, (*ii*) ; 33).]

Example 1. *Integrate* $\int \frac{dx}{(x+2)\sqrt{x+3}}$.

$$\left[\text{Form} \int \frac{1}{\text{linear}\ \sqrt{\text{linear}}}\,dx\right]$$

Put $\sqrt{x+3} = z$ ...(1) [$\sqrt{\text{linear}} = z$ (Art. 43)]

then $\frac{1}{2\sqrt{x+3}}\,dx = dz$, or $\frac{1}{\sqrt{x+3}}\,dx = 2dz.$

***How to write this step.** Write in the numerator the quadratic whose square root occurs in the denominator, *viz.*, x^2+2x+3, and adjust the coefficient of x^3 by multiplying x^2+2x+3 by x, thus getting $x(x^2+2x+3)$; now adjust the other terms by adding $-x^2-2x+1$ [$\because$ we have got the other terms $2x^2+3x$ in $x(x^2+2x+3)$ but we require x^2+x+1 as the other terms in x^3+x^2+x+1], thus getting

$$x^3+x^2+x+1 = x(x^2+2x+3)-x^2-2x+1.$$

† While taking constant coefficients a, b, c, etc., d is generally omitted to avoid confusion with d in dx. Thus here we have taken $\sqrt{cx+e}$ and not $\sqrt{cx+d}$.

$\therefore \quad \int \frac{1}{(x+2)\sqrt{x+3}}\,dx$ | From (1), squaring, $x + 3 = z^2$, or $x = z^2 - 3$

$$= \int \frac{1}{z^2 - 3 + 2}\, 2\,dz = 2\int \frac{1}{z^2 - 1}\,dz = 2\int \frac{1}{z^2 - (1)^2}\,dz$$

$$= 2\,\frac{1}{2(1)} \log \frac{z-1}{z+1} = \log \frac{\sqrt{x+3} - 1}{\sqrt{x+3} + 1}. \qquad \text{[From (1)]}$$

[**Extension.** *If in* $\int \frac{1}{\textit{linear}\sqrt{\textit{linear}}}\,dx$, *instead of 1 in the numerator there is a polynomial in x, whose degree is equal to, or greater than, that of the rational factor of the denominator, first divide the numerator by this rational factor, and then use the rule, where necessary.*]

Example 2. *Integrate* $\int \frac{x^3}{(x-1)\sqrt{x+2}}\,dx.$

[Here the degree of the num. x^3 > that of the rational factor $(x - 1)$ of the denom., $\therefore$ divide x^3 by $x - 1$.]

$$\int \frac{x^3}{(x-1)\sqrt{x+2}}\,dx = \int \left[x^2 + x + 1 + \frac{1}{x-1}\right] \frac{1}{\sqrt{x+2}}\,dx$$

$$= \int \frac{x^2 + x + 1}{\sqrt{x+2}}\,dx + \int \frac{1}{(x-1)\sqrt{x+2}}\,dx \qquad ...(1)$$

$x - 1)\; x^3 \;(x^2 + x + 1$
$\quad x^3 - x^2$
$\quad\quad x^2$
$\quad\quad x^2 - x$
$\quad\quad\quad x$
$\quad\quad\quad x - 1$
$\quad\quad\quad\quad 1$

Put $\sqrt{x+2} = z$* ...(2) $[\sqrt{\text{linear}} = z]$

then $\frac{1}{2\sqrt{x+2}}\,dx = dz$, or $\frac{1}{\sqrt{x+2}}\,dx = 2dz$.

$\therefore \quad \int \frac{x^2 + x + 1}{\sqrt{x+2}}\,dx$ | From (2), squaring, $x + 2 = z^2$, or $x = z^2 - 2$

$$= \int [(z^2 - 2)^2 + (z^2 - 2) + 1]\, 2dz$$

$$= 2\int [z^4 - 4z^2 + 4 + z^2 - 2 + 1]\,dz = 2\int (z^4 - 3z^2 + 3)dz = 2\left[\frac{z^5}{5} - \frac{3z^3}{3} + 3z\right]$$

$$= 2\left[\frac{(x+2)^{\frac{5}{2}}}{5} - (x+2)^{\frac{3}{2}} + 3(x+2)^{\frac{1}{2}}\right], \qquad \text{[From (2)]}$$

and $\int \frac{1}{(x-1)\sqrt{x+2}}\,dx = \int \frac{1}{z^2 - 2 - 1}\, 2dz = 2\int \frac{1}{z^2 - 3}\,dz$

$$= 2\int \frac{1}{z^2 - (\sqrt{3})^2}\,dz = 2\,\frac{1}{2\sqrt{3}} \log \frac{z - \sqrt{3}}{z + \sqrt{3}} = \frac{1}{\sqrt{3}} \log \frac{\sqrt{x+2} - \sqrt{3}}{\sqrt{x+2} + \sqrt{3}}.$$

***Why this step. Rule to integrate if only one radical $\sqrt{\text{linear}}$ occurs in the integrand :**

Put linear = z^2 (Art. 48), *i.e.*, $\sqrt{\text{linear}} = z$ which substitution enables us also to integrate

$\int \frac{1}{\text{linear}\sqrt{\text{linear}}}\,dx$ (Art. 43).

$\therefore$ from (1), $\displaystyle\int \frac{x^3}{(x-1)\sqrt{x+2}}\,dx$

$$= 2\left[\frac{(x+2)^{\frac{5}{2}}}{5} - (x+2)^{\frac{3}{2}} + 3(x+2)^{\frac{1}{2}}\right] + \frac{1}{\sqrt{3}}\log\frac{\sqrt{x+2}-\sqrt{3}}{\sqrt{x+2}+\sqrt{3}}.$$

EXAMPLES

1. Evaluate

(*i*) $\displaystyle\int \frac{1}{2\sqrt{x}(1+x)}\,dx.$ (*ii*) $\displaystyle\int \frac{1}{x\sqrt{x-1}}\,dx.$

2. Integrate (*i*) $\displaystyle\int \frac{1}{x\sqrt{x+1}}\,dx.$ (*ii*) $\displaystyle\int (1+x)^{-1}x^{-1/2}\,dx.$

3. Evaluate (*i*) $\displaystyle\int (2+x)^{-1}(1+2x)^{-1/2}\,dx.$ (*ii*) $\displaystyle\int_8^{15} \frac{dx}{(x-3)\sqrt{x+1}}.$

4. Integrate (*i*) $\displaystyle\int \frac{x-1}{(x+1)\sqrt{x+3}}\,dx.$ (*ii*) $\displaystyle\int \frac{x^2-x+1}{(x+1)\sqrt{x-2}}\,dx.$

44. Case II. Rule to integrate $\displaystyle\int \frac{1}{\text{linear}\sqrt{\text{quadratic}}}\,dx.$

Put linear $= \dfrac{1}{z}$.

[Proof.** To integrate $\displaystyle\int \frac{1}{(ax+b)\sqrt{cx^2+ex+f}}\,dx.$*

Put $ax+b = \dfrac{1}{z}$ ...(1), then $a\,dx = -\dfrac{1}{z^2}\,dz$, or $dx = -\dfrac{1}{az^2}\,dz.$

From (1), $ax = \dfrac{1}{z} - b.$ or $x = \dfrac{1}{a}\left(\dfrac{1}{z} - b\right)$

$$\therefore \int \frac{1}{(ax+b)\sqrt{cx^2+ex+f}}\,dx$$

$$= \int \frac{1}{\frac{1}{z}\sqrt{c\frac{1}{a^2}\left(\frac{1}{z}-b\right)^2 + e\frac{1}{a}\left(\frac{1}{z}-b\right)+f}}\left(-\frac{1}{az^2}\,dz\right)$$

$$= -\frac{1}{a}\int \frac{1}{\sqrt{\frac{A}{z^2}+\frac{B}{z}+C}}\cdot\frac{1}{z}\,dz \text{ (say)}$$

$$= -\frac{1}{a}\int \frac{1}{\sqrt{A+Bz+Cz^2}}\,dz,$$

which can be integrated (Art. 41).

*See the footnote† on p. 80.

Example 1. *Integrate* $\int \frac{dx}{(x+2)\sqrt{x^2+6x+7}}$.

$\left[\text{Form} \int \frac{1}{\text{linear}\sqrt{\text{quadratic}}}\,dx\right]$

Put $x + 2 = \frac{1}{z}$, ...(1) $\left[\text{Linear} = \frac{1}{z} \text{ (Art. 44)}\right]$

then $dx = -\frac{1}{z^2}\,dz.$

$\therefore \int \frac{1}{(x+2)\sqrt{x^2+6x+7}}\,dx$ | From (1), $x = \frac{1}{z} - 2$

$$= \int \frac{1}{\frac{1}{z}\sqrt{\left(\frac{1}{z}-2\right)^2 + 6\left(\frac{1}{z}-2\right)+7}}\left(-\frac{1}{z^2}\,dz\right)$$

$$= -\int \frac{1}{\sqrt{\frac{1}{z^2} - \frac{4}{z} + 4 + \frac{6}{z} - 12 + 7}}\,\frac{1}{z}\,dz$$

$$= -\int \frac{1}{\sqrt{\frac{1}{z^2} + \frac{2}{z} - 1}}\,\frac{1}{z}\,dz = -\int \frac{1}{\sqrt{1+2z-z^2}}\,dz$$

$$= -\int \frac{1}{\sqrt{1-(z^2-2z)}}\,dz = -\int \frac{1}{\sqrt{1-(z^2-2z+1)+1}}\,dz$$

$$= -\int \frac{1}{\sqrt{2-(z-1)^2}}\,dz = -\int \frac{1}{\sqrt{(\sqrt{2})^2-(z-1)^2}}\,dz$$

$$= -\sin^{-1}\frac{z-1}{\sqrt{2}} = -\sin^{-1}\frac{\frac{1}{x+2}-1}{\sqrt{2}}$$ $\left[\because \text{ from (1), } z = \frac{1}{x+2}\right]$

$$= -\sin^{-1}\frac{1-x-2}{(x+2)\sqrt{2}} = -\sin^{-1}\frac{-x-1}{(x+2)\sqrt{2}}.$$

[**Extension.** *If in* $\int \frac{1}{\textit{linear}\sqrt{\textit{quadratic}}}\,dx$, *instead of 1 in the numerator, there is a polynomial in x, whose degree is equal to, or greater than, that of the rational factor of the denominator, first divide the numerator by this rational factor, and then use the rule, where necessary.*]

Example 2. *Integrate* $\int \frac{x^2+2x+3}{(x+1)\sqrt{x^2+1}}\,dx.$

[Here the degree of the num. $x^2 + 2x + 3 >$ that of the rational factor $(x + 1)$ of the denom.,
$\therefore$ divide $x^2 + 2x + 3$ by $x + 1$.]

$$\int \frac{x^2+2x+3}{(x+1)\sqrt{x^2+1}}\,dx = \int \left(x+1+\frac{2}{x+1}\right)\frac{1}{\sqrt{x^2+1}}\,dx$$

$$\begin{array}{r|l} x+1) & x^2+2x+3(x+1 \\ & \underline{x^2+x} \\ & x+3 \\ & \underline{x+1} \\ & 2 \end{array}$$

$$= \int \frac{x}{\sqrt{x^2+1}}\,dx + \int \frac{1}{\sqrt{x^2+1}}\,dx + 2\int \frac{1}{(x+1)\sqrt{x^2+1}}\,dx \qquad \text{...(1)}$$

Now $\displaystyle\int \frac{x}{\sqrt{x^2+1}}\,dx = \tfrac{1}{2}\int (x^2+1)^{-\frac{1}{2}}(2x)\,dx = \tfrac{1}{2}\,\frac{(x^2+1)^{\frac{1}{2}}}{\frac{1}{2}} = \sqrt{x^2+1}$,

$$\int \frac{1}{\sqrt{x^2+1}}\,dx = \int \frac{1}{\sqrt{x^2+(1)^2}}\,dx = \sinh^{-1}\frac{x}{1} = \sinh^{-1}x,$$

and $\displaystyle\int \frac{1}{(x+1)\sqrt{x^2+1}}\,dx$ $\left[\text{Put } x+1=\frac{1}{z} \text{ ...(2), then } dx = -\frac{1}{z^2}\,dz.\right.$ From (2), $\left. x = \frac{1}{z}-1\right]$

$$= \int \frac{1}{\frac{1}{z}\sqrt{\left(\frac{1}{z}-1\right)^2+1}}\left(-\frac{1}{z^2}\,dz\right) = -\int \frac{1}{\sqrt{\frac{1}{z^2}-\frac{2}{z}+1+1}}\cdot\frac{1}{z}\,dz$$

$$= -\int \frac{1}{\sqrt{\frac{1}{z^2}-\frac{2}{z}+2}}\cdot\frac{1}{z}\,dz = -\int \frac{1}{\sqrt{1-2z+2z^2}}\,dz$$

$$= -\int \frac{1}{\sqrt{2}\sqrt{z^2-z+\frac{1}{2}}}\,dz = -\frac{1}{\sqrt{2}}\int \frac{1}{\sqrt{(z^2-z+\frac{1}{4})-\frac{1}{4}+\frac{1}{2}}}\,dz$$

$$= -\frac{1}{\sqrt{2}}\int \frac{1}{\sqrt{(z-\frac{1}{2})^2+\frac{1}{4}}}\,dz = -\frac{1}{\sqrt{2}}\int \frac{1}{\sqrt{(z-\frac{1}{2})^2+(\frac{1}{2})^2}}\,dz$$

$$= -\frac{1}{\sqrt{2}}\sinh^{-1}\frac{z-\frac{1}{2}}{\frac{1}{2}} = -\frac{1}{\sqrt{2}}\sinh^{-1}(2z-1) = -\frac{1}{\sqrt{2}}\sinh^{-1}\left(\frac{2}{x+1}-1\right)$$

$\left[\because \text{ from (2), } z = \frac{1}{x+1}\right]$

$$= -\frac{1}{\sqrt{2}}\sinh^{-1}\frac{2-x-1}{x+1} = -\frac{1}{\sqrt{2}}\sinh^{-1}\frac{1-x}{1+x}.$$

$\therefore$ from (1), $\displaystyle\int \frac{x^2+2x+3}{(x+1)\sqrt{x^2+1}}\,dx$

$$= \sqrt{x^2+1}+\sinh^{-1}x+2\left(-\frac{1}{\sqrt{2}}\sinh^{-1}\frac{1-x}{1+x}\right)$$

$$= \sqrt{x^2+1}+\sinh^{-1}x-\sqrt{2}\sinh^{-1}\frac{1-x}{1+x}.$$

EXAMPLES

1. Integrate

(i) $\int \frac{dx}{x\sqrt{1+x^2}}$. (ii) $\int \frac{dx}{(x-1)\sqrt{x^2+1}}$.

[(ii) **Note. Case when x < 1. *It is assumed that $x > 1$, so that $(x-1)$ is positive. If, however, $x < 1$, so that $(x-1)$ is negative, then*

$$\int \frac{1}{(x-1)\sqrt{x^2+1}}\,dx = \int \frac{1}{-(1-x)\sqrt{x^2+1}}\,dx = -\int \frac{1}{(1-x)\sqrt{x^2+1}}\,dx.$$

Now put $1 - x = \frac{1}{z}$. It will be found that

$$\int \frac{1}{(x-1)\sqrt{x^2+1}}\,dx = -\frac{1}{\sqrt{2}}\sinh^{-1}\frac{1+x}{1-x}.]$$

2. Integrate (i) $\int x^{-1}(3-4x+x^2)^{-1/2}\,dx$. (ii) $\int \frac{dx}{(x+1)^{\frac{3}{2}}(x-1)^{\frac{1}{2}}}$.

3. Evaluate (i) $\int_0^1 \frac{dx}{(x+1)\sqrt{x^2+2x}}$.

(ii) $\int \frac{dx}{(1+x)\sqrt{1+x+x^2}}$.

4. Integrate

(i) $\int \frac{x\,dx}{(x-1)\sqrt{x^2+1}}$. (ii) $\int \frac{x^2+3x+3}{(x+2)\sqrt{x^2+4x+6}}\,dx$.

5. Integrate

(i) $\int \frac{x^2-x-1}{(x-1)\sqrt{x^2-1}}\,dx$. (ii) $\int \frac{\sqrt{1+x+x^2}}{1+x}\,dx$.

6. Integrate $\int \log\frac{a}{x}\sin^{-1}x\,dx$.

[**Extension. Rule to integrate** $\int \frac{1}{(\text{linear})^2\sqrt{\text{quadratic}}}\,dx$.

Put linear $= \frac{1}{z}$.]

[Proof.** To integrate $\int \frac{1}{(ax+b)^2\sqrt{cx^2+ex+f}}\,dx$.

Put $ax + b = \frac{1}{z}$ ...(1). Proceed as in Art. 44.

It will be found that

$$\int \frac{1}{(ax+b)^2\sqrt{cx^2+ex+f}}\,dx = -\frac{1}{a}\int \frac{z}{\sqrt{A+Bz+Cz^2}}\,dz,$$

which can be integrated (Art. 42).]

7. Integrate $\int \frac{dx}{x^2\sqrt{x^2 - a^2}}$.

****8.** Integrate $\int \frac{1}{\sin\theta\sqrt{a\cos^2\theta + b\sin^2\theta + c}}\,d\theta.$

$$\left[\int \frac{1}{\sin\theta\sqrt{a\cos^2\theta + b\sin^2\theta + c}}\,d\theta = \int \frac{1}{\sin\theta\sqrt{a\cos^2\theta + b\sin^2\theta + c(\cos^2\theta + \sin^2\theta)}}\,d\theta\right.$$

$$= \int \frac{1}{\sin\theta\sqrt{(a+c)\cos^2\theta + (b+c)\sin^2\theta}}\,d\theta.$$

(Denom. is a homogeneous function of the second degree in $\cos\theta$ and $\sin\theta$.)

(As suggested by the factor $\sin\theta$ in the denom.) Divide the num. and denom. by $\sin^2\theta$, and put $\cot\theta = z$.]

****9.** Integrate $\int \frac{1-x}{1+x}\cdot\frac{1}{\sqrt{x + x^2 + x^3}}\,dx.$

$$\left[\int \frac{1-x}{1+x}\cdot\frac{1}{\sqrt{x+x^2+x^3}}\,dx = \int \frac{1-x}{1+x}\cdot\frac{1}{\sqrt{x}\sqrt{1+x+x^2}}\,dx.\right.$$

$\left(\text{As suggested by } \frac{1}{\sqrt{x}}\,dx\right)$ Put $\sqrt{x} = z$, then it will be found that

$$\int \frac{1-x}{1+x}\cdot\frac{1}{\sqrt{x+x^2+x^3}}\,dx = 2\int \frac{1-z^2}{(1+z^2)\sqrt{1+z^2+z^4}}\,dz$$

(Divide the num. and denom. by z^2)

$$\left.= 2\int \frac{\frac{1}{z^2} - 1}{\left(z + \frac{1}{z}\right)\sqrt{z^2 + \frac{1}{z^2} + 1}}\,dz. \text{ Put } z + \frac{1}{z} = u.\right]$$

45. Case III. Rule to integrate $\int \frac{1}{\text{quadratic}\sqrt{\text{linear}}}\,dx$.

Put $\sqrt{linear} = z$.

[Proof.** To integrate $\int \frac{1}{(ax^2 + bx + c)\sqrt{ex + f}}\,dx$.*

Put $\sqrt{ex + f} = z$ ...(1), then $\frac{1}{2\sqrt{ex+f}}\,e\,dx = dz$, or $\frac{1}{\sqrt{ex+f}}\,dx = \frac{2dz}{e}$.

From (1), squaring, $ex + f = z^2$, or $x = \frac{z^2 - f}{e}$

$\therefore \int \frac{1}{(ax^2 + bx + c)\sqrt{ex + f}}\,dx$

*See the footnote† on p. 80.

$$= \int \frac{1}{\left[a\left(\frac{z^2-f}{e}\right)^2 + b\left(\frac{z^2-f}{e}\right) + c\right]} 2\frac{dz}{e}$$

$$= \frac{2}{e}\int \frac{1}{Az^4 + Bz^2 + C} dz \text{ (say)},$$

which can be integrated by resolving the integrand into partial fractions.]

Note 1. *The rule to integrate* $\int \frac{1}{quadratic\sqrt{linear}} dx$ *(Art. 45) is the same as the rule to integrate* $\int \frac{1}{linear\sqrt{linear}} dx$ *(Art. 43), viz., put* $\sqrt{linear} = z$.

Note 2. *If in* $\int \frac{1}{quadratic\sqrt{linear}} dx$, *instead of 1 in the numerator, there is a linear in x, the same rule is applicable, viz., put* $\sqrt{linear} = z$. *(See the following solved Ex.)*

Example. *Integrate* $\int \frac{x+2}{(x^2+3x+3)\sqrt{x+1}} dx$.

Put $\sqrt{x+1} = z$ ...(1), then $\frac{1}{2\sqrt{x+1}} dx = dz$, or $\frac{1}{\sqrt{x+1}} dx = 2\, dz$.

$$\therefore \int \frac{x+2}{(x^2+3x+3)\sqrt{x+1}} dx$$

From (1), squaring, $x+1 = z^2$, or $x = z^2 - 1$.

$$= \int \frac{z^2-1+2}{(z^2-1)^2+3(z^2-1)+3} 2dz = 2\int \frac{z^2+1}{z^4-2z^2+1+3z^2-3+3} dz$$

$$= 2\int \frac{z^2+1}{z^4+z^2+1} dz$$ [Divide the num. and denom. by z^2]

$$= 2\int \frac{1+\frac{1}{z^2}}{z^2+1+\frac{1}{z^2}} dz$$ [Put $z - \frac{1}{z} = u$ ...(2), then $\left(1+\frac{1}{z^2}\right) dz = du$.

From (2), squaring, $z^2 + \frac{1}{z^2} - 2 = u^2$, or $z^2 + \frac{1}{z^2} = u^2 + 2$]

$$= 2\int \frac{du}{u^2+2+1} = 2\int \frac{1}{u^2+3} du = 2\int \frac{1}{u^2+(\sqrt{3})^2} du$$

$$= 2 \cdot \frac{1}{\sqrt{3}} \tan^{-1}\frac{u}{\sqrt{3}} = \frac{2}{\sqrt{3}} \tan^{-1}\frac{z-\frac{1}{z}}{\sqrt{3}}$$ [From (2)]

$$= \frac{2}{\sqrt{3}} \tan^{-1}\frac{z^2-1}{z\sqrt{3}} = \frac{2}{\sqrt{3}} \tan^{-1}\frac{x+1-1}{\sqrt{x+1}\sqrt{3}}$$ [From (1)]

$$= \frac{2}{\sqrt{3}} \tan^{-1}\frac{x}{\sqrt{3(x+1)}}.$$

EXAMPLES

1. Integrate (*i*) $\displaystyle\int \frac{1}{(x^2+1)\sqrt{x}}\,dx.$ (*ii*) $\displaystyle\int \frac{x+1}{(x^2+4x+5)\sqrt{x+2}}\,dx.$

[**Extension.** *If in* $\displaystyle\int \frac{1}{\text{quadratic}\sqrt{\text{linear}}}\,dx$*, instead of 1 in the numerator, there is a polynomial in x whose degree is equal to, or greater than, that of the rational factor of the denominator, first divide the numerator by this rational factor, and then use the rule, where necessary.*]

****2.** Integrate $\displaystyle\int \frac{x^2+1}{(x^2-2x+2)\sqrt{x-1}}\,dx.$

46. Case IV. Rule to integrate

$$\int \frac{\mathbf{1}}{\textbf{pure quadratic}^* \sqrt{\textbf{pure quadratic}}}\,\mathbf{dx}.$$

Put $x = \dfrac{1}{z}$ *and then in the resulting integral, put* $\sqrt{\textit{pure quadratic}} = u.$

** [**Proof.** To integrate $\displaystyle\int \frac{1}{(ax^2+b)\sqrt{cx^2+e}}\,dx.$ †

Put $x = \dfrac{1}{z}$ ...(1), then $dx = -\dfrac{1}{z^2}\,dz.$

$$\therefore \int \frac{1}{(ax^2+b)\sqrt{cx^2+e}}\,dx = \int \frac{1}{\left(\dfrac{a}{z^2}+b\right)\sqrt{\dfrac{c}{z^2}+e}}\left(-\frac{1}{z^2}\,dz\right)$$

$$= -\int \frac{z}{(a+bz^2)\sqrt{c+ez^2}}\,dz \quad ...(2) \quad [\text{Put } \sqrt{c+ez^2} = u] \quad ...(3),$$

then $\dfrac{1}{2\sqrt{c+ez^2}}(2ez)\,dz = du,$ or $\dfrac{z\,dz}{\sqrt{c+ez^2}} = \dfrac{du}{e}.$

From (3), squaring, $c + ez^2 = u^2$, or $z^2 = \dfrac{u^2-c}{e}$]

$$= -\int \frac{1}{\left[a + b\dfrac{(u^2-c)}{e}\right]}\frac{du}{e} = -\int \frac{1}{ae+bu^2-bc}\,du$$

$$= -\int \frac{1}{bu^2+ae-bc}\,du, \text{ which can be integrated (Arts. 13, (ii) ; 33).]}$$

Cor. Rule to integrate $\displaystyle\int \frac{\mathbf{x}}{\textbf{pure quadratic}\sqrt{\textbf{pure quadratic}}}\,\mathbf{dx}.$

***Pure quadratic. Def.** A **pure quadratic** is one which does not contain the term in x. Thus $ax^2 + b$ is a pure quadratic.

† See the footnote † on p. 80.

Put $\sqrt{\textit{pure quadratic}} = z.$

[Proof.** See (2) above.]

Example. *Integrate* $\int \dfrac{dx}{(x^2+a^2)\sqrt{x^2+b^2}}$.

$$\left[\text{Form} \int \frac{1}{\text{pure quadratic}\sqrt{\text{pure quadratic}}}\,dx\right]$$

Put $x = \dfrac{1}{z}$ ...(1), then $dx = -\dfrac{1}{z^2}\,dz$.

$$\therefore \int \frac{1}{(x^2+a^2)\sqrt{x^2+b^2}}\,dx = \int \frac{1}{\left(\frac{1}{z^2}+a^2\right)\sqrt{\frac{1}{z^2}+b^2}}\left(-\frac{1}{z^2}\,dz\right)$$

$$= -\int \frac{z}{(1+a^2z^2)\sqrt{1+b^2z^2}}\,dz \qquad \left[\text{Form} \int \frac{x}{\text{pure quad.}\sqrt{\text{pure quad.}}}\,dx\right.$$

$\Big[$ Put $\sqrt{1+b^2z^2} = u$ ...(2), $[\sqrt{\text{pure quad.}} = z]$

then $\dfrac{1}{2\sqrt{1+b^2z^2}}(2b^2z)dz = du,$ or $\dfrac{z\,dz}{\sqrt{1+b^2z^2}} = \dfrac{du}{b^2}$.

From (2), squaring, $1 + b^2z^2 = u^2$, or $z^2 = \dfrac{u^2-1}{b^2}\Big]$

$$= -\int \frac{1}{\left[1+a^2\left(\frac{u^2-1}{b^2}\right)\right]}\frac{du}{b^2} = -\int \frac{1}{b^2+a^2u^2-a^2}\,du$$

$$= -\int \frac{1}{a^2\left(u^2+\frac{b^2-a^2}{a^2}\right)}\,du = -\frac{1}{a^2}\int \frac{1}{u^2+\frac{b^2-a^2}{a^2}}\,du \qquad \text{...(3)}$$

Case I. If $a^2 > b^2$, then from (3),

$$\int \frac{1}{(x^2+a^2)\sqrt{x^2+b^2}}\,dx = -\frac{1}{a^2}\int \frac{1}{u^2-\frac{a^2-b^2}{a^2}}\,du$$

$$= -\frac{1}{a^2}\int \frac{1}{u^2-\left(\frac{\sqrt{a^2-b^2}}{a}\right)^2}\,du = -\frac{1}{a^2}\cdot\frac{1}{2\frac{\sqrt{a^2-b^2}}{a}}\log\frac{u-\frac{\sqrt{a^2-b^2}}{a}}{u+\frac{\sqrt{a^2-b^2}}{a}}$$

$$= -\frac{1}{2a\sqrt{a^2-b^2}}\log\frac{au-\sqrt{a^2-b^2}}{au+\sqrt{a^2-b^2}} \quad \text{which, from (2),}$$

$$= -\frac{1}{2a\sqrt{a^2-b^2}}\log\frac{a\sqrt{1+b^2z^2}-\sqrt{a^2-b^2}}{a\sqrt{1+b^2z^2}+\sqrt{a^2-b^2}} \quad \text{which, from (1),}$$

$$= -\frac{1}{2a\sqrt{a^2-b^2}} \log \frac{a\sqrt{1+\dfrac{b^2}{x^2}} - \sqrt{a^2-b^2}}{a\sqrt{1+\dfrac{b^2}{x^2}} + \sqrt{a^2-b^2}}$$

$$= -\frac{1}{2a\sqrt{a^2-b^2}} \log \frac{a\dfrac{\sqrt{x^2+b^2}}{x} - \sqrt{a^2-b^2}}{a\dfrac{\sqrt{x^2+b^2}}{x} + \sqrt{a^2-b^2}}$$

$$= -\frac{1}{2a\sqrt{a^2-b^2}} \log \frac{a\sqrt{x^2+b^2} - x\sqrt{a^2-b^2}}{a\sqrt{x^2+b^2} + x\sqrt{a^2-b^2}}.$$

Case II. If $a^2 < b^2$, then from (3),

$$\int \frac{1}{(x^2+a^2)\sqrt{x^2+b^2}}\,dx = -\frac{1}{a^2}\int \frac{1}{u^2 + \left(\dfrac{\sqrt{b^2-a^2}}{a}\right)^2}\,du$$

$$= -\frac{1}{a^2}\cdot\frac{1}{\dfrac{\sqrt{b^2-a^2}}{a}}\tan^{-1}\frac{u}{\dfrac{\sqrt{b^2-a^2}}{a}} = -\frac{1}{a\sqrt{b^2-a^2}}\tan^{-1}\frac{au}{\sqrt{b^2-a^2}}$$

which, from (2),

$$= -\frac{1}{a\sqrt{b^2-a^2}}\tan^{-1}\frac{a\sqrt{1+b^2z^2}}{\sqrt{b^2-a^2}} \text{ which, from (1),}$$

$$= -\frac{1}{a\sqrt{b^2-a^2}}\tan^{-1}\frac{a\sqrt{1+\dfrac{b^2}{x^2}}}{\sqrt{b^2-a^2}} = -\frac{1}{a\sqrt{b^2-a^2}}\tan^{-1}\frac{a\sqrt{x^2+b^2}}{x\sqrt{b^2-a^2}}.$$

EXAMPLES

1. (*i*) Integrate $\displaystyle\int \frac{1}{(x^2-1)\sqrt{x^2+1}}\,dx.$

(*ii*) Prove that $\displaystyle\int_0^1 \frac{dx}{(1+x^2)\sqrt{1-x^2}} = \frac{\pi}{2\sqrt{2}}.$

Integrate (Exs. 2—5) :

2. $\displaystyle\int \frac{1}{(x^2+4)\sqrt{(x+2)(x-2)}}\,dx.$

3. $\displaystyle\int \frac{x\,dx}{(x^2+a^2)\sqrt{x^2+b^2}}.$

$$\left[\left(\text{Form} \int \frac{x}{\text{pure quadratic}\sqrt{\text{pure quadratic}}}\,dx\right)\right.$$

Put $\sqrt{x^2+b^2} = z.$] [$\sqrt{\text{pure quadratic}} = z$ (Art. 46, Cor.)]

**4. $\displaystyle\int \frac{x+a}{(x^2+b^2)\sqrt{x^2+c^2}}\,dx.$

[**Extension.** *If in* $\displaystyle\int \frac{1}{\text{pure quadratic}\sqrt{\text{pure quadratic}}}\,dx$, *instead of 1 in the numerator, there is a polynomial in x, whose degree is equal to, or greater than, that of the rational factor of the denominator, first divide the numerator by this rational factor, and then use the rule, where necessary.*]

5. $\displaystyle\int \frac{\sqrt{1+x^2}}{1-x^2}\,dx.$

[Rule to integrate $\displaystyle\int \frac{1}{\text{quadratic}\sqrt{\text{quadratic}}}\,dx \left\{\text{or} \int \frac{\text{linear}}{\text{quadratic}\sqrt{\text{quadratic}}}\,dx\right\}$, if the quadratic not under the square root can be resolved into real linear factors :

Resolve $\dfrac{1}{\text{quadratic}}\left\{\text{or } \dfrac{\text{linear}}{\text{quadratic}}\right\}$ *into partial fractions, and split up the integral into two, each of which is of the form* $\displaystyle\int \frac{1}{\text{linear}\sqrt{\text{quadratic}}}\,dx$.]

**6. Integrate $\displaystyle\int \frac{x-1}{(x^2-2x)\sqrt{2x^2-2x+1}}\,dx.$

Three standard cases of $\int x^m (a+bx^n)^p\,dx$.

47. To integrate $\int x^m(a+bx^n)^p\,dx$.

Case I. If p is a + ve integer, we can expand $(a+bx^n)^p$ in a finite series by the binomial theorem, multiply each term by x^m, and integrate term by term.

Case II. If p is a fraction $= \dfrac{r}{s}$ (say), where r and s are integers, and s +ve, then

$$\int x^m(a+bx^n)^p\,dx = \int (a+bx^n)^{\frac{r}{s}}\,x^m\,dx$$

[Put $a+bx^n = z^s$* ...(1), then $bn\,x^{n-1}\,dx = s\,z^{s-1}\,dz$, or $dx = \dfrac{s}{bn}\dfrac{z^{s-1}}{x^{n-1}}\,dz$

$$= \int x^m z^r \frac{s}{bn}\frac{z^{s-1}}{x^{n-1}}\,dz = \frac{s}{bn}\int z^{r+s-1}\,x^{m-n+1}\,dz$$

From (1), $x^n = \dfrac{z^s-a}{b}$, or $x = \left(\dfrac{z^s-a}{b}\right)^{\frac{1}{n}}$

---

*__Why this step.__ To rid $(a+bx^n)^{\frac{r}{s}}$ of fractional power, we put $a+bx^n = z^s$, so that $(a+bx^n)^{\frac{r}{s}} = (z^s)^{\frac{r}{s}} = z^r$.

$$= \frac{s}{bn}\int z^{r+s-1}\left(\frac{z^s-a}{b}\right)^{\frac{m-n+1}{n}} dz = \frac{s}{nb^{\frac{m+1}{n}}}\int z^{r+s-1}(z^s-a)^{\frac{m+1}{n}-1}\,dz \qquad \text{...(1)}$$

Sub-case (i). If $\frac{m+1}{n}$ is a +ve integer, the integral on the R.H.S. of (1) can be evaluated (Case I).

Sub-case (ii). If $\frac{m+1}{n}$ is a –ve integer, or 0, the integral on the R.H.S. of (1)

$$= \frac{s}{nb^{\frac{m+1}{n}}}\int \frac{z^{r+s-1}}{(z^s-a)^{-\frac{m+1}{n}+1}}\,dz,$$

which can be integrated by resolving the integrand into partial fractions (Chap. IV).

Case III. If p is a fraction $= \frac{r}{s}$ (say), where r and s are integers, and s +ve, then

$$\int x^m (a+bx^n)^p\,dx = \int x^m (a+bx)^{n\frac{r}{s}}\,dx \qquad \text{[Take } x^n \text{ outside the binomial]}$$

$$= \int x^{m+\frac{nr}{s}}(ax^{-n}+b)^{\frac{r}{s}}\,dx, \qquad \left[\text{Compare with } \int x^m (a+bx^n)^p\,dx,\right.$$

$$\left.\text{here "}m\text{"} = m + \frac{nr}{s}, \text{ "}n\text{"} = -n\right]$$

which can be integrated (Case II), if $\dfrac{m+\frac{nr}{s}+1}{-n}$ $\qquad \left[\dfrac{m+1}{n}\right]$

is a + ve or –ve integer, or 0, *i.e.*, if $\frac{m+1}{n}+\frac{r}{s}$ is a – ve or + ve integer, or 0, *i.e.*, if $\frac{m+1}{n}+p$ is a + ve or –ve integer, or 0.

[Rule to integrate $\int x^m (a+bx^n)^p\,dx$:

Case I. *If p is a positive integer, expand.*

Case II. *If p is a fraction* $= \frac{r}{s}$ *(say), where r and s are integers, and s positive, then if* $\frac{m+1}{n}$ *is a positive or negative integer, or 0, put* $a+bx^n = z^s$*, and then expand, or resolve into partial fractions, as the case may require.*

Case III. *If p is a fraction* $= \frac{r}{s}$ *(say), where r and s are integers, and s positive, then, if* $\frac{m+1}{n} + p$ *is a positive or negative integer, or 0, take* x^n *outside the binomial, and put* $ax^{-n} + b = z^s$, *and then expand, or resolve into partial fractions, as the case may require.*]

Example 1. *Integrate* $\int x^5 (1 + x^3)^{1/2}\, dx.$

[Compare $\int x^5 (1 + x^3)^{1/2}\, dx$ with $\int x^m (a + bx^n)^p\, dx,$

here $m = 5$, $n = 3$, $p = \frac{1}{2}$, a fraction

$$\therefore \quad \frac{m+1}{n} = \frac{5+1}{3} = \frac{6}{3} = 2, \text{ a +ve integer.}]$$

Put $1 + x^3 = z^2$* ...(1), then $3x^2\, dx = 2z\, dz$, or $dx = \frac{2}{3}\frac{z}{x^2}\, dz.$

$$\therefore \quad \int x^5 (1 + x^3)^{1/2}\, dx = \int x^5 z \frac{2}{3}\frac{z}{x^2}\, dz = \frac{2}{3}\int z^2 x^3\, dz \qquad | \text{ From (1), } x^3 = z^2 - 1$$

$$= \frac{2}{3}\int z^2 (z^2 - 1)\, dz = \frac{2}{3}\int (z^4 - z^2)\, dz = \frac{2}{3}\left(\frac{z^5}{5} - \frac{z^3}{3}\right)$$

$$= \frac{2}{3}\left[\frac{1}{5}(1 + x^3)^{\frac{5}{2}} - \frac{1}{3}(1 + x^3)^{\frac{3}{2}}\right]. \qquad [\because \text{ from (1), } z = (1 + x^3)^{\frac{1}{2}}]$$

****Example 2.** *Integrate* $\int x^{\frac{1}{2}} (1 + x^3)^{\frac{1}{2}}\, dx.$

[Compare $\int x^{\frac{1}{2}} (1 + x^3)^{\frac{1}{2}}\, dx$ with $\int x^m (a + bx^n)^p\, dx,$

here $m = \frac{1}{2}$, $n = 3$, $p = \frac{1}{2}$, a fraction

$$\therefore \quad \frac{m+1}{n} = \frac{\frac{1}{2}+1}{3} = \frac{\frac{3}{2}}{3} = \frac{1}{2}, \text{ a fraction } \frac{m+1}{n} + p = \frac{1}{2} + \frac{1}{2} = 1, \text{ a +ve integer.}]$$

$$\int x^{\frac{1}{2}} (1 + x^3)^{\frac{1}{2}}\, dx \qquad [\text{Take } x^3 \text{ outside the binomial}]$$

$$= \int x^2 (x^{-3} + 1)^{\frac{1}{2}}\, dx \qquad [\text{Put } x^{-3} + 1 = z^2 \quad ...(1),$$

then $-3x^{-4}\, dx = 2z\, dz$, or $dx = -\frac{2}{3}\frac{z}{x^{-4}}\, dz$]

$$= \int x^2 z \left(-\frac{2}{3}\frac{z}{x^{-4}}\, dz\right) = -\frac{2}{3}\int \frac{z^2}{x^{-6}}\, dz \qquad | \text{ From (1), } x^{-3} = z^2 - 1$$

$$= -\frac{2}{3}\int \frac{z^2}{(z^2 - 1)^2}\, dz = -\frac{2}{3}\int z \cdot \frac{z}{(z^2 - 1)^2}\, dz \qquad [\text{Integrate by parts}]$$

---

*__Why this step.__ To rid $(1 + x^3)^{\frac{1}{2}}$ of fractional power, we put $1 + x^3 = z^2$, so that

$$(1 + x^3)^{\frac{1}{2}} = (z^2)^{\frac{1}{2}} = z.$$

$$= -\frac{2}{3}\left[z\left(-\frac{1}{2(z^2-1)}\right) - \int 1\left(-\frac{1}{2(z^2-1)}\right)dz\right]$$

$$= -\frac{2}{3}\left[-\frac{z}{2(z^2-1)} + \frac{1}{2}\int \frac{1}{z^2-(1)^2}\,dz\right]$$

$$= -\frac{2}{3}\left[-\frac{z}{2(z^2-1)} + \frac{1}{2}\cdot\frac{1}{2(1)}\log\frac{z-1}{z+1}\right]$$

$$= -\frac{2}{3}\left[-\frac{z}{2(z^2-1)} + \frac{1}{4}\log\frac{z-1}{z+1}\right]$$

$$\int \frac{z}{(z^2-1)^2}\,dz = \frac{1}{2}\int (z^2-1)^{-2}(2z)\,dz = \frac{1}{2}\frac{(z^2-1)^{-1}}{-1} = -\frac{1}{2(z^2-1)}$$

$$\left[\text{From (1), } z = (x^{-3}+1)^{\frac{1}{2}} = \frac{(1+x^3)^{\frac{1}{2}}}{x^{\frac{3}{2}}}\right]$$

$$= -\frac{2}{3}\left[-\frac{(1+x^3)^{\frac{1}{2}}}{2x^{\frac{3}{2}}\,\frac{1}{x^3}} + \frac{1}{4}\log\frac{\frac{(1+x^3)^{\frac{1}{2}}}{x^{\frac{3}{2}}}-1}{\frac{(1+x^3)^{\frac{1}{2}}}{x^{\frac{3}{2}}}+1}\right]$$

$$= -\frac{2}{3}\left[-\frac{1}{2}x^{\frac{3}{2}}(1+x^3)^{\frac{1}{2}} + \frac{1}{4}\log\frac{(1+x^3)^{\frac{1}{2}}-x^{\frac{3}{2}}}{(1+x^3)^{\frac{1}{2}}+x^{\frac{3}{2}}}\right].$$

EXAMPLES

1. Show that $\int x^{m-1}(a+bx^n)^p\,dx$ is directly integrable if

(*i*) p is a positive integer, (*ii*) $\frac{m}{n}$ is an integer,

(*iii*) $\frac{m}{n}+p$ is an integer.

[Proceed as in Art. 47.]

2. Integrate $\int x^3(1+x^5)^3\,dx$.

3. Show that (*i*) $\int_{\sqrt{3}}^{2\sqrt{3}} z^3(z^2-3)^{\frac{3}{2}}\,dz = \frac{16038}{35}$.

(*ii*) $\int_a^\infty \frac{dx}{x^4(a^2+x^2)^{\frac{1}{2}}} = \frac{2-\sqrt{2}}{3a^4}$

4. Integrate

(*i*) $\int \frac{dx}{(ax^2+c)^{\frac{3}{2}}}$; (*ii*) $\int \frac{dx}{x\sqrt{a^n+x^n}}$.

Three standard cases of rationalization by an algebraic substitution

48. Case I. Rule to integrate if only one radical $\sqrt{\text{linear}}$ occurs in the integrand : *Put linear* $= z^2$.

Generalization. Rule to integrate if only one radical (linear)$^{1/n}$ occurs in the integrand : *Put linear* $= z^n$.

** **Proof.** To integrate $\int f[x, (ax+b)^{\frac{1}{n}}]\, dx$.

Put $ax + b = z^n$ ...(1), then $a\, dx = nz^{n-1}\, dz$, or $dx = \dfrac{n}{a} z^{n-1}\, dz$.

$\therefore \int f[x, (ax+b)^{\frac{1}{n}}]\, dx$ | From (1), $x = \dfrac{z^n - b}{a}$

$$= \int f\left[\frac{z^n - b}{a}, z\right] \frac{n}{a} z^{n-1}\, dz = \frac{n}{a} \int f\left[\frac{z^n - b}{a}, z\right] z^{n-1}\, dz,$$

which can be integrated by resolving the integrand into partial fractions (Chap. IV) [$\because$ the integrand is a rational algebraic function of z].

EXAMPLES

1. Evaluate $\displaystyle\int_0^{\infty} \frac{\sqrt{x}\, dx}{(1+x)(2+x)(3+x)}$.

2. Integrate (*i*) $\displaystyle\int \frac{dx}{x + \sqrt[3]{x}}$. (*ii*) $\displaystyle\int \frac{dx}{\sqrt{a(x-b)} + \sqrt{b(x-a)}}$.

49. Case II. Rule to integrate if fractional powers of only one linear occur in the integrand :

Put linear $= z^l$, *where* l *is the L.C.M. of the denominators of the fractional indices.*

For, the substitution transforms the given integral into the integral of a rational algebraic function of z, which can be integrated by resolving the integrand into partial fractions (Chap. IV).

EXAMPLES

1. Integrate $\displaystyle\int \frac{dx}{(1+x)^{\frac{1}{3}} + (1+x)^{\frac{1}{2}}}$.

[The fractional indices of $(1 + x)$ are $\frac{1}{3}, \frac{1}{2}$.

The L.C.M. of the denoms. 3 and 2 is 6.

$\therefore$ Put $1 + x = z^6$ [Rule (Art. 49)]. Cancel the common factor z^2 from the num. and denom.]

2. Evaluate $\int_0^{16} \frac{x^{\frac{1}{4}}}{1+x^{\frac{1}{2}}}\,dx.$

****50. Case III. Rule to integrate if only one radical $\sqrt{\text{quadratic}}$ occurs in the integrand :**

(i) If $\sqrt{x^2+ax+b}$ *occurs in the integrand,*

put $\sqrt{x^2+ax+b}=z-x.$

(ii) If $\sqrt{-x^2+ax+b}$ *occurs in the integrand,*

and $\sqrt{-x^2+ax+b}$* $=\sqrt{(x-\alpha)(\beta-x)}$,

put $\sqrt{\dfrac{\text{one factor}}{\text{other factor}}}$, *i.e.*, $\sqrt{\dfrac{x-\alpha}{\beta-x}}=z.$

[**Proof of (*i*). To integrate $\int f(x, \sqrt{x^2+ax+b})\,dx.$

Put $\sqrt{x^2+ax+b}=z-x$

or, squaring, $x^2+ax+b=z^2-2zx+x^2$, or $x(2z+a)=z^2-b$,

or
$$x=\frac{z^2-b}{2z+a} \qquad ...(1)$$

$\therefore$ $\sqrt{x^2+ax+b}=z-x$ [*Note this step*]

$$=z-\frac{z^2-b}{2z+a}=\frac{z^2+az+b}{2z+a}.$$

From (1),
$$dx=\frac{(2z+a)2z-(z^2-b)2}{(2z+a)^2}\,dz=\frac{2z^2+2az+2b}{(2z+a)^2}\,dz$$
$$=\frac{2(z^2+az+b)}{(2z+a)^2}\,dz.$$

$$\therefore \int f(x, \sqrt{x^2+ax+b})\,dx=\int f\left[\frac{z^2-b}{2z+a}, \frac{z^2+az+b}{2z+a}\right]\frac{2(z^2+az+b)}{(2z+a)^2}\,dz$$
$$=2\int f\left[\frac{z^2-b}{2z+a}, \frac{z^2+az+b}{2z+a}\right]\frac{z^2+az+b}{(2z+a)^2}\,dz,$$

which can be integrated by resolving the integrand into partial fractions (Chap. IV) [$\because$ the integrand is a rational algebraic function of z].

---

*The quadratic $-x^2+ax+b$ must have real factors [$\because$ if it has no real factors, then, for all values of x, $-x^2+ax+b$ has the same sign as the coefficient of x^2 (From Intermediate Algebra) and is $\therefore$ negative, $\therefore$ $\sqrt{-x^2+ax+b}$ and $\therefore$ the integrand is imaginary].

Proof of (*ii*). To integrate $\int f(x, \sqrt{-x^2+ax+b})\,dx$.

$$\sqrt{-x^2+ax+b} = \sqrt{(x-\alpha)(\beta-x)}\ .$$

Put $\sqrt{\dfrac{x-\alpha}{\beta-x}} = z$ or, squaring, $\dfrac{x-\alpha}{\beta-x} = z^2$,

or $$x(1+z^2) = \alpha+\beta z^2, \text{ or } x = \frac{\alpha+\beta z^2}{1+z^2} \qquad ...(1)$$

Then $$x-\alpha = \frac{\alpha+\beta z^2}{1+z^2} - \alpha = \frac{(\beta-\alpha)z^2}{1+z^2},$$

$$\beta - x = \beta - \frac{\alpha+\beta z^2}{1+z^2} = \frac{\beta-\alpha}{1+z^2}.$$

$$\therefore \quad \sqrt{(x-\alpha)(\beta-x)} = \sqrt{\frac{(\beta-\alpha)z^2}{1+z^2}\cdot\frac{(\beta-\alpha)}{1+z^2}} = \frac{(\beta-\alpha)z}{1+z^2}.$$

From (1), $$dx = \frac{(1+z^2)\,2\beta z - (\alpha+\beta z^2)2z}{(1+z^2)^2}\,dz = \frac{2(\beta-\alpha)z}{(1+z^2)^2}\,dz.$$

$$\therefore \quad \int f(x, \sqrt{-x^2+ax+b})\,dx = \int f[x, \sqrt{(x-\alpha)(\beta-x)}]\,dx$$

$$= \int f\left[\frac{\alpha+\beta z^2}{1+z^2}, \frac{(\beta-\alpha)z}{1+z^2}\right] \frac{2(\beta-\alpha)z}{(1+z^2)^2}\,dz,$$

$$= 2(\beta-\alpha)\int f\left[\frac{\alpha+\beta z^2}{1+z^2}, \frac{(\beta-\alpha)z}{1+z^2}\right] \frac{z}{(1+z^2)^2}\,dz,$$

which can be integrated by resolving the integrand into partial fractions (Chap. IV) [$\because$ the integrand is a rational algebraic function of z].

****Note.** *If (instead of* $\sqrt{x^2+ax+b}$ *or* $\sqrt{-x^2+ax+b}$*),* $\sqrt{ax^2+bx+c}$ *occurs in the integrand, two cases arise :*

Case I. If a is + ve, then $\sqrt{ax^2+bx+c} = \sqrt{a}\sqrt{x^2+\dfrac{b}{a}x+\dfrac{c}{a}}$ $\qquad \left[\text{Put } \dfrac{b}{a} = a', \dfrac{c}{a} = b'\right]$

$= \sqrt{a}\sqrt{x^2+a'x+b'}$, so we proceed as in Art. 50, (*i*).

Case II. If a is – ve = – A (say, where A is + ve), then

$$\sqrt{ax^2+bx+c} = \sqrt{-Ax^2+bx+c} = \sqrt{A}\sqrt{-x^2+\frac{b}{A}x+\frac{c}{A}} \qquad \left[\text{Put } \frac{b}{A} = a', \frac{c}{A} = b'\right]$$

$= \sqrt{A}\sqrt{-x^2+a'x+b'}$, so we proceed as in Art. 50, (*ii*).]

EXAMPLES

****1.** Integrate $\int (x+\sqrt{a^2+x^2})^n\,dx$.

[Here $\sqrt{a^2+x^2}$ occurs in the integrand.

Put $\sqrt{a^2+x^2} = z - x$ [Rule, Art. 50, (*i*)]

or, squaring, $a^2 + x^2 = z^2 - 2zx + x^2$

or $2zx = z^2 - a^2$, or $x = \dfrac{z^2 - a^2}{2z}$.]

****2.** If n is an integer greater than 1, prove that

$$\int_0^{\infty} \frac{dx}{(x+\sqrt{1+x^2})^n} = \frac{n}{n^2-1}.$$

Rationalization by a trigonometric substitution.

51. Rule to integrate an irrational algebraic function by rationalizing by a trigonometric substitution :

(*i*) *If* $\sqrt{a^2-x^2}$ *occurs in the integrand, put* $x = a \sin \theta$. [Cf. Art. 13, (*i*).]

(*ii*) *If* $\sqrt{a^2+x^2}$ *occurs in the integrand, put* $x = a \tan \theta$.

(*iii*) *If* $\sqrt{x^2-a^2}$ *occurs in the integrand, put* $x = a \sec \theta$. [Cf. Art. 13, (*iii*).]

For, by the substitution the square root sign is in each case eliminated.

Thus in (*i*), $\sqrt{a^2-x^2} = \sqrt{a^2 - a^2 \sin^2\theta} = \sqrt{a^2(1-\sin^2\theta)} = a\cos\theta.$

Similarly in (*ii*) and (*iii*).

EXAMPLES

1. Integrate

(*i*) $\displaystyle\int \frac{dx}{(a^2-x^2)^{\frac{3}{2}}}$ (*ii*) $\displaystyle\int \frac{dx}{(a^2+x^2)^{\frac{3}{2}}}$.

2. Show that $\displaystyle\int_0^{\frac{1}{2}} \frac{dx}{(1-2x^2)\sqrt{1-x^2}} = \frac{1}{2}\log(2+\sqrt{3})$.

3. Integrate $\displaystyle\int \frac{dx}{(2ax+x^2)^{\frac{3}{2}}}$.

$$\left[\int \frac{1}{(x^2+2ax)^{\frac{3}{2}}}dx = \int \frac{1}{[(x^2+2ax+a^2)-a^2]^{\frac{3}{2}}}dx = \int \frac{1}{[(x+a)^2-a^2]^{\frac{3}{2}}}dx.\right]$$

MISCELLANEOUS EXAMPLES ON CHAPTER V

1. Evaluate $\displaystyle\int_2^3 \frac{dx}{\sqrt{5x-6-x^2}}$.

2. Evaluate (*i*) $\int \frac{dx}{\sqrt{(x-a)(x-b)}}$.

(*ii*) $\int_{\alpha}^{\beta} \frac{dx}{\sqrt{(x-\alpha)(\beta-x)}}$, $\beta > \alpha$.

[(*i*) Put the expression whose square root occurs in the denom., say, $x - \alpha = z^2$ or Put $x = \alpha \cos^2 \theta + \beta \sin^2 \theta$]

3. Evaluate (*i*) $\int \sqrt{\frac{x-a}{x-b}}\, dx$. (*ii*) $\int \sqrt{\frac{a+x}{x}}\, dx$.

(*iii*) $\int_0^a x \sqrt{\frac{a^2-x^2}{a^2+x^2}}\, dx$.

4. Integrate $\int \sqrt{e^{2x} + ae^x}\, dx$.

5. Integrate

(*i*) $\int \frac{x^2+2x+3}{\sqrt{x^2+x+1}}\, dx$. (*ii*) $\int \frac{2x^2+3}{\sqrt{3-2x-x^2}}\, dx$.

6. Evaluate $\int_0^1 \frac{1-4x+2x^2}{\sqrt{2x-x^2}}\, dx$.

7. Integrate

(*i*) $\int \frac{1}{(x+b)\sqrt{x+a}}\, dx$. (*ii*) $\int \frac{2-3x}{x\sqrt{1+x}}\, dx$.

8. (*a*) Integrate (*i*) $\int \frac{dx}{(1+x)\sqrt{1-x^2}}$.

(*ii*) $\int \frac{dx}{(x-a)\sqrt{x^2-a^2}}$.

(*b*) Show that $\int_1^2 \frac{dx}{(x+1)\sqrt{x^2-1}} = \frac{1}{\sqrt{3}}$.

9. Integrate (*i*) $\int \frac{dx}{(1+x)\sqrt{1+x-x^2}}$.

(*ii*) $\int \frac{dx}{(x+1)\sqrt{1+2x-x^2}}$.

(*iii*) $\int \frac{1}{(x-a)\sqrt{(x-a)(b-x)}}\, dx$.

10. Integrate

(i) $\displaystyle\int \frac{\sqrt{x^2-a^2}}{x}\,dx.$ (ii) $\displaystyle\int \frac{dx}{x^2\sqrt{1+x^2}}.$

11. Integrate

(i) $\displaystyle\int \frac{1-x^2}{(1+x^2)\sqrt{1+x^4}}\,dx.$ (ii) $\displaystyle\int \frac{\cos^{-1}x}{x^3}\,dx.$

12. Integrate $\displaystyle\int \frac{\sin^3 x}{(1+\cos^2 x)\sqrt{1+\cos^2 x+\cos^4 x}}\,dx.$

[Put $\cos x = z$.]

****13.** Integrate $\displaystyle\int \frac{x\,dx}{(a^2+b^2-x^2)\sqrt{(a^2-x^2)(x^2-b^2)}}.$

$$\int \frac{x\,dx}{(a^2+b^2-x^2)\sqrt{(a^2-x^2)(x^2-b^2)}} = \int \frac{x}{(a^2+b^2-x^2)\sqrt{(x^2-b^2)(a^2-x^2)}}\,dx.$$

$\left[\text{Put } \sqrt{\dfrac{x^2-b^2}{a^2-x^2}} = z.\right]$

14. Integrate $\displaystyle\int \frac{d\theta}{\cos\theta\sqrt{a^2\cos^2\theta+b^2\sin^2\theta+c^2}}.$

$$\left[\int \frac{1}{\cos\theta\sqrt{a^2\cos^2\theta+b^2\sin^2\theta+c^2}}\,d\theta\right.$$

$$= \int \frac{1}{\cos\theta\sqrt{a^2\cos^2\theta+b^2\sin^2\theta+c^2(\cos^2\theta+\sin^2\theta)}}\,d\theta$$

$$= \int \frac{1}{\cos\theta\sqrt{(a^2+c^2)\cos^2\theta+(b^2+c^2)\sin^2\theta}}\,d\theta.$$

(Denom. is a homogeneous function of the second degree in $\cos\theta$ and $\sin\theta$.)

(As suggested by the factor $\cos\theta$ in the denom.) divide the num. and denom. by $\cos^2\theta$, and put $\tan\theta = z$. (Compare Ex. 8, Art. 44.)]

****15.** Integrate $\displaystyle\int \frac{1}{(x^2+2ax+b^2)\sqrt{x^2+2ax+c^2}}\,dx.$

$$\left[\int \frac{1}{(x^2+2ax+b^2)\sqrt{x^2+2ax+c^2}}\,dx\right.$$

$$= \int \frac{1}{[(x^2+2ax+a^2)-a^2+b^2]\sqrt{(x^2+2ax+a^2)-a^2+c^2}}\,dx$$

$$= \int \frac{1}{[(x+a)^2+(b^2-a^2)]\sqrt{(x+a)^2+(c^2-a^2)}}\,dx$$

[Put $x + a = z$ ...(1), then $dx = dz$.

Let $a < b < c$.

Then $b^2 - a^2$ is + ve = A^2 (say), $c^2 - a^2$ is + ve = B^2 (say). Also $A^2 < B^2$ ($\because$ $b < c$)]

$$= \int \frac{1}{(z^2 + A^2)\sqrt{z^2 + B^2}}\, dz, \quad \text{where } A^2 < B^2.]$$

16. Integrate $\displaystyle\int \frac{x+1}{(x^2+4)\sqrt{x^2+9}}\, dx.$

17. Integrate (*i*) $\displaystyle\int x^3 (1 + x^2)^{1/3}\, dx.$

(*ii*) $\displaystyle\int \frac{dx}{x\sqrt{x^3+1}}.$

18. Integrate (*i*) $\displaystyle\int \frac{\sqrt{x}\, dx}{1+x}.$

(*ii*) $\displaystyle\int \frac{(x+3)\, dx}{(x-3)^{\frac{1}{3}}}.$

(*iii*) $\displaystyle\int \frac{dx}{x + \sqrt{x-1}}.$

****19.** Integrate $\displaystyle\int \frac{\log(\cos\theta + \sqrt{\cos 2\theta})}{1-\cos^2\theta}\, d\theta.$

$$\left[\int \frac{\log(\cos\theta + \sqrt{\cos 2\theta})}{1-\cos^2\theta}\, d\theta = \int \log(\cos\theta + \sqrt{\cos 2\theta})\, \text{cosec}^2\,\theta\, d\theta\right.$$

[Integrate by parts]

$$= \log(\cos\theta + \sqrt{\cos 2\theta})(-\cot\theta)$$

$$-\int \frac{1}{\cos\theta + \sqrt{\cos 2\theta}}\left[-\sin\theta + \frac{1}{2\sqrt{\cos 2\theta}}(-\sin 2\theta)\, 2\right](-\cot\theta)\, d\theta.$$

The integral on the R.H.S.

$$= -\int \frac{1}{\cos\theta + \sqrt{\cos 2\theta}} \frac{(\sin\theta\sqrt{\cos 2\theta} + 2\sin\theta\cos\theta)}{\sqrt{\cos 2\theta}} \frac{\cos\theta}{\sin\theta}\, d\theta$$

[*Cancel* $\sin\theta$ *from the num. and denom.*]

$$= -\int \frac{1}{\cos\theta + \sqrt{\cos 2\theta}} \frac{(\sqrt{\cos 2\theta} + 2\cos\theta)\cos\theta}{\sqrt{\cos 2\theta}}\, d\theta$$

[*Multiply the num. and denom. by* $\cos\theta - \sqrt{\cos 2\theta}$ *to rationalize the factor* $\cos\theta + \sqrt{\cos 2\theta}$ *in the denom.* (*Note this step*)]

$$= -\int \frac{(\cos\theta - \sqrt{\cos 2\theta})(\sqrt{\cos 2\theta} + 2\cos\theta)\cos\theta}{(\cos^2\theta - \cos 2\theta)\sqrt{\cos 2\theta}}\,d\theta$$

$$= -\int \frac{(-\cos\theta\sqrt{\cos 2\theta} + 2\cos^2\theta - \cos 2\theta)\cos\theta}{\sin^2\theta\sqrt{\cos 2\theta}}\,d\theta$$

$$= -\int \frac{(-\cos\theta\sqrt{\cos 2\theta} + 1)\cos\theta}{\sin^2\theta\sqrt{\cos 2\theta}}\,d\theta$$

$$= -\int \left[-\cot^2\theta + \frac{\cos\theta}{\sin^2\theta\sqrt{\cos 2\theta}}\right]d\theta$$

$$= \int \cot^2\theta\,d\theta - \int \frac{\cos\theta}{\sin^2\theta\sqrt{1-2\sin^2\theta}}\,d\theta.$$

To evaluate the second integral, put $\sin\theta = z$, and then $z = \dfrac{1}{u}$.]

CHAPTER VI

INTEGRATION OF TRIGONOMETRIC FUNCTIONS

Two standard cases of $\int \sin^n x\, dx$ or $\int \cos^n x\, dx$

52. Case I. Rule to integrate $\int \sin^n x\, dx$ or $\int \cos^n x\, dx$, where n is a positive odd integer :

(i) If the index of sin x is a positive odd integer, put cos x = z.

(ii) If the index of cos x is a positive odd integer, put sin x = z.

[**Proof of** (*i*). To integrate $\int \sin^n x\, dx$, where n is a + ve odd integer.

Let $n = 2k + 1$, where k is a + ve integer, or 0.

Put $\cos x = z$ ...(1), then $-\sin x\, dx = dz$, or $dx = -\dfrac{dz}{\sin x}$.

$$\therefore \quad \int \sin^n x\, dx = \int \sin^{2k+1} x\, dx = \int \sin^{2k+1} x \left(-\frac{dz}{\sin x}\right)$$

$$= -\int \sin^{2k} x\, dz = -\int (\sin^2 x)^k\, dz = -\int (1 - \cos^2 x)^k\, dz$$

$$= -\int (1 - z^2)^k\, dz,$$

which can be integrated by expanding $(1 - z^2)^k$ in a finite series by the binomial theorem, and integrating term by term.

Similarly $\int \cos^n x\, dx$, where n is a positive odd integer, can be integrated by putting $\sin x = z$.]

[**Aid to memory.** Of the two functions sin x and cos x, if the index of one, say, sin x is a positive odd integer, put the other, *i.e.*, cos $x = z$.]

Example. *Integrate* $\int \cos^3 x\, dx$.

[Here the index of cos x is 3, a +ve odd integer.]

Put $\sin x = z$ ...(1), then $\cos x\, dx = dz$, or $dx = \dfrac{dz}{\cos x}$.

$$\therefore \quad \int \cos^3 x\, dx = \int \cos^3 x \frac{dz}{\cos x} = \int \cos^2 x\, dz = \int (1 - \sin^2 x)\, dz$$

$$= \int (1 - z^2)\, dz = z - \frac{z^3}{3} = \sin x - \frac{\sin^3 x}{3}. \qquad \text{[From (1)]}$$

EXAMPLES

1. Evaluate (*i*) $\int \sin^3 x\, dx$. (*ii*) $\int \sin^5 x\, dx$.

(*iii*) $\int_0^{\frac{\pi}{2}} \sin^7 x\, dx$.

2. Evaluate (*i*) $\int_0^{\frac{\pi}{4}} \cos^5 \theta \, d\theta$. (*ii*) $\int \cos^7 x \, dx$.

3. Prove that $\int \cos^n x \, dx = \sin x - p_1 \dfrac{\sin^3 x}{3} + p_2 \dfrac{\sin^5 x}{5} - p_3 \dfrac{\sin^7 x}{7} + \ldots$ etc., $p_1, p_2, p_3, \ldots$ being the coefficients of the expansion $(1+\alpha)^{\frac{n-1}{2}}$, and n having any real value positive or negative.

[**De Moivre's theorem.** *For all values of n, positive or negative, integral or fractional, the value, or one of the values, of*

$$(\cos \theta + i \sin \theta)^n = \cos n\theta + i \sin n\theta \text{ (where } i = \sqrt{-1}\text{)}. \qquad \text{[From Trigonometry]}$$

Thus, if n is a positive integer,

$$(\cos \theta + i \sin \theta)^n = \cos n\theta + i \sin n\theta.$$

$$(\cos \theta + i \sin \theta)^{-1} = \cos(-\theta) + i \sin(-\theta) = \cos \theta - i \sin \theta.$$

$$(\cos \theta + i \sin \theta)^{-n} = \cos(-n\theta) + i \sin(-n\theta) = \cos n\theta - i \sin n\theta.]$$

53. Case II. Rule to integrate $\int \sin^n x \, dx$ or $\int \cos^n x \, dx$, where n is a positive even integer :

[**Multiple-angle method.**] (*i*) *Express the integrand in a series of sines or cosines of multiple angles.*

[(*a*) *If the index is a positive even integer but* **small** (*e.g.*, *2, 4*), *use the double-angle formulae* : $\sin^2 x = \dfrac{1-\cos 2x}{2}$, $\cos^2 x = \dfrac{1+\cos 2x}{2}$.

(*b*) *If the index is a positive even integer but* **large** (*e.g.*, *6, 8*), *use De Moivre's theorem.*]

(*ii*) *Integrate term by term.*

Example 1. *Integrate* (*i*) $\int \cos^2 x \, dx$.

(*ii*) $\int \cos^4 x \, dx$.

(*i*) [Here the index of $\cos x$ is 2, a + ve even integer but small.]

$$\int \cos^2 x \, dx = \int \frac{1+\cos 2x}{2} dx \qquad \left| \begin{array}{l} \cos 2x = 2\cos^2 x - 1 \\ \therefore \cos^2 x = \dfrac{1+\cos 2x}{2} \end{array} \right.$$

$$= \frac{1}{2}\left[x + \frac{\sin 2x}{2}\right].$$

(*ii*) [Here the index of $\cos x$ is 4, a + ve even integer but small.]

$$\int \cos^4 x \, dx = \int (\cos^2 x)^2 \, dx = \int \left(\frac{1+\cos 2x}{2}\right)^2 dx \qquad \left| \begin{array}{l} \cos 2x = 2\cos^2 x - 1 \\ \therefore \cos^2 x = \dfrac{1+\cos 2x}{2} \end{array} \right.$$

$$= \int \frac{1}{4}(1 + 2\cos 2x + \cos^2 2x) \, dx = \int \frac{1}{4}\left[1 + 2\cos 2x + \frac{1+\cos 4x}{2}\right] dx$$

$$= \int \frac{1}{8}(2 + 4\cos 2x + 1 + \cos 4x)\,dx = \int \frac{1}{8}(3 + 4\cos 2x + \cos 4x)^{*}\,dx$$

$$= \frac{1}{8}\left[3x + 4\frac{\sin 2x}{2} + \frac{\sin 4x}{4}\right] = \frac{1}{8}\left[3x + 2\sin 2x + \frac{\sin 4x}{4}\right].$$

Example 2. *Integrate* $\int \sin^6 x\,dx$.

[Here the index of sin x is 6, a + ve even integer but large.]

Let $\cos x + i\sin x = y$ ...(1) [Raise each side to the power (– 1)]

Then $(\cos x + i\sin x)^{-1} = y^{-1}$

or $\cos x - i\sin x = \dfrac{1}{y}$ ...(2)

Subtracting (2) from (1)†, $2i\sin x = y - \dfrac{1}{y}$.

$$\therefore \quad (2i\sin x)^6 = \left(y - \frac{1}{y}\right)^6 = y^6 - {}^6C_1 y^5\left(\frac{1}{y}\right) + {}^6C_2 y^4\left(\frac{1}{y^2}\right) - {}^6C_3 y^3\left(\frac{1}{y^3}\right)$$

$$+ {}^6C_4 y^2\left(\frac{1}{y^4}\right) - {}^6C_5 y\left(\frac{1}{y^5}\right) + {}^6C_6\frac{1}{y^6}$$

$$= y^6 - 6y^4 + 15y^2 - 20 + 15\frac{1}{y^2} - 6\frac{1}{y^4} + \frac{1}{y^6}$$

$$= \left(y^6 + \frac{1}{y^6}\right) - 6\left(y^4 + \frac{1}{y^4}\right) + 15\left(y^2 + \frac{1}{y^2}\right) - 20$$

$\Big[$ Now $y^n + \dfrac{1}{y^n} = y^n + y^{-n} = (\cos x + i\sin x)^n + (\cos x + i\sin x)^{-n}$

$$= \cos nx + i\sin nx + \cos nx - i\sin nx = 2\cos nx.$$

Put $n = 6, 4, 2.$]

$\therefore \quad -2^6\sin^6 x$ $[\because i^6 = [(-1)^{1/2}]^6 = (-1)^3 = -1]$

$$= 2\cos 6x - 6(2\cos 4x) + 15(2\cos 2x) - 20$$ [Cancel 2]

---

*__Check.__ Taking the integrands on both sides, $\cos^4 x = \frac{1}{8}(3 + 4\cos 2x + \cos 4x)$.

Put $x = 0$ on both sides, we get $1 = \frac{1}{8}(3 + 4 + 1)$, or $1 = 1$, which is true.

Note. To check the power of a cosine expressed in a series of cosines of multiple angles, put $x = 0$ on both sides.

Similarly to check the power of a sine, expressed in a series of sines or cosines of multiple angles, put $x = \dfrac{\pi}{2}$ on both sides (to avoid zero on the L.H.S.). See the footnote on p. 106.

And to check the product of the powers of a sine and a cosine, put $x = \dfrac{\pi}{4}$ on both sides. See the footnote on p. 109.

† **Why this step.** In order to get $2i\sin x$ (as we have to express a power of sin x in a series of sines or cosines of multiple angles) we have subtracted (2) from (1). Had we to express a power of cos x in a series of cosines of multiple angles, we would have added (1) and (2) to get $2\cos x$.

$\therefore \quad \sin^6 x = -\frac{1}{2^5}[\cos 6x - 6\cos 4x + 15\cos 2x - 10]$*

$$\therefore \quad \int \sin^6 x\, dx = -\frac{1}{2^5}\int [\cos 6x - 6\cos 4x + 15\cos 2x - 10]\, dx$$

$$= -\frac{1}{2^5}\left[\frac{\sin 6x}{6} - 6\frac{\sin 4x}{4} + 15\frac{\sin 2x}{2} - 10x\right]$$

$$= -\frac{1}{2^5}\left[\frac{\sin 6x}{6} - \frac{3}{2}\sin 4x + \frac{15}{2}\sin 2x - 10x\right].$$

EXAMPLES

1. Evaluate (*i*) $\int \sin^2 x\, dx.$ (*ii*) $\int \sin^4 x\, dx.$

(*iii*) $\int_0^{\frac{\pi}{4}} \sin^4 \theta\, d\theta.$

2. Evaluate $\int_0^{\frac{\pi}{4}} \cos^2 x\, dx.$

3. Evaluate (*i*) $\int_0^{\frac{\pi}{4}} \cos^6 x\, dx.$ (*ii*) $\int \sin^8 x\, dx.$

Three standard cases of $\int \sin^p x \cos^q x\, dx$.

54. Case I. Rule to integrate $\int \sin^p x \cos^q x\, dx$, where either p or q is a positive odd integer, whatever the other may be :

(*i*) *If the index of sin x is a positive odd integer, put cos x = z.*

(*ii*) *If the index of cos x is a positive odd integer, put sin x = z.*

[Proof of** (*i*). To integrate $\int \sin^p x \cos^q x\, dx$, where p is a positive odd integer.

Let $p = 2k + 1$, where k is a + ve integer, or 0.

Put $\cos x = z$ ...(1), then $-\sin x\, dx = dz$, or $dx = -\frac{dz}{\sin x}$.

$$\therefore \quad \int \sin^p x \cos^q x\, dx = \int \sin^{2k+1} x \cos^q x\, dx$$

$$= \int \sin^{2k+1} x \,.\, z^q . \left(-\frac{dz}{\sin x}\right) = -\int z^q \sin^{2k} x\, dz = -\int z^q (\sin^2 x)^k\, dz$$

---

*__Check.__ (See Note in the footnote on p. 105.)

Put $x = \frac{\pi}{2}$ on both sides, we get

$$1 = -\frac{1}{2^5}[\cos 3\pi - 6\cos 2\pi + 15\cos \pi - 10]$$

or $\quad 1 = -\frac{1}{32}[-1 - 6 - 15 - 10]$, or $1 = 1$, which is true.

$$= -\int z^q (1-\cos^2 x)^k \, dz = -\int z^q (1-z^2)^k \, dz,$$

which can be integrated by expanding $(1 - z^2)^k$ in a finite series by the binomial theorem, multiplying each term by z^q, and integrating term by term.

Similarly $\int \sin^p x \cos^q x \, dx$, where q is a positive odd integer, can be integrated by putting $\sin x = z$.]

Example 1. *Integrate* $\int \sin^5 x \cos^4 x \, dx$.

[Here the index of $\sin x$ is 5, a + ve odd integer.]

Put $\cos x = z$ ...(1), then $-\sin x \, dx = dz$, or $dx = -\dfrac{dz}{\sin x}$.

$$\therefore \quad \int \sin^5 x \cos^4 x \, dx = \int \sin^5 x \,.\, z^4 \left(-\frac{dz}{\sin x}\right) = -\int z^4 \sin^4 x \, dz$$

$$= -\int z^4 (\sin^2 x)^2 \, dz = -\int z^4 (1-\cos^2 x)^2 \, dz = -\int z^4 (1-z^2)^2 \, dz$$

$$= -\int z^4 (1 - 2z^2 + z^4) \, dz = -\int (z^4 - 2z^6 + z^8) \, dz$$

$$= -\left[\frac{z^5}{5} - 2\frac{z^7}{7} + \frac{z^9}{9}\right] = -\left[\frac{\cos^5 x}{5} - 2\frac{\cos^7 x}{7} + \frac{\cos^9 x}{9}\right]. \qquad \text{[From (1)]}$$

Example 2. *Integrate* $\int \sin^{2/3} x \cos^3 x \, dx$.

[Here the index of $\cos x$ is 3, a + ve odd integer.]

Put $\sin x = z$ ...(1), then $\cos x \, dx = dz$, or $dx = \dfrac{dz}{\cos x}$.

$$\therefore \quad \int \sin^{\frac{2}{3}} x \cos^3 x \, dx = \int z^{\frac{2}{3}} \cos^3 x \frac{dx}{\cos x} = \int z^{\frac{2}{3}} \cos^2 x \, dz$$

$$= \int z^{\frac{2}{3}} (1 - \sin^2 x) \, dz = \int z^{\frac{2}{3}} (1 - z^2) \, dz = \int (z^{\frac{2}{3}} - z^{\frac{8}{3}}) \, dz$$

$$= \frac{z^{\frac{5}{3}}}{\frac{5}{3}} - \frac{z^{\frac{11}{3}}}{\frac{11}{3}} = \frac{3}{5} z^{\frac{5}{3}} - \frac{3}{11} z^{\frac{11}{3}} = \frac{3}{5} \sin^{\frac{5}{3}} x - \frac{3}{11} \sin^{\frac{11}{3}} x. \qquad \text{[From (1)]}$$

EXAMPLES

1. Integrate (*i*) $\int \sin^2 x \cos^3 x \, dx$. (*ii*) $\int \sin^3 x \cos^3 x \, dx$.

2. Write down the indefinite integrals of :

(*i*) $\int \sin^3 x \cos^4 x \, dx$. (*ii*) $\int \sin^2 \theta \cos^7 \theta \, d\theta$.

3. Integrate (*i*) $\int \sin^5 \theta \cos^3 \theta \, d\theta$. (*ii*) $\int \sin^5 \theta \cos^7 \theta \, d\theta$.

[(*i*) **Short cut.** If in $\int \sin^p x \cos^q x \, dx$, the indices of both $\sin x$ and $\cos x$ are positive odd integers, we may put either $\cos x = z$, or $\sin x = z$. But *the solution is shorter if we consider the* **smaller** *index as a positive odd integer, and choose the proper substitution accordingly.* Thus in $\int \sin^5 \theta \cos^3 \theta \, d\theta$, the indices of both $\sin \theta$ and $\cos \theta$, *viz.*, 5 and 3 are positive odd integers. But 3 being < 5, we consider the index of $\cos \theta$ as a positive odd integer, and so put $\sin \theta = z$.]

4. (*i*) Integrate $\int \sin^8 \theta \cos^3 \theta \, d\theta$ and evaluate $\int_0^{\frac{\pi}{2}} \sin^8 \theta \cos^3 \theta \, d\theta$.

(*ii*) Integrate $\int \sec x \tan^3 x \, dx$.

5. (*i*) Evaluate $\int_0^{\frac{\pi}{4}} \sin^5 \theta \cos^2 \theta \, d\theta$.

(*ii*) Integrate $\int \cos 3x \sin^3 x \, dx$.

6. Integrate $\int \dfrac{x^3 \sin^{-1} x}{(1-x^2)^{\frac{3}{2}}} dx$.

55. Case II. Rule to integrate $\int$ sin^p x cos^q x dx, where both p and q are positive even integers :

[Multiple-angle method.] (*i*) *Express the integrand in a series of sines or cosines of multiple angles.*

[(*a*) *If the indices are positive even integers but* **small** (*e.g.*, *2, 4*), *use the double-angle formulae* :

$$\sin^2 x = \frac{1-\cos 2x}{2}, \cos^2 x = \frac{1+\cos 2x}{2}, \sin x \cos x = \frac{\sin 2x}{2}.$$

(*b*) *If the indices are positive even integers but* **large** (*e.g.*, 6, 8), *use De Moivre's theorem.*]

(*ii*) *Integrate term by term.*

****Example.** *Evaluate* $\int_0^{\frac{\pi}{4}} \sin^6 x \cos^2 x \, dx$.

[Here the index of sin x is 6, a + ve even integer but large.]

Let $\cos x + i \sin x = y$ ...(1) [Raise each side to the power (– 1)]

Then $(\cos x + i \sin x)^{-1} = y^{-1}$

or $$\cos x - i \sin x = \frac{1}{y} \quad ...(2)$$

From (1) and (2), subtracting and adding,

$$2 i \sin x = y - \frac{1}{y}, \quad \text{and} \quad 2 \cos x = y + \frac{1}{y}$$

$$\therefore \quad (2 i \sin x)^6 (2 \cos x)^2 = \left(y - \frac{1}{y}\right)^6 \left(y + \frac{1}{y}\right)^2$$

[Coeffs. in $(1-t)^6$ are $1 - 6 + 15 - 20 + 15 - 6 + 1$* ...(3)

***How to write this step.** Coefficients in $(1-t)^6$ are
$1 - {}^6C_1 + {}^6C_2 - {}^6C_3 + {}^6C_4 - {}^6C_5 + {}^6C_6$.

Coeffs. in $(1 - t)^6(1 + t)$ are [Take upper + L.H.*]

$$1 - 5 + 9 - 5 - 5 + 9 - 5 + 1 \qquad ...(4)$$

Coeffs. in $(1 - t)^6 (1 + t)^2$ are [Take upper + L.H.]

$$1 - 4 + 4 + 4 - 10 + 4 + 4 - 4 + 1]$$

$$= y^8 - 4y^6 + 4y^4 + 4y^2 - 10 + 4\frac{1}{y^2} + 4\frac{1}{y^4} - 4\frac{1}{y^6} + \frac{1}{y^8}$$

$$= \left(y^8 + \frac{1}{y^8}\right) - 4\left(y^6 + \frac{1}{y^6}\right) + 4\left(y^4 + \frac{1}{y^4}\right) + 4\left(y^2 + \frac{1}{y^2}\right) - 10$$

$\Big[$ Now $y^n + \dfrac{1}{y^n} = y^n + y^{-n} = (\cos x + i \sin x)^n + (\cos x + i \sin x)^{-n}$

$$= \cos nx + i \sin nx + \cos nx - i \sin nx = 2 \cos nx.$$

Put $n = 8, 6, 4, 2.]$

$\therefore \quad - 2^8 \sin^6 x \cos^2 x$ $[\because \; i^6 = [(-1)^{1/2}]^6 = (-1)^3 = -1]$

$$= 2 \cos 8x - 4(2 \cos 6x) + 4(2 \cos 4x) + 4(2 \cos 2x) - 10 \quad \text{[Cancel 2]}$$

$$\therefore \quad \sin^6 x \cos^2 x = -\frac{1}{2^7} [\cos 8x - 4 \cos 6x + 4 \cos 4x + 4 \cos 2x - 5†]$$

$$\therefore \quad \int_0^{\frac{\pi}{4}} \sin^6 x \cos^2 x \, dx = -\frac{1}{2^7} \int_0^{\frac{\pi}{4}} [\cos 8x - 4 \cos 6x + 4 \cos 4x + 4 \cos 2x - 5] \, dx$$

$$= -\frac{1}{2^7} \left[\frac{\sin 8x}{8} - 4\frac{\sin 6x}{6} + 4\frac{\sin 4x}{4} + 4\frac{\sin 2x}{2} - 5x\right]_0^{\frac{\pi}{4}}$$

***How to write this step.** To find the numbers in (4), take the upper number in (3) + the number on L.H. Thus from (3), below 1 we get 1 + 0, below – 6 we get – 6 + 1, below 15 we get 15 – 6, and so on.

Reason. Coefficients in $a + bt + ct^2 + ...$ are $a + b + c + ...$.

Now $(a + bt + ct^2 + ...) (1 + t) = a + (b + a)t + (c + b)t^2 + ...$

$\therefore$ coefficients in $(a + bt + ct^2 + ...) (1 + t)$ are $a + (b + a) + (c + b) + ...$.

†**Check.** (See Note in the footnote on p. 105.)

Put $x = \dfrac{\pi}{4}$ on both sides, we get

$$\frac{1}{2^3} \cdot \frac{1}{2} = -\frac{1}{2^7}\left[\cos 2\pi - 4 \cos \frac{3\pi}{2} + 4 \cos \pi + 4 \cos \frac{\pi}{2} - 5\right]$$

or $$\frac{1}{16} = -\frac{1}{128} [1 - 0 - 4 + 0 - 5], \text{ or } \frac{1}{16} = \frac{1}{16}, \text{ which is true.}$$

$$= -\frac{1}{2^7}\left[\left(\frac{1}{8}\sin 2\pi - \frac{2}{3}\sin\frac{3\pi}{2} + \sin\pi + 2\sin\frac{\pi}{2} - 5\frac{\pi}{4}\right) - 0\right]$$

$$= -\frac{1}{2^7}\left[0 - \frac{2}{3}(-1) + 0 + 2(1) - 5\frac{\pi}{4}\right] = -\frac{1}{2^7}\left[\frac{8}{3} - \frac{5\pi}{4}\right] = \frac{1}{2^7}\left[\frac{5\pi}{4} - \frac{8}{3}\right].$$

EXAMPLES

1. Integrate $\int \sin^2 x \cos nx\, dx$.

2. (*i*) Evaluate $\int_0^{\frac{\pi}{4}} \sin^2 x \cos^4 x\, dx$. (*ii*) $\int \sin^4 x \cos^4 x\, dx$.

****3.** Integrate $\int \sin^4 x \cos^6 x\, dx$.

[Here the index of cos x is 6, a + ve even integer but large.]

Let $\cos x + i \sin x = y$ ...(1) [Raise each side to the power (– 1)]

Then $(\cos x + i \sin x)^{-1} = y^{-1}$

or $\cos x - i \sin x = \frac{1}{y}$ ...(2)

From (1) and (2), subtracting and adding,

$$2i \sin x = y - \frac{1}{y}, \text{ and } 2\cos x = y + \frac{1}{y}$$

$$\therefore \quad (2i\sin x)^4 (2\cos x)^6 = \left(y - \frac{1}{y}\right)^4 \left(y + \frac{1}{y}\right)^6 = \left(y + \frac{1}{y}\right)^6 \left(y - \frac{1}{y}\right)^4$$

[Take factor with larger index first (*Note this step*)]

[Coeffs. in $(1 + t)^6$ are 1 + 6 + 15 + 20 + 15 + 6 + 1 ...(3)

Coeffs. in $(1 + t)^6 (1 - t)$ are [Take upper – L.H.*]

1 + 5 + 9 + 5 – 5 – 9 – 5 – 1 ...(4)

Coeffs. in $(1 + t)^6(1 - t)^2$ are [Take upper – L.H.]

1 + 4 + 4 – 4 – 10 – 4 + 4 + 4 + 1

Coeffs. in $(1 + t)^6 (1 - t)^3$ are [Take upper – L.H.]

1 + 3 + 0 – 8 – 6 + 6 + 8 + 0 – 3 – 1

Coeffs. in $(1 + t)^6(1 - t)^4$ are [Take upper – L.H.]

1 + 2 – 3 – 8 + 2 + 12 + 2 – 8 – 3 + 2 + 1.]

***How to write this step.** To find the numbers in (4), take the upper number in (3) – the number on L.H. Thus from (3), below 1 we get 1 – 0, below 6 we get 6 – 1, below 15 we get 15 – 6, and so on.

Reason. Coefficients in $a + bt + ct^2 + \ldots$ are $a + b + c + \ldots$.

Now $(a + bt + ct^2 + \ldots)(1 - t) = a + (b - a)t + (c - b)t^2 + \ldots$

$\therefore$ coefficients in $(a + bt + ct^2 + \ldots)(1 - t)$ are $a + (b - a) + (c - b) + \ldots$.

56. Case III. Rule to integrate $\int \sin^p x \cos^q x\, dx$, where p + q is a negative even integer (p, q are not necessarily integers) :

Express the integrand as the product of the powers of tan x and sec x, and put tan x = z.

[Proof.** Let $p + q = -2k$, where k is a + ve integer.

$$\int \sin^p x \cos^q x\, dx = \int \left(\frac{\sin^p x}{\cos^p x}\right) \cos^p x \cos^q x\, dx$$

$$= \int \tan^p x \cos^{p+q} x\, dx = \int \tan^p x \cos^{-2k} x\, dx = \int \tan^p x \sec^{2k} x\, dx$$

$$\left[\text{Put } \tan x = z \text{ ...(1), then } \sec^2 x\, dx = dz, \text{ or } dx = \frac{dz}{\sec^2 x}\right]$$

$$= \int z^p \sec^{2k} x \frac{dz}{\sec^2 x} = \int z^p \sec^{2k-2} x\, dz$$

$$= \int z^p (\sec^2 x)^{k-1} dz = \int z^p (1 + \tan^2 x)^{k-1} dz = \int z^p (1 + z^2)^{k-1} dz,$$

which can be integrated by expanding $(1 + z^2)^{k-1}$ in a finite series by the binomial theorem, multiplying each term by z^p, and integrating term by term.]

Example 1. *Integrate* $\int \frac{\sin^2 x}{\cos^6 x} dx.$

$\left[\text{Compare } \int \frac{\sin^2 x}{\cos^6 x} dx \text{ with } \int \sin^p x \cos^q x\, dx,\right.$

here $p = 2, q = -6, \quad \therefore \quad p + q = 2 - 6 = -4$, a – ve even integer.]

$$\int \frac{\sin^2 x}{\cos^6 x} dx = \int \left(\frac{\sin^2 x}{\cos^2 x}\right) \frac{\cos^2 x}{\cos^6 x} dx = \int \tan^2 x \frac{1}{\cos^4 x} dx = \int \tan^2 x \sec^4 x\, dx$$

$$\left[\text{Put } \tan x = z \text{ ...(1), then } \sec^2 x\, dx = dz, \text{ or } dx = \frac{dz}{\sec^2 x}\right]$$

$$= \int z^2 \sec^4 x \frac{dz}{\sec^2 x} = \int z^2 \sec^2 x\, dz = \int z^2 (1 + \tan^2 x)\, dz$$

$$= \int z^2 (1 + z^2)\, dz = \int (z^2 + z^4)\, dz = \frac{z^3}{3} + \frac{z^5}{5} = \frac{\tan^3 x}{3} + \frac{\tan^5 x}{5}. \quad \text{[From (1)]}$$

Example 2. *Integrate* $\int \sec^{\frac{3}{4}} x \operatorname{cosec}^{\frac{5}{4}} x\, dx.$

$$\int \sec^{\frac{3}{4}} x \operatorname{cosec}^{\frac{5}{4}} x\, dx = \int \frac{1}{\cos^{\frac{3}{4}} x \sin^{\frac{5}{4}} x} dx = \int \frac{1}{\sin^{\frac{5}{4}} x \cos^{\frac{3}{4}} x} dx.$$

$\left[\text{Compare } \int \frac{1}{\sin^{\frac{5}{4}} x \cos^{\frac{3}{4}} x} dx \text{ with } \int \sin^p x \cos^q x\, dx.\right.$

here $p = -\frac{5}{4}, q = -\frac{3}{4}.$

$$\therefore \qquad p+q=-\frac{5}{4}-\frac{3}{4}=-\frac{8}{4}=-2, \text{ a} - \text{ve even integer.}]$$

$$\int \sec^{\frac{3}{4}} x \operatorname{cosec}^{\frac{5}{4}} x\, dx = \int \frac{1}{\sin^{\frac{5}{4}} x \cos^{\frac{3}{4}} x}\, dx$$

$$= \int \frac{1}{\left[\dfrac{\sin^{\frac{5}{4}} x}{\cos^{\frac{5}{4}} x}\right] \cos^{\frac{5}{4}} x \cos^{\frac{3}{4}} x}\, dx = \int \frac{1}{\tan^{\frac{5}{4}} x \cos^2 x}\, dx$$

$$= \int \frac{1}{\tan^{\frac{5}{4}} x} \sec^2 x\, dx = \int \tan^{-\frac{5}{4}} x\, (\sec^2 x)\, dx = \frac{\tan^{-\frac{1}{4}} x}{-\frac{1}{4}}$$

$$= -4 \frac{1}{\tan^{\frac{1}{4}} x} = -4 \cot^{\frac{1}{4}} x.$$

EXAMPLES

1. Integrate (*i*) $\displaystyle\int \frac{\sin^2 x}{\cos^4 x}\, dx.$ (*ii*) $\displaystyle\int \frac{\cos^2 x}{\sin^4 x}\, dx.$

2. Integrate (*i*) $\displaystyle\int \frac{dx}{\sin^2 x \cos^2 x}.$ (*ii*) $\displaystyle\int \frac{dx}{\sin^4 x \cos^2 x}.$

(*iii*) $\displaystyle\int \frac{dx}{\sin^4 x \cos^4 x}.$

3. Integrate $\displaystyle\int \frac{\sqrt{\tan x}}{\sin x \cos x}\, dx.$

One standard case of $\int \sec^n x\, dx$ or $\int \operatorname{cosec}^n x\, dx$

57. Rule to integrate $\int \sec^n x\, dx$ or $\int \operatorname{cosec}^n x\, dx$, where n is a positive even integer :

(*i*) *If the index of sec x is a positive even integer, put tan x = z.*
(*ii*) *If the index of cosec x is a positive even integer, put cot x = z.*

[**Proof of** (*i*). To integrate $\int \sec^n x\, dx$, where n is a + ve even integer.

Let $n = 2k$, where k is a + ve integer.

Put $\tan x = z$ ...(1), then $\sec^2 x\, dx = dz$, or $dx = \dfrac{dz}{\sec^2 x}$.

$$\therefore \qquad \int \sec^n x\, dx = \int \sec^{2k} x\, dx = \int \sec^{2k} x \frac{dz}{\sec^2 x}$$

$$= \int \sec^{2k-2} x\, dz = \int (\sec^2 x)^{k-1}\, dz = \int (1+\tan^2 x)^{k-1}\, dz = \int (1+z^2)^{k-1}\, dz,$$

which can be integrated by expanding $(1+z^2)^{k-1}$ in a finite series by the binomial theorem, and integrating term by term.

Similarly $\int \operatorname{cosec}^n x\, dx$, where n is a positive even integer, can be integrated by putting $\cot x = z$.]

Example 1. *Integrate* (*i*) $\int \sec^4 x\, dx$.

(*ii*) $\int \sec^6 x\, dx$.

(*i*) [Here the index of sec x is 4, a + ve even integer.]

Put $\tan x = z$ ...(1) then $\sec^2 x\, dx = dz$, or $dx = \dfrac{dz}{\sec^2 x}$.

$$\therefore \quad \int \sec^4 x\, dx = \int \sec^4 x \frac{dz}{\sec^2 x} = \int \sec^2 x\, dz = \int (1 + \tan^2 x)\, dz$$

$$= \int (1 + z^2)\, dz = z + \frac{z^3}{3} = \tan x + \frac{\tan^3 x}{3}. \qquad \text{[From (1)]}$$

(*ii*) [Here the index of sec x is 6, a + ve even integer.]

Put $\tan x = z$ ...(1), then $\sec^2 x\, dx = dz$, or $dx = \dfrac{dz}{\sec^2 x}$.

$$\therefore \quad \int \sec^6 x\, dx = \int \sec^6 x \frac{dz}{\sec^2 x} = \int \sec^4 x\, dz = \int (\sec^2 x)^2\, dz$$

$$= \int (1 + \tan^2 x)^2\, dz = \int (1 + z^2)^2 dz = \int (1 + 2z^2 + z^4)\, dz$$

$$= z + 2\frac{z^3}{3} + \frac{z^5}{5} = \tan x + 2\frac{\tan^3 x}{3} + \frac{\tan^5 x}{5}. \qquad \text{[From (1)]}$$

Example 2. *Integrate* $\int \operatorname{cosec}^4 x\, dx$.

[Here the index of cosec x is 4, a + ve even integer.]

Put $\cot x = z$ ...(1), then $-\operatorname{cosec}^2 x\, dx = dz$, or $dx = -\dfrac{dz}{\operatorname{cosec}^2 x}$.

$$\therefore \quad \int \operatorname{cosec}^4 x\, dx = \int \operatorname{cosec}^4 x \left(-\frac{dz}{\operatorname{cosec}^2 x}\right) = -\int \operatorname{cosec}^2 x\, dz$$

$$= -\int (1 + \cot^2 x)\, dz = -\int (1 + z^2)\, dz = -\left[z + \frac{z^3}{3}\right]$$

$$= -\left[\cot x + \frac{\cot^3 x}{3}\right]. \qquad \text{[From (1)]}$$

EXAMPLE

(*i*) Prove that $\int_0^{\frac{\pi}{4}} \sec^4 \theta\, d\theta = \dfrac{4}{3}$.

(*ii*) Integrate $\int \dfrac{d\theta}{\sin^4 \frac{1}{2}\theta}$.

58. To integrate $\int \dfrac{\mathbf{dx}}{\mathbf{a + b \cos x}}$.

$$\int \frac{1}{a+b\cos x}\,dx = \int \frac{1}{a\left(\cos^2\frac{x}{2}+\sin^2\frac{x}{2}\right)+b\left(\cos^2\frac{x}{2}-\sin^2\frac{x}{2}\right)}\,dx$$

$$\left[\because \cos^2\frac{x}{2}+\sin^2\frac{x}{2}=1\right]$$

$$= \int \frac{1}{(a+b)\cos^2\frac{x}{2}+(a-b)\sin^2\frac{x}{2}}\,dx \qquad \left[\text{Divide the num. and denom. by } \cos^2\frac{x}{2}^{*}\right]$$

$$= \int \frac{\sec^2\frac{x}{2}}{(a+b)+(a-b)\tan^2\frac{x}{2}}\,dx$$

$$\left[\text{Put } \tan\frac{x}{2}=z \ \ldots(1), \text{ then } \sec^2\frac{x}{2}\cdot\frac{1}{2}\,dx = dz, \text{ or } \sec^2\frac{x}{2}\,dx = 2dz\right]$$

$$= \int \frac{2dz}{(a+b)+(a-b)z^2} = 2\int \frac{1}{(a+b)+(a-b)z^2}\,dz \qquad \ldots(2)$$

[*Assume that* $a + b$ *is + ve*]

Case I. If $a^2 > b^2$,

[$\because$ $a^2 - b^2$ is + ve, and $a + b$ is + ve, $\therefore$ dividing, $a - b$ is + ve] then from (2),

$$\int \frac{1}{a+b\cos x}\,dx = 2\int \frac{1}{(a-b)\left[\frac{a+b}{a-b}+z^2\right]}\,dz = \frac{2}{a-b}\int \frac{1}{\left(\sqrt{\frac{a+b}{a-b}}\right)^2+z^2}\,dz$$

$$= \frac{2}{a-b}\cdot\frac{1}{\sqrt{\frac{a+b}{a-b}}}\tan^{-1}\frac{z}{\sqrt{\frac{a+b}{a-b}}}$$

$$= \frac{2}{\sqrt{a^2-b^2}}\tan^{-1}\left[\sqrt{\frac{a-b}{a+b}}\tan\frac{x}{2}\right] \qquad \text{[From (1)]}$$

$$= \frac{1}{\sqrt{a^2-b^2}}\cos^{-1}\frac{b+a\cos x}{a+b\cos x}. \qquad \text{[From Trigonometry]}$$

***Why this step.** In $\int \frac{1}{(a+b)\cos^2\frac{x}{2}+(a-b)\sin^2\frac{x}{2}}\,dx$, the denominator is a homogeneous function of the second degree in $\cos\frac{x}{2}$ and $\sin\frac{x}{2}$,

$\therefore$ we divide the numerator and denominator by $\cos^2\frac{x}{2}$, and put $\tan\frac{x}{2} = z$.

[Rule (Ex. 7, Art. 21)]

Case II. If $a^2 < b^2$,

[$\because$ $a^2 - b^2$ is – ve, and $a + b$ is + ve, $\therefore$ dividing, $a - b$ is –ve] then from (2),

$$\int \frac{1}{a+b\cos x}\,dx = 2\int \frac{1}{(b+a)-(b-a)z^2}\,dz$$

$$= 2\int \frac{1}{(b-a)\left[\dfrac{b+a}{b-a} - z^2\right]}\,dz = \frac{2}{b-a}\int \frac{1}{\left(\sqrt{\dfrac{b+a}{b-a}}\right)^2 - z^2}\,dz$$

$$= \frac{2}{b-a}\cdot\frac{1}{2\sqrt{\dfrac{b+a}{b-a}}}\log\frac{\sqrt{\dfrac{b+a}{b-a}}+z}{\sqrt{\dfrac{b+a}{b-a}}-z}$$

$$= \frac{1}{\sqrt{b^2-a^2}}\log\frac{\sqrt{b+a}+\sqrt{b-a}\tan\dfrac{x}{2}}{\sqrt{b+a}-\sqrt{b-a}\tan\dfrac{x}{2}} \qquad \text{[From (1)]}$$

$$= \frac{1}{\sqrt{b^2-a^2}}\cosh^{-1}\frac{b+a\cos x}{a+b\cos x}. \qquad \text{[From Trigonometry]}$$

Hence we have

A standard integral.

If a + b is positive*, then

(*i*) **if** $\mathbf{a^2 > b^2}$, $\displaystyle\int \frac{1}{a+b\cos x}\,dx = \frac{1}{\sqrt{a^2-b^2}}\cos^{-1}\frac{b+a\cos x}{a+b\cos x}$ † ;

(*ii*) **if** $\mathbf{a^2 < b^2}$, $\displaystyle\int \frac{1}{a+b\cos x}\,dx = \frac{1}{\sqrt{b^2-a^2}}\cosh^{-1}\frac{b+a\cos x}{a+b\cos x}$.

***Caution.** It is assumed that *a + b is positive.* If, however, in a numerical example, $a + b$ is negative, take the negative sign outside the integral, and use then the proper formula. (See Ex. 4 following.)

† [**Aid to memory.** In order to write $\dfrac{b+a\cos x}{a+b\cos x}$, it is convenient to write the denominator $a + b\cos x$ $\left(\text{the denominator of the integrand } \dfrac{1}{a+b\cos x}\right)$ first thus, $\dfrac{1}{a+b\cos x}$, and then to write the numerator $b + a\cos x$ thus, $\dfrac{b+a\cos x}{a+b\cos x}$.]

Example 1. *Integrate* $\displaystyle\int \frac{dx}{5+4\cos x}$.

$\displaystyle\int \frac{1}{5+4\cos x}\,dx$ $\left[\text{Compare with } \displaystyle\int \frac{1}{a+b\cos x}\,dx, \text{ here } a = 5, b = 4\right.$

$\therefore$ $a + b = 5 + 4 = 9$, + ve ; $a^2 = 25$, $b^2 = 16$, $\therefore$ $a^2 > b^2$ ($\because$ 25 > 16)]

$$= \frac{1}{\sqrt{(5)^2 - (4)^2}} \cos^{-1} \frac{4 + 5\cos x}{5 + 4\cos x}$$

$$\left[\frac{1}{\sqrt{a^2 - b^2}} \cos^{-1} \frac{b + a\cos x}{a + b\cos x} \text{ (Art. 58, Case I)} \right]$$

$$= \frac{1}{3} \cos^{-1} \frac{4 + 5\cos x}{5 + 4\cos x}.$$

Example 2. *Evaluate* $\int_0^{\frac{\pi}{2}} \frac{dx}{1 + 2\cos x}$.

$\int_0^{\frac{\pi}{2}} \frac{1}{1 + 2\cos x}\, dx$ $\left[\text{Compare with } \int \frac{1}{a + b\cos x}\, dx, \text{ here } a = 1, b = 2 \right.$

$\therefore \quad a + b = 1 + 2 = 3, + \text{ve}\,;\ a^2 = 1, b^2 = 4,\ \therefore\ a^2 < b^2\ (\because\ 1 < 4)]$

$$= \left[\frac{1}{\sqrt{(2)^2 - (1)^2}} \cosh^{-1} \frac{2 + 1\cos x}{1 + 2\cos x} \right]_0^{\frac{\pi}{2}}$$

$$\left[\frac{1}{\sqrt{b^2 - a^2}} \cosh^{-1} \frac{b + a\cos x}{a + b\cos x} \text{ (Art. 58, Case II)} \right]$$

$$\frac{1}{\sqrt{3}} [\cosh^{-1} 2 - \cosh^{-1} 1] = \frac{1}{\sqrt{3}} [\cosh^{-1} 2 - 0] = \frac{1}{\sqrt{3}} \log (2 + \sqrt{(2)^2 - 1})$$

$$[\because\ \cosh^{-1} x = \log (x + \sqrt{x^2 - 1}) \text{ (Art. 8, (d), (2))}]$$

$$= \frac{1}{\sqrt{3}} \log (2 + \sqrt{3}).$$

Note. *While evaluating definite integrals, if the answer turns to be in the inverse hyperbolic form, it is usual to express it in logarithmic form.*

EXAMPLES

1. Evaluate (*i*) $\int_0^{\frac{\pi}{2}} \frac{dx}{2 + \cos x}$. (*ii*) $\int \frac{dx}{4 + 5\cos x}$.

2. (*a*) Prove that $\int_0^{\theta} \frac{dx}{1 - \cos\theta\cos x} = \frac{\pi}{2} \operatorname{cosec} \theta$.

(*b*) Evaluate $\int_0^{\pi} \frac{dx}{1-2a\cos x+a^2}$.

3. Integrate

(*i*) $\int \frac{dx}{\cos\alpha+\cos x}$. (*ii*) $\int \frac{\cos\alpha\cos x+1}{\cos\alpha+\cos x}\,dx$.

4. Integrate $\int \frac{d\theta}{2\sin^2\theta-3\cos\theta}$.

[Replace $\sin^2\theta$ by $1-\cos^2\theta$, and resolve the integrand into partial fractions by putting $\cos\theta = y$. It will be found that

$$\frac{1}{2\sin^2\theta-3\cos\theta} = \frac{1}{5}\,\frac{1}{2+\cos\theta}+\frac{2}{5}\,\frac{1}{1-2\cos\theta}$$

$$\therefore \int \frac{1}{2\sin^2\theta-3\cos\theta}\,d\theta = \frac{1}{5}\int \frac{1}{2+\cos\theta}\,d\theta+\frac{2}{5}\int\frac{1}{1-2\cos\theta}\,d\theta$$

$$= \frac{1}{5}\int \frac{1}{2+\cos\theta}\,d\theta-\frac{2}{5}\int\frac{1}{-1+2\cos\theta}\,d\theta$$

[To make "$a+b$" + ve.

(*Note this step*) (See 'Caution' in Art. 58)]

59. To integrate $\int \frac{dx}{a+b\sin x}$.

Put $x = \frac{\pi}{2}-z$ ...(1), then $dx = -\,dz$.

$$\therefore \int \frac{1}{a+b\sin x}\,dx = \int \frac{1}{a+b\sin\left(\frac{\pi}{2}-z\right)}(-\,dz)$$

$$= -\int \frac{1}{a+b\cos z}\,dz \quad ...(2) \quad [\textit{Assume that } a+b \textit{ is } +ve]$$

Case I. If $a^2 > b^2$, then from (2),

$$\int \frac{1}{a+b\sin x}\,dx = -\frac{1}{\sqrt{a^2-b^2}}\cos^{-1}\frac{b+a\cos z}{a+b\cos z}$$

$$= -\frac{1}{\sqrt{a^2-b^2}}\cos^{-1}\frac{b+a\cos\left(\frac{\pi}{2}-x\right)}{a+b\cos\left(\frac{\pi}{2}-x\right)} \quad \left[\because \text{ from (1), } z=\frac{\pi}{2}-x\right]$$

$$= -\frac{1}{\sqrt{a^2-b^2}}\cos^{-1}\frac{b+a\sin x}{a+b\sin x}.$$

Case II. If $a^2 < b^2$, then from (2),

$$\int \frac{1}{a+b\sin x}\,dx = -\frac{1}{\sqrt{b^2-a^2}}\cosh^{-1}\frac{b+a\cos z}{a+b\cos z}$$

$$= -\frac{1}{\sqrt{b^2 - a^2}} \cosh^{-1} \frac{b + a\cos\left(\frac{\pi}{2} - x\right)}{a + b\cos\left(\frac{\pi}{2} - x\right)}$$

$$= -\frac{1}{\sqrt{b^2 - a^2}} \cosh^{-1} \frac{b + a\sin x}{a + b\sin x}.$$

EXAMPLE

Evaluate

(*i*) $\int_0^{\frac{\pi}{2}} \frac{dx}{5 + 4\sin x}$. (*ii*) $\int_0^{\frac{\pi}{2}} \frac{dx}{4 + 5\sin x}$.

60. To integrate $\int \frac{\mathbf{dx}}{\mathbf{a + b\cos x + c\sin x}}$.

Put $b = r\cos\alpha$ ...(1), $c = r\sin\alpha$ ...(2)

[so that squaring (1) and (2), and adding,

$$b^2 + c^2 = r^2 \,.\, 1, \quad \therefore \quad r = \sqrt{b^2 + c^2} \qquad ...(3)$$

and dividing (2) by (1),

$$\frac{c}{b} = \frac{\sin\alpha}{\cos\alpha} = \tan\alpha, \quad \therefore \quad \alpha = \tan^{-1}\frac{c}{b} \qquad ...(4)]$$

Then $\int \frac{1}{a + b\cos x + c\sin x}\, dx = \int \frac{1}{a + r\cos\alpha\cos x + r\sin\alpha\sin x}\, dx$

$$= \int \frac{1}{a + r(\cos x\cos\alpha + \sin x\sin\alpha)}\, dx = \int \frac{1}{a + r\cos(x - \alpha)}\, dx$$

[Substitute the values of r and α from (3) and (4)]

$$= \int \frac{1}{a + \sqrt{b^2 + c^2}\cos\left(x - \tan^{-1}\frac{c}{b}\right)}\, dx \qquad ...(5)$$

[*Assume that* $a + \sqrt{b^2 + c^2}$ *is* + *ve*]

Case I. If $a^2 > b^2 + c^2$, then from (5),

$$\int \frac{1}{a + b\cos x + c\sin x}\, dx = \frac{1}{\sqrt{a^2 - (b^2 + c^2)}} \cos^{-1} \frac{\sqrt{b^2 + c^2} + a\cos\left(x - \tan^{-1}\frac{c}{b}\right)}{a + \sqrt{b^2 + c^2}\cos\left(x - \tan^{-1}\frac{c}{b}\right)}.$$

Case II. If $a^2 < b^2 + c^2$, then from (5),

$$\int \frac{1}{a + b\cos x + c\sin x}\, dx$$

$$= \frac{1}{\sqrt{(b^2+c^2)-a^2}} \cosh^{-1} \frac{\sqrt{b^2+c^2} + a\cos\left(x - \tan^{-1}\frac{c}{b}\right)}{a + \sqrt{b^2+c^2}\cos\left(x - \tan^{-1}\frac{c}{b}\right)}.$$

Example. *Integrate* $\int \frac{dx}{13 + 3\cos x + 4\sin x}$.

Put $3 = r\cos\alpha$ ...(1), $4 = r\sin\alpha$ ...(2)

[so that squaring (1) and (2), and adding,

$$(3)^2 + (4)^2 = r^2 \cdot 1, \quad \therefore \quad r = 5 \qquad ...(3)$$

and dividing (2) by (1),

$$\frac{4}{3} = \frac{\sin\alpha}{\cos\alpha} = \tan\alpha, \quad \therefore \quad \alpha = \tan^{-1}\frac{4}{3} \qquad ...(4)]$$

Then $\int \frac{dx}{13 + 3\cos x + 4\sin x} dx$

$$= \int \frac{1}{13 + r\cos\alpha\cos x + r\sin\alpha\sin x} dx$$

$$= \int \frac{1}{13 + r(\cos x\cos\alpha + \sin x\sin\alpha)} dx = \int \frac{1}{13 + r\cos(x-\alpha)} dx$$

[Substitute the values of r and α from (3) and (4)]

$$= \int \frac{1}{13 + 5\cos(x - \tan^{-1}\frac{4}{3})} dx$$

$$= \frac{1}{\sqrt{(13)^2 - (5)^2}} \cos^{-1} \frac{5 + 13\cos(x - \tan^{-1}\frac{4}{3})}{13 + 5\cos(x - \tan^{-1}\frac{4}{3})}$$

$$= \frac{1}{12} \cos^{-1} \frac{5 + 13\cos(x - \tan^{-1}\frac{4}{3})}{13 + 5\cos(x - \tan^{-1}\frac{4}{3})}$$

EXAMPLES

1. Integrate

(*i*) $\int \frac{\sqrt{2}\,dx}{2\sqrt{2} + \cos x + \sin x}$. (*ii*) $\int \frac{dx}{3(1 - \sin x) - \cos x}$.

2. Prove that $\int_0^{\pi} \frac{dx}{3 + 2\sin x + \cos x} = \frac{\pi}{4}$.

Put $\tan\frac{x}{2} = t$ ...(1)

then, $\sec^2\frac{x}{2} \cdot \frac{1}{2} dx = dt$, or $dx = \frac{2dt}{\sec^2\frac{x}{2}} = \frac{2dt}{1 + \tan^2\frac{x}{2}} = \frac{2dt}{1+t^2}$, and from (1), when $x = 0, t = 0$; when $x \to \pi, t \to \infty$;

$$\cos x = \frac{1 - \tan^2 \frac{x}{2}}{1 + \tan^2 \frac{x}{2}} = \frac{1 - t^2}{1 + t^2}, \sin x = \frac{2 \tan \frac{x}{2}}{1 + \tan^2 \frac{x}{2}} = \frac{2t}{1 + t^2},$$

$$\therefore \int_0^{\pi} \frac{1}{3 + 2 \sin x + \cos x} dx = \int_0^{\infty} \frac{1}{3 + 2\left(\frac{2t}{1 + t^2}\right) + \frac{1 - t^2}{1 + t^2}} \cdot \frac{2dt}{1 + t^2}$$

$$= 2 \int_0^{\infty} \frac{1}{3(1 + t^2) + 4t + 1 - t^2} dt$$

$$= 2 \int_0^{\infty} \frac{1}{2t^2 + 4t + 4} dt = \int_0^{\infty} \frac{1}{t^2 + 2t + 2} dt.]$$

Note 1. Comparison of the two methods. *For evaluating the indefinite integral* $\int \frac{1}{a + b \cos x + c \sin x} dx$, *the method of Art. 60 [by reducing it to the form* $\int \frac{1}{a + r \cos (x - \alpha)} dx$*] is shorter. But, for evaluating the definite integral* $\int \frac{1}{a + b \cos x + c \sin x} dx$ *between given limits the method used in the above Ex. 2 is better. For, the trigonometrical calculations in the answer are then easier.*

Note 2. Rule to integrate a rational function of cos x and sin x.

Put $\tan \frac{x}{2} = t$,

then $\sec^2 \frac{x}{2} \cdot \frac{1}{2} dx = dt$, or $dx = \frac{2dt}{\sec^2 \frac{x}{2}} = \frac{2dt}{1 + \tan^2 \frac{x}{2}} = \frac{2dt}{1 + t^2}$,

$$\cos x = \frac{1 - \tan^2 \frac{x}{2}}{1 + \tan^2 \frac{x}{2}} = \frac{1 - t^2}{1 + t^2}, \sin x = \frac{2 \tan \frac{x}{2}}{1 + \tan^2 \frac{x}{2}} = \frac{2t}{1 + t^2},$$

$\therefore$ *the substitution transforms the given integral into the integral of a rational algebraic function of t, which can be integrated by resolving the integrand into partial fractions (Chap. IV).*

Caution. Though the above rule is applicable to integrate *any* rational function of $\cos x$ and $\sin x$, yet it should be used only as the last resort. For, in practice, in the resulting integrand, the denominator is generally of very high degree in t and, therefore, this method is not generally the simplest.

3. Integrate $\int \dfrac{1+\sin x}{\sin x\,(1+\cos x)}\,dx.$

61. To integrate $\displaystyle\int \frac{\mathbf{a\cos x + b\sin x}}{\mathbf{c\cos x + e\sin x}}\,\mathbf{dx}.$

[*Let num.* = l *(denom.)* + m*(diff. coeff. of denom.) i.e.,*]

Let $a\cos x + b\sin x = l(c\cos x + e\sin x) + m(-c\sin x + e\cos x)$

[so that equating coeffs. of $\cos x$, $a = cl + em$ ...(1) $\;c \mid e$

equating coeffs. of $\sin x$, $b = el - cm$ ...(2) $\;e \mid -c$

[To find the values of l, m.]

Multiplying (1) by c, (2) by e, and adding,

$$ac + be = l(c^2 + e^2), \quad \therefore \quad l = \frac{ac + be}{c^2 + e^2} \qquad ...(3)$$

Multiplying (1) by e, (2) by $(-c)$, and adding,

$$ae - bc = m(e^2 + c^2), \quad \therefore \quad m = \frac{ae - bc}{c^2 + e^2} \qquad ...(4)]$$

Then $\displaystyle\int \frac{a\cos x + b\sin x}{c\cos x + e\sin x}\,dx$

$$= \int \frac{l(c\cos x + e\sin x) + m(-c\sin x + e\cos x)}{c\cos x + e\sin x}\,dx$$

$$= \int \left[l + m\frac{-c\sin x + e\cos x}{c\cos x + e\sin x}\right]dx$$

$$= lx + m\log(c\cos x + e\sin x)$$

[Substitute the values of l and m from (3) and (4)]

$$= \frac{ac + be}{c^2 + e^2}\,x + \frac{ae - bc}{c^2 + e^2}\log(c\cos x + e\sin x).$$

EXAMPLES

1. Integrate

(i) $\displaystyle\int \frac{a\sin\theta + b\cos\theta}{c\sin\theta + e\cos\theta}\,d\theta.$ (ii) $\displaystyle\int \frac{d\theta}{a + b\tan\theta}.$

2. Integrate $\int \cos 2\theta \log(1 + \tan\theta)\,d\theta.$

****62. To integrate** $\displaystyle\int \frac{\mathbf{a + b\cos x + c\sin x}}{\mathbf{a' + b'\cos x + c'\sin x}}\,\mathbf{dx}.$

[*Let num.* = l*(denom.)* + m*(diff. coeff. of denom.)* + n, *i.e.,*]

Let $a + b\cos x + c\sin x$

$$= l(a' + b'\cos x + c'\sin x) + m(-b'\sin x + c'\cos x) + n$$

[so that equating coeffs. of $\cos x$, $b = b'l + c'm$,

equating coeffs. of $\sin x$, $c = c'l + b'm$,

equating constant terms, $a = a'l + n$.

Solving these equations, we get the values of l, m, n.]

Then $\displaystyle\int \frac{a + b\cos x + c\sin x}{a' + b'\cos x + c'\sin x}\,dx$

$$= \int \frac{l(a' + b'\cos x + c'\sin x) + m(-b'\sin x + c'\cos x) + n}{a' + b'\cos x + c'\sin x}\,dx$$

$$= \int \left[l + m\frac{-b'\sin x + c'\cos x}{a' + b'\cos x + c'\sin x} + \frac{n}{a' + b'\cos x + c'\sin x} \right] dx$$

$$= lx + m\log(a' + b'\cos x + c'\sin x) + n\int \frac{1}{a' + b'\cos x + c'\sin x}\,dx \qquad ...(1)$$

Now $\displaystyle\int \frac{1}{a' + b'\cos x + c'\sin x}\,dx$ can be integrated (Art. 60),

$\therefore$ from (1), $\displaystyle\int \frac{a + b\cos x + c\sin x}{a' + b'\cos x + c'\sin x}\,dx$ can be integrated.

Substituting the values of l, m, n found above, we get the required answer.

EXAMPLE

**Integrate $\displaystyle\int \frac{\sin x + \cos x}{3\sin x + 4\cos x + 1}\,dx$.

63. To integrate $\displaystyle\int \frac{dx}{a + b\cosh x}$.

$$\int \frac{1}{a + b\cosh x}\,dx$$

$$= \int \frac{1}{a\left(\cosh^2 \frac{x}{2} - \sinh^2 \frac{x}{2}\right) + b\left(\cosh^2 \frac{x}{2} + \sinh^2 \frac{x}{2}\right)}\,dx$$

[Examples, Art. 8 (c)]

$$= \int \frac{1}{(a + b)\cosh^2 \frac{x}{2} + (b - a)\sinh^2 \frac{x}{2}}\,dx$$

$\left[\text{Divide the num. and denom. by } \cosh^2 \frac{x}{2}\right]$

$$= \int \frac{\operatorname{sech}^2 \frac{x}{2}}{(a + b) + (b - a)\tanh^2 \frac{x}{2}}\,dx$$

$\left[\text{Put } \tanh \frac{x}{2} = z \quad ...(1), \quad \text{then } \operatorname{sech}^2 \frac{x}{2} \cdot \frac{1}{2}\,dx = dz, \text{ or } \operatorname{sech}^2 \frac{x}{2}\,dx = 2dz\right]$

$$= \int \frac{2dz}{(a+b)+(b-a)z^2} = 2\int \frac{1}{(a+b)+(b-a)z^2}\,dz \qquad ...(2)$$

[*Assume that* $a + b$ *is* + *ve*]

Case I. If $a^2 > b^2$,

[$\because$ $a^2 - b^2$ is + ve, and $a + b$ is + ve, $\therefore$ dividing, $a - b$ is + ve, or $b - a$ is – ve] then from (2),

$$\int \frac{1}{a+b\cosh x}\,dx = 2\int \frac{1}{(a+b)-(a-b)z^2}\,dz = 2\int \frac{1}{(a-b)\left[\dfrac{a+b}{a-b} - z^2\right]}\,dz$$

$$= \frac{2}{a-b}\int \frac{1}{\left(\sqrt{\dfrac{a+b}{a-b}}\right)^2 - z^2}\,dz = \frac{2}{a-b}\,\frac{1}{2\sqrt{\dfrac{a+b}{a-b}}}\log \frac{\sqrt{\dfrac{a+b}{a-b}}+z}{\sqrt{\dfrac{a+b}{a-b}}-z}$$

$$= \frac{1}{\sqrt{a^2-b^2}}\log \frac{\sqrt{a+b}+\sqrt{a-b}\tanh\dfrac{x}{2}}{\sqrt{a+b}-\sqrt{a-b}\tanh\dfrac{x}{2}}. \qquad \text{[From (1)]}$$

Case II. If $a^2 < b^2$,

[$\because$ $a^2 - b^2$ is – ve, and $a + b$ is + ve, $\therefore$ dividing, $a - b$ is – ve, or $b - a$ is + ve] then from (2),

$$\int \frac{1}{a+b\cosh x}\,dx = 2\int \frac{1}{(b-a)\left[\dfrac{b+a}{b-a} + z^2\right]}\,dz$$

$$= \frac{2}{b-a}\int \frac{1}{\left(\sqrt{\dfrac{b+a}{b-a}}\right)^2 + z^2}\,dz = \frac{2}{b-a}\,\frac{1}{\sqrt{\dfrac{b+a}{b-a}}}\tan^{-1}\frac{z}{\sqrt{\dfrac{b+a}{b-a}}}$$

$$= \frac{2}{\sqrt{b^2-a^2}}\tan^{-1}\left[\sqrt{\frac{b-a}{b+a}}\tanh\frac{x}{2}\right]. \qquad \text{[From (1)]}$$

EXAMPLES

1. Integrate $\displaystyle\int \frac{dx}{3+4\cosh x}$.

2. Integrate $\displaystyle\int \frac{dx}{a+b\sinh x}$.

$$\left[\int \frac{1}{a+b\sinh x}\,dx = \int \frac{1}{a\left(\cosh^2\dfrac{x}{2} - \sinh^2\dfrac{x}{2}\right) + b\,2\sinh\dfrac{x}{2}\cosh\dfrac{x}{2}}\,dx\right.$$

$$= \int \frac{1}{a \cosh^2 \frac{x}{2} - a \sinh^2 \frac{x}{2} + 2b \sinh \frac{x}{2} \cosh \frac{x}{2}} dx.$$

Divide the num. and denom. by $\cosh^2 \frac{x}{2}$, and put $\tanh \frac{x}{2} = z.$]

MISCELLANEOUS EXAMPLES ON CHAPTER VI

1. Evaluate

(a) $\int \sin^5 \theta \sqrt{\cos \theta}\, d\theta.$ (b) $\int_0^{\frac{\pi}{2}} \sin^{\frac{3}{2}} x \cos^3 x\, dx.$

2. Evaluate

(i) $\int_0^{\frac{\pi}{2}} \cos^2 x\, dx.$ (ii) $\int_0^a \frac{x^4\, dx}{(a^2 + x^2)^4}.$

3. Prove that (i) $\int_0^{\pi} \frac{d\theta}{5 + 3\cos\theta} = \frac{\pi}{4}.$

(ii) $\int_0^{\alpha} \frac{dx}{\cos a + \cos x} = \text{cosec } \alpha \log \sec \alpha.$

4. Prove that $\int_0^{\frac{\pi}{2}} \frac{dx}{1 - \sin x \cos \lambda} = (\pi - \lambda) \text{ cosec } \lambda \; (0 < \lambda < \pi).$

5. Evaluate $\int_0^{\frac{\pi}{2}} \frac{dx}{2\cos x + \sin x + a}$ in the two cases when $a = 1$ and $a = 3$.

6. Integrate (i) $\int \frac{2\cos x + 3\sin x}{3\cos x + 2\sin x} dx.$

**(ii) $\int \frac{a + b\cos\theta + c\sin\theta}{1 + \sin\theta} d\theta.$

7. Evaluate

(i) $\int \frac{\cos x}{a + b\cos x} dx.$ (ii) $\int_{-1}^{+1} \frac{\sqrt{1 - x^2}}{a - x} dx.$

CHAPTER VII

REDUCTION FORMULAE

Reduction formulae play an important role in systematic integration and it is well to treat them as an independent subject. —*Osgood.*

Reduction formulae found by integrating by parts

64. Reduction formula. Def. A reduction formula for an integral is a formula which connects the integral with one or two other integrals in which the integrand is of the same type but of lower degree or lower order.

Integration by successive reduction. Integration by repeated application of a reduction formula is called *integration by successive reduction.*

Example. Obtain the reduction formula

$$\int \mathbf{x^m (\log x)^n \, dx = \frac{x^{m+1} (\log x)^n}{m+1} - \frac{n}{m+1} \int x^m (\log x)^{n-1} \, dx.}$$

Integrate out $\int \mathbf{x^m (\log x)^3 \, dx}$**.**

(*a*) $\int x^m (\log x)^n \, dx = \int (\log x)^n \, . \, x^m \, dx$ [Rule (Note, p. 37, top)]

[Integrate by parts]

$$= (\log x)^n \, . \, \frac{x^{m+1}}{m+1} - \int n(\log x)^{n-1} \frac{1}{x} \, . \, \frac{x^{m+1}}{m+1} \, dx$$

$$= \frac{x^{m+1}}{m+1} (\log x)^n - \frac{n}{m+1} \int x^m (\log x)^{n-1} \, dx \qquad ...(1)$$

(*b*) Putting $n = 3$ in (1),

$$\int x^m (\log x)^3 \, dx = \frac{x^{m+1}}{m+1} (\log x)^3 - \frac{3}{m+1} \int x^m (\log x)^2 \, dx \qquad ...(2)$$

Putting $n = 2$ in (1),

$$\int x^m (\log x)^2 \, dx = \frac{x^{m+1}}{m+1} (\log x)^2 - \frac{2}{m+1} \int x^m \log x \, dx \qquad ...(3)$$

Putting $n = 1$ in (1),

$$\int x^m \log x \, dx = \frac{x^{m+1}}{m+1} \log x - \frac{1}{m+1} \int x^m (\log x)^0 \, dx$$

$$\left[\text{But} \int x^m (\log x)^0 \, dx = \int x^m \, dx = \frac{x^{m+1}}{m+1}\right]$$

$$= \frac{x^{m+1}}{m+1} \log x - \frac{1}{m+1}\frac{x^{m+1}}{m+1} = \frac{x^{m+1}}{m+1} \log x - \frac{x^{m+1}}{(m+1)^2} \quad ...(4)$$

Substituting this value of $\int x^m \log x \, dx$ in (3),

$$\int x^m (\log x)^2 \, dx = \frac{x^{m+1}}{m+1} (\log x)^2 - \frac{2}{m+1}\left[\frac{x^{m+1}}{m+1} \log x - \frac{x^{m+1}}{(m+1)^2}\right]$$

$$= \frac{x^{m+1}}{m+1} (\log x)^2 - \frac{2}{(m+1)^2} x^{m+1} \log x + \frac{2}{(m+1)^3} x^{m+1}.$$

Substituting this value of $\int x^m (\log x)^2 \, dx$ in (2),

$$\int x^m (\log x)^3 \, dx = \frac{x^{m+1}}{m+1} (\log x)^3$$

$$- \frac{3}{m+1}\left[\frac{x^{m+1}}{m+1} (\log x)^2 - \frac{2}{(m+1)^2} x^{m+1} \log x + \frac{2}{(m+1)^3} x^{m+1}\right]$$

$$= \frac{x^{m+1}}{m+1} (\log x)^3 - 3\frac{x^{m+1}}{(m+1)^2} (\log x)^2 + 6\frac{x^{m+1}}{(m+1)^3} \log x - 6\frac{x^{m+1}}{(m+1)^4}$$

$$= \frac{x^{m+1}}{m+1}\left[(\log x)^3 - \frac{3}{m+1} (\log x)^2 + \frac{6}{(m+1)^2} \log x - \frac{6}{(m+1)^3}\right].$$

[*Note this method of substituting backwards from (4) in (3), and then from the result in (2).*]

Note 1. What is the use of a reduction formula ? *The use of a reduction formula is that the index of the troublesome factor [here* $(\log x)^n$*] is decreased, and by repeated application of the same formula the integral is reduced to one which is either a standard form, or which can be easily integrated.*

Note 2. Method to find a reduction formula for $\int x^n e^{ax} \, dx$ or $\int x^m (\log x)^n \, dx$. *Integrate by parts.*

Note 3. Abridged notation for the reduction formula.

If $I_{m, n} = \int x^m (\log x)^n \, dx$, *then changing n to n – 1,*

$$I_{m, n-1} = \int x^m (\log x)^{n-1} \, dx.$$

$\therefore$ *from (1),*

$$I_{m, n} = \frac{x^{m+1}}{m+1} (\log x)^n - \frac{n}{m+1} I_{m, n-1}.$$

EXAMPLES

1. Prove that

$$\int x^m (\log x)^n \, dx = \frac{x^{m+1} (\log x)^n}{m+1} - \frac{n}{m+1}\int x^m (\log x)^{n-1} \, dx.$$

Integrate out $\int x^m \log x \, dx$, $\int x^m (\log x)^2 \, dx$.

[Two important limits. (1) $\mathbf{\underset{x \to \infty}{Lt} \dfrac{x^n}{e^x} = 0}$**, (2)** $\mathbf{\underset{x \to 0}{Lt} x^m (\log x)^n = 0}$**, where m, n are positive.** (From Differential Calculus)]

2. Evaluate $\int_0^1 x^m (\log x)^n \, dx$, when $m \geq 0$ and n is a positive integer.

$$\int_0^1 x^m (\log x)^n \, dx = \int_0^1 (\log x)^n \, . \, x^m \, dx \qquad \text{[Integrate by parts]}$$

$$= \left[(\log x)^n \, . \, \frac{x^{m+1}}{m+1} \right]_0^1 - \int_0^1 n \, (\log x)^{n-1} \, \frac{1}{x} \, . \, \frac{x^{m+1}}{m+1} \, dx$$

$$\left[\text{But } \underset{x \to 0}{\text{Lt}} \, x^{m+1} (\log x)^n = 0 \right.$$

$$\left. \left\{ \because \underset{x \to 0}{\text{Lt}} \, x^m (\log x)^n = 0 \text{ (From Diff. Calculus)} \right\} \right]$$

$$\therefore \quad \int_0^1 x^m (\log x)^n \, dx = -\frac{n}{m+1} \int_0^1 x^m (\log x)^{n-1} \, dx \qquad \text{...(1)}$$

Changing n to $n - 1$,

$$\int_0^1 x^m (\log x)^{n-1} \, dx = -\frac{n-1}{m+1} \int_0^1 x^m (\log x)^{n-2} \, dx.$$

Substituting this value of $\int_0^1 x^m (\log x)^{n-1} \, dx$ in (1),

$$\int_0^1 x^m (\log x)^n \, dx = \frac{n(n-1)}{(m+1)^2} \int_0^1 x^m (\log x)^{n-2} \, dx \qquad \text{...(2)}$$

Generalizing from (1) and (2),

$$\int_0^1 x^m (\log x)^n \, dx = (-1)^n{}^* \, \frac{n(n-1)(n-2) \ldots \text{ to } n \text{ factors}}{(m+1)^n} \int_0^1 x^m (\log x)^{n-n} \, dx$$

$$\left[\text{But } \int_0^1 x^m (\log x)^0 \, dx = \int_0^1 x^m \, dx = \left[\frac{x^{m+1}}{m+1} \right]_0^1 = \frac{1}{m+1} \right]$$

---

*__How to write this step.__ In (1), we have (–), *i.e.*, $(-1)^1$, in (2), we have (+), *i.e.*, $(-1)^2$, ..., $\therefore$ in general, we have $(-1)^n$.

$$= \frac{(-1)^n\, n(n-1)(n-2)\ldots 1}{(m+1)^n}\,\frac{1}{m+1} = \frac{(-1)^n\, n!}{(m+1)^{n+1}}.$$

****3.** Prove that $\int_0^1 v^{vx}\, dv = 1 - \frac{x}{2^2} + \frac{x^2}{3^3} - \frac{x^3}{4^4} + \frac{x^4}{5^5} - \ldots$.

[Expand the integrand in an infinite series in powers of x by using the expansion :

$$a^x = 1 + x\log a + \frac{x^2}{2!}(\log a)^2 + \frac{x^3}{3!}(\log a)^3 + \ldots .\Big]$$

4. Find a reduction formula for the integral $\int \frac{x^m}{(\log x)^n}\, dx$.

$$\left[\int \frac{x^m}{(\log x)^n}\, dx = \int x^{m+1}\left[\frac{1}{(\log x)^n}\,\frac{1}{x}\right]^* dx = \int x^{m+1}\left[(\log x)^{-n}\,\frac{1}{x}\right] dx.\right.$$

Integrate by parts.]

5. Evaluate $\int_0^\infty e^{-x} x^n\, dx$, where n is a positive integer.

Reduction formulae found by integrating twice by parts

Reduction formulae for $\int x^m \sin nx\, dx$ and $\int x^m \cos nx\, dx$

65. To find a reduction formula for $\int x^m \sin nx\, dx$.

$\int x^m \sin nx\, dx$ [Integrate by parts]

$$= x^m\left(-\frac{\cos nx}{n}\right) - \int m\,x^{m-1}\left(-\frac{\cos nx}{n}\right) dx$$

$$= -\frac{x^m \cos nx}{n} + \frac{m}{n}\int x^{m-1}\cos nx\, dx \quad \text{[Integrate again by parts]}$$

$$= -\frac{x^m \cos nx}{n} + \frac{m}{n}\left[x^{m-1}\frac{\sin nx}{n} - \int (m-1)\,x^{m-2}\frac{\sin nx}{n}\, dx\right]$$

$$= -\frac{x^m \cos nx}{n} + \frac{m}{n}\left[\frac{x^{m-1}\sin nx}{n} - \frac{m-1}{n}\int x^{m-2}\sin nx\, dx\right]$$

$$= -\frac{x^m \cos nx}{n} + \frac{m}{n^2}x^{m-1}\sin nx - \frac{m(m-1)}{n^2}\int x^{m-2}\sin nx\, dx \quad \ldots(1)$$

which is the required reduction formula.

Note 1. To evaluate the integral completely, if m is a positive integer.

***Why this step.** In order to make $\int \frac{1}{(\log x)^n}\, dx$ integrable, we take the differential coefficient of $\log x$, *viz.*, $\frac{1}{x}$ with it, [From $\int [f(x)]^n f'(x)\, dx$]

and adjust the power of x, thus getting $\int \frac{x^m}{(\log x)^n}\, dx = \int x^{m+1}\left[\frac{1}{(\log x)^n}\,\frac{1}{x}\right] dx.$

Case I. If m is a positive odd integer, then from (1), ultimately we are left with $\int x^1 \sin nx\, dx$* $= \int x \sin nx\, dx$ [Integrate by parts]

$$= x\left(-\frac{\cos nx}{n}\right) - \int 1\left(-\frac{\cos nx}{n}\right) dx = -\frac{x \cos nx}{n} + \frac{1}{n}\int \cos nx\, dx$$

$$= -\frac{x \cos nx}{n} + \frac{1}{n}\frac{\sin nx}{n} = -\frac{x \cos nx}{n} + \frac{\sin nx}{n^2}.$$

Case II. If m is a positive even integer, then from (1), ultimately we are left with $\int x^0 \sin nx\, dx$† $= \int \sin nx\, dx = -\frac{\cos nx}{n}$.

Note 2. Method to find a reduction formula for $\int x^m \sin nx\, dx$ or $\int x^m \cos nx\, dx$. *Integrate twice by parts.*

Note 3. Important. *In order to find a reduction formula when a sine or a cosine or a power of a sine or of a cosine is a factor of the product, it is generally useful to integrate twice by parts* (Art. 65, 66).

Note 4. Abridged notation for the reduction formula.

If $I_{m,n} = \int x^m \sin nx\, dx$, *then changing* m to $m - 2$,

$$I_{m-2,n} = \int x^{m-2} \sin nx\, dx,$$

$\therefore$ from (1),

$$I_{m,n} = -\frac{x^m \cos nx}{n} + \frac{m}{n^2} x^{m-1} \sin nx - \frac{m(m-1)}{n^2} I_{m-2,n}.$$

EXAMPLES

1. Find a reduction formula for $\int x^n \sin x\, dx$.

2. If $u_n = \int_0^{\frac{\pi}{2}} x^n \sin x\, dx$, and $n > 1$, prove that

$$u_n + n(n-1)\, u_{n-2} = n\left(\frac{\pi}{2}\right)^{n-1}.$$

Hence evaluate (*a*) $u_3 \int_0^{\frac{\pi}{2}} x^3 \sin x\, dx$

(*b*) $\int_0^{\frac{\pi}{2}} x^5 \sin x\, dx$.

***How to write this step.** m being a positive odd integer, the indices of x in the integrals on the R.H.S., *viz.*, $m - 2$ (and subtracting 2 each time), $m - 4$, ..., are all positive odd integers,

$\therefore$ ultimately we are left with $\int x^1 \sin nx\, dx$.

†How to write this step. m being a positive even integer, the indices of x in the integrals on the R.H.S., *viz.*, $m - 2$ (and subtracting 2 each time), $m - 4$, ..., are all positive even integers,

$\therefore$ ultimately we are left with $\int x^0 \sin nx\, dx$.

3. Find a reduction formula for $\int x^m \cos nx\, dx$.

**Show how to evaluate the integral completely, if m is a positive integer. [Proceed as in Art. 65 and Note 1.]

Reduction formulae for $\int e^{ax} \sin^n bx\, dx$ and $\int e^{ax} \cos^n bx\, dx$

****66. Reduction formula for $\int e^{ax} \sin^n bx\, dx$. To prove that $\int e^{ax} \sin^n bx\, dx$**

$$= \frac{a \sin bx - nb \cos bx}{a^2 + n^2b^2} e^{ax} \sin^{n-1} bx + \frac{n(n-1)b^2}{a^2 + n^2 b^2} \int e^{ax} \sin^{n-2} bx\, dx.$$

$\int e^{ax} \sin^n bx\, dx = \int \sin^n bx\, e^{ax}\, dx$ [Integrate by parts]

$$= \sin^n bx \,.\, \frac{e^{ax}}{a} - \int (n \sin^{n-1} bx \cos bx \,.\, b) \frac{e^{ax}}{a}\, dx$$

$$= \frac{e^{ax} \sin^n bx}{a} - \frac{nb}{a} \int (\sin^{n-1} bx \cos bx)\, e^{ax}\, dx$$

[Integrate again by parts]

$$= \frac{e^{ax} \sin^n bx}{a} - \frac{nb}{a} \left[(\sin^{n-1} bx \cos bx) . \frac{e^{ax}}{a} - \int \{(n-1) \sin^{n-2} bx \cos bx \right.$$

$$\left. . \, b (\cos bx) + \sin^{n-1} bx\, (-\sin bx \,.\, b)\} \frac{e^{ax}}{a}\, dx \right] = \frac{e^{ax} \sin^n bx}{a} - \frac{nb}{a}$$

$$\left[\frac{e^{ax} \sin^{n-1} bx \cos bx}{a} - \int \{(n-1)\, b \sin^{n-2} bx \cos^2 bx - b \sin^n bx\} \frac{e^{ax}}{a}\, dx \right]$$

$$= \frac{e^{ax} \sin^n bx}{a} - \frac{nb}{a} \left[\frac{e^{ax} \sin^{n-1} bx \cos bx}{a} \right.$$

$$\left. - \int \{(n-1)\, b \sin^{n-2} bx (1 - \sin^2 bx)^* - b \sin^n bx\} \frac{e^{ax}}{a}\, dx \right]$$

$$= \frac{e^{ax} \sin^n bx}{a} - \frac{nb}{a} \left[\frac{e^{ax} \sin^{n-1} bx \cos bx}{a} \right.$$

$$\left. - \int \{(n-1) b \sin^{n-2} bx - nb \sin^n bx\} \frac{e^{ax}}{a}\, dx \right]$$

***Why this step.** In the two expressions $e^{ax} \sin^n bx$ and $e^{ax} \sin^{n-2} bx$, whose integrals are to be connected, we don't require $\cos^2 bx$, but require sines only,

$\therefore$ we replace $\cos^2 bx$ by $1 - \sin^2 bx$.

$$= \frac{e^{ax}\sin^n bx}{a} - \frac{nb}{a}\left[\frac{e^{ax}\sin^{n-1} bx\cos bx}{a} - \frac{(n-1)b}{a}\int e^{ax}\sin^{n-2} bx\, dx + \frac{nb}{a}\int e^{ax}\sin^n bx\, dx\right]$$

$$= \frac{e^{ax}\sin^n bx}{a} - \frac{nb}{a^2}e^{ax}\sin^{n-1} bx\cos bx + \frac{n(n-1)b^2}{a^2}\int e^{ax}\sin^{n-2} bx\, dx - \frac{n^2b^2}{a^2}\int e^{ax}\sin^n bx\, dx.$$

Transposing,

$$\left(1+\frac{n^2b^2}{a^2}\right)\int e^{ax}\sin^n bx\, dx$$

$$= \frac{ae^{ax}\sin^n bx - nbe^{ax}\sin^{n-1} bx\cos bx}{a^2} + \frac{n(n-1)b^2}{a^2}\int e^{ax}\sin^{n-2} bx\, dx$$

$$\therefore \quad \int e^{ax}\sin^n bx\, dx$$

$$= \frac{e^{ax}\sin^{n-1} bx\,(a\sin bx - nb\cos bx)}{a^2\left(1+\dfrac{n^2b^2}{a^2}\right)} + \frac{n(n-1)b^2}{a^2\left(1+\dfrac{n^2b^2}{a^2}\right)}\int e^{ax}\sin^{n-2} bx\, dx$$

$$= \frac{e^{ax}\sin^{n-1} bx\,(a\sin bx - nb\cos bx)}{a^2+n^2b^2} + \frac{n(n-1)b^2}{a^2+n^2b^2}\int e^{ax}\sin^{n-2} bx\, dx.$$

Note. Method to find a reduction formula for $\int e^{ax}\sin^n bx\, dx$.

Integrate twice by parts, replace $\cos^2 bx$ by $1 - \sin^2 bx$, and transpose.

Similarly we have the

Method to find a reduction formula for $\int e^{ax}\cos^n bx\, dx$.

Integrate twice by parts, replace $\sin^2 bx$ by $1 - \cos^2 bx$, and transpose.

EXAMPLES

1. Find a reduction formula for $\int_0^{\pi} e^{-x}\sin^n x\, dx$.

2. (*i*) Find a formula of reduction for $\int e^{ax}\sin^n x\, dx$.

(*ii*) Hence evaluate $\int e^{ax}\sin^3 x\, dx$.

****3.** Reduction formula for $\int e^{ax}\cos^n bx\, dx$. Prove that $\int e^{ax}\cos^n bx\, dx$

$$= \frac{a\cos bx + nb\sin bx}{a^2+n^2b^2}\, e^{ax}\cos^{n-1} bx + \frac{n(n-1)b^2}{a^2+n^2b^2}\int e^{ax}\cos^{n-2} bx\, dx.$$

4. (*i*) Find a reduction formula for $\int e^{ax}\cos^n x\, dx$, where n is a positive integer.

(*ii*) Evaluate $\int e^{ax}\cos^4 x\, dx$.

**5. (*i*) Reduction formula for $\int \cos^m x \cos nx\, dx$.

If $\quad I_{m,n} = \int \cos^m x \cos nx\, dx$, prove that

$$I_{m,n} = -\frac{\cos^{m-1} x\,(n \cos x \sin nx - m \sin x \cos nx)}{m^2 - n^2} + \frac{m(m-1)}{m^2 - n^2} I_{m-2,n}.$$

(*ii*) Show that

$$\Big[I_{m,n} \Big]_0^{\frac{\pi}{2}} = \frac{m(m-1)}{m^2 - n^2} \Big[I_{m-2,n} \Big]_0^{\frac{\pi}{2}}.$$

[(*i*) Integrate twice by parts as in Art. 66.]

Note. *The above reduction formula connects* $I_{m,n}$ *with* $I_{m-2,n}$. *Another reduction formula connecting* $I_{m,n}$ *with* $I_{m-1,n-1}$ *is given in Ex. 2, (i), Art. 76-I.*

Reduction formulae found by the rule of "smaller index + 1"

67. Rule of "smaller index + 1" to connect $\int x^m (a + bx^n)^p\, dx$ with a given integral of the same type :

(*i*) *Let* $P = x^{\lambda+1} (a + bx^n)^{\mu+1}$, *where* λ *and* μ *are the smaller indices of x and* $a + bx^n$ *respectively in the two expressions whose integrals are to be connected.*

(*ii*) *Find* $\frac{dP}{dx}$, *and rearrange it as a linear function of the two expressions whose integrals are to be connected.*

(*iii*) *Integrate both sides w.r.t. x, transpose, and solve for the integral given to be connected.*

Example 1. *Connect* $\int x^m (a + bx^n)^p\, dx$ *with* $\int x^m (a + bx^n)^{p-1}\, dx$.

We have to connect $\int x^m (a + bx^n)^p\, dx$ with $\int x^m (a + bx^n)^{p-1}\, dx$.

(*i*) Let $P = x^{m+1} (a + bx^n)^{p-1+1}$ [Rule of "smaller index + 1" (Art. 67)]

$= x^{m+1} (a + bx^n)^p$.

(*ii*) $\frac{dP}{dx} = (m+1)x^m \cdot (a + bx^n)^p + x^{m+1} \cdot p(a + bx^n)^{p-1}\, bnx^{n-1}$

$= (m+1)\, x^m (a + bx^n)^p + pbnx^{m+n} (a + bx^n)^{p-1}$

$= (m+1)\, x^m (a + bx^n)^p + pbnx^m \cdot x^{n*} (a + bx^n)^{p-1}$

---

*__Why this step.__ In the two expressions $x^m (a + bx^n)^p$ and $x^m (a + bx^n)^{p-1}$, whose integrals are to be connected, we don't require x^{m+n}, but require x^m, $\therefore$ we replace x^{m+n} by $x^m \cdot x^n$; again we don't require x^n but require $a + bx^n$, $\therefore$ we replace bx^n by $\overline{a + bx^n} - a$.

$$= (m + 1)\, x^m (a + bx^n)^p + pnx^m(\overline{a + bx^n - a})(a + bx^n)^{p-1} \quad \textit{[Note this step]}$$

$$= (m + 1)\, x^m (a + bx^n)^p + pnx^m(a + bx^n)^p - pnax^m(a + bx^n)^{p-1}$$

$$= (m + 1 + pn)x^m (a + bx^n)^p - pnax^m(a + bx^n)^{p-1}.$$

(*iii*) Integrating both sides w.r.t. x,

$$P = (m + 1 + pn) \int x^m (a + bx^n)^p \, dx - pna \int x^m (a + bx^n)^{p-1} \, dx.$$

Transposing,

$$(m + 1 + pn) \int x^m (a + bx^n)^p \, dx = P + pna \int x^m(a + bx^n)^{p-1} \, dx.$$

$$\therefore \quad \int x^m (a + bx^n)^p \, dx$$

$$= \frac{P}{m+1+pn} + \frac{pna}{m+1+pn} \int x^m (a + bx^n)^{p-1} \, dx \quad \text{[Substitute the value of P]}$$

$$= \frac{x^{m+1} (a + bx^n)^p}{m+1+pn} + \frac{pna}{m+1+pn} \int x^m (a + bx^n)^{p-1} \, dx \quad \ldots(1)$$

Note. *Changing m to $m - 1$,*

$$\int x^{m-1} (a + bx^n)^p \, dx = \frac{x^m (a + bx^n)^p}{m + pn} + \frac{pna}{m + pn} \int x^{m-1} (a + bx^n)^{p-1} \, dx,$$

in which the coefficients (involving $m + pn$) are simpler than the corresponding coefficients in (1). Hence sometimes $\int x^{m-1} (a + bx^n)^p \, dx$ is discussed instead of $\int x^m(a + bx^n)^p \, dx$. (See solved Ex. 2 following and Exs. 1—6, Art. 67.)

Example 2. *What are the integrals with which $\int x^{m-1}(a + bx^n)^p \, dx$ can be connected by reduction formulae ?*

$\int x^{m-1} (a + bx^n)^p \, dx$ can be connected with any of the following six integrals :

(1) with $\int x^{m-1}(a + bx^n)^{p-1} \, dx$, *i.e., the index of $a + bx^n$ can be decreased by 1, that of x remaining unchanged ;*

or (2) with $\int x^{m-1}(a + bx^n)^{p+1} \, dx$, *i.e., the index of $a + bx^n$ can be increased by 1, that of x remaining unchanged ;*

or (3) with $\int x^{m-n-1}(a + bx^n)^p \, dx$, *i.e., the index of x can be decreased by n, that of $a + bx^n$ remaining unchanged ;*

or (4) with $\int x^{m+n-1}(a + bx^n)^p \, dx$, *i.e., the index of x can be increased by n, that of $a + bx^n$ remaining unchanged ;*

or (5) with $\int x^{m+n-1}(a + bx^n)^{p-1} \, dx$, *i.e., the index of $a + bx^n$ can be decreased by 1, and that of x increased by n at the same time ;*

or (6) with $\int x^{m-n-1}(a + bx^n)^{p+1} \, dx$, *i.e., the index of $a + bx^n$ can be increased by 1, and that of x decreased by n at the same time.*

[**Aid to memory.** In finding a reduction formula for $\int x^m(a+bx^n)^p\,dx$ or $\int x^{m-1}(a+bx^n)^p\,dx$,

(*i*) the index of $a + bx^n$ can be decreased by 1 or increased by 1, that of x remaining unchanged (this gives two reduction formulae) ;

or (*ii*) the index of x can be decreased by n or increased by n, that of $(a + bx^n)$ remaining unchanged (this gives two other reduction formulae) ;

or (*iii*) the index of $a + bx^n$ can be decreased by 1 and that of x increased by n, or the index of $a + bx^n$ can be increased by 1 and that of x decreased by n (this gives the remaining two reduction formulae).]

Caution. The indices of $a + bx^n$ and x cannot both be decreased (by 1 and n respectively) or both increased (by 1 and n respectively) in the same operation.

Example 3. *Find* $\int (x^2+a^2)^{\frac{5}{2}}\,dx$.

Let us connect $\int (x^2+a^2)^{\frac{n}{2}}\,dx$ with $\int (x^2+a^2)^{\frac{n}{2}-1}\,dx$, * *i.e.*,

$$\int x^0\,(x^2+a^2)^{\frac{n}{2}}\,dx \qquad \left[\text{Form} \int x^m\,(a+bx^n)^p\,dx\right]$$

with
$$\int x^0\,(x^2+a^2)^{\frac{n}{2}-1}\,dx.$$

(*i*) Let $P = x^{0+1}\,(x^2+a^2)^{\frac{n}{2}-1+1}$ [Rule of "smaller index + 1" (Art. 67)]

$$= x(x^2+a^2)^{\frac{n}{2}}.$$

(*ii*)
$$\frac{dP}{dx} = 1\,.\,(x^2+a^2)^{\frac{n}{2}} + x\,.\,\frac{n}{2}\,(x^2+a^2)^{\frac{n}{2}-1}\,(2x)$$

$$= (x^2+a^2)^{\frac{n}{2}} + nx^2\,(x^2+a^2)^{\frac{n}{2}-1}$$

$$= (x^2+a^2)^{\frac{n}{2}} + n(\overline{x^2+a^2}-a^2)\dagger\,(x^2+a^2)^{\frac{n}{2}-1} \qquad [\textit{Note this step}]$$

***Why this step.** In order to find $\int (x^2+a^2)^{\frac{5}{2}}\,dx$, we may connect $\int (x^2+a^2)^{\frac{5}{2}}\,dx$ with [$\int (x^2+a^2)^{\frac{5}{2}-1}\,dx$, *i.e.*, with] $\int (x^2+a^2)^{\frac{3}{2}}\,dx$, and this again with $\int (x^2+a^2)^{\frac{1}{2}}\,dx$, which is a standard form. *But instead of finding two reduction formulae,* we find only one (the general one) connecting $\int (x^2+a^2)^{\frac{n}{2}}\,dx$ with $\int (x^2+a^2)^{\frac{n}{2}-1}\,dx$, and *use it twice,* putting $n = 5$, and then $n = 3$.

† **Why this step.** In the two expressions $(x^2+a^2)^{\frac{n}{2}}$ and $(x^2+a^2)^{\frac{n}{2}-1}$, whose integrals are to be connected, we don't require x^2 but require $x^2 + a^2$,

$\therefore$ we replace x^2 by $\overline{x^2+a^2}-a^2$.

$$= (x^2 + a^2)^{\frac{n}{2}} + n(x^2 + a^2)^{\frac{n}{2}} - na^2(x^2 + a^2)^{\frac{n}{2}-1}$$

$$= (n + 1)(x^2 + a^2)^{\frac{n}{2}} - na^2(x^2 + a^2)^{\frac{n}{2}-1}.$$

(*iii*) Integrating both sides w.r.t. x,

$$P = (n + 1) \int (x^2 + a^2)^{\frac{n}{2}} dx - na^2 \int (x^2 + a^2)^{\frac{n}{2}-1} dx.$$

Transposing,

$$(n + 1) \int (x^2 + a^2)^{\frac{n}{2}} dx = P + na^2 \int (x^2 + a^2)^{\frac{n}{2}-1} dx$$

$$\therefore \quad \int (x^2 + a^2)^{\frac{n}{2}} dx = \frac{P}{n+1} + \frac{na^2}{n+1} \int (x^2 + a^2)^{\frac{n}{2}-1} dx$$

[Substitute the value of P]

$$= \frac{x(x^2 + a^2)^{\frac{n}{2}}}{n+1} + \frac{na^2}{n+1} \int (x^2 + a^2)^{\frac{n}{2}-1} dx \qquad \text{...(1)}$$

Putting $n = 5$ in (1),

$$\int (x^2 + a^2)^{\frac{5}{2}} dx = \frac{x(x^2 + a^2)^{\frac{5}{2}}}{6} + \frac{5a^2}{6} \int (x^2 + a^2)^{\frac{3}{2}} dx \qquad \text{...(2)}$$

Putting $n = 3$ in (1),

$$\int (x^2 + a^2)^{\frac{3}{2}} dx = \frac{x(x^2 + a^2)^{\frac{3}{2}}}{4} + \frac{3a^2}{4} \int (x^2 + a^2)^{\frac{1}{2}} dx \qquad \text{...(3)}$$

Now $\int (x^2 + a^2)^{\frac{1}{2}} dx$ $\qquad \left[\int \sqrt{x^2 + a^2}\, dx\right]$

$$= \frac{x(x^2 + a^2)^{\frac{1}{2}}}{2} + \frac{a^2}{2} \sinh^{-1} \frac{x}{a} \qquad \text{...(4)}$$

Substituting this value of $\int (x^2 + a^2)^{\frac{1}{2}} dx$ in (3),

$$\int (x^2 + a^2)^{\frac{3}{2}} dx = \frac{x(x^2 + a^2)^{\frac{3}{2}}}{4} + \frac{3a^2}{4} \left[\frac{x(x^2 + a^2)^{\frac{1}{2}}}{2} + \frac{a^2}{2} \sinh^{-1} \frac{x}{a}\right]$$

$$= \frac{x(x^2 + a^2)^{\frac{3}{2}}}{4} + \frac{3a^2}{4.2}\, x\, (x^2 + a^2)^{\frac{1}{2}} + \frac{3a^4}{4.2} \sinh^{-1} \frac{x}{a}.$$

Substituting this value of $\int (x^2 + a^2)^{\frac{3}{2}} dx$ in (2),

$$\int (x^2 + a^2)^{\frac{5}{2}} dx = \frac{x(x^2 + a^2)^{\frac{5}{2}}}{6} + \frac{5a^2}{6} \left[\frac{x(x^2 + a^2)^{\frac{3}{2}}}{4} + \frac{3a^2}{4.2} x(x^2 + a^2)^{\frac{1}{2}} + \frac{3a^4}{4.2} \sinh^{-1} \frac{x}{a}\right]$$

$$= \frac{x(x^2+a^2)^{\frac{5}{2}}}{6} + \frac{5}{6.4} a^2 x\,(x^2+a^2)^{\frac{3}{2}} + \frac{5.3}{6.4.2} a^4 x\,(x^2+a^2)^{\frac{1}{2}}$$

$$+ \frac{5.3}{6.4.2} a^6 \sinh^{-1}\frac{x}{a}.$$

Example 4. *Calculate the value of* $\int_0^{2a} x^m \sqrt{2ax - x^2}\, dx$ *(m being a positive integer) by the use of reduction formula.*

(*a*) [To find a reduction formula for $\int x^m \sqrt{2ax - x^2}\, dx$.]

$$\int x^m \sqrt{2ax-x^2}\, dx = \int x^{m+\frac{1}{2}} (2a-x)^{\frac{1}{2}}\, dx. \qquad \text{[Form } \textstyle\int x^m (a+bx^n)^p\, dx]$$

Let us connect $\int x^{m+\frac{1}{2}} (2a-x)^{\frac{1}{2}}\, dx$ with $\int x^{m-\frac{1}{2}*} (2a-x)^{\frac{1}{2}}\, dx$.

(*i*) Let $\quad P = x^{m-\frac{1}{2}+1} (2a-x)^{\frac{1}{2}+1}$ [Rule of "smaller index + 1" (Art. 67)]

$$= x^{m+\frac{1}{2}} (2a-x)^{\frac{3}{2}}.$$

(*ii*)
$$\frac{dP}{dx} = (m+\tfrac{1}{2})\, x^{m-\frac{1}{2}} . (2a-x)^{\frac{3}{2}} + x^{m+\frac{1}{2}} . \tfrac{3}{2} (2a-x)^{\frac{1}{2}} (-1)$$

$$= (m+\tfrac{1}{2})\, x^{m-\frac{1}{2}} (2a-x)(2a-x)^{\frac{1}{2}\dagger} - \tfrac{3}{2} x^{m+\frac{1}{2}} (2a-x)^{\frac{1}{2}} \qquad [\textit{Note this step}]$$

$$= (2m+1) a x^{m-\frac{1}{2}} (2a-x)^{\frac{1}{2}} - (m+\tfrac{1}{2})\, x^{m+\frac{1}{2}} (2a-x)^{\frac{1}{2}} - \tfrac{3}{2} x^{m+\frac{1}{2}} (2a-x)^{\frac{1}{2}}$$

$$= (2m+1) a x^{m-\frac{1}{2}} (2a-x)^{\frac{1}{2}} - (m+2) x^{m+\frac{1}{2}} (2a-x)^{\frac{1}{2}}.$$

(*iii*) Integrating both sides w.r.t. x,

$$P = (2m+1)a \int x^{m-\frac{1}{2}} (2a-x)^{\frac{1}{2}}\, dx - (m+2) \int x^{m+\frac{1}{2}} (2a-x)^{\frac{1}{2}}\, dx.$$

Transposing,

$$(m+2) \int x^{m+\frac{1}{2}} (2a-x)^{\frac{1}{2}}\, dx = -P + (2m+1)a \int x^{m-\frac{1}{2}} (2a-x)^{\frac{1}{2}} dx$$

***How to write this step.** Compare $\int x^{m+\frac{1}{2}} (2a-x)^{\frac{1}{2}}\, dx$ with $\int x^m (a+bx^n)^p\, dx$, here $n = 1$. Decrease the index of x by (n =)1, *i.e.*, connect $\int x^{m+\frac{1}{2}} (2a-x)^{\frac{1}{2}}\, dx$ with $\int x^{m+\frac{1}{2}-1} (2a-x)^{\frac{1}{2}}\, dx$.

[Solved Ex. 2, (3)]

† **Why this step.** In the two expressions $x^{m+\frac{1}{2}} (2a-x)^{\frac{1}{2}}$ and $x^{m-\frac{1}{2}} (2a-x)^{\frac{1}{2}}$, whose integrals are to be connected, we don't require $(2a-x)^{\frac{3}{2}}$ but require $(2a-x)^{\frac{1}{2}}$,

$\therefore$ we replace $(2a-x)^{\frac{3}{2}}$ by $(2a-x)\,(2a-x)^{\frac{1}{2}}$.

or $$\int x^{m+\frac{1}{2}}(2a-x)^{\frac{1}{2}}\,dx = -\frac{P}{m+2}+\frac{(2m+1)\,a}{m+2}\int x^{m-\frac{1}{2}}(2a-x)^{\frac{1}{2}}\,dx$$

[Substitute the value of P]

$$= -\frac{x^{m+\frac{1}{2}}(2a-x)^{\frac{3}{2}}}{m+2}+\frac{(2m+1)a}{m+2}\int x^{m-\frac{1}{2}}(2a-x)^{\frac{1}{2}}\,dx$$

[But $2a - x$ (Multiply and divide by x)

$$= \frac{2ax-x^2}{x}\ *\ \Big]$$

or $$\int x^{m+\frac{1}{2}}\frac{(2ax-x^2)^{\frac{1}{2}}}{x^{\frac{1}{2}}}\,dx = -\frac{x^{m+\frac{1}{2}}(2ax-x^2)^{\frac{3}{2}}}{(m+2)x^{\frac{3}{2}}}+\frac{(2m+1)a}{m+2}\int x^{m-\frac{1}{2}}\frac{(2ax-x^2)^{\frac{1}{2}}}{x^{\frac{1}{2}}}\,dx$$

$$\therefore\quad \int x^m\sqrt{2ax-x^2}\,dx = -\frac{x^{m-1}(2ax-x^2)^{\frac{3}{2}}}{m+2}+\frac{(2m+1)a}{m+2}\int x^{m-1}\sqrt{2ax-x^2}\,dx,$$

which is the required reduction formula.

(b) $\int_0^{2a} x^m\sqrt{2ax-x^2}\,dx$

$$= \left[-\frac{x^{m-1}(2ax-x^2)^{\frac{3}{2}}}{m+2}\right]_0^{2a}+\frac{(2m+1)a}{m+2}\int_0^{2a} x^{m-1}\sqrt{2ax-x^2}\,dx$$

$$= \frac{(2m+1)a}{m+2}\int_0^{2a} x^{m-1}\sqrt{2ax-x^2}\,dx \qquad \text{...(1)}$$

Changing m to $m-1$,

$$\int_0^{2a} x^{m-1}\sqrt{2ax-x^2}\,dx = \frac{(2m-1)a}{m+1}\int_0^{2a} x^{m-2}\sqrt{2ax-x^2}\,dx.$$

Substituting this value of $\int_0^{2a} x^{m-1}\sqrt{2ax-x^2}\,dx$ in (1),

$$\int_0^{2a} x^m\sqrt{2ax-x^2}\,dx = \frac{(2m+1)(2m-1)}{(m+2)(m+1)}a^2\int_0^{2a} x^{m-2}\sqrt{2ax-x^2}\,dx \qquad \text{...(2)}$$

***Why this step.** To put the reduction formula in the same form as the original integral $\int x^m\sqrt{2ax-x^2}\,dx$, we replace $2a-x$ by $\dfrac{2ax-x^2}{x}$.

Generalizing from (1) and (2),

$$\int_0^{2a} x^m \sqrt{2ax - x^2}\, dx$$

$$= \frac{(2m+1)(2m-1)\ldots \text{to } m \text{ factors}}{(m+2)(m+1)\ldots \text{to } m \text{ factors}}\, a^m \int_0^{2a} x^{m-m} \sqrt{2ax - x^2}\, dx$$

[$2m + 1, 2m - 1, \ldots$ is an A.P., $\therefore$ mth term $= 2m + 1 + (m - 1)(-2) = 3$,
$m + 2, m + 1, \ldots$ is an A.P., $\therefore$ mth term $= m + 2 + (m - 1)(-1) = 3$]

$$= \frac{(2m+1)(2m-1)\ldots 3}{(m+2)(m+1)\ldots 3}\, a^m \int_0^{2a} \sqrt{2ax - x^2}\, dx \qquad \ldots(3)$$

Now $\displaystyle\int_0^{2a} \sqrt{2ax - x^2}\, dx = \int_0^{2a} \sqrt{-(x^2 - 2ax)}\, dx$

$$= \int_0^{2a} \sqrt{-(x^2 - 2ax + a^2) + a^2}\, dx = \int_0^{2a} \sqrt{a^2 - (x-a)^2}\, dx$$

$$= \left[\frac{(x-a)\sqrt{a^2 - (x-a)^2}}{2} + \frac{a^2}{2}\sin^{-1}\frac{x-a}{a}\right]_0^{2a}$$

$$= \frac{a^2}{2}\sin^{-1} 1 - \left[\frac{a^2}{2}\sin^{-1}(-1)\right] = \frac{a^2}{2}\frac{\pi}{2} - \left[\frac{a^2}{2}\left(-\frac{\pi}{2}\right)\right] = \frac{a^2}{2}\pi.$$

$\therefore$ from (3),

$$\int_0^{2a} x^m \sqrt{2ax - x^2}\, dx$$

$$= \frac{(2m+1)(2m-1)\ldots 3}{(m+2)(m+1)\ldots 3}\, a^m \frac{a^2}{2}\pi = \frac{(2m+1)(2m-1)\ldots 3}{(m+2)(m+1)\ldots 3}\, a^{m+2}\frac{\pi}{2}.$$

EXAMPLES

1. If $X = a + bx^n$, connect $\int x^{m-1} X^p\, dx$ with $\int x^{m-1} X^{p-1}\, dx$.

[Substitute the value of X. We have to connect

$\int x^{m-1}(a + bx^n)^p\, dx$ with $\int x^{m-1}(a + bx^n)^{p-1}\, dx$.

Proceed as in solved Ex. 1, Art. 67. In the answer put $a + bx^n = X$.]

Note. *This reduction formula is useful when p is positive.*

2. Obtain the following reduction formula (where $X = a + bx^n$) :

$$\int x^{m-1} X^p\, dx = -\frac{x^m X^{p+1}}{na(p+1)} + \frac{m + pn + n}{na(p+1)} \int x^{m-1} X^{p+1}\, dx.$$

Note. *This reduction formula is useful when p is negative.*

3. Connect $\int x^{m-1}(a + bx^n)^p\, dx$ with $\int x^{m-n-1}(a + bx^n)^p\, dx$ and evaluate $\displaystyle\int \frac{x^8\, dx}{(1 - x^3)^{\frac{1}{3}}}$.

Note. *This reduction formula is useful when m is positive.*

4. If Z = $a + bx^n$, then show that

$$\int x^{m-1} Z^p \, dx = \frac{x^m Z^{p+1}}{am} - \frac{b(m+np+n)}{am} \int x^{m+n-1} Z^p \, dx.$$

Note. *This reduction formula is useful when m is negative.*

Obtain the following reduction formulae (Exs. 5—6) (where X = $a + bx^n$) :

5. $$\int x^{m-1} X^p \, dx = \frac{x^m X^p}{m} - \frac{pbn}{m} \int x^{m+n-1} X^{p-1} \, dx.$$

Note. *This reduction formula is useful when p is positive and m negative.*

6. $$\int x^{m-1} X^p \, dx = \frac{x^{m-n} X^{p+1}}{bn(p+1)} - \frac{m-n}{bn(p+1)} \int x^{m-n-1} X^{p+1} \, dx.$$

Note. *This reduction formula is useful when p is negative and m positive.*

7. Prove that

$$\int (a^2+x^2)^{\frac{2n+1}{2}} \, dx = \frac{x}{2n+2}(a^2+x^2)^{\frac{2n+1}{2}} + \frac{2n+1}{2n+2} a^2 \int (a^2+x^2)^{\frac{2n-1}{2}} \, dx.$$

8. (*i*) If m be a positive integer, find a reduction formula for $\int x^m \sqrt{2ax - x^2} \, dx.$

(*ii*) Hence obtain the value of $\int_0^{2a} x^3 \sqrt{2ax - x^2} \, dx.$

(*iii*) Use the formula obtained in (*i*) to show that, if m is a positive integer,

$$\int_0^{2a} x^m \sqrt{2ax - x^2} \, dx = \pi a^{m+2} \frac{2m+1!}{2^m \, m! \, m+2!}.$$

[As in solved Ex. 4,

$$\int_0^{2a} x^m \sqrt{2ax - x^2} \, dx = \pi a^{m+2} \frac{(2m+1)(2m-1)\ldots 3}{(m+2)(m+1)\ldots 3.2}$$

[Multiply the num. and denom. by $2m(2m-2)\ldots 2$*]

---

*__Why this step.__ As suggested by the factor $(2m+1)!$ in the numerator of the R.H.S. of the result to be proved, we make $(2m+1)(2m-1)\ldots 3$ also $(2m+1)!$ multiplying the numerator by $2m(2m-2)\ldots 2$, and multiply the denominator also these factors to adjust.

$$= \pi\, a^{m+2} \frac{(2m+1)!}{(m+2)(m+1)\ldots 2 \,.\, (2m)(2m-2)\ldots 2}$$

$$= \pi\, a^{m+2} \frac{(2m+1)!}{(m+2)(m+1)\ldots 2 \,.\, 2^m\, m(m-1)\ldots 1} = \pi\, a^{m+2} \frac{(2m+1)!}{2^m\, m!\,(m+2)!} \,. \Bigg]$$

9. Investigate a formula of reduction applicable to

$$\int x^m (1+x^2)^{\frac{n}{2}}\, dx$$

when m and n are positive integers, and complete the integration if $m = 5$, $n = 7$.

10. (*a*) Find a reduction formula for $\displaystyle\int \frac{x^m\, dx}{(a^3+x^3)^{\frac{n}{3}}}$.

(*b*) Find a reduction formula for $\displaystyle\int \frac{x^m\, dx}{(x^3-1)^{\frac{1}{3}}}$ and obtain the value of $\displaystyle\int x^8 (x^3-1)^{-\frac{1}{3}}\, dx$.

11. Obtain a reduction formula for $\displaystyle\int \frac{x^m}{\sqrt{ax^2+2bx+c}}\, dx$.

[(*i*) Let $P = x^{m-1}\sqrt{ax^2+2bx+c}$.*

(*ii*) $$\frac{dP}{dx} = (m-1)\, x^{m-2}\sqrt{ax^2+2bx+c} + x^{m-1} \frac{1}{2\sqrt{ax^2+2bx+c}}(2ax+2b)$$

$$= (m-1) \frac{x^{m-2}(ax^2+2bx+c)\dagger}{\sqrt{ax^2+2bx+c}} + \frac{x^{m-1}(ax+b)}{\sqrt{ax^2+2bx+c}}$$

$$= (m-1)a \frac{x^m}{\sqrt{ax^2+2bx+c}} + 2(m-1)b \frac{x^{m-1}}{2\sqrt{ax^2+2bx+c}}$$

***How to write this step.** In order to find a reduction formula for $\displaystyle\int \frac{x^m}{\sqrt{ax^2+c}}\, dx$, we would connect it with $\displaystyle\int \frac{x^{m-2}}{\sqrt{ax^2+c}}\, dx$ (solved Ex. 2, (3), Art. 67) and so we would let $P = x^{m-1}\sqrt{ax^2+c}$ (Rule of "smaller index + 1").

So here we let $P = x^{m-1}\sqrt{ax^2+2bx+c}$.

† **Why this step.** In the two expressions $\dfrac{x^m}{\sqrt{ax^2+2bx+c}}$ and $\dfrac{x^{m-2}}{\sqrt{ax^2+2bx+c}}$, whose integrals are to be connected, we don't require $\sqrt{ax^2+2bx+c}$ but require $\dfrac{1}{\sqrt{ax^2+2bx+c}}$,

$\therefore$ we replace $\sqrt{ax^2+2bx+c}$ by $\dfrac{ax^2+2bx+c}{\sqrt{ax^2+2bx+c}}$.

$$+ (m-1)c \frac{x^{m-2}}{\sqrt{ax^2+2bx+c}} + a \frac{x^m}{\sqrt{ax^2+2bx+c}} + b \frac{x^{m-1}}{\sqrt{ax^2+2bx+c}}$$

$$= ma \frac{x^m}{\sqrt{ax^2+2bx+c}} + (2m-1)b \frac{x^{m-1}}{\sqrt{ax^2+2bx+c}}$$

$$+ (m-1)c \frac{x^{m-2}}{\sqrt{ax^2+2bx+c}}.$$

(*iii*) Integrate both sides w.r.t. x.]

68. Rule of "smaller index + 1" to connect $\int \sin^p x \cos^q x \, dx$ with a given integral of the same type :

(i) Let $P = \sin^{\lambda+1} x \cos^{\mu+1} x$, *where* λ *and* μ *are the smaller indices of* $\sin x$ *and* $\cos x$ *respectively in the two expressions whose integrals are to be connected.*

(ii) Find $\frac{dP}{dx}$, *and rearrange it as a linear function of the two expressions whose integrals are to be connected.*

(iii) Integrate both sides w.r.t. x, transpose, and solve for the integral given to be connected.

Example 1. *Connect* $\int \sin^p x \cos^q x \, dx$ *with* $\int \sin^{p-2} x \cos^q x \, dx$.

We have to connect $\int \sin^p x \cos^q x \, dx$ with $\int \sin^{p-2} x \cos^q x \, dx$.

(*i*) Let $P = \sin^{p-2+1} x \cos^{q+1} x$ [Rule of "smaller index + 1" (Art. 68)]

$$= \sin^{p-1} x \cos^{q+1} x.$$

(*ii*)
$$\frac{dP}{dx} = (p-1)\sin^{p-2} x \cos x \,.\, \cos^{q+1} x + \sin^{p-1} x \,.\, (q+1)\cos^q x(-\sin x)$$

$$= (p-1)\sin^{p-2} x \cos^{q+2} x - (q+1)\sin^p x \cos^q x$$

$$= (p-1)\sin^{p-2} x \cos^q x (1-\sin^2 x)^* - (q+1)\sin^p x \cos^q x$$

$$= (p-1)\sin^{p-2} x \cos^q x - (p-1)\sin^p x \cos^q x - (q+1)\sin^p x \cos^q x$$

$$= (p-1)\sin^{p-2} x \cos^q x - (p+q)\sin^p x \cos^q x.$$

(*iii*) Integrating both sides w.r.t. x,

$$P = (p-1)\int \sin^{p-2} x \cos^q x \, dx - (p+q)\int \sin^p x \cos^q x \, dx.$$

Transposing,

$$(p+q)\int \sin^p x \cos^q x \, dx = -P + (p-1)\int \sin^{p-2} x \cos^q x \, dx.$$

***Why this step.** In the two expressions $\sin^p x \cos^q x$ and $\sin^{p-2} x \cos^q x$, whose integrals are to be connected, we don't require $\cos^{q+2} x$ but require $\cos^q x$,

$\therefore$ we replace $\cos^{q+2} x$ by $(\cos^q x . \cos^2 x =) \cos^q x (1 - \sin^2 x)$.

$$\therefore \int \sin^p x \cos^q x\, dx = -\frac{P}{p+q} + \frac{p-1}{p+q}\int \sin^{p-2} x \cos^q x\, dx \quad \text{[Substitute the value of P]}$$

$$= -\frac{\sin^{p-1} x \cos^{q+1} x}{p+q} + \frac{p-1}{p+q}\int \sin^{p-2} x \cos^q x\, dx \qquad ...(1)$$

Note 1. *This reduction formula is useful when p is positive.*

Note 2. *Abridged notation for the reduction formula.*

If $I_{p,q} = \int \sin^p x \cos^q x\, dx$, then changing p to $p-2$,

$$I_{p-2,q} = \int \sin^{p-2} x \cos^q x\, dx,$$

$\therefore$ from (1),

$$I_{p,q} = -\frac{\sin^{p-1} x \cos^{q+1} x}{p+q} + \frac{p-1}{p+q} I_{p-2,q}.$$

Example 2. *What are the integrals with which* $\int \sin^p x \cos^q x\, dx$ *can be connected by reduction formulae ?*

$\int \sin^p x \cos^q x\, dx$ can be connected with any of the following six integrals :

(1) with $\int \sin^{p-2} x \cos^q x\, dx$, *i.e., the index of sin x can be decreased by 2, that of cos x remaining unchanged ;*

or (2) with $\int \sin^{p+2} x \cos^q x\, dx$, *i.e., the index of sin x can be increased by 2, that of cos x remaining unchanged ;*

or (3) with $\int \sin^p x \cos^{q-2} x\, dx$, *i.e., the index of cos x can be decreased by 2, that of sin x remaining unchanged ;*

or (4) with $\int \sin^p x \cos^{q+2} x\, dx$, *i.e., the index of cos x can be increased by 2, that of sin x remaining unchanged ;*

or (5) with $\int \sin^{p-2} x \cos^{q+2} x\, dx$, *i.e., the index of sin x can be decreased by 2 and that of cos x increased by 2 ; at the same time ;*

or (6) with $\int \sin^{p+2} x \cos^{q-2} x\, dx$, *i.e., the index of sin x can be increased by 2 and that of cos x decreased by 2 ; at the same time.*

[**Aid to memory.** In finding a reduction formula for $\int \sin^p x \cos^q x\, dx$,

(*i*) the index of sin x can be decreased by 2 or increased by 2, that of cos x remaining unchanged (this gives two reduction formulae) ;

or (*ii*) the index of cos x can be decreased by 2 or increased by 2, that of sin x remaining unchanged (this gives two other reduction formulae) ;

or (*iii*) the index of sin x can be decreased by 2 and that of cos x increased by 2, or the index of sin x can be increased by 2 and that of cos x decreased by 2 (this gives the remaining two reduction formulae).]

Caution. The indices of sin x and cos x cannot both be decreased by 2 or both increased by 2 in the same operation.

Note. Rule to integrate $\int \sin^p x \cos^q x \, dx$:

(i) If either p or q is a positive odd integer, use the substitution method, viz., put cos x = z or sin x = z as the case may require (Art. 54).

(ii) If both p and q are positive even integers, use the multiple-angle method (Art. 55) or the reduction formula (Art. 68).

(iii) If p + q is a negative even integer, use the substitution method, viz., put tan x = z (Art. 56).

(iv) In all other cases use the reduction formula (Art. 68).

Caution. We must warn the student against the stupidity of applying a reduction formula when there is an obvious short cut ; to use a reduction formula in such a case to evaluate the integral would be much like multiplying 700 and 800 together by logarithms.

—*Osgood.*

EXAMPLES

1. (*a*) Obtain a reduction formula for $\int \sin^m x \cos^n x \, dx$, where m and n are positive integers.

Hence integrate $\int \sin^4 x \cos^2 x \, dx$.

(*b*) Integrate $\int \dfrac{\sin^4 x}{\cos^2 x} dx$.

2. Connect $\int \sin^p x \cos^q x \, dx$ with $\int \sin^{p+2} x \cos^q x \, dx$.

Note. *This reduction formula is useful when p is negative.*

3. Connect $\int \sin^p x \cos^q x \, dx$ with $\int \sin^p x \cos^{q-2} x \, dx$.

Note. *This reduction formula is useful when q is positive.*

4. Connect $\int \sin^p x \cos^q x \, dx$ with $\int \sin^p x \cos^{q+2} x \, dx$.

Note. *This reduction formula is useful when q is negative.*

5. Connect $\int \sin^p x \cos^q x \, dx$ with $\int \sin^{p-2} x \cos^{q+2} x \, dx$.

Note. *This reduction formula is useful when p is positive and q negative.*

6. Show that

$$\int \sin^p x \cos^q x \, dx = \frac{\sin^{p+1} x \cos^{q-1} x}{p+1} + \frac{q-1}{p+1} \int \sin^{p+2} x \cos^{q-2} x \, dx.$$

Note. *This reduction formula is useful when p is negative and q positive.*

7. Evaluate $\int \dfrac{dx}{\sin x \cos^4 x}$.

Reduction formulae for $\int \sin^n x \, dx$ and $\int \cos^n x \, dx$

69-I. To find a reduction formula for $\int \sin^n x \, dx$.

Let us connect $\int \sin^n x \, dx$ with $\int \sin^{n-2} x \, dx$,

i.e., $\int \sin^n x \cos^0 x \, dx$ [Form $\int \sin^p x \cos^q x \, dx$]

with $\int \sin^{n-2} x \cos^0 x \, dx$.

(*i*) Let $P = \sin^{n-2+1} x \cos^{0+1} x$. [Rule of "smaller index + 1" (Art. 68)]

$= \sin^{n-1} x \cos x$.

(*ii*) $$\frac{dP}{dx} = (n-1) \sin^{n-2} x \cos x \cdot \cos x + \sin^{n-1} x (-\sin x)$$

$$= (n-1) \sin^{n-2} x \cos^2 x - \sin^n x$$

$$= (n-1) \sin^{n-2} x (1 - \sin^2 x)^* - \sin^n x$$

$$= (n-1) \sin^{n-2} x - (n-1) \sin^n x - \sin^n x$$

$$= (n-1) \sin^{n-2} x - n \sin^n x.$$

(*iii*) Integrating both sides w.r.t. x,

$$P = (n-1) \int \sin^{n-2} x \, dx - n \int \sin^n x \, dx.$$

Transposing,

$$n \int \sin^n x \, dx = -P + (n-1) \int \sin^{n-2} x \, dx$$

$$\therefore \quad \int \sin^n x \, dx = -\frac{P}{n} + \frac{n-1}{n} \int \sin^{n-2} x \, dx$$ [Substitute the value of P]

$$= -\frac{\sin^{n-1} x \cos x}{n} + \frac{n-1}{n} \int \sin^{n-2} x \, dx \quad ...(1)$$

which is the required reduction formula.

Note 1. To evaluate the integral completely, if n is a positive even integer.

∵ *n is a positive even integer,*

∴ *from (1), ultimately we are left with*

$$\int \sin^0 x \, dx = \int 1 \, dx = x.$$

Note 2. Rule to integrate $\int \sin^n x \, dx$, where n is a positive integer :

(*i*) *If the index of sin x is a positive odd integer, use the substitution method, viz., put cos x = z (Art. 52).*

(*ii*) *If the index of sin x is a positive even integer, use the multiple-angle method (Art. 53) or the reduction formula (Art. 69-I).*

Note 3. Abridged notation for the reduction formula.

If $I_n = \int \sin^n x \, dx$, *then changing n to n – 2,*

$$I_{n-2} = \int \sin^{n-2} x \, dx,$$

∴ *from (1),*

$$I_n = -\frac{\sin^{n-1} x \cos x}{n} + \frac{n-1}{n} I_{n-2}.$$

***Why this step.** In the two expressions $\sin^n x$ and $\sin^{n-2} x$, whose integrals are to be connected, we don't require $\cos^2 x$ but require sines only,

∴ we replace $\cos^2 x$ by $1 - \sin^2 x$.

EXAMPLES

1. Prove that

$$\int \sin^n x\,dx = -\frac{\cos x \sin^{n-1} x}{n} + \frac{n-1}{n}\int \sin^{n-2} x\,dx.$$

Hence integrate (*i*) $\int \sin^4 x\,dx$. (*ii*) $\int \sin^6 x\,dx$.

2. Integrate $\int \sin^8 x\,dx$.

3. Establish a formula of reduction for $\int \cos^n x\,dx$.

[Let us connect $\int \cos^n x\,dx$ with $\int \cos^{n-2} x\,dx$,

i.e., $\int \sin^0 x \cos^n x\,dx$ $\left[\text{Form} \int \sin^p x \cos^q x\,dx\right]$

with $\int \sin^0 x \cos^{n-2} x\,dx$.

Use the rule of "smaller index + 1" (Art. 68).]

4. Show by the method of integration by parts,

$$\int \sin^n x\,dx = -\frac{\sin^{n-1} x \cos x}{n} + \frac{n-1}{n}\int \sin^{n-2} x\,dx.$$

$\left[\left[\int \sin^n x\,dx = \int \sin^{n-1} x \sin x\,dx\right.\right.$ [Integrate by parts]

$$= \sin^{n-1} x(-\cos x) - \int (n-1)\sin^{n-2} x \cos x(-\cos x)\,dx$$

$$= -\sin^{n-1} x \cos x + (n-1)\int \sin^{n-2} x \cos^2 x\,dx$$

$$= -\sin^{n-1} x \cos x + (n-1)\int \sin^{n-2} x(1-\sin^2 x)^{*}\,dx$$

$$= -\sin^{n-1} x \cos x + (n-1)\int \sin^{n-2} x\,dx - (n-1)\int \sin^n x\,dx.$$

Transpose.

Note 1. Method to find a reduction formula for $\int \sin^n x\,dx$ by integration by parts.

Separate sin x, integrate by parts, replace $\cos^2 x$ by $1 - \sin^2 x$, and transpose, and solve for the integral given to be connected.]

Note 2. *This method is also applicable to find a reduction formula for $\int x \sin^n x\,dx$. (See Misc. Ex. 14, Chap. VII.)]*

69-II. To evaluate $\int_0^{\frac{\pi}{2}} \sin^n x\,dx$, where n is a positive integer.

Let us connect $\int \sin^n x\,dx$ with $\int \sin^{n-2} x\,dx$, *i.e.*, $\int \sin^n x \cos^0 x\,dx$

[Form $\int \sin^p x \cos^q x\,dx$]

with $\int \sin^{n-2} x \cos^0 x\,dx$.]

(*i*) Let $P = \sin^{n-2+1} x \cos^{0+1} x$ [Rule of "smaller index + 1" (Art. 68)]

$= \sin^{n-1} x \cos x.$

***Why this step.** See the footnote on p. 144.

(*ii*)
$$\frac{dP}{dx} = (n-1)\sin^{n-2}x\cos x.\cos x + \sin^{n-1}x(-\sin x)$$
$$= (n-1)\sin^{n-2}x\cos^2 x - \sin^n x$$
$$= (n-1)\sin^{n-2}x(1-\sin^2 x)^* - \sin^n x$$
$$= (n-1)\sin^{n-2}x - (n-1)\sin^n x - \sin^n x$$
$$= (n-1)\sin^{n-2}x - n\sin^n x.$$

(*iii*) **Integrating both sides w.r.t. x,**

$$P = (n-1)\int \sin^{n-2}x\,dx - n\int \sin^n x\,dx.$$

Transposing,

$$n\int \sin^n x\,dx = -P + (n-1)\int \sin^{n-2}x\,dx$$

or
$$\int \sin^n x\,dx = -\frac{P}{n} + \frac{n-1}{n}\int \sin^{n-2}x\,dx \qquad \text{[Substitute the value of P]}$$
$$= -\frac{\sin^{n-1}x\cos x}{n} + \frac{n-1}{n}\int \sin^{n-2}x\,dx.$$

$\therefore$
$$\int_0^{\frac{\pi}{2}} \sin^n x\,dx = \left[-\frac{\sin^{n-1}x\cos x}{n}\right]_0^{\frac{\pi}{2}} + \frac{n-1}{n}\int_0^{\frac{\pi}{2}} \sin^{n-2}x\,dx$$

[But when $x = \frac{\pi}{2}$, $\sin^{n-1}x\cos x = 0$, and when $x = 0$, $\sin^{n-1}x\cos x = 0$, if n is a + ve integer > 1]

$$= \frac{n-1}{n}\int_0^{\frac{\pi}{2}} \sin^{n-2}x\,dx \qquad ...(1)$$

Changing n to $n-2$,

$$\int_0^{\frac{\pi}{2}} \sin^{n-2}x\,dx = \frac{n-3}{n-2}\int_0^{\frac{\pi}{2}} \sin^{n-4}x\,dx.$$

Substituting this value of $\int_0^{\frac{\pi}{2}} \sin^{n-2}x\,dx$ in (1),

$$\int_0^{\frac{\pi}{2}} \sin^n x\,dx = \frac{(n-1)(n-3)}{n(n-2)}\int_0^{\frac{\pi}{2}} \sin^{n-4}x\,dx \qquad ...(2)$$

---

*__Why this step.__ See the footnote on p. 144.

Generalizing from (1) and (2),

Case I. If n is a + ve odd integer,

$$\int_0^{\frac{\pi}{2}} \sin^n x\, dx = \frac{(n-1)(n-3)\ldots 2^*}{n(n-2)\ldots 3} \int_0^{\frac{\pi}{2}} \sin^1 x\, dx$$

$$\left[\text{But } \int_0^{\frac{\pi}{2}} \sin^1 x\, dx = \int_0^{\frac{\pi}{2}} \sin x\, dx = \Big[-\cos x\Big]_0^{\frac{\pi}{2}} = -[0-1] = 1\right]$$

$$= \frac{(n-1)(n-3)\ldots 2}{n(n-2)\ldots 3}.$$

Case II. If n is a + ve even integer,

$$\int_0^{\frac{\pi}{2}} \sin^n x\, dx = \frac{(n-1)(n-3)\ldots 1\dagger}{n(n-2)\ldots 2} \int_0^{\frac{\pi}{2}} \sin^0 x\, dx$$

$$\left[\text{But } \int_0^{\frac{\pi}{2}} \sin^0 x\, dx = \int_0^{\frac{\pi}{2}} 1\, dx = \Big[x\Big]_0^{\frac{\pi}{2}} = \frac{\pi}{2}\right]$$

$$= \frac{(n-1)(n-3)\ldots 1}{n(n-2)\ldots 2} \cdot \frac{\pi}{2}.$$

[**Rule to write down the value of $\int_0^{\frac{\pi}{2}} \sin^n x\, dx$, where n is a positive integer :**

---

*__How to write this step.__ n being a positive odd integer, $n-1$, $n-3$, ..., are all positive even integers,

$\therefore$ the last factor in the numerator = the least positive even integer = 2.

Now in the numerator, the first factor $n-1$ has become the last factor = 1

($\because$ $n = 3$), $\therefore$ in the denominator, the first factor n will become the last factor = 3.

Again, from (1) and (2), the index of sin x in the integral on the R.H.S. is less than the last factor in the numerator and is $\therefore$ = 2 – 1 = 1.

† **How to write this step.** n being a positive even integer, $n-1$, $n-3$, ..., are all positive odd integers,

$\therefore$ the last factor in the numerator = the least positive odd integer = 1.

Now in the numerator, the first factor $n-1$ has become the last factor = 1 ($\because$ $n = 2$), $\therefore$ in the denominator, the first factor n will become the last factor = 2.

Again, from (1) and (2), the index of sin x in the integral on the R.H.S. is less than the last factor in the numerator and is $\therefore$ = 1 – 1 = 0.

$$\int_0^{\frac{\pi}{2}} \sin^n x\,dx = \frac{(n-1)\times \text{go on diminishing by } 2\ldots}{n\times \text{go on diminishing by } 2\ldots}\cdot\frac{\pi}{2} \text{ only if } n \text{ is a positive even integer}$$

$\left(\textit{otherwise no } \frac{\pi}{2}\right)$, *each series of factors being continued so long as the factors are positive (i.e., omitting zero and negative factors).*]

Example. *Write down the values of*

(*i*) $\int_0^{\frac{\pi}{2}} \sin^8 \theta\,d\theta$. (*ii*) $\int_0^{\frac{\pi}{2}} \sin^7 \theta\,d\theta$.

(*i*) $\int_0^{\frac{\pi}{2}} \sin^8 \theta\,d\theta$ $\left[\text{Compare with } \int_0^{\frac{\pi}{2}} \sin^n \theta\,d\theta, \text{ here } n = 8\right]$

$= \frac{7.5.3.1}{8.6.4.2}\cdot\frac{\pi}{2}$ $\left[\frac{\pi}{2} (\because\ n = 8, \text{ a + ve even integer})\right]$

$\left[\frac{(n-1)\times \text{go on diminishing by } 2\ldots}{n\times \text{go on diminishing by } 2\ldots}\cdot\frac{\pi}{2} \text{ only if } n \text{ is a + ve even integer}\right]$

$= \frac{35}{256}\pi$.

(*ii*) $\int_0^{\frac{\pi}{2}} \sin^7 \theta\,d\theta$ $\left[\text{Compare with } \int_0^{\frac{\pi}{2}} \sin^n \theta\,d\theta, \text{ here } n = 7\right]$

$= \frac{6.4.2}{7.5.3.1}$ $\left[\text{no } \frac{\pi}{2} (\because\ n = 7, \text{ not a + ve even integer})\right]$

$\left[\frac{(n-1)\times \text{go on diminishing by } 2\ldots}{n\times \text{go on diminishing by } 2\ldots}\cdot\frac{\pi}{2} \text{ only if } n \text{ is a + ve even integer}\right]$

$= \frac{16}{35}$.

EXAMPLES

Write down the values of (Exs. 1 — 2) :

1. (*i*) $\int_0^{\frac{\pi}{2}} \sin^2 x\,dx$. (*ii*) $\int_0^{\frac{\pi}{2}} \sin^3 \theta\,d\theta$.

(*iii*) $\int_0^{\frac{\pi}{2}} \sin^4 x\,dx$.

2. (*i*) $\int_0^{\frac{\pi}{2}} \sin^5 x\, dx.$ (*ii*) $\int_0^{\frac{\pi}{2}} \sin^{10} x\, dx.$

3. Find the value of $\int_0^{\frac{\pi}{2}} \sin^{2m} \theta\, d\theta$ for + ve integral values of m.

Integrals reducible to $\int_0^{\frac{\pi}{2}} \sin^n x\, dx$ by a trigonometric substitution.

4. Show that (*i*) $\int_0^a \frac{x^4}{\sqrt{a^2 - x^2}}\, dx = \frac{3a^4 \pi}{16}$.

(*ii*) Find the value of $\int_0^1 \frac{x^4}{\sqrt{1-x^2}}\, dx$. Put $x = a \sin \theta$.

[**Rule to evaluate** $\int_0^a \frac{x^n}{\sqrt{a^2 - x^2}}\, dx$ **:** *Put* $x = a \sin \theta$.]

5. Investigate a formula of reduction for $\int \frac{x^{2n+1}\, dx}{(1-x^2)^{\frac{1}{2}}}$, and by means of this integral show that

$$\frac{1}{2n+2} + \frac{1}{2} \cdot \frac{1}{2n+4} + \frac{1.3}{2.4} \cdot \frac{1}{2n+6} + \frac{1.3.5}{2.4.6} \cdot \frac{1}{2n+8} + \ldots \text{ ad. inf.}$$
$$= \frac{2.4.6 \ldots 2n}{3.5.7 \ldots 2n+1}.$$

Sum also the series

$$\frac{1}{2n+1} + \frac{1}{2} \cdot \frac{1}{2n+3} + \frac{1.3}{2.4} \cdot \frac{1}{2n+5} + \frac{1.3.5}{2.4.6} \cdot \frac{1}{2n+7} + \ldots \text{ ad. inf.}$$

70. To evaluate $\int_0^{\frac{\pi}{2}} \cos^n x\, dx$, where n is a positive integer.

Let us connect $\int \cos^n x\, dx$ with $\int \cos^{n-2} x\, dx$,

i.e., $\int \sin^0 x \cos^n x\, dx$ [Form $\int \sin^p x \cos^q x\, dx$]

with $\int \sin^0 x \cos^{n-2} x\, dx.$

(*i*) Let $P = \sin^{0+1} x \cos^{n-2+1} x$ [Rule of "smaller index + 1" (Art. 68)]

$= \sin x \cos^{n-1} x.$

(*ii*) $\frac{dP}{dx} = \cos x \,.\, \cos^{n-1} x + \sin x.\, (n-1) \cos^{n-2} x\, (-\sin x)$

$= \cos^n x - (n-1) \sin^2 x \cos^{n-2} x$

$= \cos^n x - (n-1)(1 - \cos^2 x)^* \cos^{n-2} x.$

---

*__Why this step.__ In the two expressions $\cos^n x$ and $\cos^{n-2} x$, whose integrals are to be connected, we don't require $\sin^2 x$ but require cosines only,

$\therefore$ we replace $\sin^2 x$ by $1 - \cos^2 x$.

$$= \cos^n x - (n-1)\cos^{n-2} x + (n-1)\cos^n x$$

$$= n\cos^n x - (n-1)\cos^{n-2} x.$$

(*iii*) Integrating both sides w.r.t. x,

$$P = n\int \cos^n x\, dx - (n-1)\int \cos^{n-2} x\, dx.$$

Transposing,

$$n\int \cos^n x\, dx = P + (n-1)\int \cos^{n-2} x\, dx$$

or
$$\int \cos^n x\, dx = \frac{P}{n} + \frac{n-1}{n}\int \cos^{n-2} x\, dx \qquad \text{[Substitute the value of P]}$$

$$= \frac{\sin x\cos^{n-1} x}{n} + \frac{n-1}{n}\int \cos^{n-2} x\, dx.$$

$$\therefore \quad \int_0^{\frac{\pi}{2}} \cos^n x\, dx = \left[\frac{\sin x\cos^{n-1} x}{n}\right]_0^{\frac{\pi}{2}} + \frac{n-1}{n}\int_0^{\frac{\pi}{2}} \cos^{n-2} x\, dx$$

[But when $x = \frac{\pi}{2}$, $\sin x\cos^{n-1} x = 0$, if n is a + ve integer > 1,

and when $x = 0$, $\sin x\cos^{n-1} x = 0$]

$$= \frac{n-1}{n}\int_0^{\frac{\pi}{2}} \cos^{n-2} x\, dx \qquad ...(1)$$

Changing n to $n-2$,

$$\int_0^{\frac{\pi}{2}} \cos^{n-2} x\, dx = \frac{n-3}{n-2}\int_0^{\frac{\pi}{2}} \cos^{n-4} x\, dx.$$

Substituting this value of $\int_0^{\frac{\pi}{2}} \cos^{n-2} x\, dx$ in (1),

$$\int_0^{\frac{\pi}{2}} \cos^n x\, dx = \frac{(n-1)(n-3)}{n(n-2)}\int_0^{\frac{\pi}{2}} \cos^{n-4} x\, dx \qquad ...(2)$$

Generalizing from (1) and (2),

Case I. If n is a + ve odd integer,

$$\int_0^{\frac{\pi}{2}} \cos^n x\, dx = \frac{(n-1)(n-3)\ldots 2^*}{n(n-2)\ldots 3}\int_0^{\frac{\pi}{2}} \cos^1 x\, dx$$

***How to write this step.** See the footnote* on p. 147.

$$\left[\text{But } \int_0^{\frac{\pi}{2}} \cos^1 x\, dx = \int_0^{\frac{\pi}{2}} \cos x\, dx = \Big[\sin x\Big]_0^{\frac{\pi}{2}} = 1 - 0 = 1\right]$$

$$= \frac{(n-1)(n-3)\ldots 2}{n(n-2)\ldots 3}.$$

Case II. If n is a + ve even integer,

$$\int_0^{\frac{\pi}{2}} \cos^n x\, dx = \frac{(n-1)(n-3)\ldots 1^*}{n(n-2)\ldots 2} \int_0^{\frac{\pi}{2}} \cos^0 x\, dx$$

$$\left[\text{But } \int_0^{\frac{\pi}{2}} \cos^0 x\, dx = \int_0^{\frac{\pi}{2}} 1\, dx = \Big[x\Big]_0^{\frac{\pi}{2}} = \frac{\pi}{2}\right]$$

$$= \frac{(n-1)(n-3)\ldots 1}{n(n-2)\ldots 2} \cdot \frac{\pi}{2}.$$

[**Rule to write down the value of $\int_0^{\frac{\pi}{2}} \cos^n x\, dx$, where n is a positive integer :** *Same as the rule to write down the value of* $\int_0^{\frac{\pi}{2}} \sin^n x\, dx$. *(Rule of Art. 69-II)*]

EXAMPLES

1. Write down the values of

(*i*) $\int_0^{\frac{\pi}{2}} \cos^6 \phi\, d\phi$. (*ii*) $\int_0^{\frac{\pi}{2}} \cos^9 x\, dx$.

2. Evaluate (*i*) $\int_0^{\frac{\pi}{4}} \cos^2 2x\, dx$. (*ii*) $\int_0^{\frac{\pi}{6}} \cos^3 3\theta\, d\theta$.

Integrals reducible to $\int_0^{\frac{\pi}{2}} \cos^n x\, dx$ by a trigonometric substitution.

3. Evaluate

(*i*) $\int_0^{\infty} \frac{dx}{(1+x^2)^2}$. (*ii*) $\int_0^{\infty} \frac{dx}{(1+x^2)^{n+\frac{1}{2}}}$.

***How to write this step.** See the footnote on p. 147.

[Rule to evaluate $\int_0^{\infty} \frac{1}{(a^2+x^2)^n}\,dx$, where n is a positive integer ; or $\int_0^{\infty} \frac{1}{(a^2+x^2)^{\frac{n}{2}}}\,dx$, where n is a positive odd integer : *Put* $x = a \tan\theta$.]

4. Prove that

$$\frac{2}{\pi}\int_0^{\frac{\pi}{2}} \frac{dx}{(1-e^2\cos^2 x)^{\frac{1}{2}}} = 1 + \frac{1^2}{2^2}e^2 + \frac{1^2 . 3^2}{2^2 . 4^2}e^4 + \frac{1^2 . 3^2 . 5^2}{2^2 . 4^2 . 6^2}e^6 + \ldots,$$

where $e < 1$.

5. Prove that

$$\int_0^1 \frac{a\,dz}{1+a^2(1-z^2)} = a - \frac{2}{3}a^3 + \frac{2.4}{3.5}a^5 - \frac{2.4.6}{3.5.7}a^7 + \ldots .$$

71. To evaluate $\int_0^{\frac{\pi}{2}} \sin^p x \cos^q x\, dx$, where p and q are positive integers.

Let us connect $\int \sin^p x \cos^q x\, dx$ with $\int \sin^{p-2} x \cos^q x\, dx$.

(*i*) Let $\quad P = \sin^{p-2+1} x \cos^{q+1} x \quad$ [Rule of "smaller index + 1" (Art. 68)]

$$= \sin^{p-1} x \cos^{q+1} x.$$

(*ii*) $$\frac{dP}{dx} = (p-1)\sin^{p-2} x \cos x . \cos^{q+1} x + \sin^{p-1} x . (q+1)\cos^q x(-\sin x)$$

$$= (p-1)\sin^{p-2} x\cos^{q+2} x - (q+1)\sin^p x \cos^q x$$

$$= (p-1)\sin^{p-2} x \cos^q x(1-\sin^2 x)^* - (q+1)\sin^p x\cos^q x$$

$$= (p-1)\sin^{p-2} x\cos^q x - (p-1)\sin^p x\cos^q x - (q+1)\sin^p x\cos^q x$$

$$= (p-1)\sin^{p-2} x\cos^q x - (p+q)\sin^p x\cos^q x.$$

(*iii*) Integrating both sides w.r.t. x,

$$P = (p-1)\int \sin^{p-2} x\cos^q x\, dx - (p+q)\int \sin^p x\cos^q x\, dx.$$

Transposing,

$$(p+q)\int \sin^p x\cos^q x\, dx = -P + (p-1)\int \sin^{p-2} x\cos^q x\, dx$$

or $$\int \sin^p x\cos^q x\, dx = -\frac{P}{p+q} + \frac{p-1}{p+q}\int \sin^{p-2} x\cos^q x\, dx$$

[Substitute the value of P]

*__Why this step.__ See the footnote on p. 141.

$$= -\frac{\sin^{p-1} x \cos^{q+1} x}{p+q} + \frac{p-1}{p+q} \int \sin^{p-2} x \cos^q x \, dx.$$

$$\therefore \quad \int_0^{\frac{\pi}{2}} \sin^p x \cos^q x \, dx$$

$$= \left[-\frac{\sin^{p-1} x \cos^{q+1} x}{p+q} \right]_0^{\frac{\pi}{2}} + \frac{p-1}{p+q} \int_0^{\frac{\pi}{2}} \sin^{p-2} x \cos^q x \, dx$$

$\left[\text{But when } x = \frac{\pi}{2}, \sin^{p-1} x \cos^{q+1} x = 0, \right.$

and when $x = 0$, $\sin^{p-1} x \cos^{q+1} x = 0$, if p is a + ve integer $> 1 \big]$

$$= \frac{p-1}{p+q} \int_0^{\frac{\pi}{2}} \sin^{p-2} x \cos^q x \, dx \qquad ...(1)$$

Changing p to $p - 2$,

$$\int_0^{\frac{\pi}{2}} \sin^{p-2} x \cos^q x \, dx = \frac{p-3}{p+q-2} \int_0^{\frac{\pi}{2}} \sin^{p-4} x \cos^q x \, dx.$$

Substituting this value of $\int_0^{\frac{\pi}{2}} \sin^{p-2} x \cos^q x \, dx$ in (1),

$$\int_0^{\frac{\pi}{2}} \sin^p x \cos^q x \, dx = \frac{(p-1)(p-3)}{(p+q)(p+q-2)} \int_0^{\frac{\pi}{2}} \sin^{p-4} x \cos^q x \, dx \qquad ...(2)$$

Generalizing from (1) and (2),

Case I. If p is a + ve odd integer,

$$\int_0^{\frac{\pi}{2}} \sin^p x \cos^q x \, dx = \frac{(p-1)(p-3) \ldots 2^*}{(p+q)(p+q-2) \ldots (q+3)} \int_0^{\frac{\pi}{2}} \sin^1 x \cos^q x \, dx$$

***How to write this step.** p being a positive odd integer, $p - 1$, $p - 3$, ..., are all positive even integers,

$\therefore$ the last factor in the numerator = the least positive even integer = 2.

Now in the numerator, the first factor $p - 1$ has become the last factor = 2

($\therefore$ $p = 3$), $\therefore$ in the denominator, the first factor $p + q$ will become the last factor = $3 + q$.

Again, from (1) and (2), the index of sin x in the integral on the R.H.S. is 1 less than the last factor in the numerator and is $\therefore$ $= 2 - 1 = 1$.

$$\left[\text{But} \int_0^{\frac{\pi}{2}} \sin^1 x \cos^q dx = -\int_0^{\frac{\pi}{2}} \cos^q x(-\sin x)\, dx\right.$$

$$\left. = \left[-\frac{\cos^{q+1} x}{q+1}\right]_0^{\frac{\pi}{2}} = -\frac{1}{q+1}[0-1] = \frac{1}{q+1}\right]$$

$$= \frac{(p-1)(p-3)\ldots 2}{(p+q)(p+q-2)\ldots(q+3)(q+1)}.$$

Case II. If p is a + ve even integer,

$$\int_0^{\frac{\pi}{2}} \sin^p x \cos^q x\, dx = \frac{(p-1)(p-3)\ldots 1^*}{(p+q)(p+q-2)\ldots(q+2)} \int_0^{\frac{\pi}{2}} \sin^0 x \cos^q x\, dx$$

$$= \frac{(p-1)(p-3)\ldots 1}{(p+q)(p+q-2)\ldots(q+2)} \int_0^{\frac{\pi}{2}} \cos^q x\, dx \qquad \ldots(3)$$

Sub-case (*i*). If q is a + ve odd integer, then from (3),

$$\int_0^{\frac{\pi}{2}} \sin^p x \cos^q x\, dx$$

$$= \frac{(p-1)(p-3)\ldots 1}{(p+q)(p+q-2)\ldots(q+2)} \cdot \frac{(q-1)(q-3)\ldots 2}{q(q-2)\ldots 3}. \qquad \text{[Art. 70, Case I]}$$

Sub-case (*ii*). If q is a + ve even integer, then from (3),

$$\int_0^{\frac{\pi}{2}} \sin^p x \cos^q x\, dx$$

$$= \frac{(p-1)(p-3)\ldots 1}{(p+q)(p+q-2)\ldots(q+2)} \cdot \frac{(q-1)(q-3)\ldots 1}{q(q-2)\ldots 2} \cdot \frac{\pi}{2}.$$

[Art. 70, Case II]

***How to write this step.** p being a positive even integer, $p-1$, $p-3$,..., are all positive odd integers,

$\therefore$ the last factor in the numerator = the least positive odd integer = 1.

Now in the numerator, the first factor $p-1$ has become the last factor = 1 ($\therefore$ $p = 2$), $\therefore$ in the denominator, the first factor $p+q$ will become the last factor = $2+q$.

Again, from (1) and (2), the index of $\sin x$ in the integral on the R.H.S. is 1 less than the last factor in the numerator and is $\therefore$ $= 1 - 1 = 0$.

[**Rule to write down the value of $\int_0^{\frac{\pi}{2}} \sin^p x \cos^q x\, dx$, where p, q are positive integers :**

$$\int_0^{\frac{\pi}{2}} \sin^p x \cos^q x\, dx$$

$$= \frac{(p-1) \times \text{go on diminishing by } 2 \ldots \times (q-1) \times \text{go on diminishing by } 2 \ldots}{(p+q) \times \text{go on diminishing by } 2 \ldots} \cdot \frac{\pi}{2}$$

only if both p and q are positive even integers $\left(\text{otherwise no } \frac{\pi}{2}\right)$, *each series of factors being continued so long as the factors are positive (i.e., omitting zero and negative factors).*]

[Proof. Case I.** If p is a + ve odd integer, then from Art. 71, Case I,

$$\int_0^{\frac{\pi}{2}} \sin^p x \cos^q x\, dx = \frac{(p-1)(p-3)\ldots 2}{(p+q)(p+q-2)\ldots(q+1)} \qquad \ldots(1)$$

Sub-case (i). If q is a + ve odd integer, then from (1), multiplying the num. and denom. by $(q-1)(q-3)\ldots 2$,

$$\int_0^{\frac{\pi}{2}} \sin^p x \cos^q x\, dx$$

$$= \frac{(p-1)(p-3)\ldots 2}{(p+q)(p+q-2)\ldots(q+1)} \cdot \frac{(q-1)(q-3)\ldots 2}{(q-1)(q-3)\ldots 2}$$

$$= \frac{(p-1) \times \text{go on diminishing by } 2 \ldots \times (q-1) \times \text{go on diminishing by } 2 \ldots}{(p+q) \times \text{go on diminishing by } 2 \ldots} \qquad \ldots(2)$$

Sub-case (ii). If q is a + ve even integer, then from (1), multiplying the num. and denom. by $(q-1)(q-3)\ldots 1$,

$$\int_0^{\frac{\pi}{2}} \sin^p x \cos^q x\, dx = \frac{(p-1)(p-3)\ldots 2}{(p+q)(p+q-2)\ldots(q+1)} \cdot \frac{(q-1)(q-3)\ldots 1}{(q-1)(q-3)\ldots 1}$$

$$= \frac{(p-1) \times \text{go on diminishing by } 2 \ldots \times (q-1) \times \text{go on diminishing by } 2 \ldots}{(p+q) \times \text{go on diminishing by } 2 \ldots} \qquad \ldots(3)$$

Case II. If p is a + ve even integer,

Sub-case (i). If q is a + ve odd integer, then from Art. 71, Case II, (*i*),

$$\int_0^{\frac{\pi}{2}} \sin^p x \cos^q x \, dx = \frac{(p-1)(p-3)\ldots 1}{(p+q)(p+q-2)\ldots(q+2)} \cdot \frac{(q-1)(q-3)\ldots 2}{q(q-2)\ldots 3}$$

$$= \frac{(p-1)\times \text{go on diminishing by 2} \ldots \times (q-1) \times \text{go on diminishing by 2} \ldots}{(p+q)\times \text{go on diminishing by 2} \ldots} \qquad \ldots(4)$$

Sub-case (ii). If q is a + ve even integer, then from Art. 71, Case II, (*ii*),

$$\int_0^{\frac{\pi}{2}} \sin^p x \cos^q x \, dx = \frac{(p-1)(p-3)\ldots 1}{(p+q)(p+q-2)\ldots(q+2)} \cdot \frac{(q-1)(q-3)\ldots 1}{q(q-2)\ldots 2} \cdot \frac{\pi}{2}$$

$$= \frac{(p-1)\times \text{go on diminishing by 2} \ldots \times (q-1) \times \text{go on diminishing by 2} \ldots}{(p+q)\times \text{go on diminishing by 2} \ldots} \cdot \frac{\pi}{2} \qquad \ldots(5)$$

Hence the rule from (2), (3), (4), (5).]

Example 1. *Write down the value of* $\int_0^{\frac{\pi}{2}} \sin^4 \theta \cos^6 \theta \, d\theta$.

$$\int_0^{\frac{\pi}{2}} \sin^4 \theta \cos^6 \theta \, d\theta \quad \left[\text{Compare with } \int_0^{\frac{\pi}{2}} \sin^p \theta \cos^q \theta \, d\theta, \text{ here } p = 4, q = 6\right]$$

$$= \frac{3.1 \times 5.3.1}{10.8.6.4.2} \cdot \frac{\pi}{2} \quad \left[\frac{\pi}{2} \because p(=4), q\ (=6) \text{ are both + ve even integers}\right]$$

$$\left[\frac{(p-1)\times \text{go on diminishing by 2} \ldots \times (q-1) \times \text{go on diminishing by 2} \ldots}{(p+q)\times \text{go on diminishing by 2} \ldots} \cdot \frac{\pi}{2}\right.$$

only if both p and q are + ve even integers (Rule of Art. 71)]

$$= \frac{3\pi}{512}.$$

Example 2. *Write down the value of* $\int_0^{\frac{\pi}{2}} \sin^5 \theta \cos^6 \theta \, d\theta$.

$$\int_0^{\frac{\pi}{2}} \sin^5 \theta \cos^6 \theta \, d\theta \quad \left[\text{Compare with } \int_0^{\frac{\pi}{2}} \sin^p \theta \cos^q \theta \, d\theta, \text{ here } p = 5, q = 6\right]$$

$$= \frac{4.2 \times 5.3.1}{11.9.7.5.3.1} \quad \left[\text{no } \frac{\pi}{2} \because p(=5), q\ (=6) \text{ are not both + ve even integers}\right]$$

$$\left[\frac{(p-1)\times \text{go on diminishing by 2} \ldots \times (q-1) \times \text{go on diminishing by 2} \ldots}{(p+q)\times \text{go on diminishing by 2} \ldots} \cdot \frac{\pi}{2}\right.$$

only if both p and q are + ve even integers (Rule of Art. 71)]

$$= \frac{8}{693}.$$

EXAMPLES

Write down the values of (Exs. 1—2) :

1. (*i*) $\int_0^{\frac{\pi}{2}} \sin^6 x \cos^4 x \, dx$. (*ii*) $\int_0^{\frac{\pi}{2}} \sin^6 x \cos^5 x \, dx$.

2. (*i*) $\int_0^{\frac{\pi}{2}} \sin^5 x \cos^4 x \, dx$. (*ii*) $\int_0^{\frac{\pi}{2}} \sin^5 x \cos^5 x \, dx$.

3. Evaluate $\int_0^{\frac{\pi}{6}} \cos^6 3\theta \sin^2 6\theta \, d\theta$.

| Put $3\theta = \phi, \theta = \frac{\phi}{3}$ so that $d\theta = \frac{d\phi}{3}$ when $\theta = 0, \phi = 0$ and when $\theta = \frac{\pi}{6}, \phi = \frac{\pi}{2}$

4. Show that $\int_0^{\frac{\pi}{2}} \sin^{2m} \theta \cos^{2m-1} \theta \, d\theta = \frac{(2m-2)(2m-4)\ldots 4.2}{(4m-1)(4m-3)\ldots(2m+1)}$, m being a positive integer.

5. Find the value of $\int_0^{\frac{\pi}{2}} \sin^{2m} \theta \cos^{2n} \theta \, d\theta$ for + ve integral values of m and n.

Integrals reducible to $\int_0^{\frac{\pi}{2}} \sin^p x \cos^q x \, dx$ by a trigonometric substitution.

Evaluate (Exs. 6—9) :

6. (*i*) $\int_0^1 x^2(1-x^2)^{\frac{3}{2}} \, dx$ (*ii*) $\int_0^1 x^5 (1-x^2)^{\frac{5}{2}} \, dx$.

[(*i*) Rule to evaluate $\int_0^a x^m (a^2 - x^2)^{\frac{n}{2}} \, dx$, where m, n are positive integers and n odd : *Put* $x = a \sin \theta$.

(*ii*) Rule to evaluate $\int_0^\infty \frac{x^m}{(a^2+x^2)^n} \, dx$, where m, n are positive integers, or $\int_0^\infty \frac{x^m}{(a^2+x^2)^{\frac{n}{2}}} \, dx$, where m, n are positive integers and n odd : *Put* $x = a \tan \theta$.]

7. $\int_0^\infty \frac{t^4}{(1+t^2)^4} \, dt$.

8. (*i*) $\int_0^1 x^{\frac{3}{2}} (1-x)^{\frac{3}{2}} \, dx$. (*ii*) $\int_0^{2a} x\sqrt{2ax - x^2} \, dx$.

[(*ii*) Rule to evaluate $\int_0^{2a} x^m \sqrt{2ax - x^2}\, dx \left(= \int_0^{2a} x^{m+\frac{1}{2}} \sqrt{2a - x}\, dx \right)$, where *m* is a positive integer : *Put* $x = 2a \sin^2 \theta$.]

9. (*i*) $\int_0^{2a} x^2 \sqrt{2ax - x^2}\, dx$. (*ii*) $\int_0^{2a} x^3 \sqrt{2ax - x^2}\, dx$.

10. $\int_0^{2a} x^m \sqrt{2ax - x^2}\, dx$, *m* being a positive integer.

Reduction formulae for

$\int \tan^n x\, dx, \int \cot^n x\, dx, \int \sec^n x\, dx, \int \operatorname{cosec}^n x\, dx$

72. To find a reduction formula for $\int \tan^n x\, dx$.

$$\int \tan^n x\, dx = \int \tan^{n-2} x \tan^2 x\, dx = \int \tan^{n-2} x (\sec^2 x - 1)\, dx$$

[Note in Ex. 4, Art. 7]

$$= \int \tan^{n-2} x \sec^2 x\, dx - \int \tan^{n-2} x\, dx$$

$$= \frac{\tan^{n-1} x}{n-1} - \int \tan^{n-2} x\, dx \qquad \text{...(1)}$$

which is the required reduction formula.

Note 1. To evaluate the integral completely, if n is a positive integer.

Case I. If n is a positive odd integer, then from (1), ultimately we are left with $\int \tan^1 x\, dx$* $= \int \tan x\, dx = -\log \cos x$.

Case II. If n is a positive even integer, then from (1), ultimately we are left with $\int \tan^0 x\, dx$† $= \int 1\, dx = x$.

Note 2. Method to find a reduction formula for $\int \tan^n x\, dx$.

Separate $\tan^2 x$, *replace it by* $\sec^2 x - 1$, *and integrate.*

Note 3. Rule to integrate $\int \tan^n x\, dx$, where n is a positive integer : *Use the reduction formula.*

Example. *Integrate* $\int \tan^5 x\, dx$.

---

*__How to write this step.__ n being a positive odd integer, the indices of $\tan x$ in the integrals on the R.H.S., *viz.*, $n - 2$ (and subtracting 2 each time), $n - 4$, ..., are all positive odd integers,

$\therefore$ ultimately we are left with $\int \tan^1 x\, dx$.

†**How to write this step.** n being a positive even integer, the indices of $\tan x$ in the integrals on the R.H.S., *viz.*, $n - 2$ (and subtracting 2 each time), $n - 4$, ..., are all positive even integers,

$\therefore$ ultimately we are left with $\int \tan^0 x\, dx$.

Let us find a reduction formula for $\int \tan^n x\, dx$*.

$$\int \tan^n x\, dx = \int \tan^{n-2} x \tan^2 x\, dx = \int \tan^{n-2} x\,(\sec^2 x - 1)\, dx$$

$$= \int \tan^{n-2} x \sec^2 x\, dx - \int \tan^{n-2} x\, dx = \frac{\tan^{n-1} x}{n-1} - \int \tan^{n-2} x\, dx \quad ...(1)$$

Putting $n = 5$ in (1),

$$\int \tan^5 x\, dx = \frac{\tan^4 x}{4} - \int \tan^3 x\, dx \quad ...(2)$$

Putting $n = 3$ in (1),

$$\int \tan^3 x\, dx = \frac{\tan^2 x}{2} - \int \tan x\, dx$$

$$= \frac{\tan^2 x}{2} - (-\log \cos x) = \frac{\tan^2 x}{2} + \log \cos x.$$

Substituting this value of $\int \tan^3 x\, dx$ in (2),

$$\int \tan^5 x\, dx = \frac{\tan^4 x}{4} - \left[\frac{\tan^2 x}{2} + \log \cos x\right] = \frac{\tan^4 x}{4} - \frac{\tan^2 x}{2} - \log \cos x.$$

EXAMPLES

1. Integrate (*i*) $\int \tan^3 x\, dx$.

(*ii*) $\int \tan^4 x\, dx$.

2. Find a reduction formula for $\int \tanh^n x\, dx$.

73. To find a reduction formula for $\int \cot^n x\, dx$.

$$\int \cot^n x\, dx = \int \cot^{n-2} x \cot^2 x\, dx = \int \cot^{n-2} x\,(\mathrm{cosec}^2 x - 1)\, dx$$

[Note in Ex. 4, Art. 7]

$$= \int \cot^{n-2} x\, \mathrm{cosec}^2 x\, dx - \int \cot^{n-2} x\, dx$$

$$= -\int \cot^{n-2} x\,(-\mathrm{cosec}^2 x)\, dx - \int \cot^{n-2} x\, dx$$

$$= -\frac{\cot^{n-1} x}{n-1} - \int \cot^{n-2} x\, dx \quad ...(1)$$

which is the required reduction formula.

Note 1. To evaluate the integral completely, if n is a positive integer.

***Why this step.** In order to integrate $\int \tan^5 x\, dx$, we may connect $\int \tan^5 x\, dx$ with $\int \tan^3 x\, dx$, and this again with $\int \tan x\, dx$, which is a standard form. *But instead of finding two reduction formulae,* we find only one (the general one) connecting $\int \tan^n x\, dx$ with $\int \tan^{n-2} x\, dx$, and *use it twice,* putting $n = 5$, and then $n = 3$.

(Compare the footnote* on p.134.)

Case I. If n is a positive odd integer, then from (1), ultimately we are left with

$$\int \cot^1 x\, dx^* = \int \cot x\, dx = \log \sin x.$$

Case II. If n is a positive even integer, then from (1), ultimately we are left with

$$\int \cot^0 x\, dx^\dagger = \int 1\, dx = x.$$

Note 2. Method to find a reduction formula for $\int \cot^n x\, dx$.

Separate $\cot^2 x$, replace it by $\operatorname{cosec}^2 x - 1$, and integrate.

Note 3. Rule to integrate $\int \cot^n x\, dx$, where n is a positive integer : *Use the reduction formula.*

EXAMPLES

1. Integrate (*i*) $\int \cot^3 x\, dx$. (*ii*) $\cot^4 x\, dx$.

2. Integrate $\int \cot^5 x\, dx$.

74. To find a reduction formula for $\int \sec^n x\, dx$.

$$\int \sec^n x\, dx = \int \sec^{n-2} x \sec^2 x\, dx \qquad \text{[Integrate by parts]}$$

$$= \sec^{n-2} x . \tan x - \int (n-2) \sec^{n-3} x \sec x \tan x . \tan x\, dx$$

$$= \sec^{n-2} x \tan x - (n-2) \int \sec^{n-2} x \tan^2 x\, dx$$

$$= \sec^{n-2} x \tan x - (n-2) \int \sec^{n-2} x (\sec^2 x - 1)^\ddagger\, dx$$

$$= \sec^{n-2} x \tan x - (n-2) \int \sec^n x\, dx + (n-2) \int \sec^{n-2} x\, dx.$$

Transposing,

$$(1 + n - 2) \int \sec^n x\, dx = \sec^{n-2} x \tan x + (n-2) \int \sec^{n-2} x\, dx$$

$$\therefore \qquad \int \sec^n x\, dx = \frac{\sec^{n-2} x \tan x}{n-1} + \frac{n-2}{n-1} \int \sec^{n-2} x\, dx \qquad ...(1)$$

which is the required reduction formula.

Note 1. *To evaluate the integral completely, if n is a positive odd integer.*

$\because$ *n is a positive odd integer,*

$\therefore$ *from (1), ultimately we are left with*

$$\int \sec^1 x\, dx^{**} = \int \sec x\, dx = \log \tan \left(\frac{p}{4} + \frac{x}{2}\right).$$

Note 2. *Method to find a reduction formula for $\int \sec^n x\, dx$.*

Separate $\sec^2 x$, integrate by parts, replace $\tan^2 x$ by $\sec^2 x - 1$, and transpose.

***How to write this step.** See the footnote* on p. 158.

†**How to write this step.** See the footnote† on p. 158.

‡**Why this step.** In the two expressions $\sec^n x$ and $\sec^{n-2} x$, whose integrals are to be connected, we don't require $\tan^2 x$ but require secants only,

$\therefore$ we replace $\tan^2 x$ by $\sec^2 x - 1$. (Note in Ex. 4, Art. 7)

****How to write this step.** See the footnote* on p. 158.

Note 3. Rule to integrate $\int \sec^n x\, dx$, where n is a positive integer :

(i) If the index of sec x is a positive odd integer, use the reduction formula (Art. 74).

(ii) If the index of sec x is a positive even integer, use the substitution method, viz., put tan x = z (Art. 57, (i)).

EXAMPLES

1. Find a reduction formula for $\int \sec^n x\, dx$, and deduce the values of

(*i*) $\int \sec^3 x\, dx$.

(*ii*) $\int \sec^5 x\, dx$.

****2.** Integrate $\int \sec^7 x\, dx$.

75. To find a reduction formula for $\int \text{cosec}^n x\, dx$.

$$\int \text{cosec}^n x\, dx = \int \text{cosec}^{n-2} x\, \text{cosec}^2 x\, dx \qquad \text{[Integrate by parts]}$$

$$= \text{cosec}^{n-2} x\,(-\cot x) - \int (n-2)\,\text{cosec}^{n-3} x\,(-\text{cosec}\, x \cot x)(-\cot x)\, dx$$

$$= -\text{cosec}^{n-2} x \cot x - (n-2) \int \text{cosec}^{n-2} x \cot^2 x\, dx$$

$$= -\text{cosec}^{n-2} x \cot x - (n-2) \int \text{cosec}^{n-2} x\,(\text{cosec}^2 x - 1)^*\, dx$$

$$= -\text{cosec}^{n-2} x \cot x - (n-2) \int \text{cosec}^n x\, dx + (n-2) \int \text{cosec}^{n-2} x\, dx.$$

Transposing,

$$(1 + n - 2) \int \text{cosec}^n x\, dx = -\text{cosec}^{n-2} x \cot x + (n-2) \int \text{cosec}^{n-2} x\, dx$$

$$\therefore \qquad \int \text{cosec}^n x\, dx = -\frac{\text{cosec}^{n-2} x \cot x}{n-1} + \frac{n-2}{n-1} \int \text{cosec}^{n-2} x\, dx \qquad ...(1)$$

which is the required reduction formula.

Note 1. *To evaluate the integral completely, if n is a positive odd integer.*

$\because$ *n is a positive odd integer,*

$\therefore$ *from (1), ultimately we are left with*

$$\int \text{cosec}^1 x\, dx^\dagger = \int \text{cosec}\, x\, dx = \log \tan \frac{x}{2}.$$

***Why this step.** In the two expressions $\text{cosec}^n x$ and $\text{cosec}^{n-2} x$, whose integrals are to be connected, we don't require $\cot^2 x$ but require cosecants only,

$\therefore$ we replace $\cot^2 x$ by $\text{cosec}^2 x - 1$. (Note in Ex. 4, Art. 7)

†**How to write this step.** See the footnote* on p. 158.

Note 2. *Method to find a reduction formula for* $\int \operatorname{cosec}^n x \, dx$.

Separate $\operatorname{cosec}^2 x$, *integrate by parts, replace* $\cot^2 x$ *by* $\operatorname{cosec}^2 x - 1$, *and transpose.*

Note 3. *This method is also applicable to find a reduction formula for* $\int x \operatorname{cosec}^n x \, dx$. *(See Misc. Ex. 25, Chap. VII.)*

Note 4. *Rule to integrate* $\int \operatorname{cosec}^n x \, dx$, *where n is a positive integer :*

(*i*) *If the index of cosec x is a positive odd integer, use the reduction formula (Art. 75).*

(*ii*) *If the index of cosec x is a positive even integer, use the substitution method, viz., put* $\cot x = z$ *(Art. 57, (ii)).*

EXAMPLE

1. Find a reduction formula for $\int \operatorname{cosec}^n x \, dx$, and deduce the values of

(*i*) $\int \operatorname{cosec}^3 x \, dx$.

(*ii*) $\int \operatorname{cosec}^5 x \, dx$.

76-I. Example. Reduction formula for $\int \cos^m x \sin nx \, dx$.

(*a*) **Prove that, if** $I_{m, n} = \int \cos^m x \sin nx \, dx$,

$$(m + n) I_{m, n} = -\cos^m x \cos nx + m \, I_{m-1, n-1}.$$

(*b*) **Prove also that**

$$\left[I_{m, m} \right]_0^{\frac{\pi}{2}} = \frac{1}{2^{m+1}} \left(2 + \frac{2^2}{2} + \frac{2^3}{3} + \ldots + \frac{2^m}{m} \right).$$

(*a*) $\int \cos^m x \sin nx \, dx$ [Integrate by parts]

$$= \cos^m x \left(-\frac{\cos nx}{n} \right) - \int m \cos^{m-1} x \, (-\sin x) \left(-\frac{\cos nx}{n} \right) dx$$

$$= -\frac{\cos^m x \cos nx}{n} - \frac{m}{n} \int \cos^{m-1} x \, (\sin x \cos nx) \, dx \qquad \ldots(1)$$

Now $\sin (n - 1) x = \sin nx \cos x - \cos nx \sin x$

$\therefore \quad \sin x \cos nx = \sin nx \cos x - \sin (n - 1)x.$

Substituting this value of $\sin x \cos nx$ in (1),

$$\int \cos^m x \sin nx \, dx = -\frac{\cos^m x \cos nx}{n} - \frac{m}{n} \int \cos^{m-1} x \, [\sin nx \cos x - \sin (n - 1)x] \, dx$$

$$= -\frac{\cos^m x \cos nx}{n} - \frac{m}{n} \int \cos^m x \sin nx \, dx + \frac{m}{n} \int \cos^{m-1} x \sin (n - 1) \, x \, dx.$$

Transposing,

$$\left(1+\frac{m}{n}\right)\int \cos^m x \sin nx\, dx$$

$$= -\frac{\cos^m x \cos nx}{n} + \frac{m}{n}\int \cos^{m-1} x \sin (n-1)\, x\, dx$$

(Multiply by n)

$\therefore$ $$(m+n)\, I_{m,n} = -\cos^m x \cos nx + m\, I_{m-1,\, n-1}.$$

(b) $$I_{m,n} = -\frac{\cos^m x \cos nx}{m+n} + \frac{m}{m+n} I_{m-1,\, n-1}.$$ (Put $n = m$)

$\therefore$ $$\left[I_{m,m}\right]_0^{\frac{\pi}{2}} = \left[-\frac{\cos^m x \cos mx}{2m}\right]_0^{\frac{\pi}{2}} + \frac{m}{2m}\left[I_{m-1,\, m-1}\right]_0^{\frac{\pi}{2}}$$

$$= \frac{1}{2m} + \frac{1}{2}\left[I_{m-1,\, m-1}\right]_0^{\frac{\pi}{2}} \qquad ...(1)$$

Changing m to $m-1$,

$$\left[I_{m-1,\, m-1}\right]_0^{\frac{\pi}{2}} = \frac{1}{2(m-1)} + \frac{1}{2}\left[I_{m-2,\, m-2}\right]_0^{\frac{\pi}{2}}.$$

Substituting this value of $\left[I_{m-1,\, m-1}\right]_0^{\frac{\pi}{2}}$ in (1),

$$\left[I_{m,m}\right]_0^{\frac{\pi}{2}} = \frac{1}{2m} + \frac{1}{2}\left\{\frac{1}{2(m-1)} + \frac{1}{2}\left[I_{m-2,\, m-2}\right]_0^{\frac{\pi}{2}}\right\}$$

$$= \frac{1}{2m} + \frac{1}{2^2(m-1)} + \frac{1}{2^2}\left[I_{m-2,\, m-2}\right]_0^{\frac{\pi}{2}} \qquad ...(2)$$

Generalizing from (1) and (2),

$$\left[I_{m,m}\right]_0^{\frac{\pi}{2}} = \frac{1}{2m} + \frac{1}{2^2(m-1)} + \frac{1}{2^3(m-2)} + \ldots \text{ to } m \text{ terms}$$

$$+ \frac{1}{2^m}\left[I_{m-m,\, m-m}\right]_0^{\frac{\pi}{2}}$$

$$\left[\text{But } \left[I_{0,0}\right]_0^{\frac{\pi}{2}} = \int_0^{\frac{\pi}{2}} \cos^0 x \sin 0\,.\,x\, dx = 0\right]$$

[2, 2^2, 2^3, ... is a G.P., $\therefore$ mth term = $2.2^{m-1} = 2^m$,
$m, m-1, m-2, ...$ is an A.P., $\therefore$ mth term = $m + (m-1)(-1) = 1$]

$$= \frac{1}{2m} + \frac{1}{2^2(m-1)} + \frac{1}{2^3(m-2)} + ... + \frac{1}{2^m(1)}$$

[Multiply and divide by 2^{m+1}*]

$$= \frac{1}{2^{m+1}}\left[\frac{2^{m+1}}{2m} + \frac{2^{m+1}}{2^2(m-1)} + \frac{2^{m+1}}{2^3(m-1)} + ... + \frac{2^{m+1}}{2^m(1)}\right]$$

[Write the series backwards]

$$= \frac{1}{2^{m+1}}\left[2 + \frac{2^2}{2} + \frac{2^3}{3} + ... + \frac{2^m}{m}\right].$$

EXAMPLES

1. Prove that $\int_0^{\frac{\pi}{2}} \cos^n x \cos nx\, dx = \frac{\pi}{2^{n+1}}$.

2. (*i*) If $I_{m,n} = \int \cos^m x \cos nx\, dx$, show that

$$I_{m,n} = \frac{\cos^m x \sin nx}{m+n} + \frac{m}{m+n} I_{m-1,n-1}.$$

Hence find the value of $\int_0^{\frac{\pi}{2}} \cos^3 x \cos 2x\, dx$.

(*ii*) If $\int_0^{\frac{\pi}{2}} \cos^m x \cos nx\, dx$ be denoted by $f(m, n)$, show that

$$f(m, n) = \frac{m}{m-n} f(m-1, n+1)$$
$$= \frac{m}{m+n} f(m-1, n-1).$$

****3.** If m and n are integers, m is greater than n and $m + n$ is even, prove that

$$\int_0^{\frac{\pi}{2}} \cos^m x \cos nx\, dx = \frac{\pi}{2^{m+1}} + \frac{m!}{\frac{m+n}{2}!\,\frac{m-n}{2}!}.$$

***Why this step.** $\because$ on the R.H.S. of the result to be proved, the factor $\frac{1}{2^{m+1}}$ occurs,

$\therefore$ we multiply and divide by 2^{m+1}.

76-II. Example. Find a reduction formula for the integral

$$\int \frac{\sin nx}{\sin x}\,dx.$$

Let us connect $\int \frac{\sin nx}{\sin x}\,dx$ with $\int \frac{\sin (n-2)x}{\sin x}\,dx$. *[Note this step]*

$$\frac{\sin nx - \sin (n-2)x^{*}}{\sin x} = \frac{2\cos (n-1)x \sin x}{\sin x} = 2\cos (n-1)x.$$

Integrating both sides w.r.t. x,

$$\int \frac{\sin nx}{\sin x}\,dx - \int \frac{\sin (n-2)x}{\sin x}\,dx = 2\,\frac{\sin (n-1)x}{(n-1)}.$$

Transposing,

$$\int \frac{\sin nx}{\sin x}\,dx = 2\,\frac{\sin (n-1)x}{(n-1)} + \int \frac{\sin (n-2)x}{\sin x}\,dx,$$

which is the required reduction formula.

EXAMPLE

1. Prove that

$$\int \sin n\theta \sec \theta \, d\theta = -\frac{2\cos (n-1)\theta}{n-1} - \int \sin (n-2)\theta \sec \theta \, d\theta.$$

**Hence or otherwise evaluate $\int_0^{\frac{\pi}{2}} \frac{\cos 5\theta \sin 3\theta}{\cos \theta}\,d\theta$.

$$\left[\ \int_0^{\frac{\pi}{2}} \frac{\cos 5\theta \sin 3\theta}{\cos \theta}\,d\theta = \int_0^{\frac{\pi}{2}} \frac{1}{2}\,\frac{2\cos 5\theta \sin 3\theta}{\cos \theta}\,d\theta\right.$$

$$= \int_0^{\frac{\pi}{2}} \frac{1}{2}\,\frac{\sin 8\theta - \sin 2\theta}{\cos \theta}\,d\theta = \frac{1}{2}\int_0^{\frac{\pi}{2}} \frac{\sin 8\theta}{\cos \theta}\,d\theta - \frac{1}{2}\int_0^{\frac{\pi}{2}} \frac{\sin 2\theta}{\cos \theta}\,d\theta.$$

$\left(\text{To evaluate } \int_0^{\frac{\pi}{2}} \frac{\sin 8\theta}{\cos \theta}\,d\theta.\right)$

***Why this step.** We have taken $\frac{\sin nx - \sin (n-2)x}{\sin x}$, $\because$ by doing so, $\sin x$ in the denominator cancels. Had $\cos x$ occurred in the denominator (see Ex. following), we would have taken $\frac{\sin nx + \sin (n-2)x}{\cos x}$.

From part (*i*),

$$\int_0^{\frac{\pi}{2}} \frac{\sin n\theta}{\cos\theta}\, d\theta = \left[-\frac{2\cos(n-1)\theta}{n-1} \right]_0^{\frac{\pi}{2}} - \int_0^{\frac{\pi}{2}} \frac{\sin(n-2)\theta}{\cos\theta}\, d\theta \qquad ...(1)$$

Put $n = 8$ in (1), and get ...(2)
Put $n = 6$ in (1), and get ...(3)
Put $n = 4$ in (1), and get ...(4)
Put $n = 2$ in (1), and get ...(5)

Substitute backwards from (5), *i.e.*, from (5) in (4), and from the result in (3), and from the result in (2).]

****76-III. Example. Reduction formula for $\int \frac{1}{(a + b\cos x)^n} dx$.**

If $u_n = \int \frac{1}{(a + b\cos x)^n}$ **, show that**

$$u_n = \frac{A\sin x}{(a + b\cos x)^{n-1}} + Bu_{n-1} + Cu_{n-2},$$

where $A = -\frac{1}{n-1}\frac{b}{a^2 - b^2}$, $B = \frac{(2n-3)a}{(n-1)(a^2 - b^2)}$, $C = -\frac{n-2}{n-1}\frac{1}{a^2 - b^2}$.

Find u_2.

(*a*) (*i*) Let $P = \frac{\sin x}{(a + b\cos x)^{n-1}}$.

[*i.e., in the integrand* $\frac{1}{(a + b\cos x)^n}$, *insert* sin *x in the numerator, and decrease the index of a* + *b* cos *x in the denominator by 1.*]

$$= \sin x\,(a + b\cos x)^{1-n}.$$

(*ii*)

$$\frac{dP}{dx} = \cos x.\,(a + b\cos x)^{1-n} + \sin x\,(1-n)(a + b\cos x)^{-n}(-b\sin x)$$

$$= \cos x\,(a + b\cos x)^{1-n} - b(1-n)\sin^2 x\,(a + b\cos x)^{-n}$$

$$= \cos x\,(a + b\cos x)^{1-n} - b(1-n)\,(1 - \cos^2 x)^*\,(a + b\cos x)^{-n}$$

$$\left[\text{But } \cos x = \frac{\overline{a + b\cos x} - a}{b}\right]$$

$$= \left(\frac{\overline{a + b\cos x} - a}{b}\right)(a + b\cos x)^{1-n}$$

***Why this step.** In the three expressions

$$\frac{1}{(a + b\cos x)^n}, \frac{1}{(a + b\cos x)^{n-1}}, \frac{1}{(a + b\cos x)^{n-2}},$$

whose integrals are to be connected, we don't require $\sin^2 x$ but require powers of $a + b\cos x$ only,

$\therefore$ we replace $\sin^2 x$ by $1 - \cos^2 x$; and then again $\cos x$ by $\frac{\overline{a + b\cos x} - a}{b}$.

$$- b(1-n)\left[1-\left(\frac{a+b\cos x-a}{b}\right)^2\right](a+b\cos x)^{-n}$$

$$= \frac{1}{b}(a+b\cos x)^{2-n} - \frac{a}{b}(a+b\cos x)^{1-n} - b(1-n)$$

$$\times\left[1-\left\{\frac{(a+b\cos x)^2}{b^2} - \frac{2a(a+b\cos x)}{b^2} + \frac{a^2}{b^2}\right\}\right](a+b\cos x)^{-n}$$

$$= \frac{1}{b}(a+b\cos x)^{2-n} - \frac{a}{b}(a+b\cos x)^{1-n} - b(1-n)$$

$$\times\left[\left(1-\frac{a^2}{b^2}\right) - \frac{(a+b\cos x)^2}{b^2} + \frac{2a}{b^2}(a+b\cos x)\right](a+b\cos x)^{-n}$$

$$= \frac{1}{b}(a+b\cos x)^{2-n} - \frac{a}{b}(a+b\cos x)^{1-n}$$

$$- \frac{(1-n)(b^2-a^2)}{b}(a+b\cos x)^{-n} + \frac{(1-n)}{b}(a+b\cos x)^{2-n}$$

$$- \frac{2(1-n)a}{b}(a+b\cos x)^{1-n} \quad \text{[Collect like terms]}$$

$$= \frac{1}{b}(2-n)(a+b\cos x)^{2-n} - \frac{a}{b}(3-2n)(a+b\cos x)^{1-n}$$

$$- \frac{(1-n)(b^2-a^2)}{b}(a+b\cos x)^{-n}$$

$$= -\frac{n-2}{b}\,\frac{1}{(a+b\cos x)^{n-2}} + \frac{(2n-3)a}{b}\,\frac{1}{(a+b\cos x)^{n-1}}$$

$$- \frac{(n-1)(a^2-b^2)}{b}\,\frac{1}{(a+b\cos x)^n}.$$

(*iii*) Integrating both sides w.r.t. x,

$$P = -\frac{n-2}{b}\int \frac{1}{(a+b\cos x)^{n-2}}\,dx + \frac{(2n-3)a}{b}\int \frac{1}{(a+b\cos x)^{n-1}}\,dx$$

$$- \frac{(n-1)(a^2-b^2)}{b}\int \frac{1}{(a+b\cos x)^n}\,dx.$$

Transposing,

$$\frac{(n-1)(a^2-b^2)}{b}\int \frac{1}{(a+b\cos x)^n}\,dx$$

$$= -P + \frac{(2n-3)a}{b}\int \frac{1}{(a+b\cos x)^{n-1}}\,dx - \frac{n-2}{b}\int \frac{1}{(a+b\cos x)^{n-2}}\,dx.$$

$$\therefore \int \frac{1}{(a+b\cos x)^n}\,dx$$

$$= -\frac{bP}{(n-1)(a^2-b^2)} + \frac{(2n-3)a}{(n-1)(a^2-b^2)}\int \frac{1}{(a+b\cos x)^{n-1}}\,dx$$

$$- \frac{n-2}{(n-1)(a^2-b^2)}\int \frac{1}{(a+b\cos x)^{n-2}}\,dx. \quad \text{[Substitute the value of P]}$$

$$= -\frac{b\sin x}{(n-1)(a^2-b^2)(a+b\cos x)^{n-1}} + \frac{(2n-3)a}{(n-1)(a^2-b^2)}\int \frac{1}{(a+b\cos x)^{n-1}}\,dx$$

$$- \frac{n-2}{(n-1)(a^2-b^2)}\int \frac{1}{(a+b\cos x)^{n-2}}\,dx$$

or $$u_n = -\frac{b\sin x}{(n-1)(a^2-b^2)(a+b\cos x)^{n-1}} + \frac{(2n-3)a}{(n-1)(a^2-b^2)}\,u_{n-1}$$

$$- \frac{n-2}{(n-1)(a^2-b^2)}\,u_{n-2} \quad ...(1)$$

(*b*) Putting $n = 2$ in (1),

$$u_2 = -\frac{b\sin x}{(a^2-b^2)(a+b\cos x)} + \frac{a}{a^2-b^2}\,u_1 - 0$$

$$= -\frac{b\sin x}{(a^2-b^2)(a+b\cos x)} + \frac{a}{a^2-b^2}\int \frac{1}{a+b\cos x}\,dx \quad ...(2)$$

Case I. If $a^2 > b^2$, then from (2),

$$u_2 = -\frac{b\sin x}{(a^2-b^2)(a+b\cos x)} + \frac{a}{a^2-b^2}\cdot\frac{1}{\sqrt{a^2-b^2}}\cos^{-1}\frac{b+a\cos x}{a+b\cos x}$$

$$= -\frac{b\sin x}{(a^2-b^2)(a+b\cos x)} + \frac{a}{(a^2-b^2)^{\frac{3}{2}}}\cos^{-1}\frac{b+a\cos x}{a+b\cos x}.$$

Case II. If $a^2 < b^2$, then from (2),

$$u_2 = -\frac{b\sin x}{-(b^2-a^2)(a+b\cos x)} + \frac{a}{-(b^2-a^2)}\cdot\frac{1}{\sqrt{b^2-a^2}}\cosh^{-1}\frac{b+a\cos x}{a+b\cos x}$$

$$= \frac{b\sin x}{(b^2-a^2)(a+b\cos x)} - \frac{a}{(b^2-a^2)^{\frac{3}{2}}}\cosh^{-1}\frac{b+a\cos x}{a+b\cos x}.$$

EXAMPLES

1. Evaluate $\displaystyle\int_0^{\pi} \frac{dx}{(5+4\cos x)^2}$.

(*i*) Let $\quad P = \dfrac{\sin x}{5+4\cos x}$.

(*ii*) Then it will be found that

$$\frac{dP}{dx} = \frac{5\cos x+4}{(5+4\cos x)^2} \qquad \left[\text{But (in the num.) } \cos x = \frac{5+4\cos x-5}{4}\right]$$

$$= \frac{5\left[\frac{5+4\cos x-5}{4}\right]+4}{(5+4\cos x)^2} = \frac{5}{4}\,\frac{1}{5+4\cos x} - \frac{9}{4}\,\frac{1}{(5+4\cos x)^2}.$$

(*iii*) Integrate both sides w.r.t. x and transpose. Substitute the value of P. It will be found that

$$\int_0^{\pi} \frac{1}{(5+4\cos x)^2}\,dx = \left[-\frac{4}{9}\frac{\sin x}{5+4\cos x}\right]_0^{\pi} + \frac{5}{9}\int_0^{\pi}\frac{1}{5+4\cos x}\,dx.$$

Evaluate the integral on the R.H.S.]

****2. Evaluate $\int \frac{dx}{(1+2\cos x)^3}$.**

[Find a reduction formula for $\int \frac{1}{(1+2\cos x)^n}\,dx$. (Proceed as in Art. 76-III.) Let

$$P = \frac{\sin x}{(1+2\cos x)^{n-1}} = \sin x\,(1+2\cos x)^{1-n}.$$

Find $\frac{dP}{dx}$, and replace $\sin^2 x$ by $1-\cos^2 x$. It will be found that $\frac{dP}{dx} = \cos x(1+2\cos x)^{1-n}$

$-2(1-n)(1-\cos^2 x)(1+2\cos x)^{-n}$ $\left[\text{But } \cos x = \frac{1+2\cos x - 1}{2}\right]$

$$= \left(\frac{1+2\cos x-1}{2}\right)(1+2\cos x)^{1-n} - 2(1-n)$$

$$\times\left[1-\left(\frac{1+2\cos x-1}{2}\right)^2\right](1+2\cos x)^{-n}$$

It will be found that the reduction formula is

$$\int \frac{1}{(1+2\cos x)^n}\,dx = \frac{2}{3(n-1)}\frac{\sin x}{(1+2\cos x)^{n-1}} - \frac{2n-3}{3(n-1)}$$

$$\times\int\frac{1}{(1+2\cos x)^{n-1}}\,dx + \frac{n-2}{3(n-1)}\int\frac{1}{(1+2\cos x)^{n-2}}\,dx.$$

First put $n = 3$, and then $n = 2$.]

MISCELLANEOUS EXAMPLES ON CHAPTER VII

1. If m and n are positive integers and

$$f(m, n) = \int_0^1 x^{n-1}(\log x)^m\,dx,$$

prove that $f(m, n) = -\frac{m}{n} f(m-1, n)$.

Deduce that $f(m, n) = \frac{(-1)^m\, m!}{n^{m+1}}$.

2. If $$I_n = \int_0^{\frac{\pi}{2}} x^n \sin(2p+1)x\,dx,$$

prove that $$I_n + \frac{n(n-1)}{(2p+1)^2} I_{n-2} = (-1)^p \frac{n}{(2p+1)^2}\left(\frac{\pi}{2}\right)^{n-1},$$

n and p being positive integers.

3. Obtain a reduction formula for $I_m = \int_0^{\infty} e^{-x} \sin^m x\, dx$, where $m \geq 2$ in the form $(1 + m^2)$ $I_m = m(m-1)\, I_{m-2}$; and hence evaluate I_4.

4. If I_n denotes $\int_0^a (a^2 - x^2)^n\, dx$, and $n > 0$, prove that

$$I_n = \frac{2na^2}{2n+1} I_{n-1}.$$

Hence evaluate $\int_0^a (a^2 - x^2)^3\, dx$.

5. If I_n denotes $\int_0^1 x^p (1 - x^q)^n\, dx$, where p, q and n are positive, prove that $(qn + p + 1)$ $I_n = qn\, I_{n-1}$.

Evaluate I_n when n is a positive integer.

****6.** Prove that $\int_0^1 x^{-\frac{1}{4}}(1 - x^{\frac{1}{2}})^{\frac{5}{2}}\, dx = \frac{5}{16}\int_0^1 x^{-\frac{1}{4}}(1 - x^{\frac{1}{2}})^{\frac{1}{2}}\, dx$.

7. Prove that if $X = x^2 + ax + a^2$,

$$\int X^{\frac{n}{2}}\, dx = \frac{2x + a}{2(n+1)} X^{\frac{n}{2}} + \frac{3n\,a^2}{4(n+1)} \int X^{\frac{n}{2}-1}\, dx.$$

8. If $I_n = \int x^n (a - x)^{1/2}\, dx$, prove that

$$(2n + 3)\, I_n = 2an\, I_{n-1} - 2x^n (a - x)^{\frac{3}{2}}.$$

Evaluate $\int_0^a x^2 \sqrt{ax - x^2}\, dx$.

9. If $U_n = \int x^2 \sqrt{a^2 - x^2}\, dx$, prove that

$$U_n = -\frac{x^{n-1}(a^2 - x^2)^{\frac{3}{2}}}{n+2} + \frac{n-1}{n+2} a^2\, U_{n-2}.$$

Evaluate $\int_0^a x^4 \sqrt{a^2 - x^2}\, dx$.

10. Find reduction formulae for

(*i*) $\int x^m (a + bx)^{\frac{2n+1}{2}}\, dx$. (*ii*) $\int x^{2m} (a^2 + x^2)^{n + \frac{1}{2}}\, dx$.

11. Obtain a reduction formula for $\int \frac{dx}{(a^2+x^2)^n}$, where n is a positive integer.

Show that $\int_0^{\infty} \frac{dx}{(1+x^2)^5} = \frac{\pi}{2}\frac{1.3.5.7}{2.4.6.8}$.

[(*a*) Let us connect $\int \frac{1}{(a^2+x^2)^n}\,dx$ with $\int \frac{1}{(a^2+x^2)^{n-1}}\,dx$,

i.e., $\int x^0 (a^2+x^2)^{-n}\,dx$ [Form $\int x^m (a+bx^n)^p\,dx$]

with $\int x^0 (a^2+x^2)^{-n+1}\,dx$.

(*i*) Let $P = x^{0+1}(a^2+x^2)^{-n+1}$ [Rule of "smaller index + 1" (Art. 67)]

$= x(a^2+x^2)^{1-n}$.

(*ii*)

$$\frac{dP}{dx} = 1.(a^2+x^2)^{1-n} + x.(1-n)(a^2+x^2)^{-n}(2x)$$
$$= (a^2+x^2)^{1-n} + 2(1-n)x^2(a^2+x^2)^{-n}$$
$$= (a^2+x^2)^{1-n} + 2(1-n)(\overline{a^2+x^2}-a^2)^{*}(a^2+x^2)^{-n}$$
$$= (a^2+x^2)^{1-n} + 2(1-n)(a^2+x^2)^{1-n} - 2(1-n)a^2(a^2+x^2)^{-n}$$
$$= (3-2n)(a^2+x^2)^{1-n} - 2(1-n)a^2(a^2+x^2)^{-n}$$
$$= -(2n-3)\frac{1}{(a^2+x^2)^{n-1}} + 2(n-1)a^2\frac{1}{(a^2+x^2)^n}.$$

(*iii*) Integrate both sides w.r.t. x.]

12. If $f(n) = a^{2n-1}\int_0^{\infty}\frac{dx}{(a^2+x^2)^n}$, prove $f(n) = \frac{2n-3}{2n-2}f(n-1)$.

13. Integrate $\int \frac{1}{(x^2+2)^3}\,dx$.

[Find a reduction formula for $\int \frac{1}{(x^2+2)^n}\,dx$. Put $n = 3$, and then $n = 2$.]

14. If $u_n = \int_0^{\frac{\pi}{2}} \theta \sin^n\theta\,d\theta$, and $n > 1$, prove that

$$u_n = \frac{n-1}{n}u_{n-2} + \frac{1}{n^2}.$$

Deduce that $u_5 = \frac{149}{225}$.

[(See Note 2 in Ex. 2, Art. 69-I).

---

*__Why this step.__ See the footnote† on p. 134.

$\int \theta \sin^n \theta \, d\theta = \int (\theta \sin^{n-1} \theta) \sin \theta \, d\theta$. Integrate by parts, replace

$\cos^2 \theta$ by $1 - \sin^2 \theta$. $\left(\text{Notice that } \int \sin^{n-1} \theta \cos \theta \, d\theta = \dfrac{\sin^n \theta}{n}. \right)$

Transpose.]

****15.** Show that

$$(m + n)(m + n - 2) \int \text{sim}^m \theta \cos^n \theta \, d\theta$$
$$= (m - 1) \sin^{m+1} \theta \cos^{n-1} \theta - (n - 1) \sin^{m-1} \theta \cos^{n+1} \theta$$
$$+ (m - 1)(n - 1) \int \sin^{m-2} \theta \cos^{n-2} \theta \, d\theta.$$

16. Prove that if $a < 1$,

$$(i) \int_0^{\frac{\pi}{2}} \sin^{-1} (a \cos x) \, dx = \frac{a}{1^2} + \frac{a^3}{3^2} + \frac{a^5}{5^2} + \frac{a^7}{7^2} + \dots .$$

$$(ii) \frac{1}{\pi} \int_0^{\frac{\pi}{2}} [\sin^{-1} (a \cos x)]^2 \, dx = \frac{a^2}{2^2} + \frac{a^4}{4^2} + \frac{a^6}{6^2} + \frac{a^8}{8^2} + \dots .$$

[(*i*) Remember : $\sin^{-1} x = x + \dfrac{1}{2} \dfrac{x^3}{3} + \dfrac{1.3}{2.4} \dfrac{x^5}{5} + \dots .$

(*ii*) Rule to write down the expansion of $\dfrac{(\sin^{-1} x)^2}{2}$ from that of $\sin^{-1} x$:

In the expansion of $\sin^{-1} x$, *viz.*,

$$\frac{(\sin^{-1} x)^1}{1} = \frac{x^1}{1} + \frac{1}{2} \frac{x^3}{3} + \frac{1.3}{2.4} \frac{x^5}{5} + \dots,$$

increase every number by 1, thus getting the expansion

$$\frac{(\sin^{-1} x)^2}{2} = \frac{x^2}{2} + \frac{2}{3} \frac{x^4}{4} + \frac{2.4}{3.5} \frac{x^6}{6} + \dots .$$

(From Diff. Calculus)]

17. Evaluate (*i*) $\displaystyle\int_0^{\frac{\pi}{2}} \sin^4 x \cos^2 x \, dx.$

(*ii*) $\displaystyle\int_0^{\frac{\pi}{2}} \sin^5 x \cos^8 x \, dx.$

18. Evaluate (*i*) $\displaystyle\int_0^{2a} x \sqrt{2ax - x^2} \, dx.$

(*ii*) $\displaystyle\int_0^{2a} x^2 \sqrt{2ax - x^2} \, dx.$

19. Evaluate (*i*) $\int_0^{2a} x^3 \sqrt{2ax - x^2}\, dx.$

(*ii*) $\int_0^{2a} x^{\frac{9}{2}} (2a - x)^{-\frac{1}{2}}\, dx.$

20. If m and n are positive integers, prove that

$$\int_a^b (x-a)^m (b-x)^n\, dx = \frac{m!\,n!}{(m+n+1)!}(b-a)^{m+n+1}.$$

[Put $x = a\cos^2\theta + b\sin^2\theta$.]

21. Evaluate (*i*) $\int_0^{\infty} \frac{x^2\, dx}{(4+x^2)^{\frac{5}{2}}}.$

(*ii*) $\int_0^{\frac{\pi}{2}} \frac{dx}{(a^2\cos^2 x + b^2\sin^2 x)^2}.$

22. Evaluate $\int_0^{\frac{\pi}{4}} \tan^5\theta\, d\theta.$

23. If $\phi(n) = \int_0^{\frac{\pi}{4}} \tan^n x\, dx$, show that

(*i*) $\phi(n) + \phi(n-2) = \frac{1}{n-1}$, and deduce the value of $\phi\,(5)$.

(*ii*) $n[\phi(n-1) + \phi(n+1)] = 1.$

24. Prove that

(*a*) $\int \cos^{2n}\phi\, d\phi = \frac{1}{2n}\tan\phi\cos^{2n}\phi + \left(1 - \frac{1}{2n}\right)\int \cos^{2n-2}\phi\, d\phi.$

(*b*) $\int \sec^{2n+1}\phi\, d\phi = \frac{1}{2n}\tan\phi\sec^{2n-1}\phi + \left(1 - \frac{1}{2n}\right)\int \sec^{2n-1}\phi\, d\phi.$

25. Find a reduction formula for $\int \frac{x}{\sin^n x}\, dx.$

[$\int \frac{x}{\sin^n x}\, dx = \int x \operatorname{cosec}^n x\, dx$ (See Note 3 in Art. 75.)

$= \int (x\operatorname{cosec}^{n-2} x)\operatorname{cosec}^2 x\, dx$. Integrate by parts, replace $\cot^2 x$ by $\operatorname{cosec}^2 x - 1$.

$$\left(\text{Notice that } \int \operatorname{cosec}^{n-2} x \cot x \, dx \right.$$

$$\left. = -\int \operatorname{cosec}^{n-3} x \, (-\operatorname{cosec} x \cot x) \, dx = -\frac{\operatorname{cosec}^{n-2} x}{n-2}. \right)$$

Transpose.]

26. (*a*) Integrate $\int \frac{1}{\cosh^3 x} dx.$

(*b*) Prove that $\int_0^\infty \operatorname{sech}^n u \, du = \frac{n-2}{n-1} \int_0^\infty \operatorname{sech}^{n-2} u \, du \; (n > 2),$

and find the value of $\int_0^\infty \frac{du}{\cosh^5 u}.$

27. Prove that

$$\int_0^{\frac{\pi}{2}} \cos^m x \sin nx \, dx = \frac{1}{m+n} + \frac{m}{m+n} \int_0^{\frac{\pi}{2}} \cos^{m-1} x \sin (n-1) x \, dx.$$

****28.** Prove that if n be a positive integer greater than unity,

$$\int_0^{\frac{\pi}{2}} \cos^{n-2} x \sin nx \, dx = \frac{1}{n-1}.$$

$$\left[\int_0^{\frac{\pi}{2}} \cos^{n-2} x \sin nx \, dx = \int_0^{\frac{\pi}{2}} \cos^{n-2} x \sin [(n-1) x + x] \, dx \right.$$

$$= \int_0^{\frac{\pi}{2}} \cos^{n-2} x \left[\sin (n-1) x \cos x + \cos (n-1) x \sin x \right] dx$$

$$= \int_0^{\frac{\pi}{2}} \cos^{n-1} x \sin (n-1) x \, dx + \int_0^{\frac{\pi}{2}} \cos^{n-2} x \sin x \cos (n-1) x \, dx$$

[Integrate only 1st integral on R.H.S. by parts]

$$= \left\{ \cos^{n-1} x \left[\frac{-\cos (n-1) x}{n-1} \right] \right\}_0^{\frac{\pi}{2}}$$

$$- \int_0^{\frac{\pi}{2}} (n-1) \cos^{n-2} x \, (-\sin x) \left[\frac{-\cos (n-1) x}{n-1} \right] dx$$

$$+ \int_0^{\frac{\pi}{2}} \cos^{n-2} x \sin x \cos (n-1) x \, dx.$$

$$= -\frac{1}{n-1} [0 - 1] \quad (\because \text{ last two integrals on R.H.S. cancel})$$

$$= \frac{1}{n-1}. \Big]$$

29. If $S_n = \int_0^{\frac{\pi}{2}} \frac{\sin (2n-1)x}{\sin x} dx, V_n = \int_0^{\frac{\pi}{2}} \left(\frac{\sin nx}{\sin x}\right)^2 dx,$

(n an integer), show that $S_{n+1} - S_n = 0, V_{n+1} - V_n = S_{n+1}$.

30. Prove that $\int_0^{\pi} \frac{\sin n\theta}{\sin \theta} d\theta$ is equal to 0 or π according as n is an even or odd integer.

By means of a reduction formula or otherwise, prove that $\int_0^{\pi} \frac{\sin^2 n\theta}{\sin^2 \theta} d\theta = n\pi$, where n is a positive integer.

[(*b*) Let us connect $\int_0^{\pi} \frac{\sin^2 n\theta}{\sin^2 \theta} d\theta$ with $\int_0^{\pi} \frac{\sin^2 (n-1)\theta}{\sin^2 \theta} d\theta$. [*Note this step*]

$$\frac{\sin^2 n\theta - \sin^2 (n-1)\theta}{\sin^2 \theta} = \frac{\sin (2n-1)\theta \sin \theta}{\sin^2 \theta} = \frac{\sin (2n-1)\theta}{\sin \theta}.$$

Integrating both sides w.r.t. θ between the limits 0 and π,

$$\int_0^{\pi} \frac{\sin^2 n\theta - \sin^2 (n-1)\theta}{\sin^2 \theta} d\theta = \int_0^{\pi} \frac{\sin (2n-1)\theta}{\sin \theta} d\theta = \pi$$

[From part (*a*), $\because$ $2n-1$ is a + ve odd integer]

or
$$\int_0^{\pi} \frac{\sin^2 n\theta}{\sin^2 \theta} d\theta - \int_0^{\pi} \frac{\sin^2 (n-1)\theta}{\sin^2 \theta} d\theta = \pi$$

$$\therefore \quad \int_0^{\pi} \frac{\sin^2 n\theta}{\sin^2 \theta} d\theta = \pi + \int_0^{\pi} \frac{\sin^2 (n-1)\theta}{\sin^2 \theta} d\theta \quad ...(1)$$

Changing n to $n-1$,

$$\int_0^{\pi} \frac{\sin^2 (n-1)\theta}{\sin^2 \theta}\, d\theta = \pi + \int_0^{\pi} \frac{\sin^2 (n-2)\theta}{\sin^2 \theta}\, d\theta.$$

Substituting this value of $\int_0^{\pi} \frac{\sin^2 (n-1)\theta}{\sin^2 \theta}\, d\theta$ in (1),

$$\int_0^{\pi} \frac{\sin^2 n\theta}{\sin^2 \theta}\, d\theta = \pi + \pi + \int_0^{\pi} \frac{\sin^2 (n-2)\theta}{\sin^2 \theta}\, d\theta$$

or

$$\int_0^{\pi} \frac{\sin^2 n\theta}{\sin^2 \theta}\, d\theta = 2\pi + \int_0^{\pi} \frac{\sin^2 (n-2)\theta}{\sin^2 \theta}\, d\theta \qquad \text{...(2)}$$

Generalizing from (1) and (2),

$$\int_0^{\pi} \frac{\sin^2 n\theta}{\sin^2 \theta}\, d\theta = n\pi + \int_0^{\pi} \frac{\sin^2 (n-n)\theta}{\sin^2 \theta}\, d\theta = n\pi + 0.\Bigg]$$

31. Show that, if n is a positive integer, then

$$\int_0^{2\pi} \frac{\cos (n-1)x - \cos nx}{1 - \cos x}\, dx = 2\pi,$$

and deduce that $\displaystyle\int_0^{2\pi} \left(\frac{\sin \frac{1}{2} nx}{\sin \frac{1}{2} x} \right)^2 dx = 2n\pi.$

$$\Bigg[(a) \int_0^{2\pi} \frac{\cos (n-1)x - \cos nx}{1 - \cos x}\, dx = \int_0^{2\pi} \frac{2 \sin \dfrac{(2n-1)x}{2} \sin \dfrac{x}{2}}{2 \sin^2 \dfrac{x}{2}}\, dx$$

$$= \int_0^{2\pi} \frac{\sin \dfrac{(2n-1)x}{2}}{\sin \dfrac{x}{2}}\, dx.$$

Put $\frac{x}{2} = z$, and proceed as in Ex. 30, part (*a*).

(*b*) $\displaystyle\int_0^{2\pi} \frac{\cos (n-1)x - \cos nx}{1 - \cos x}\, dx = 2\pi.$

Put $n = 1, 2, 3, \ldots, n$, and add vertically. Cancel diagonally.
It will be found that

$$\int_0^{2\pi} \frac{1-\cos nx}{1-\cos x}\,dx = 2n\pi, \quad \text{or} \quad \int_0^{2\pi} \frac{2\sin^2 \frac{nx}{2}}{2\sin^2 \frac{x}{2}}\,dx = 2n\pi.]$$

32. (*a*) Evaluate $\displaystyle\int_0^{\pi/2} \frac{1+2\cos x}{(2+\cos x)^2}\,dx.$

**(*b*) Use the substitution $\tan\dfrac{\theta}{2} = \sqrt{\dfrac{1+e}{1-e}}\tan\dfrac{u}{2}$ to evaluate

$$\int \frac{d\theta}{(1+e\cos\theta)^2}.$$

[(*a*) Proceed as in finding a reduction formula for $\displaystyle\int \frac{1}{(2+\cos x)^2}\,dx$.

(*i*) Let $P = \dfrac{\sin x}{2+\cos x}$.

(*ii*) Find $\dfrac{dP}{dx}$, which will be found to be $= \dfrac{1+2\cos x}{(2+\cos x)^2}$.

(*iii*) Integrate both sides w.r.t. x.]

CHAPTER VIII

DEFINITE INTEGRALS

SECTION I

GEOMETRICAL INTERPRETATION OF A DEFINITE INTEGRAL

[**Rectangular axes.** *Throughout this book we shall suppose that the co-ordinate axes are rectangular.*]

77. Area formula for Cartesian equation. To prove that the area bounded by the curve y = f(x), the x-axis and the ordinates x = a, x = b is $\int_a^b y\,dx$.

Let AB be the curve $y = f(x)$, and CA, DB the ordinates $x = a$, $x = b$.

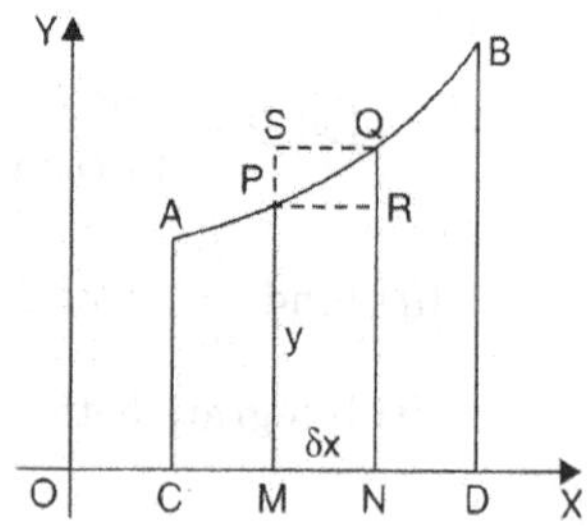

Let P (x, y) be any pt. on the curve and MP its ordinate. Then the area ACMP is clearly a function of x = A (say).

Let Q$(x + \delta x, y + \delta y)$ be a neighbouring pt. on the curve and NQ its ordinate. Then the area PMNQ = δA.

From P, Q draw PR, QS ⊥ s respectively on NQ, MP produced.

Then the area PMNQ lies between the areas of the rects. PMNR and SMNQ,

i.e., δA lies between $y\,\delta x$ and $(y + \delta y)\,\delta x$

$\therefore$ $\dfrac{\delta A}{\delta x}$ lies between y and $y + \delta y$

$\therefore$ taking the limits when $\delta x \to 0$ (and $\therefore$ $\delta y \to 0$),

$$\frac{dA}{dx} = y \qquad \text{...(1)}$$

$$\therefore \qquad \int_a^b y\,dx = \int_a^b \frac{dA}{dx}\,dx = \Big[A\Big]_a^b$$

= (value of A when $x = b$) – (value of A when $x = a$)

= area ACDB – 0 = area ACDB ...(2)

Cor. 1. From (1), $\mathbf{A = \int y\, dx.}$

Cor. 2. To give the geometrical interpretation of a definite integral.

From (2), $\because \quad y = f(x)$,

$$\therefore \quad \int_a^b f(x)\, dx = \text{area ACDB.}$$

Hence the **geometrical interpretation :**

$$\int_a^b f(x)\, dx = \textit{area bounded by the curve } y = f(x), \textit{ the x-axis and the ordinates } x = a,\ x = b.$$

Note 1. Assumptions. *(i) $f(x)$ is continuous and single-valued in the interval (a, b).*

(ii) a, b are finite, and $a < b$.

(iii) The integrand $f(x)$ does not change sign in the interval (a, b), i.e., the curve does not cross the x-axis.

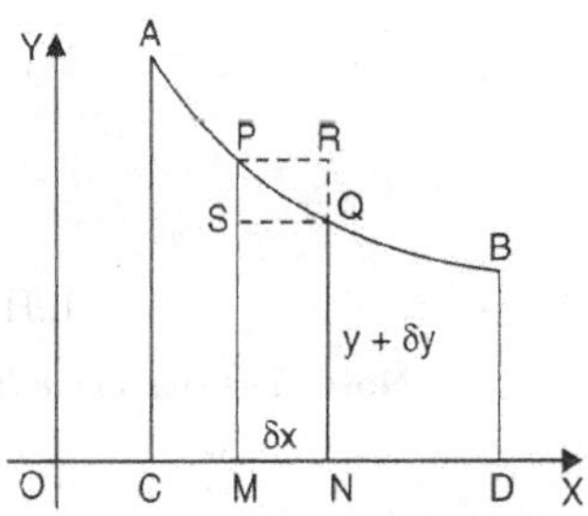

Note 2. *In the Fig. of Art. 77, it has been assumed that as a point moves along the curve from A to B, the ordinate constantly increases. If, however, the ordinate constantly decreases as in the adjoining Fig., the proposition is still true. For, the area PMNQ still lies between the areas of the rects. SMNQ and PMNR,*

i.e., δA lies between $(y + \delta y)\, \delta x$ and $y\, \delta x$, and the proof holds as before.

Note 3. *If, however, the ordinate constantly increases in some parts of the interval and constantly decreases in the others as in the adjoining Fig., the proposition is still true.*

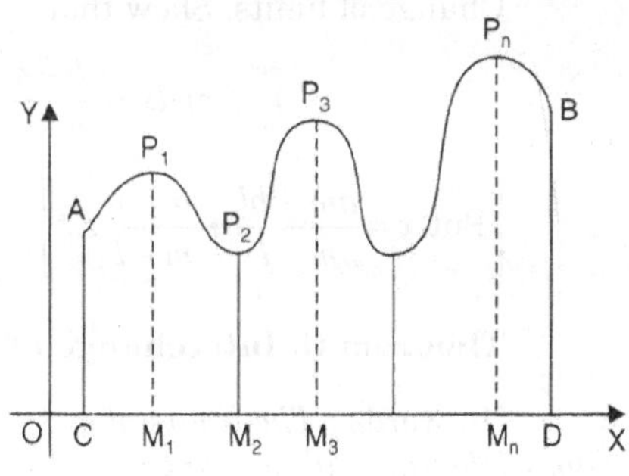

For, if the curve has a finite number of maximum and minimum ordinates M_1P_1, M_2P_2,..., $M_n P_n$, then

area ACDB = area ACM_1P_1 + area $P_1M_1M_2P_2$ + ... + area P_nM_nDB, and the proposition is true for each of the areas on the R.H.S.

SECTION II

SEVEN GENERAL THEOREMS ON DEFINITE INTEGRALS

78. Theorem I. $\displaystyle\int_a^b \mathbf{f(x)\, dx} = \int_a^b \mathbf{f(z)\, dz.}$

[In words : *The value of a definite integral depends only on the limits and not on the variable of integration (x or z).*]

To prove that $\int_a^b f(x)\,dx = \int_a^b f(z)\,dz$.

Let $\int f(x)\,dx = \phi(x)^*$ ...(1) [*Note this step*]

Then $\text{L.H.S.} = \Big[\phi(x)\Big]_a^b = \phi(b) - \phi(a).$

$\text{R.H.S.} = \Big[\phi(z)\Big]_a^b$

$\left[\because \text{ from (1), putting } x = z \text{ (and } \therefore\ dx = dz), \int f(z)\,dz = \phi(z)\right]$

$= \phi(b) - \phi(a).$

$\therefore$ L.H.S. = R.H.S.

*__Note.__ *The step (1) will also be the starting point in the proofs of theorems II and III.*

EXAMPLE

Change of limits. Show that

$$\int_a^b f(x)\,dx = \frac{b-a}{m-l}\int_l^m f\left(\frac{am-bl}{m-l} + \frac{b-a}{m-l}\,x\right) dx.$$

$$\left[\text{Put } x = \frac{am-bl}{m-l} + \frac{b-a}{m-l}\,z.^*\right]$$

Theorem II. Interchange of limits. $\int_a^b f(x)\,dx = -\int_b^a f(x)\,dx.$

[In words : *The limits of a definite integral can be interchanged if at the same time the sign of the integral is changed.*]

To prove that $\int_a^b f(x)\,dx = -\int_b^a f(x)\,dx$.

***How to write this step.** Comparing the integrands on the L.H.S. and R.H.S. of the result to be proved, *viz.*, $f(x)$ and $f\left(\frac{am-bl}{m-l} + \frac{b-a}{m-l}\,x\right)$,

we get $x = \frac{am-bl}{m-l} + \frac{b-a}{m-l}\,x$

[Change x to z on the R.H.S. to avoid confusion with x on the L.H.S.]

$= \frac{am-bl}{m-l} + \frac{b-a}{m-l}\,z.$

Let $\int f(x)\,dx = \phi(x)$.

Then $\quad$ L.H.S. $= \Big[\phi(x)\Big]_a^b = \phi(b) - \phi(a)$.

R.H.S. $= -\Big[\phi(x)\Big]_b^a = -[\phi(a) - \phi(b)] = \phi(b) - \phi(a)$.

$\therefore$ $\quad$ L.H.S. = R.H.S.

EXAMPLES

1. Prove that $\int_a^b f(a+b-x)\,dx = \int_a^b f(x)\,dx$.

2. Prove that $\int_a^b \dfrac{f(x-a)}{f(b-x)}\,dx = \int_a^b \dfrac{f(b-x)}{f(x-a)}\,dx$, provided $f(x)$ remains finite when x vanishes.

3. Prove that

$$\int_{-\infty}^{+\infty} \frac{dx}{(x^2 \pm ax + a^2)(x^2 \pm bx + b^2)} = \frac{2\pi}{\sqrt{3}}\,\frac{a+b}{ab\,(a^2+ab+b^2)}.$$

[We have to prove that

(1) taking upper signs throughout,

$$\int_{-\infty}^{+\infty} \frac{dx}{(x^2 + ax + a^2)(x^2 + bx + b^2)} = \frac{2\pi}{\sqrt{3}}\,\frac{a+b}{ab\,(a^2+ab+b^2)};$$

(2) taking lower signs throughout,

$$\int_{-\infty}^{+\infty} \frac{dx}{(x^2 - ax + a^2)(x^2 - bx + b^2)} = \frac{2\pi}{\sqrt{3}}\,\frac{a+b}{ab\,(a^2+ab+b^2)}.$$

(1) $$\frac{1}{(x^2 + ax + a^2)(x^2 + bx + b^2)} = \frac{1}{a^3 - b^3}\left[\frac{x-b}{x^2+ax+a^2} - \frac{x-a}{x^2+bx+b^2}\right].^{*}$$

***How to find these partial fractions by inspection.**

We write the denominators $\overline{x^2 + ax + a^2}\;\overline{x^2 + bx + b^2}$.

Now $(x-b)(x^2+bx+b^2) = x^3 - b^3$, and $(x-a)(x^2+ax+a^2) = x^3 - a^3$, and the *difference* of $x^3 - b^3$ and $x^3 - a^3$ is a *constant*,

$\therefore$ we write $x - b$ over $x^2 + ax + a^2$ and $x - a$ over $x^2 + bx + b^2$, thus getting

$$\frac{x-b}{x^2+ax+a^2} - \frac{x-a}{x^2+bx+b^2}.$$

But this $= \dfrac{x^3 - b^3 - (x^3 - a^3)}{(x^2+ax+a^2)(x^2+bx+b^2)} = \dfrac{a^3 - b^3}{(x^2+ax+a^2)(x^2+bx+b^2)}$.

$\therefore$ we adjust by writing $\dfrac{1}{a^3 - b^3}$ outside $\left[\dfrac{x-b}{x^2+ax+a^2} - \dfrac{x-a}{x^2+bx+b^2}\right]$.

(2) Put $x = -z$, then $dx = -dz$;

when $x \to -\infty, z \to +\infty$, when $x \to +\infty, z \to -\infty$.

$$\therefore \int_{-\infty}^{+\infty} \frac{1}{(x^2 - ax + a^2)(x^2 - bx + b^2)} dx = \int_{+\infty}^{-\infty} \frac{1}{(z^2 + az + a^2)(z^2 + bz + b^2)} (-dz)$$

$$= -\int_{+\infty}^{-\infty} \frac{1}{(z^2 + az + a^2)(z^2 + bz + b^2)} dz$$

$$= \int_{-\infty}^{+\infty} \frac{1}{(z^2 + az + a^2)(z^2 + bz + b^2)} dz \qquad \text{[Theorem II]}$$

$$= \int_{-\infty}^{+\infty} \frac{1}{(x^2 + ax + a^2)(x^2 + bx + b^2)} dx \text{ (Theorem I) = ? (Proved in (1))]}$$

Theorem III. Decomposition of the range of integration.

$$\int_a^b \mathbf{f(x)\,dx} = \int_a^c \mathbf{f(x)\,dx} + \int_c^b \mathbf{f(x)\,dx}.$$

To prove that $\int_a^b f(x)\,dx = \int_a^c f(x)\,dx + \int_c^b f(x)\,dx$.

Let $\int f(x)\,dx = \phi(x)$.

Then $\text{L.H.S.} = \Big[\phi(x)\Big]_a^b = \phi(b) - \phi(a)$.

$$\text{R.H.S.} = \Big[\phi(x)\Big]_a^c + \Big[\phi(x)\Big]_c^b$$

$$= \phi(c) - \phi(a) + \phi(b) - \phi(c) = \phi(b) - \phi(a).$$

$\therefore$ L.H.S. = R.H.S.

To illustrate the theorem geometrically.

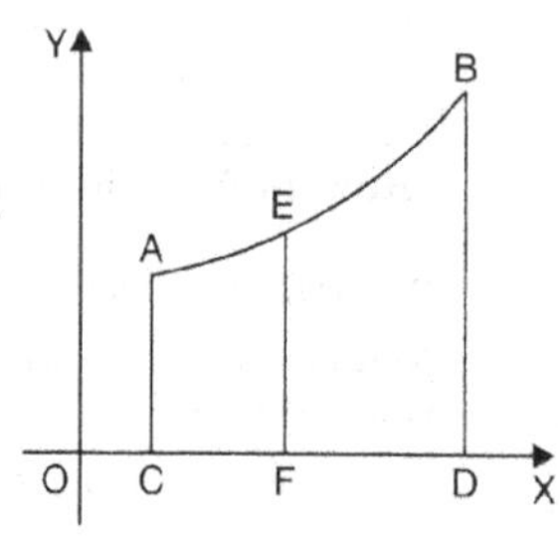

Let AB be the curve $y = f(x)$, and CA, DB the ordinates $x = a, x = b$.

Let FE be the ordinate $x = c$.

Then the theorem

$$\int_a^b f(x)\,dx = \int_a^c f(x)\,dx + \int_c^b f(x)\,dx$$

expresses that

area ACDB = area ACFE + area EFDB, which is obvious.

****Note. Generalization of theorem III.**

$$\int_a^b f(x)\,dx = \int_a^c f(x)\,dx + \int_c^d f(x)\,dx + \int_d^e f(x)\,dx + \ldots + \int_k^b f(x)\,dx.$$

To prove that

$$\int_a^b f(x)\,dx = \int_a^c f(x)\,dx + \int_c^d f(x)\,dx + \int_d^e f(x)\,dx + \ldots + \int_k^b f(x)\,dx.$$

Let $\int f(x)\,dx = \phi(x)$.

Then $$\text{L.H.S.} = \Big[\phi(x)\Big]_a^b = \phi(b) - \phi(a).$$

$$\text{R.H.S.} = \Big[\phi(x)\Big]_a^c + \Big[\phi(x)\Big]_c^d + \Big[\phi(x)\Big]_d^e + \ldots + \Big[\phi(x)\Big]_k^b$$

$$= \phi(c) - \phi(a)$$
$$+ \phi(d) - \phi(c)$$
$$+ \phi(e) - \phi(d)$$
$$+ \ldots \quad \ldots$$
$$+ \phi(b) - \phi(k) \qquad \text{[Cancel diagonally]}$$
$$= \phi(b) - \phi(a).$$

$\therefore$ L.H.S. = R.H.S.

EXAMPLE

Odd and even functions. Prove that

(*i*) if $f(x)$ is odd, then $\int_{-a}^{a} f(x)\,dx = 0$;

(*ii*) if f(x) is even, then $\int_{-a}^{a} f(x)\,dx = 2\int_0^a f(x)\,dx$.

$$\int_{-a}^{a} f(x)\,dx = \int_{-a}^{0} f(x)\,dx + \int_0^a f(x)\,dx \qquad \ldots(1)$$

$$\left[\because \int_a^b f(x)\,dx = \int_a^c f(x)\,dx + \int_c^b f(x)\,dx \quad \text{(Theorem III)}\right]$$

In the first integral on the R.H.S., put $x = -z$, then $dx = -dz$;

when $x = -a$, $z = a$, when $x = 0$, $z = 0$.

$$\therefore \quad \int_{-a}^{0} f(x)\,dx = \int_a^0 f(-z)(-dz) = -\int_a^0 f(-z)\,dz$$

$$= \int_0^a f(-z)\,dz \qquad \left[\because \int_a^b f(x)\,dx = -\int_b^a f(x)\,dx \quad \text{(Theorem II)}\right]$$

$$= \int_0^a f(-x)\,dx \qquad \left[\because \int_a^b f(x)\,dx = \int_a^b f(z)\,dz \quad \text{(Theorem I)}\right]$$

Substituting this value of $\int_{-a}^{0} f(x)\,dx$ in (1),

$$\int_{-a}^{a} f(x)\,dx = \int_{0}^{a} f(-x)\,dx + \int_{0}^{a} f(x)\,dx \qquad ...(2)$$

(*i*) If $f(x)$ is odd, $f(-x) = -f(x)$*

$\therefore$ from (2), $\int_{-a}^{a} f(x)\,dx = -\int_{0}^{a} f(x)\,dx + \int_{0}^{a} f(x)\,dx = 0.$

(*ii*) If $f(x)$ is even, $f(-x) = f(x)$*

$\therefore$ from (2), $\int_{-a}^{a} f(x)\,dx = \int_{0}^{a} f(x)\,dx + \int_{0}^{a} f(x)\,dx = 2\int_{0}^{a} f(x)\,dx.$

[**Aid to memory.** (*i*) If $f(x)$ is odd, then $\int f(x)\,dx$ is even,

$\therefore$ if we substitute $x = a$, and then $x = -a$, we get the same results, and

$\therefore$ subtracting the second result from the first, we get 0.]

Theorem IV. Important. $\int_{0}^{a} \mathbf{f(x)\,dx} = \int_{0}^{a} \mathbf{f(a-x)\,dx}$.

[In words : *If the lower limit of a definite integral is zero, then in the integrand, x can be changed to (upper limit – x), the limits remaining unchanged.*]

Put $x = a - z$†, then $dx = -dz$;

when $x = 0$, $z = a$, when $x = a$, $z = 0$.

$$\therefore \quad \int_{0}^{a} f(x)\,dx = \int_{a}^{0} f(a-z)(-dz) = -\int_{a}^{0} f(a-z)\,dz = \int_{0}^{a} f(a-z)\,dz$$

$$\left[\because \int_{a}^{b} f(x)\,dx = -\int_{b}^{a} f(x)\,dx \text{ (Theorem II)}\right]$$

$$= \int_{0}^{a} f(a-x)\,dx. \qquad \left[\because \int_{a}^{b} f(x)\,dx = \int_{a}^{b} f(z)\,dz \text{ (Theorem I)}\right]$$

***Odd and even functions. Defs.**

(*i*) $f(x)$ is said to be **odd,** if $f(-x) = -f(x)$.

(*ii*) $f(x)$ is said to be **even,** if $f(-x) = f(x)$.

Examples. $\sin x$ is an odd function and $\cos x$ an even function.

For, let $f(x) = \sin x$, then $f(-x) = \sin(-x) = -\sin x = -f(x)$,

$\therefore$ $f(x)$, *i.e.*, $\sin x$ is an odd function.

Similarly $\cos x$ is an even function.

†**How to write this step.** Comparing the integrands on the L.H.S. and R.H.S. of the result to be proved, *viz.*, $f(x)$ and $f(a-x)$, we get

$x = a - x$ [Change x to z on the R.H.S. to avoid confusion with x on the L.H.S.]

$= a - z.$

Note. *The above theorem is extremely important. It enables us, in certain cases, to evaluate a definite integral without first finding the corresponding indefinite integral which it may be difficult or even impossible to find. (See solved Exs. 1 and 2 following.)*

To illustrate the theorem geometrically.

Let AB be the curve $y = f(x)$ ...(1),

and OA, O′B the ordinates $x = 0, x = a$.

Let P(x, y) be any pt. on the curve and MP its ordinate.

Take O′O and O′B as the new axes, and let (x', y') be the co-ordinates of P referred to them.

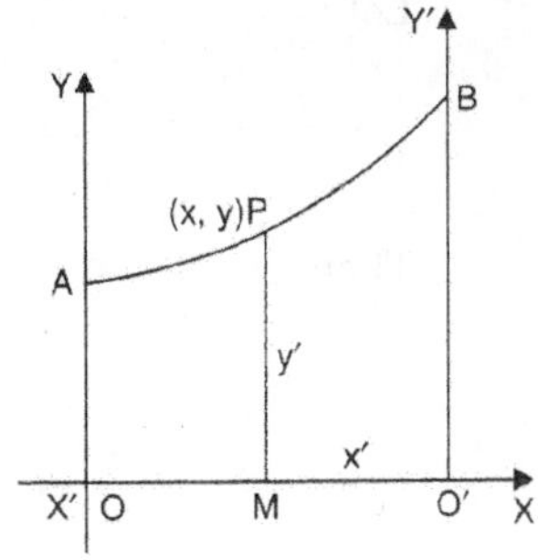

Then $x = \text{OM} = \text{OO}' - \text{MO}' = a - x'$,

$y = \text{MP} = y'$

$\therefore$ from (1), the equation of the curve AB becomes $y' = f(a - x')$

or, dropping dashes, $y = f(a - x)$ ...(2)

and the ordinates O′B, OA are $x = 0, x = a$.

$\therefore$ from (1) and (2), the theorem $\int_0^a f(x)\,dx = \int_0^a f(a-x)\,dx$

expresses that area AOO′B = area O′BAO,

which is obvious.

Cor. The rule to write down the value of $\int_0^{\frac{\pi}{2}} \cos^n x\,dx$ is the same as that to write down the value of $\int_0^{\frac{\pi}{2}} \sin^n x\,dx$. [See rule of Art. 70.]

For, $$\int_0^{\frac{\pi}{2}} \cos^n x\,dx = \int_0^{\frac{\pi}{2}} \cos^n\left(\frac{\pi}{2} - x\right) dx$$

$$\left[\because \int_0^a f(x)\,dx = \int_0^a f(a-x)\,dx \text{ (Theorem IV)}\right]$$

$$= \int_0^{\frac{\pi}{2}} \sin^n x\,dx.$$

Example 1. *Evaluate* $\int_0^{\frac{\pi}{2}} \frac{\sqrt{\sin x}}{\sqrt{\sin x}+\sqrt{\cos x}}\,dx.$

Let $$I = \int_0^{\frac{\pi}{2}} \frac{\sqrt{\sin x}}{\sqrt{\sin x}+\sqrt{\cos x}}\,dx \quad ...(1)$$

Then $$I = \int_0^{\frac{\pi}{2}} \frac{\sqrt{\sin\left(\frac{\pi}{2}-x\right)}}{\sqrt{\sin\left(\frac{\pi}{2}-x\right)}+\sqrt{\cos\left(\frac{\pi}{2}-x\right)}}\,dx$$

$$\left[\because \int_0^a f(x)\,dx = \int_0^a f(a-x)\,dx \text{ (Theorem IV)}\right]$$

$$= \int_0^{\frac{\pi}{2}} \frac{\sqrt{\cos x}}{\sqrt{\cos x}+\sqrt{\sin x}}\,dx \quad ...(2)$$

Adding (1) and (2),

$$2I = \int_0^{\frac{\pi}{2}} \left[\frac{\sqrt{\sin x}}{\sqrt{\sin x}+\sqrt{\cos x}} + \frac{\sqrt{\cos x}}{\sqrt{\cos x}+\sqrt{\sin x}}\right] dx$$

$$= \int_0^{\frac{\pi}{2}} \frac{\sqrt{\sin x}+\sqrt{\cos x}}{\sqrt{\sin x}+\sqrt{\cos x}}\,dx$$

$$= \int_0^{\frac{\pi}{2}} 1\,dx = \Big[x\Big]_0^{\frac{\pi}{2}} = \frac{\pi}{2}$$

$$\therefore \quad I = \frac{\pi}{4}.$$

Note. When to use the theorem IV. *The theorem IV is generally used*

(1) for evaluating a definite integral when the integrand is a trigonometric function of x, and the limits of integration are 0 and $\frac{\pi}{2}$ *(solved Ex. 1 above),*

or (2) for evaluating a definite integral when the integrand is the product of x and a trigonometric function of x, and the limits of integration are 0 and π (solved Ex. 2 below).

Example 2. *Prove that* $\int_0^{\pi} \frac{x \tan x \, dx}{\sec x + \cos x} = \frac{\pi^2}{4}$.

Let $$I = \int_0^{\pi} \frac{x \tan x}{\sec x + \cos x} \, dx \qquad \text{...(1)}$$

Then $$I = \int_0^{\pi} \frac{(\pi - x) \tan (\pi - x)}{\sec (\pi - x) + \cos (\pi - x)} \, dx$$

$$\left[\because \int_0^a f(x)\, dx = \int_0^a f(a - x)\, dx \text{ (Theorem IV)}\right]$$

$$= \int_0^{\pi} \frac{(\pi - x)(-\tan x)}{-\sec x - \cos x} \, dx = \int_0^{\pi} \frac{(\pi - x) \tan x}{\sec x + \cos x} \, dx$$

$$= \int_0^{\pi} \frac{\pi \tan x}{\sec x + \cos x} \, dx - \int_0^{\pi} \frac{x \tan x}{\sec x + \cos x} \, dx \qquad \text{...(2)}$$

Adding (1) and (2),

$$2I = \int_0^{\pi} \frac{\pi \tan x}{\sec x + \cos x} \, dx = \pi \int_0^{\pi} \frac{\frac{\sin x}{\cos x}}{\frac{1}{\cos x} + \cos x} \, dx$$

$$= \pi \int_0^{\pi} \frac{\sin x}{1 + \cos^2 x} \, dx$$

[Put $\cos x = z$, then $-\sin x \, dx = dz$, or $\sin x \, dx = -dz$;
when $x = 0, z = 1$, when $x = \pi, z = -1$]

$$= \pi \int_1^{-1} \frac{-dz}{1 + z^2} = -\pi \int_1^{-1} \frac{1}{(1)^2 + z^2} \, dz = -\pi \left[\frac{1}{1} \tan^{-1} \frac{z}{1}\right]_1^{-1}$$

$$= -\pi [\tan^{-1}(-1) - \tan^{-1} 1] = -\pi \left[-\frac{\pi}{4} - \frac{\pi}{4}\right] = \frac{\pi^2}{2}$$

$\therefore$ $$I = \frac{\pi^2}{4}.$$

EXAMPLES

1. (*a*) Prove that $\int_0^a f(x)\, dx = \int_0^a f(a - x)\, dx$.

(*b*) Show that $\int_0^{\frac{\pi}{2}} \log \tan x \, dx = 0$.

(*c*) Evaluate $\int_0^{\frac{\pi}{2}} \sin 2x \log \tan x \, dx$.

2. Show that $\int_0^{\frac{\pi}{2}} \frac{\sin^2 x}{\sin x + \cos x} dx = \frac{1}{\sqrt{2}} \log(\sqrt{2} + 1)$.

[It will be found that

$$I = \frac{1}{\sqrt{2}} \log \frac{\tan \frac{3\pi}{8}}{\tan \frac{\pi}{8}} = \frac{1}{\sqrt{2}} \log \frac{\sin \frac{3\pi}{8} \cos \frac{\pi}{8}}{\cos \frac{3\pi}{8} \sin \frac{\pi}{8}}$$

[Multiply the num. and denom. by 2]

$$= \frac{1}{\sqrt{2}} \log \frac{2 \sin \frac{3\pi}{8} \cos \frac{\pi}{8}}{2 \cos \frac{3\pi}{8} \sin \frac{\pi}{8}}.$$

Use the "A, B" formulae of Trigonometry for the num. and denom.]

3. Show that $\int_0^{\pi} \frac{x \sin x \, dx}{1 + \cos^2 x} = \frac{\pi^2}{4}$.

4. Evaluate $\int_0^{\pi} \frac{x \, dx}{1 + \sin x}$.

Theorem V. $\int_0^{2a} \mathbf{f(x)\,dx} = \int_0^{a} \mathbf{f(x)\,dx} + \int_0^{a} \mathbf{f(2a - x)\,dx}$.

$$\int_0^{2a} f(x)\,dx = \int_0^{a} f(x)\,dx + \int_a^{2a} f(x)\,dx \qquad ...(1)$$

In the second integral on the R.H.S., put $x = 2a - z$*,

then $dx = -dz$; when $x = a, z = a$, when $x = 2a, z = 0$.

$$\therefore \quad \int_0^{2a} f(x)\,dx = \int_a^{0} f(2a - z)(-dz)$$

$$= -\int_a^{0} f(2a - z)\,dz$$

$$= \int_0^{a} f(2a - z)\,dz$$

$$= \int_0^{a} f(2a - x)\,dx.$$

***How to write this step.** Comparing the integrands in the second integrals on the R.H.S.'s of (1) and the result to be proved, *viz.*, $f(x)$ and $f(2a - x)$, we get

$x = 2a - x$ [Change x to z on the R.H.S. to avoid confusion with x on the L.H.S.]

$= 2a - z$.

Substituting this value of $\int_a^{2a} f(x)\,dx$ in (1),

$$\int_0^{2a} f(x)\,dx = \int_0^a f(x)\,dx + \int_0^a f(2a-x)\,dx.$$

To illustrate the theorem geometrically.

Let AB be the curve $y = f(x)$ ...(1),

and OA, O′B the ordinates $x = 0, x = 2a$.

Let FE be the ordinate $x = a$.

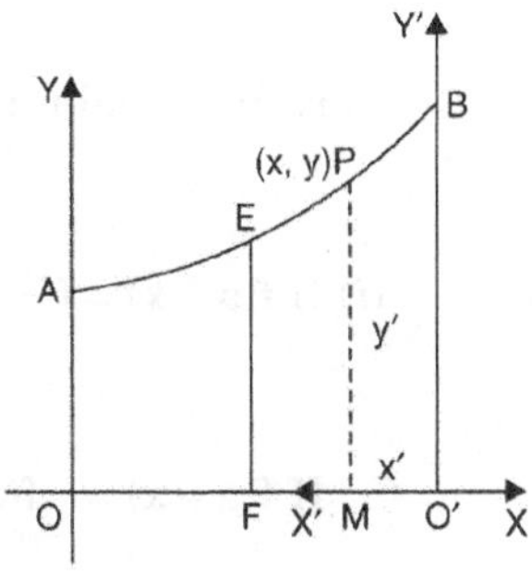

Let P(x, y) be any pt. *on the curve EB* and MP its ordinate. [*Note this step*]

For the area EFO′B, take O′F and O′B as the new axes, and let (x', y') be the co-ordinates of P referred to them.

Then $x = \text{OM} = \text{OO}' - \text{MO}' = 2a - x'$,

$y = \text{MP} = y'$

∴ from (1), the equation of the curve EB becomes $y' = f(2a - x')$

or, dropping dashes, $y = f(2a - x)$ ...(2),

and the ordinates O′B, FE are $x = 0, x = a$.

∴ from (1) and (2), the theorem

$$\int_0^{2a} f(x)\,dx = \int_0^a f(x)\,dx + \int_0^a f(2a-x)\,dx$$

expresses that

area AOO′B = area AOFE + area O′BEF,

which is obvious.

EXAMPLE

Prove that $\int_0^{2a} \phi(x)\,dx = \int_0^a \{\phi(x) + \phi(2a-x)\}\,dx$, and illustrate the theorem geometrically.

Theorem VI. (*i*) **If f(2a − x) = f(x), then** $\int_0^{2a} f(x)\,dx = 2\int_0^a f(x)\,dx$.

(*ii*) **If f(2a − x) = − f(x), then** $\int_0^{2a} f(x)\,dx = 0$.

$$\int_0^{2a} f(x)\,dx = \int_0^a f(x)\,dx + \int_0^a f(2a-x)\,dx \quad \text{...(1)} \quad \text{[Theorem V]}$$

(*i*) If $f(2a - x) = f(x)$, then from (1),

$$\int_0^{2a} f(x)\,dx = \int_0^a f(x)\,dx + \int_0^a f(x)\,dx = 2\int_0^a f(x)\,dx.$$

(*ii*) If $f(2a - x) = -f(x)$, then from (1),

$$\int_0^{2a} f(x)\,dx = \int_0^a f(x)\,dx - \int_0^a f(x)\,dx = 0.$$

Cor. Practical form of theorem VI. From theorem VI, changing a to $\frac{a}{2}$, we get

(*i*) **If f(a − x) = f(x), then** $\int_0^a f(x)\,dx = 2\int_0^{\frac{a}{2}} f(x)\,dx.$

(*ii*) **If f(a − x) = − f(x), then** $\int_0^a f(x)\,dx = 0.$

[In words : *If the lower limit of a definite integral is zero, then*

(*i*) *if on changing x to (upper limit − x), the integrand remains unchanged, the integral is equal to twice the integral of the same function between half the limits ;*

(*ii*) *if on changing x to (upper limit − x), the integrand changes only in sign but not in magnitude, the integral is equal to zero.*]

Note. When to use the theorem VI (Practical form). *The theorem VI (Practical form) is generally used for evaluating a definite integral when the integrand is a trigonometric function of x, and the limits of integration are 0 and π (or 0 and a multiple of π).*

Example 1. *Evaluate* $\int_0^{\pi} \sin^n x\,dx.$

$$\sin^n(\pi - x) = \sin^n x \qquad [f(a - x) = f(x)]$$

$$\therefore \qquad \int_0^{\pi} \sin^n x\,dx = 2\int_0^{\frac{\pi}{2}} \sin^n x\,dx \qquad ...(1)$$

$$\left[\int_0^a f(x)\,dx = 2\int_0^{\frac{a}{2}} f(x)\,dx \text{ (Theorem VI, Cor. (i)')}\right]$$

Case I. If n is a + ve odd integer, then from (1),

$$\int_0^{\pi} \sin^n x \, dx = 2 \frac{(n-1)(n-3)\ldots 2^*}{n(n-2)\ldots 3}.$$

Case II. If n is a + ve even integer, then from (1),

$$\int_0^{\pi} \sin^n x \, dx = 2 \frac{(n-1)(n-3)\ldots 1}{n(n-2)\ldots 2} \cdot \frac{\pi \dagger}{2} = \frac{(n-1)(n-3)\ldots 1}{n(n-2)\ldots 2} \pi.$$

Example 2. *Evaluate*

(i) $\int_0^{\pi} \cos^{2n+1} x \, dx.$ (ii) $\int_0^{\pi} \cos^{2n} x \, dx.$

(i) $\cos^{2n+1} (\pi - x) = (-\cos x)^{2n+1} = -\cos^{2n+1} x$ $[f(a-x) = -f(x)]$

$\therefore$ $\int_0^{\pi} \cos^{2n+1} x = 0.$ $\left[\int_0^a f(x)\, dx = 0 \text{ (Theorem VI, Cor. } (ii)')\right]$

(ii) $\cos^{2n} (\pi - x) = (-\cos x)^{2n} = \cos^{2n} x$ $[f(a-x) = f(x)]$

$$\therefore \int_0^{\pi} \cos^{2n} x \, dx = 2 \int_0^{\frac{\pi}{2}} \cos^{2n} x \, dx$$

$$\left[\int_0^a f(x)\, dx = 2 \int_0^{\frac{a}{2}} f(x)\, dx \text{ (Theorem VI, Cor. } (i)')\right]$$

$$= 2 \frac{(2n-1)(2n-3)\ldots 1}{2n(2n-2)\ldots 2} \cdot \frac{\pi \ddagger}{2} = \frac{(2n-1)(2n-3)\ldots 1}{2n(2n-2)\ldots 2} \pi.$$

---

*__How to write this step.__ n being a positive odd integer, $n-1$, $n-3$, ..., are all positive even integers,

$\therefore$ the last factor in the numerator = the least positive even integer = 2.

Now in the numerator, the first factor $n-1$ has become the last factor = 2 ($\therefore$ $n = 3$), $\therefore$ in the denominator, the first factor n will become the last factor = 3.

Again $\because$ n is not a positive even integer,

$\therefore$ we do not multiply the result by $\frac{\pi}{2}$ (Rule of Art. 69-II).

†**How to write this step.** n being a positive even integer, $n-1$, $n-3$, ...,are all positive odd integers,

$\therefore$ the last factor in the numerator = the least positive odd integer = 1.

Now in the numerator, the first factor $n-1$ has become the last factor = 1 ($\therefore$ $n = 2$), $\therefore$ in the denominator, the first factor n will become the last factor = 2.

Again $\because$ n is a positive even integer,

$\therefore$ we multiply the result by $\frac{\pi}{2}$ (Rule of Art. 69-II).

‡**How to write this step.** Proceed as in the footnote† above, $2n$ being a positive even integer etc.

Example 3. *Evaluate* $\int_0^{2\pi} \cos^{2n} x\, dx.$

$$\cos^{2n}(2\pi - x) = \cos^{2n} x \qquad [f(a-x) = f(x)]$$

$$\therefore \quad \int_0^{2\pi} \cos^{2n} x\, dx = 2\int_0^{\pi} \cos^{2n} x\, dx \qquad \left[\int_0^a f(x)\,dx = 2\int_0^{\frac{a}{2}} f(x)\,dx\right.$$

(Theorem VI, Cor. $(i)'$)]

[Again $\cos^{2n}(\pi - x) = (-\cos x)^{2n} = \cos^{2n} x$]

$$= 2.2\int_0^{\frac{\pi}{2}} \cos^{2n} x\, dx = 4\,\frac{(2n-1)(2n-3)\ldots 1}{2n(2n-2)\ldots 2}\cdot\frac{\pi}{2}$$

$$= 2\pi\,\frac{(2n-1)(2n-3)\ldots 1}{2n(2n-2)\ldots 2}.$$

Example 4. *Evaluate* $\int_0^{\frac{\pi}{2}} \log \sin x\, dx.$

Let $$I = \int_0^{\frac{\pi}{2}} \log \sin x\, dx \qquad \ldots(1)$$

Then $$I = \int_0^{\frac{\pi}{2}} \log \sin\left(\frac{\pi}{2} - x\right) dx = \int_0^{\frac{\pi}{2}} \log \cos x\, dx \qquad \ldots(2)$$

Adding (1) and (2),

$$2I = \int_0^{\frac{\pi}{2}} (\log \sin x + \log \cos x)\, dx$$

$$= \int_0^{\frac{\pi}{2}} \log \sin x \cos x\, dx = \int_0^{\frac{\pi}{2}} \log \frac{\sin 2x}{2}\, dx$$

$$= \int_0^{\frac{\pi}{2}} \log \sin 2x\, dx - \int_0^{\frac{\pi}{2}} \log 2\, dx \qquad \ldots(3)$$

Now $\int_0^{\frac{\pi}{2}} \log \sin 2x\, dx$ $\qquad$ [Put $2x = z$, then $2dx = dz$, or $dx = \frac{dz}{2}$; when $x = 0$, $z = 0$, when $x = \frac{\pi}{2}$, $z = \pi$]

$$\int_0^{\pi} \log \sin z \,.\, \frac{dz}{2} = \frac{1}{2}\int_0^{\pi} \log \sin z\, dz = \frac{1}{2}\int_0^{\pi} \log \sin x\, dx$$

[But $\log \sin(\pi - x) = \log \sin x$] $\qquad [f(a-x) = f(x)]$

$$= \frac{1}{2} \cdot 2 \int_0^{\frac{\pi}{2}} \log \sin x \, dx$$

$$\left[\int_0^a f(x)\, dx = 2 \int_0^{\frac{a}{2}} f(x)\, dx \text{ (Theorem VI, Cor. } (i)') \right]$$

$$= \int_0^{\frac{\pi}{2}} \log \sin x \, dx = \text{I},$$

$$\int_0^{\frac{\pi}{2}} \log 2 \, dx = \log 2 \int_0^{\frac{\pi}{2}} 1 \, dx$$

$$= \log 2 \cdot \Big[x \Big]_0^{\frac{\pi}{2}} = \frac{\pi}{2} \log 2.$$

$\therefore$ from (3), $\quad 2\text{I} = \text{I} - \frac{\pi}{2} \log 2, \quad \therefore \quad \text{I} = -\frac{\pi}{2} \log 2.$

Hence we have an important integral.

$$\int_0^{\frac{\pi}{2}} \log \sin x \, dx = -\frac{\pi}{2} \log 2.$$

Example 5. *Evaluate* $\int_0^\infty \log \left(x + \frac{1}{x} \right) \frac{dx}{1 + x^2}$.

Put $x = \tan \theta$, then $dx = \sec^2 \theta \, d\theta$;

when $\quad x = 0, \theta = 0$, when $x \to \infty, \theta \to \frac{\pi}{2}$.

$$\therefore \int_0^\infty \log \left(x + \frac{1}{x} \right) \frac{dx}{1 + x^2} = \int_0^{\frac{\pi}{2}} \log \left(\tan \theta + \frac{1}{\tan \theta} \right) \frac{\sec^2 \theta \, d\theta}{1 + \tan^2 \theta}$$

$$= \int_0^{\frac{\pi}{2}} \log \left(\tan \theta + \frac{1}{\tan \theta} \right) \frac{\sec^2 \theta \, d\theta}{\sec^2 \theta}$$

$$= \int_0^{\frac{\pi}{2}} \log \left(\frac{\sin \theta}{\cos \theta} + \frac{\cos \theta}{\sin \theta} \right) d\theta$$

$$= \int_0^{\frac{\pi}{2}} \log \frac{1}{\sin \theta \cos \theta} \, d\theta$$

$$= -\int_0^{\frac{\pi}{2}} \log \sin \theta \, d\theta - \int_0^{\frac{\pi}{2}} \log \cos \theta \, d\theta$$

$$\left[\text{But} \int_0^{\frac{\pi}{2}} \log \cos \theta \, d\theta = \int_0^{\frac{\pi}{2}} \log \cos \left(\frac{\pi}{2} - \theta \right) d\theta = \int_0^{\frac{\pi}{2}} \log \sin \theta \, d\theta \right]$$

$$= -\int_0^{\frac{\pi}{2}} \log \sin \theta \, d\theta - \int_0^{\frac{\pi}{2}} \log \sin \theta \, d\theta = -2\int_0^{\frac{\pi}{2}} \log \sin \theta \, d\theta$$

$$= -2\left(-\frac{\pi}{2}\log 2\right)$$ [Solved Ex. 4, theorem VI]

$$= \pi \log 2.$$

EXAMPLES

1. If $f(x) = f(a - x)$, prove that $\int_0^a f(x)\,dx = 2\int_0^{a/2} f(x)\,dx$.

Hence or otherwise evaluate $\int_0^{\pi} \sin^3 \theta \cos^2 \theta \, d\theta$.

2. (*a*) Evaluate $\int_0^{\pi} \sin^4 x \, dx$.

**(*b*) Prove that $\int_0^{\pi} \theta \sin^2 \theta \cos \theta \, d\theta = -\frac{4}{9}$.

[(*b*) $\int_0^{\pi} \theta \sin^2 \theta \cos \theta \, d\theta = \int_0^{\pi} \theta(\sin^2 \theta \cos \theta)\, d\theta$. Integrate by parts.]

3. If $f(x) = -f(2a - x)$, $\int_b^{2a} f(x)\,dx = -\int_0^b f(x)\,dx$.

4. If $\phi(x) = \phi(2a - x)$, show that $\int_0^{2a} \phi(x)\,dx = 2\int_0^a \phi(x)\,dx$.

Prove that $\int_0^{\pi} \cos^n x \, dx$ is equal to zero or $\pi\,(n\,!)/2^n\,(\frac{1}{2}\,n\,!)^2$ according as n is odd or even.

5. Prove that $\int_0^{\pi} \frac{x\,dx}{a^2 - \cos^2 x} = \frac{\pi^2}{2a\sqrt{a^2 - 1}}$, a being supposed greater than unity.

6. Evaluate $\int_0^{\frac{\pi}{2}} \log \cos x \, dx$.

7. (*i*) Evaluate $\int_0^{\frac{\pi}{2}} x \cot x \, dx$.

(*ii*) Prove that $\int_0^{\frac{\pi}{2}} \left(\frac{\theta}{\sin\theta}\right)^2 d\theta = \pi \log 2.$

Theorem VII. Periodic function. If f(x) = f(a + x), then

$$\int_0^{na} f(x)\, dx = n \int_0^a f(x)\, dx.$$

$$\int_0^{na} f(x)\, dx = \int_0^a f(x)\, dx + \int_a^{2a} f(x)\, dx + \int_{2a}^{3a} f(x)\, dx + \ldots + \int_{(n-1)a}^{na} f(x)\, dx \qquad \ldots(1)$$

[Generalization of theorem III]

In the second integral on the R.H.S. of (1), put $x = a + z$,* then $dx = dz$; when $x = a$, $z = 0$, when $x = 2a$, $z = a$.

$$\therefore \quad \int_a^{2a} f(x)\, dx = \int_0^a f(a+z)\, dz = \int_0^a f(a+x)\, dx = \int_0^a f(x)\, dx.$$

[$\because$ $f(a + x) = f(x)$ (Given)]

Again, in the third integral on the R.H.S. of (1), put $x = a + z$, then $dx = dz$; when $x = 2a$, $z = a$, when $x = 3a$, $z = 2a$.

$$\therefore \quad \int_{2a}^{3a} f(x)\, dx = \int_a^{2a} f(a+z)\, dz = \int_a^{2a} f(a+x)\, dx = \int_a^{2a} f(x)\, dx$$

[$\because$ $f(a + x) = f(x)$ (Given)]

$$= \int_0^a f(x)\, dx, \qquad \text{[Proved above]}$$

and so on.

$$\therefore \quad \text{from (1)}, \int_0^{na} f(x)\, dx = \int_0^a f(x)\, dx + \int_0^a f(x)\, dx + \int_0^a f(x)\, dx + \ldots \text{ to } n \text{ terms}$$

$$= n \int_0^a f(x)\, dx.$$

***To illustrate the theorem geometrically.*

Let $P_0P_1P_2 \ldots P_n$ be the curve $y = f(x)$, and OP_0, M_1P_1, M_2P_2, ..., M_nP_n the ordinates $x = 0$, $x = a$, $x = 2a$, ..., $x = na$.

$\because$ $f(x) = f(a + x)$ [Given]

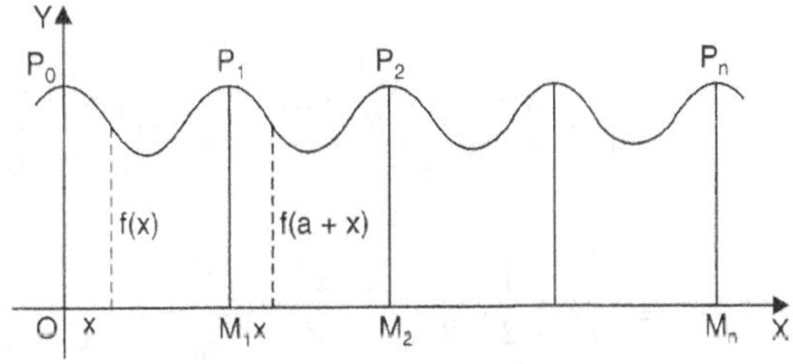

*As suggested by $f(x) = f(a + x)$ (Given).

∴ the corresponding ordinates shown dotted in the Fig. are equal

∴ the curve consists of the portion P_0P_1 from $x = 0$ to $x = a$, and an endless repetition of that portion, P_1P_2, ...

∴ area $P_0OM_1P_1$ = area $P_1M_1M_2P_2$ = ...

∴ the theorem $\int_0^{na} f(x)\,dx = n\int_0^a f(x)\,dx$

expresses that area $P_0OM_nP_n = n$. area $P_0OM_1P_1$,

which is obvious.

EXAMPLE

If $f(x) = f(a + x)$, prove that

$$\int_a^{na} f(x)\,dx = (n-1)\int_0^a f(x)\,dx,$$

and illustrate geometrically.

79. Rule to evaluate a definite integral by expansion and summation :

(*i*) Expand the integrand (a function of x) in powers of x, and integrate term by term.

(*ii*) Sum up the resulting infinite series.

Note. When to use the above rule. *The above rule is generally used for evaluating a definite integral when the integrand is the product of $\frac{1}{x}$ and a logarithm (which can be expanded in powers of x), and the limits of integration are 0 and 1.*

Example. *Evaluate* $\int_0^1 \frac{\log(1-x)}{x}\,dx.$

$$\int_0^1 \frac{\log(1-x)}{x}\,dx \qquad \text{[Expand } \log(1-x) \text{ in powers of } x\text{]}$$

$$= \int_0^1 \frac{-x - \frac{x^2}{2} - \frac{x^3}{3} - \dots}{x}\,dx = -\int_0^1 \left(1 + \frac{x}{2} + \frac{x^2}{3} + \dots\right) dx$$

$$= -\left[x + \frac{1}{2}\cdot\frac{x^2}{2} + \frac{1}{3}\cdot\frac{x^3}{3} + \dots\right]_0^1 = -\left[\frac{1}{1^2} + \frac{1}{2^2} + \frac{1}{3^2} + \dots\right]$$

$$= -\frac{\pi^2}{6}. \qquad \text{[From Trigonometry]}$$

*Note. Remember :

$$\left.\begin{aligned} \frac{1}{1^2}+\frac{1}{2^2}+\frac{1}{3^2}+\ldots &= \frac{\pi^2}{6} \\ \frac{1}{1^2}-\frac{1}{2^2}+\frac{1}{3^2}-\frac{1}{4^2}+\ldots &= \frac{\pi^2}{12} \end{aligned}\right\} \text{(From Trigonometry)}$$

EXAMPLES

1. Prove $\int_0^1 \frac{\log x}{1+x}\,dx = -\frac{\pi^2}{12}$.

$$\left[\int_0^1 \frac{\log x}{1+x}\,dx = \int_0^1 \log x \cdot \frac{1}{1+x}\,dx \qquad \text{[Integrate by parts]}\right.$$

$$= \Big[\log x \cdot \log(1+x)\Big]_0^1 - \int_0^1 \frac{1}{x}\log(1+x)\,dx = -\int_0^1 \frac{\log(1+x)}{x}\,dx.$$

Now proceed as in Solved Ex., Art. 79.

It will be found that

$$\left.\int_0^1 \frac{\log x}{1+x}\,dx = -\left(\frac{1}{1^2}-\frac{1}{2^2}+\frac{1}{3^2}-\frac{1}{4^2}+\ldots\right) = -\frac{\pi^2}{12}. \quad \text{(From Trigonometry)}\right]$$

2. Evaluate $\int_0^1 \frac{\log x}{1-x}\,dx$.

3. Evaluate (*i*) $\int_0^{\frac{\pi}{2}} \tan x \log \sin x\,dx$.

(*ii*) $\int_0^{\frac{\pi}{2}} \sin x \log \sin x\,dx$.

[(*i*) Put $\cos x = z$*, and expand the logarithm.

(*ii*) Put $\cos x = z$, and expand the logarithm.

It will be found that

$$\int_0^{\frac{\pi}{2}} \sin x \log \sin x\,dx = -\frac{1}{2}\left[\frac{1}{3}+\frac{1}{2.5}+\frac{1}{3.7}+\ldots\right]$$

---

*As suggested by the factor $\sin x$ of the numerator. Compare Note in Ex. 7, Art. 36.

$$= -\left[\frac{1}{2.3} + \frac{1}{4.5} + \frac{1}{6.7} + \ldots\right] = -\left[\frac{1}{2} - \frac{1}{3} + \frac{1}{4} - \frac{1}{5} + \ldots\right]$$

$$= -\frac{1}{2} + \frac{1}{3} - \frac{1}{4} + \frac{1}{5} - \ldots \qquad \text{[Add 1 and subtract 1]}$$

$$= \left(1 - \frac{1}{2} + \frac{1}{3} - \frac{1}{4} + \ldots\right) - 1 = \log 2 - 1.$$

$$\left[\because \log(1+x) = x - \frac{x^2}{2} + \frac{x^3}{3} - \frac{x^4}{4} + \ldots,\right.$$

$$\left.\therefore \text{ putting } x = 1, \log 2 = 1 - \frac{1}{2} + \frac{1}{3} - \frac{1}{4} + \ldots\right]$$

SECTION III

DEFINITE INTEGRAL AS THE LIMIT OF A SUM

80. Two definitions of integration.

(1) Integration may be defined as the inverse of differentiation. Thus an integral of $f(x)$ may be defined as the function of x whose differential coefficient w.r.t. x is $f(x)$ (see Art. 2).

(2) Integration may also be defined as a process of summation.

Thus a definite integral and, therefore, also an indefinite integral may be defined as the limit of the sum of a finite series when the number of terms of the series tends to infinity whilst each term of the series tends to zero (see Art. 81 and Note 4 after it).

81. To define a definite integral as the limit of a sum. Def.

If (*i*) $f(x)$ *is continuous and single-valued in the interval* (a, b),

(*ii*) a, b *are finite, and* $a < b$, *then*

$$\underset{h \to 0}{Lt}\ h\,[f(a) + f(a+h) + f(a+2h) + \ldots + f(a + \overline{n-1}h)] \qquad (\textit{where } nh = b - a)$$

is called the **definite integral** *of* $f(x)$ *between the limits a and b, and is written* $\int_a^b f(x)\,dx$ [read "integral, from a to b, of $f(x)\,dx$"]. Thus

$$\int_a^b \mathbf{f(x)\,dx} = \underset{h \to 0}{\mathbf{Lt}}\ \mathbf{h[f(a) + f(a+h) + f(a+2h) + \ldots + f(a + \overline{n-1}h)]},$$

where nh = b − a.

Note 1. Important. *Integration by summation* is also called **integration from definition** or **integration from first principles** or **ab initio integration.**

****Note 2. A check on the answer.** *If* $f(x)$ *is positive in the interval* (a, b), *then* $\int_a^b f(x)\,dx$ *is also positive.*

$$\int_a^b f(x)\,dx = \underset{h\to 0}{\text{Lt}}\ h[f(a) + f(a+h) + f(a+2h) + \ldots + f(a+\overline{n-1}h)] \qquad \ldots(1)$$

where $nh = b - a$.

Now each term of the series on the R.H.S. of (1) is + ve.

$\therefore$ their sum and the limit of their sum is also + ve.

Similarly, *if $f(x)$ is negative in the interval (a, b), then* $\int_a^b f(x)\,dx$ *is also negative.*

The above Note enables us to say whether the value of a definite integral should be positive or negative and, therefore, serves as a check on the answer.

****Note 3. Why the notation $\int_a^b f(x)\,dx$.**

$$\underset{h\to 0}{\text{Lt}}\ h[f(a) + f(a+h) + f(a+2h) + \ldots + f(a+\overline{n-1}h)], \qquad \text{where } nh = b - a$$

$$= \underset{h\to 0}{\text{Lt}} \sum_{r=0}^{r=n-1} hf(a+rh) = \underset{h\to 0}{\text{Lt}} \underset{r=0}{\overset{r=n-1}{\text{S}}} f(a+rh).h$$

(where S (like Σ) denotes the sum)

$$= \underset{h\to 0}{\text{Lt}} \underset{a+rh=a}{\overset{a+rh=a+\overline{n-1}h=b-h}{\text{S}}} f(a+rh).h \qquad [\because\ nh = b - a]$$

[Regard $a + rh$ as a variable x, so that the infinitesimal increment $h = \delta x$]

$$= \underset{\delta x\to 0}{\text{Lt}} \underset{x=a}{\overset{x=b}{\text{S}}} f(x)\,\delta x \qquad \left[\because\ \underset{h\to 0}{\text{Lt}}\ (b-h) = b\right]$$

[On taking the limit, it is customary, for convenience, to replace the sign S by the sign $\int$ and δx by dx]

$$= \int_a^b f(x)\,dx.$$

****Note 4. An indefinite integral can be written as a definite integral.**

Let $\int f(x)\,dx = \phi(x) + c.$

Then $$\int_a^x f(x)\,dx = \Big[\phi(x) + c\Big]_a^x = \phi(x) + c - [\phi(a) + c] = \phi(x) - \phi(a)$$

$$= \phi(x) + c, \quad \text{where } c = -\phi(a).$$

$$\therefore \qquad \int f(x)\,dx = \int_a^x f(x)\,dx.$$

Example 1. *Find by summation the value of* $\int_a^b x\,dx$.

$\left[\text{Compare } \int_a^b x\,dx \text{ with } \int_a^b f(x)\,dx, \text{ here } f(x) = x,\right.$

$\therefore \quad f(a) = a, f(a+h) = a+h, f(a+2h) = a+2h, \ldots, f(a+\overline{n-1}h) = a+\overline{n-1}h.]$

$$\int_a^b x\,dx = \underset{h\to 0}{\text{Lt}}\; h[a + (a+h) + (a+2h) + \ldots + (a+\overline{n-1}h)], \qquad \text{where } nh = b-a$$

$$\left[\underset{h\to 0}{\text{Lt}}\; h\{f(a) + f(a+h) + f(a+2h) + \ldots + f(a+\overline{n-1}h)\} \text{ (Art. 81)}\right]$$

$$= \underset{h\to 0}{\text{Lt}}\; h\,[a + a + a + \ldots + a(n \text{ terms}) + h(1 + 2 + \ldots + \overline{n-1})]$$

$$\left[\text{But } 1 + 2 + \ldots + n = \frac{n(n+1)}{2}, \;\therefore \text{ changing } n \text{ to } n-1, \right.$$

$$\left. 1 + 2 + \ldots + \overline{n-1} = \frac{(n-1)n}{2}\right]$$

$$= \underset{h\to 0}{\text{Lt}}\; h\left[na + h\frac{(n-1)n}{2}\right] = \underset{h\to 0}{\text{Lt}}\left[nha + \frac{(nh-h)nh}{2}\right]$$

[Put $nh = b - a$* (*Note this step*)]

$$= \underset{h\to 0}{\text{Lt}}\left[(b-a)a + \frac{(b-a-h)(b-a)}{2}\right] = (b-a)\,a + \frac{(b-a)^2}{2}$$

$$= (b-a)\left[a + \frac{b-a}{2}\right] = \frac{(b-a)(b+a)}{2} = \frac{b^2-a^2}{2}.$$

$\left[\textbf{Check. } \int_a^b x\,dx = \left[\frac{x^2}{2}\right]_a^b = \frac{1}{2}(b^2 - a^2).\right.$

Similarly in each case of integration by summation, the answer should be checked.]

*__Note. Important step.__ *First put $nh = b - a$, and then take the limit when $h \to 0$.*

Example 2. *Calculate from first principles* $\int_a^b e^x\,dx$.

$\left[\text{Compare } \int_a^b e^x\,dx \text{ with } \int_a^b f(x)\,dx, \text{ here } f(x) = e^x,\right.$

$\therefore \quad f(a) = e^a, f(a+h) = e^{a+h}, f(a+2h) = e^{a+2h}, \ldots, f(a+\overline{n-1}h) = e^{a+\overline{n-1}h}.]$

$$\int_a^b e^x\,dx = \underset{h\to 0}{\text{Lt}}\; h[e^a + e^{a+h} + e^{a+2h} + \ldots + e^{a+\overline{n-1}h}], \quad \text{where } nh = b-a$$

$$\left[\underset{h \to 0}{\text{Lt}} \; h\{f(a) + f(a+h) + f(a+2h) + \ldots + f(a + \overline{n-1}h)\} \text{ (Art. 81)} \right]$$

$$= \underset{h \to 0}{\text{Lt}} \; h \left[\frac{e^a (e^{nh} - 1)^*}{e^h - 1} \right] \qquad \left[\frac{a(r^n - 1)}{r - 1} \right] \quad \text{[Put } nh = b - a \text{ (Note this step)]}$$

$$= \underset{h \to 0}{\text{Lt}} \; h \left[\frac{e^a (e^{b-a} - 1)}{e^h - 1} \right] = \underset{h \to 0}{\text{Lt}} \; \frac{h}{e^h - 1} (e^b - e^a)$$

$$\left[\text{But } \underset{h \to 0}{\text{Lt}} \frac{h}{e^h - 1} = \underset{h \to 0}{\text{Lt}} \frac{h}{1 + h + \frac{h^2}{2!} + \ldots - 1} = \underset{h \to 0}{\text{Lt}} \frac{1}{1 + \frac{h}{2!} + \ldots} = 1 \right.$$

$$= 1\,(e^b - e^a) = e^b - e^a.$$

$$\left[\textbf{Check. } \int_a^b e^x \, dx = \left[e^x \right]_a^b = e^b - e^a. \right]$$

[**Note 1. Two important results from Trigonometry :**

(1) **sin a + sin (a + h) + sin (a + 2h) + ... + sin** $(\mathbf{a} + \overline{\mathbf{n-1}}\mathbf{h})$

$$= \frac{\sin\left(a + \frac{n-1}{2}h\right) \sin \frac{nh}{2}}{\sin \frac{h}{2}}.$$

(In words : *The sum of the sines of a series of n angles in A.P.*

$$= \left. \frac{\sin \frac{\textit{first angle} + \textit{last angle}}{2} \sin \frac{n \textit{ diff.}}{2}}{\sin \frac{\textit{diff.}}{2}} \right)$$

Note that "*diff.*" stands for the common "*difference*" of the angles in A.P.

(2) **cos a + cos (a + h) + cos (a + 2h) + ... + cos** $(\mathbf{a} + \overline{\mathbf{n-1}}\mathbf{h})$

$$= \frac{\cos\left(a + \frac{n-1}{2}h\right) \sin \frac{nh}{2}}{\sin \frac{h}{2}}.$$

(In words : *The sum of the cosines of a series of n angles in A.P.*

$$= \left. \frac{\cos \frac{\textit{first angle} + \textit{last angle}}{2} \sin \frac{n \textit{ diff.}}{2}}{\sin \frac{\textit{diff.}}{2}} . \right)$$

Note 2. To deduce the sum of the series

***How to write this step.** $e^a + e^{a+h} + e^{a+2h} + \ldots + e^{a + \overline{n-1}h}$ is a G.P. (of n terms), whose first term is e^a, and common ratio is e^h.

$$\therefore \qquad \text{sum} = \frac{e^a (e^{nh} - 1)}{e^h - 1}. \qquad \left[\frac{a(r^n - 1)}{r - 1} \right]$$

$\cos a + \cos (a + h) + \cos (a + 2h) + ... + \cos (a + \overline{n-1}h)$
from the sum of the series

$\sin a + \sin (a + h) + \sin (a + 2h) + ... + \sin (a + \overline{n-1}h)$.

We remember that

$$\sin a + \sin (a + h) + \sin (a + 2h) + ... + \sin (a + \overline{n-1}h)$$

$$= \frac{\sin\left(a + \frac{n-1}{2}h\right)\sin\frac{nh}{2}}{\sin\frac{h}{2}}.$$

Differentiating both sides w.r.t. a,

$$\cos a + \cos (a + h) + \cos (a + 2h) + ... + \cos (a + \overline{n-1}h)$$

$$\left.= \frac{\cos\left(a + \frac{n-1}{2}h\right)\sin\frac{nh}{2}}{\sin\frac{h}{2}}.\right]$$

Example 3. *Prove ab initio that*

$$\int_a^b \sin x\, dx = \cos a - \cos b.$$

$\left[\text{Compare } \int_a^b \sin x\, dx \text{ with } \int_a^b f(x)\, dx, \text{ here } f(x) = \sin x,\right.$

$\therefore$ $f(a) = \sin a,\ f(a + h) = \sin (a + h),\ f(a + 2h) = \sin (a + 2h), ...,$

$f(a + \overline{n-1}\,h) = \sin (a + \overline{n-1}\,h).]$

$$\sin x\, dx = \underset{h \to 0}{\text{Lt}}\ h[\sin a + \sin (a + h) + \sin (a + 2h) + ... + \sin (a + \overline{n-1}\,h)]$$

where $nh = b - a$

$$\left[\underset{h \to 0}{\text{Lt}}\ h\{f(a) + f(a + h) + f(a + 2h) + ... + f(a + \overline{n-1}h)\}\ \text{(Art. 81)}\right]$$

$$= \underset{h \to 0}{\text{Lt}}\ h\left[\frac{\sin\left(a + \frac{n-1}{2}h\right)\sin\frac{nh}{2}}{\sin\frac{h}{2}}\right]$$

$$= \underset{h \to 0}{\text{Lt}}\ \frac{\frac{h}{2}}{\sin\frac{h}{2}}\left[2\sin\left(a + \frac{n-1}{2}h\right)\sin\frac{nh}{2}\right]$$

$$= \underset{h \to 0}{\text{Lt}}\ \frac{\frac{h}{2}}{\sin\frac{h}{2}}\left[\cos\left(a + \frac{n-1}{2}h - \frac{nh}{2}\right) - \cos\left(a + \frac{n-1}{2}h + \frac{nh}{2}\right)\right]$$

$$= \underset{h \to 0}{\text{Lt}} \frac{\frac{h}{2}}{\sin \frac{h}{2}} \left[\cos\left(a - \frac{h}{2}\right) - \cos\left(a + nh - \frac{h}{2}\right) \right]$$

[Put $nh = b - a$ (*Note this step*)]

$$= \underset{h \to 0}{\text{Lt}} \frac{\frac{h}{2}}{\sin \frac{h}{2}} \left[\cos\left(a - \frac{h}{2}\right) - \cos\left(a + b - a - \frac{h}{2}\right) \right]$$

$$= 1 . [\cos a - \cos b] = \cos a - \cos b.$$

EXAMPLES

1. Prove by summation that $\int_a^b x^2 \, dx = \dfrac{b^3 - a^3}{3}$.

****2.** Find $\int_a^b x^3 \, dx$ from its definition as the limit of a sum.

3. Prove by summation that $\int_a^b e^{-x} \, dx = e^{-a} - e^{-b}$.

4. Prove *ab initio* that $\int_a^b \sinh x \, dx = \cosh b - \cosh a$.

5. (*a*) Prove by summation that $\int_a^b \cos \theta \, d\theta = \sin b - \sin a$.

(*b*) Evaluate from first principles $\int_0^{\frac{\pi}{2}} \cos x \, dx$.

**82. General definition of a definite integral as the limit of a sum.

If (i) $f(x)$ is continuous and single-valued in the interval (a, b),

(ii) a, b are finite, and $a < b$,

(iii) the interval (a, b) is divided into n unequal sub-intervals by means of the points of division $x_1, x_2, x_3, \ldots, x_{n-1}$, then

$$Lt\ [f(a)(x_1 - a) + f(x_1)(x_2 - x_1) + f(x_2)(x_3 - x_2) + \ldots + f(x_{n-1})(b - x_{n-1})],$$

when $x_1 - a \to 0$, $x_2 - x_1 \to 0$, $x_3 - x_2 \to 0$, ..., $b - x_{n-1} \to 0$, is called the

definite integral *of $f(x)$ between the limits a and b, and is written* $\int_a^b f(x) \, dx$ [read "integral, from a to b, of $f(x) \, dx$"]. Thus

$$\int_a^b \mathbf{f(x)\,dx = Lt\ [f(a)(x_1 - a) + f(x_1)(x_2 - x_1) + f(x_2)(x_3 - x_2) + \ldots + f(x_{n-1})(b - x_{n-1})]} \quad \ldots(1)$$

$$\text{when } x_1 - a \to 0,\ x_2 - x_1 \to 0,\ x_3 - x_2 \to 0,\ \ldots,\ b - x_{n-1} \to 0.$$

Cor. Two important particular cases.

Case I. If the points of division $x_1, x_2, x_3, \ldots, x_{n-1}$ are such that $a, x_1, x_2, x_3, \ldots, x_{n-1}, b$ is the A.P.

$a, a + h, a + 2h, a + 3h, \ldots, a + \overline{n-1}\,h, a + nh$, so that $b = a + nh$, then from (1),

$$\int_a^b f(x)\,dx = \text{Lt}\,[f(a)h + f(a+h)h + f(a+2h)h + \ldots + f(a + \overline{n-1}\,h)h] \text{ when } h \to 0$$

i.e.,
$$\int_a^b \mathbf{f(x)\,dx} = \underset{h \to 0}{\mathbf{Lt}}\ \mathbf{h[f(a) + f(a+h) + f(a+2h) + \ldots + f(a + \overline{n-1}h)]}$$

where nh = b – a.

This is the usual definition (Art. 81).

Case II. If the points of division $x_1, x_2, x_3, \ldots, x_{n-1}$ are such that $a, x_1, x_2, x_3, \ldots, x_{n-1}, b$ is the G.P.

$$a, ar, ar^2, ar^3, \ldots, ar^{n-1}, ar^n, \text{ so that } b = ar^n,$$

then from (1),

$$\int_a^b f(x)\,dx = \text{Lt}\,[f(a)(ar - a) + f(ar)(ar^2 - ar) + f(ar^2)(ar^3 - ar^2) + \ldots$$
$$+ f(ar^{n-1})(ar^n - ar^{n-1})],$$
$$\text{when } ar - a \to 0,\ ar^2 - ar \to 0,\ ar^3 - ar^2 \to 0,\ \ldots,\ ar^n - ar^{n-1} \to 0,$$

i.e.,
$$\int_a^b \mathbf{f(x)\,dx} = \underset{r \to 1}{\mathbf{Lt}}\ \mathbf{(r-1)\,[f(a).a + f(ar).ar + f(ar^2).ar^2 + \ldots + f(ar^{n-1}).ar^{n-1}]},$$

$$\textbf{where } \mathbf{r^n = \frac{b}{a}}.$$

Hence we have another definition of a definite integral as the limit of a sum (see Art. 83).

****83. Another definition of a definite integral as the limit of the sum, suitable for evaluating $\int_a^b x^m\,dx$ by summation.**

If (i) f(x) is continuous and single-valued in the interval (a, b),

(ii) a, b are finite, and a < b, then

$$\underset{r \to 1}{Lt}\ (r-1)\,[f(a)\,.\,a + f(ar).ar + f(ar^2).ar^2 + \ldots + f(ar^{n-1}).ar^{n-1}]$$

$$\left(\text{where } r^n = \frac{b}{a}\right)$$

is called the **definite integral** *of f(x) between the limits a and b, and is written* $\int_a^b f(x)\,dx$ [read "integral, from a to b, of $f(x)\,dx$"]. Thus

$$\int_a^b f(x)\,dx = \operatorname*{Lt}_{r\to 1} (r-1)\,[f(a).a + f(ar).ar + f(ar^2).ar^2 + \ldots + f(ar^{n-1}).ar^{n-1}],$$

$$\text{where } r^n = \frac{b}{a}.$$

****Example 1.** *Find by summation* $\int_a^b x^m\,dx$.

$\left[\text{Compare } \int_a^b x^m\,dx \text{ with } \int_a^b f(x)\,dx, \text{ here } f(x) = x^m,\right.$

$\therefore \qquad f(a) = a^m,\ f(ar) = (ar)^m,\ f(ar^2) = (ar^2)^m,\ \ldots,\ f(ar^{n-1}) = (ar^{n-1})^m.]$

$$\int_a^b x^m\,dx = \operatorname*{Lt}_{r\to 1} (r-1)\,[a^m\,.\,a + (ar)^m.ar + (ar^2)^m\,.\,ar^2 + \ldots + (ar^{n-1})^m\,.ar^{n-1}],$$

$$\text{where } r^n = \frac{b}{a}$$

$$\left[\operatorname*{Lt}_{r\to 1} (r-1)\,\{f(a).a + f(ar).ar + f(ar^2).ar^2 + \ldots + f(ar^{n-1}).ar^{n-1}\}\ \text{(Art. 83)}\right]$$

$$= \operatorname*{Lt}_{r\to 1} (r-1)\,[a^{m+1} + (ar)^{m+1} + (ar^2)^{m+1} + \ldots + (ar^{n-1})^{m+1}]$$

$$= \operatorname*{Lt}_{r\to 1} (r-1)\,a^{m+1}\,\frac{[(r^{m+1})^n - 1]}{r^{m+1}-1} \qquad \left[\frac{a(r^n-1)}{r-1}\right]$$

$$= \operatorname*{Lt}_{r\to 1} \frac{(r-1)\,a^{m+1}}{r^{m+1}-1}\,[\,(r^n)^{m+1} - 1] \qquad \left[\text{Put } r^n = \frac{b}{a}^{*}\ \textit{(Note this step)}\right]$$

$$= \operatorname*{Lt}_{r\to 1} \frac{(r-1)}{r^{m+1}-1}\,a^{m+1}\left[\left(\frac{b}{a}\right)^{m+1} - 1\right]$$

$$= \operatorname*{Lt}_{r\to 1} \frac{r-1}{r^{m+1}-1}\,(b^{m+1} - a^{m+1})$$

$$\left[\text{But } \operatorname*{Lt}_{r\to 1} \frac{r-1}{r^{m+1}-1}\ \text{(Put } r = 1+h, \text{ so that when } r\to 1,\ h\to 0)\right.$$

$$= \operatorname*{Lt}_{h\to 0} \frac{1+h-1}{(1+h)^{m+1}-1} = \operatorname*{Lt}_{h\to 0} \frac{h}{1+(m+1)h+\dfrac{(m+1)m}{2!}h^2+\ldots-}$$

$$\left.= \operatorname*{Lt}_{h\to 0} \frac{1}{m+1+\dfrac{(m+1)m}{2!}h+\ldots} = \frac{1}{m+1}\right]$$

$$= \frac{1}{m+1}\,(b^{m+1} - a^{m+1}).$$

$$\left[\textbf{Check.}\ \int_a^b x^m\,dx = \left[\frac{x^{m+1}}{m+1}\right]_a^b = \frac{1}{m+1}(b^{m+1} - a^{m+1}).\right]$$

*Note. **Important step.** *First put* $r^n = \dfrac{b}{a}$, *and then take the limit when* $r \to 1$.

Example 2. *Find by summation* $\displaystyle\int_a^b \frac{1}{x^2}\,dx.$

$$\left[\text{Compare} \int_a^b \frac{1}{x^2}\,dx \text{ with} \int_a^b f(x)\,dx, \text{ here } f(x) = \frac{1}{x^2},\right.$$

$$\left.\therefore \quad f(a) = \frac{1}{a^2},\ f(ar) = \frac{1}{(ar)^2},\ f(ar^2) = \frac{1}{(ar^2)^2},\ \ldots,\ f(ar^{n-1}) = \frac{1}{(ar^{n-1})^2}.\right]$$

$$\int_a^b \frac{1}{x^2}\,dx = \operatorname*{Lt}_{r \to 1} (r-1) \left[\frac{1}{a^2}.a + \frac{1}{(ar)^2}.ar + \frac{1}{(ar^2)^2}.ar^2 + \ldots + \frac{1}{(ar^{n-1})^2}.ar^{n-1}\right],$$

where $r^n = \dfrac{b}{a}$

$$\left[\operatorname*{Lt}_{r \to 1} (r-1)\,\{f(a).a + f(ar).ar + f(ar^2).ar^2 + \ldots + f(ar^{n-1}).ar^{n-1}\}\ \text{(Art. 83)}\right]$$

$$= \operatorname*{Lt}_{r \to 1} (r-1)\left[\frac{1}{a} + \frac{1}{ar} + \frac{1}{ar^2} + \ldots + \frac{1}{ar^{n-1}}\right]$$

$$= \operatorname*{Lt}_{r \to 1} (r-1)\left[\frac{\frac{1}{a}\left(\frac{1}{r^n} - 1\right)}{\frac{1}{r} - 1}\right] \qquad \left[\frac{a(r^n - 1)}{r-1}\right] \qquad \left[\text{Put } r^n = \frac{b}{a}\ \textit{(Note this step)}\right]$$

$$= \operatorname*{Lt}_{r \to 1} (r-1)\left[\frac{\frac{1}{a}\left(\frac{a}{b} - 1\right)}{\frac{1-r}{r}}\right] = \operatorname*{Lt}_{r \to 1} \frac{(r-1)r}{1-r}\left[\frac{1}{b} - \frac{1}{a}\right]$$

[Cancel $(r-1)$ from the num. and denom.]

$$= \operatorname*{Lt}_{r \to 1} -r\left[\frac{1}{b} - \frac{1}{a}\right] = -\left[\frac{1}{b} - \frac{1}{a}\right] = \frac{1}{a} - \frac{1}{b}.$$

$$\left[\textbf{Check.}\ \int_a^b \frac{1}{x^2}\,dx = \int_a^b x^{-2}\,dx = \left[\frac{x^{-1}}{-1}\right]_a^b = -\left[\frac{1}{x}\right]_a^b = -\left[\frac{1}{b} - \frac{1}{a}\right] = \frac{1}{a} - \frac{1}{b}.\right]$$

EXAMPLE

**Prove by summation that $\int_a^b \frac{1}{\sqrt{x}}\,dx = 2(\sqrt{b} - \sqrt{a})$.

Fundamental theorem of Integral Calculus

84. Equivalence of the two definitions of a definite integral.

If (*i*) f(x) is continuous and single-valued in the interval (a, b),

(*ii*) a, b are finite, and a < b,

then $$\int_a^b f(x)\,dx = \phi(b) - \phi(a), \text{ where } \frac{d}{dx}[\phi(x)] = f(x).$$

$$\frac{d}{dx}[\phi(x)] = f(x), \text{ i.e., } \underset{h\to 0}{\text{Lt}} \frac{\phi(x+h)-\phi(x)}{h} = f(x)$$

[Def. of diff. coeff. (From Diff. Calculus)]

$\therefore$ $$\frac{\phi(x+h)-\phi(x)}{h} = f(x) + \alpha, \text{ where } \alpha \to 0 \text{ when } h \to 0$$

or $$\phi(x+h) - \phi(x) = hf(x) + h\alpha.$$

Put $x = a, a + h, a + 2h, ..., a + \overline{n-1}\,h$, then

$$\phi(a+h) - \phi(a) = hf(a) + h\alpha_1,$$

$$\phi(a+2h) - \phi(a+h) = hf(a+h) + h\alpha_2,$$

$$\phi(a+3h) - \phi(a+2h) = hf(a+2h) + h\alpha_3,$$

$$\ldots \quad \ldots \quad \ldots \quad \ldots$$

$$\phi(a+nh) - \phi(a+\overline{n-1}\,h) = hf(a+\overline{n-1}\,h) + h\alpha_n,$$

where $\alpha_1, \alpha_2, \alpha_3, \ldots\ldots, \alpha_n$ all $\to 0$ when $h \to 0$.

Adding vertically, [Cancelling terms on the L.H.S. diagonally]

$$\phi(a+nh) - \phi(a) = h[f(a) + f(a+h) + f(a+2h) + \ldots + f(a+\overline{n-1}\,h)]$$
$$+ h(\alpha_1 + \alpha_2 + \alpha_3 + \ldots + \alpha_n)$$

[Put $nh = b - a$]

or $$\phi(a+b-a) - \phi(a) = h[f(a) + f(a+h) + f(a+2h) + \ldots + f(a+\overline{n-1}\,h)]$$
$$+ h(\alpha_1 + \alpha_2 + \alpha_3 + \ldots + \alpha_n).$$

Taking the limits when $h \to 0$,

$$\phi(b) - \phi(a) = \underset{h\to 0}{\text{Lt}}\, h[f(a) + f(a+h) + f(a+2h) + \ldots + f(a+\overline{n-1}\,h)]$$
$$+ \underset{h\to 0}{\text{Lt}} \cdot h\,(\alpha_1 + \alpha_2 + \alpha_3 + \ldots + \alpha_n)$$

$$= \int_a^b f(x)\,dx + \underset{h\to 0}{\text{Lt}}\, h\,(\alpha_1 + \alpha_2 + \alpha_3 + \ldots + \alpha_n) \qquad \ldots(1)$$

Now $| h(\alpha_1 + \alpha_2 + \alpha_3 + ... + \alpha_n) |$* $= h | \alpha_1 + \alpha_2 + \alpha_3 + ... + \alpha_n |$

$\leq h [|\alpha_1| + |\alpha_2| + |\alpha_3| + ... + |\alpha_n|]$†

(Let $|\alpha'|$ be the greatest of $|\alpha_1|, |\alpha_2|, |\alpha_3|, ..., |\alpha_n|$)

$< h [|\alpha'| + |\alpha'| + |\alpha'| + ... + |\alpha'|$ (n terms)]

i.e., $< hn |\alpha'|$ [Put $nh = b - a$]

i.e., $< (b - a) |\alpha'|$, which $\to 0$ when $h \to 0$ [$\because$ then $\alpha' \to 0$]

$\therefore \quad \underset{h \to 0}{\text{Lt}}\, h(\alpha_1 + \alpha_2 + \alpha_3 + ... + \alpha_n) = 0$

$\therefore$ from (1),

$$\phi(b) - \phi(a) = \int_a^b f(x)\, dx, \quad \text{or} \quad \int_a^b f(x)\, dx = \phi(b) - \phi(a).$$

Note. Fundamental theorem of Integral Calculus. *The above theorem is called the* **fundamental theorem of Integral Calculus.** *It proves that the two definitions of a definite integral, viz.,*

(1) $\int_a^b f(x)\, dx = \phi(b) - \phi(a)$, *where* $\frac{d}{dx}[\phi(x)] = f(x)$, *(Art. 10)*

and (2) $\int_a^b f(x)\, dx = \underset{h \to 0}{\text{Lt}}\, h[f(a) + f(a + h) + f(a + 2h) + ... + f(a + \overline{n-1}\, h)]$,

where $nh = b - a$, *(Art. 81)*

are equivalent.

Hence in order to find

$$\underset{h \to 0}{\text{Lt}}\, h[f(a) + f(a + h) + f(a + 2h) + ... + f(a + \overline{n-1}\, h)] \quad ...(1)$$

where $nh = b - a$,

which occurs also in many problems of Geometry and Physics, we need not use Algebra (as in solved Exs. 1, 2, Art. 81) or Trigonometry (as in solved Ex. 3, Art. 81), which is generally tedious, and which was done before the invention of Integral Calculus.

Now we find a function $\phi(x)$ *such that* $\frac{d}{dx}[\phi(x)] = f(x)$, *and then find* $\phi(b) - \phi(a)$. *This is the required limit of the sum of the series (1).*

***Modulus. Def. and notation.** The **modulus** of a real number x is defined as its numerical value, and is denoted by $|x|$ [read "mod x"].

Thus $|2| = 2, |-2| = 2.$

Theorem. *The modulus of the product of any (finite) number of numbers is equal to the product of their moduli.*

Thus $|abc ... k| = |a|\,|b|\,|c| ... |k|.$

†**Theorem.** *The modulus of the sum of any (finite) number of numbers is less than, or equal to, the sum of their moduli.*

Thus $|a + b + c + ... + k| \leq |a| + |b| + |c| + ... + |k|.$ [From Algebra]

EXAMPLE

If $\int f(x)\,dx = F(x) + C$, then

$$\underset{n \to \infty}{\text{Lt}}\; h[f(a) + f(a+h) + f(a+2h) + \ldots + f(a + \overline{n-1}\,h)] = F(b) - F(a)$$

where $h = \dfrac{b-a}{n}$, $f(x)$ is a continuous function of x in the domain (a, b), and a and b are fixed finite numbers.

$$\left[\int f(x)\,dx = F(x) + C, \quad \therefore \quad \frac{d}{dx}\left[\int f(x)\,dx\right] = \frac{d}{dx}[F(x) + C]\right.$$

or $$f(x) = \frac{d}{dx}[F(x)], \text{ i.e., } \frac{d}{dx}[F(x)] = f(x).$$

When $n \to \infty$, $h\left(=\dfrac{b-a}{n}\right) \to 0$. Now proceed as in Art. 84.]

85-1. The definition of a definite integral as the limit of a sum enables us to find the limits of the sums of series of a certain type and also the limits of products of a certain type.

Rule to find, by integration, the limit, when n → ∞, of the sum of a series of n* terms :

(i) Find the rth term ; if it involves r but not r – 1, it is the general term. But if it involves r – 1, find the (r + 1)th term, which is then the general term.

(ii) In the general term, put $n = \dfrac{1}{h}$, *and simplify it so as to get h as a factor of the numerator.*

(iii) Put rh = x, h = dx.

(iv) Integrate the resulting expression between the limits 0 and 1. This is the required limit.

[**Proof. $\displaystyle\int_a^b f(x)\,dx = \underset{h \to 0}{\text{Lt}}\; h[f(a) + f(a+h) + f(a+2h) + \ldots + f(a + \overline{n-1}\,h)],$

where $nh = b - a$.

Put $a = 0$, $b = 1$, then

$$\int_0^1 f(x)\,dx = \underset{h \to 0}{\text{Lt}}\; h[f(0) + f(h) + f(2h) + \ldots + f(\overline{n-1}\,h)],$$

where $nh = 1 - 0 = 1$

$\left[\text{Substitute the value of } h\left(=\dfrac{1}{n}\right)\right]$

$$\underset{n \to \infty}{\text{Lt}}\; \frac{1}{n}\left[f(0) + f\left(\frac{1}{n}\right) + f\left(\frac{2}{n}\right) + \ldots + f\left(\frac{n-1}{n}\right)\right] \quad \left[\because \text{ when } h \to 0, n\left(=\frac{1}{h}\right) \to \infty\right]$$

i.e., $$\underset{n \to \infty}{\text{Lt}}\; \frac{1}{n}\left[f(0) + f\left(\frac{1}{n}\right) + f\left(\frac{2}{n}\right) + \ldots + f\left(\frac{n-1}{n}\right)\right] = \int_0^1 f(x)\,dx \qquad \ldots(1)$$

The series on the L.H.S. of (1) is

$$\frac{1}{n}\left[f(0)+f\left(\frac{1}{n}\right)+f\left(\frac{2}{n}\right)+\ldots+f\left(\frac{n-1}{n}\right)\right].$$

(*i*) rth term $= \dfrac{1}{n} f\left(\dfrac{r-1}{n}\right)$, which involves $r - 1$; changing r to $r + 1$, $(r + 1)$th term $= \dfrac{1}{n} f\left(\dfrac{r}{n}\right)$, which is the general term.

(*ii*) Putting $n = \dfrac{1}{h}$, this $= hf(rh)$.

(*iii*) Putting $rh = x$, $h = dx$, this $= f(x)\, dx$.

(*iv*) Integrating this between the limits 0 and 1, we get $\int_0^1 f(x)\, dx$, which is the R.H.S. of (1).]

***Note 1.** *If one term (or a finite number of terms) is added to (or removed from) the series of n terms, the limit, when* $n \to \infty$, *of the sum of the series, found by integration, remains unchanged.*

For, each term of the series on the L.H.S. of (1) $\to 0$ when $n \to \infty$.

$\left(\because \text{ then the factor } \dfrac{1}{n} \to 0\right)$

Note 2. *If the number of terms of the series (instead of being n) is 2n or 3n, the limits of integration are not 0 and 1.* See Misc. Ex. 20, Chap. VIII, in which it is shown how to find the limits of integration in such a case.

Example 1. *Determine by integration the limiting value of the sum of the following series when n is indefinitely great :*

$$\frac{1}{n+1}+\frac{1}{n+2}+\frac{1}{n+3}+\ldots+\frac{1}{n+n}.$$

The series is $\dfrac{1}{n+1}+\dfrac{1}{n+2}+\dfrac{1}{n+3}+\ldots+\dfrac{1}{n+n}$.

(*i*) rth term $= \dfrac{1}{n+r}$, which involves r but not $r - 1$, and is, therefore, the general term.

(*ii*) Putting $n = \dfrac{1}{h}$, this $= \dfrac{1}{\frac{1}{h}+r} = \dfrac{h}{1+rh}$. [$h$ a factor of the num.]

(*iii*) Putting $rh = x$, $h = dx$, this $= \dfrac{dx}{1+x}$.

(*iv*) Integrating this between the limits 0 and 1,

$$\int_0^1 \frac{dx}{1+x} = \int_0^1 \frac{1}{1+x}\, dx = \Big[\log(1+x)\Big]_0^1 = \log 2 - \log 1 = \log 2,$$

which is the required limit.

Example 2. *Determine by integration the limit to which the sum of the following series tends as n tends to infinity :*

$$\frac{n}{n^2}+\frac{n}{n^2+1^2}+\frac{n}{n^2+2^2}+\ldots+\frac{n}{n^2+(n-1)^2}.$$

The series is $\dfrac{n}{n^2}+\dfrac{n}{n^2+1^2}+\dfrac{n}{n^2+2^2}+\ldots+\dfrac{n}{n^2+(n-1)^2}$.

(*i*) rth term $=\dfrac{n}{n^2+(r-1)^2}$, which involves $r-1$; changing r to $r+1$ $(r+1)$th term $=\dfrac{n}{n^2+r^2}$, which is the general term.

(*ii*) Putting $n=\dfrac{1}{h}$, this $=\dfrac{\frac{1}{h}}{\frac{1}{h^2}+r^2}=\dfrac{h}{1+(rh)^2}$. [$h$ a factor of the num.]

(*iii*) Putting $rh=x$, $h=dx$, this $=\dfrac{dx}{1+x^2}$.

(*iv*) Integrating this between the limits 0 and 1,

$$\int_0^1\frac{dx}{1+x^2}=\int_0^1\frac{1}{(1)^2+x^2}\,dx=\left[\frac{1}{1}\tan^{-1}\frac{x}{1}\right]_0^1=\tan^{-1}1-\tan^{-1}0$$

$$=\frac{\pi}{4}-0=\frac{\pi}{4},$$

which is the required limit.

EXAMPLES

1. Determine by integration the limiting value of the sums of the following series when n is indefinitely great :

(*i*) $\dfrac{n}{n^2+1^2}+\dfrac{n}{n^2+2^2}+\dfrac{n}{n^2+3^2}+\ldots+\dfrac{n}{n^2+n^2}$.

(*ii*) $\dfrac{1}{\sqrt{2n-1^2}}+\dfrac{1}{\sqrt{4n-2^2}}+\dfrac{1}{\sqrt{6n-3^2}}+\ldots+\dfrac{1}{\sqrt{2n^2-n^2}}$.

(*iii*) $\dfrac{1}{n}\left[\sin^{2k}\dfrac{\pi}{2n}+\sin^{2k}\dfrac{2\pi}{2n}+\sin^{2k}\dfrac{3\pi}{2n}+\ldots+\sin^{2k}\dfrac{\pi}{2}\right]$, k being an integer.

[(*ii*) It will be found that the general term $\left(\text{putting } n=\dfrac{1}{h}\right)$

$$=\frac{\sqrt{h}}{\sqrt{2r-r^2h}}$$

[Multiply the num. and denom. by $\sqrt{h}$ (to get h as a factor of the num.)
(*Note this step*)]

$$= \frac{h}{\sqrt{2rh-(rh)^2}}.$$

(*iii*) To evaluate $\int_0^1 \sin^{2k} \frac{\pi}{2} x\, dx$. Put (the angle, *i.e.,*) $\frac{\pi}{2} x = z$.

Note. *It is assumed that k is a positive integer.]*

2. Prove that

(*i*) $\underset{n\to\infty}{\text{Lt}} \frac{1}{n}\left[\left(\frac{1}{n}\right)^m + \left(\frac{2}{n}\right)^m + \ldots + \left(\frac{n-1}{n}\right)^m\right] = \frac{1}{m+1}$ or ∞ according as $m+1$ is positive or negative.

(*ii*) $\underset{n\to\infty}{\text{Lt}} \frac{1^m+2^m+3^m+\ldots+n^m}{n^{m+1}} = \frac{1}{m+1}$ or ∞ according as $m+1$ is positive or negative.

3. Determine by integration the limit to which the sum of the following series tends as n tends to infinity :

(*i*) $\frac{1}{n} + \frac{1}{n+1} + \frac{1}{n+2} + \ldots + \frac{1}{n+(n-1)}$.

(*ii*) $\frac{1}{n} + \frac{1}{\sqrt{n^2-1^2}} + \frac{1}{\sqrt{n^2-2^2}} + \ldots + \frac{1}{\sqrt{n^2-(n-1)^2}}$.

4. (*i*) Evaluate $\underset{n\to\infty}{\text{Lt}} \sum_{r=0}^{n-1} \frac{1}{\sqrt{n^2-r^2}}$.

(*ii*) Prove that, as $n\to\infty$, $n \sum_{p=0}^{n-1} \frac{1}{n^2+p^2} \to \frac{\pi}{4}$.

[(*i*) The general term $= \frac{1}{\sqrt{n^2-r^2}}$.]

5. (*a*) Evaluate $\underset{n\to\infty}{\text{Lt}} \left(\sum_{r=1}^{n} \frac{r^3}{r^4+n^4}\right)$.

(*b*) Show that the limit when n is increased indefinitely of

$$\frac{(n-m)^{\frac{1}{3}}}{n} + \frac{(2^2 n-m)^{\frac{1}{3}}}{2n} + \frac{(3^2 n-m)^{\frac{1}{3}}}{3n} + \ldots + \frac{(n^3-m)^{\frac{1}{3}}}{n^2} \text{ is } \frac{3}{2}.$$

[(*i*) rth term $= \frac{(r^2 n-m)^{\frac{1}{3}}}{rn}$, which involves r but not $r-1$, and is, therefore, the general term.

(*ii*) Putting $n = \frac{1}{h}$, this $= \dfrac{\left(\dfrac{r^2}{h} - m\right)^{\frac{1}{3}}}{\dfrac{r}{h}} = \dfrac{(r^2 - mh)^{\frac{1}{3}}}{h^{\frac{1}{3}} r} h$ which, when $h \to 0$, is ultimately

$$= \frac{(r^2 - 0)^{\frac{1}{3}}}{h^{\frac{1}{3}} r} h \qquad (\because \ m \text{ is finite and } h \to 0, \ \therefore \ mh \to 0)$$

$$= \frac{1}{(rh)^{\frac{1}{3}}} h. \ (h \text{ a factor of the num.})]$$

85-II. Rule to find, by integration, the limit, when n → ∞, of a product of n factors :

(*i*) *Let* $l = \underset{n \to \infty}{\text{Lt}}$ *product.*

(*ii*) *Take logs of both sides, and find, by integration, the limit, when* $n \to \infty$, *of the sum of the series of n terms on the R.H.S. [Rule (Art. 85-I)]. This gives log l.*

(*iii*) *Now find l.*

Example. *Find the limiting value of* $(n\,!)^{\frac{1}{n}}/n$ *when n is infinite.*

(*i*) Let
$$l = \underset{n \to \infty}{\text{Lt}} \frac{(n\,!)^{\frac{1}{n}}}{n} = \underset{n \to \infty}{\text{Lt}} \left(\frac{n\,!}{n^n}\right)^{\frac{1}{n}}$$

$$= \underset{n \to \infty}{\text{Lt}} \left[\frac{1.2.3 \ldots n}{n.n.n \ldots n}\right]^{\frac{1}{n}}$$

$$= \underset{n \to \infty}{\text{Lt}} \left(\frac{1}{n} \cdot \frac{2}{n} \cdot \frac{3}{n} \cdots \frac{n}{n}\right)^{\frac{1}{n}}.$$

(*ii*) Taking logs of both sides,

$$\log l = \underset{n \to \infty}{\text{Lt}} \frac{1}{n}\left[\log \frac{1}{n} + \log \frac{2}{n} + \log \frac{3}{n} + \ldots + \log \frac{n}{n}\right] \qquad \ldots(1)$$

The series on the R.H.S. is

$$\frac{1}{n}\left[\log \frac{1}{n} + \log \frac{2}{n} + \log \frac{3}{n} + \ldots + \log \frac{n}{n}\right].$$

rth term $= \frac{1}{n} \log \frac{r}{n}$, which involves r but not $r - 1$, and is, therefore, the general term.

Putting $n = \frac{1}{h}$, this $= h \log rh$.

Putting $rh = x$, $h = dx$, this $= \log x \, dx$.

Integrating this between the limits 0 and 1,

$$\int_0^1 \log x \, dx = \int_0^1 \log x \, . \, 1 \, dx \qquad \text{[Integrate by parts]}$$

$$= \Big[\log x . x \Big]_0^1 - \int_0^1 \frac{1}{x} . x \, dx = - \int_0^1 1 \, dx = - \Big[x \Big]_0^1 = -1.$$

$\therefore \qquad \log l = -1.$

(*iii*) $\therefore \qquad l = e^{-1}.$

EXAMPLES

1. Prove that $\lim\limits_{n \to \infty} \left\{ \left(1+\frac{1}{n^2}\right)\left(1+\frac{2^2}{n^2}\right)\left(1+\frac{3^2}{n^2}\right)...\left(1+\frac{n^2}{n^2}\right) \right\}^{\frac{1}{n}}$

is equal to $2e^{\frac{\pi-4}{2}}$.

2. If na is always equal to unity and n is indefinitely great, show that the limiting value of the product

$$\{1+a^4\}\{1+(2a)^4\}^{\frac{1}{2}}\{1+(3a)^4\}^{\frac{1}{3}}\{1+(4a)^4\}^{\frac{1}{4}}...\{1+(na)^4\}^{\frac{1}{n}} \text{ is } e^{\frac{\pi^2}{48}}.$$

$\Big[\because \; na = 1, \; \therefore \; a = \frac{1}{n}.$

$$\therefore \quad \underset{n \to \infty}{\text{Lt}} \{1+a^4\}\{1+(2a)^4\}^{\frac{1}{2}}\{1+(3a)^4\}^{\frac{1}{3}}\{1+(4a)^4\}^{\frac{1}{4}}...\{1+(na)^4\}^{\frac{1}{n}}$$

$$= \underset{n \to \infty}{\text{Lt}} \left\{1+\frac{1}{n^4}\right\}\left\{1+\left(\frac{2}{n}\right)^4\right\}^{\frac{1}{2}}\left\{1+\left(\frac{3}{n}\right)^4\right\}^{\frac{1}{3}}\left\{1+\left(\frac{4}{n}\right)^4\right\}^{\frac{1}{4}}...\left\{1+\left(\frac{n}{n}\right)^4\right\}^{\frac{1}{n}}. \Bigg]$$

3. Find the limiting value when n is infinite of the nth part of the sum of the n quantities

$$\frac{n+1}{n}, \frac{n+2}{n}, \frac{n+3}{n}, ..., \frac{n+n}{n};$$

and show that it is to the limiting value of the nth root of the product of the same quantities as $3e : 8$, where e is the base of the Naperian logarithms.

**SECTION IV

DEFINITE INTEGRALS WITH DISCONTINUOUS INTEGRANDS

****86. In the definition of $\int_a^b f(x)\,dx$ (Art. 81), it is assumed that f(x) is continuous in the interval (a, b). If, however, f(x) is continuous in the interval (a, b) except that it becomes infinite at some point of**

the interval, we have the following definitions :

Def. 1. [*Integrand infinite at lower limit.*]

If $f(x)$ is continuous in the interval (a, b) except that $f(x) \to \infty$ when $x \to a$, then

$$\int_a^b f(x)\,dx = \underset{h \to 0}{\text{Lt}} \int_{a+h}^b f(x)\,dx, \text{ where } h \text{ is positive,}$$

provided the limit is a definite number.

Def. 2. [*Integrand infinite at upper limit.*]

If $f(x)$ is continuous in the interval (a, b) except that $f(x) \to \infty$ when $x \to b$, then

$$\int_a^b f(x)\,dx = \underset{h \to 0}{\text{Lt}} \int_a^{b-h} f(x)\,dx, \text{ where } h \text{ is positive,}$$

provided the limit is a definite number.

Def. 3. [*Integrand infinite at a point between the limits of integration.*]

If $f(x)$ is continuous in the interval (a, b) except that $f(x) \to \infty$ when $x \to c$, where $a < c < b$, then

$$\int_a^b f(x)\,dx = \underset{h \to 0}{\text{Lt}} \int_a^{c-h} f(x)\,dx + \underset{h' \to 0}{\text{Lt}} \int_{c+h'}^b f(x)\,dx^*,$$

where h, h' are positive,

provided each limit is a definite number.

* [**Aid to memory.** $\int_a^b f(x)\,dx = \int_a^c f(x)\,dx + \int_c^b f(x)\,dx$. Now use defs. 1 and 2.

****Note 1. Integral does not exist. Def.** *If the limit is not a definite number, the integral is said **not to exist**. [Cf. Note 2, Art. 11.]*

Note 2. *The above definitions (1 – 3) hold also if $f(x) \to -\infty$ when $x \to a$, or when $x \to b$, or when $x \to c$ (where $a < c < b$), or if the limits when $x \to c$ from the right and when $x \to c$ from the left are infinite but of opposite signs.*

****Example 1.** *Find* $\int_0^1 \frac{dx}{\sqrt{x}}$.

Here $\frac{1}{\sqrt{x}} \to \infty$ when $x \to 0$.

$$\therefore \quad \int_0^1 \frac{1}{\sqrt{x}}\,dx = \underset{h \to 0}{\text{Lt}} \int_{0+h}^1 \frac{1}{\sqrt{x}}\,dx \qquad \text{[Def. 1, Art. 86]}$$

$$= \underset{h\to 0}{\text{Lt}} \int_h^1 x^{-\frac{1}{2}}\, dx = \underset{h\to 0}{\text{Lt}} \left[\frac{x^{\frac{1}{2}}}{\frac{1}{2}}\right]_h^1 = \underset{h\to 0}{\text{Lt}}\ 2(1 - h^{\frac{1}{2}}) = 2.$$

****Example 2.** *Find* $\int_0^1 \frac{dx}{\sqrt{1-x^2}}$.

Here $\frac{1}{\sqrt{1-x^2}} \to \infty$ when $x \to 1$.

$$\therefore \quad \int_0^1 \frac{1}{\sqrt{1-x^2}}\, dx = \underset{h\to 0}{\text{Lt}} \int_0^{1-h} \frac{1}{\sqrt{1-x^2}}\, dx \qquad \text{[Def. 2, Art. 86]}$$

$$= \underset{h\to 0}{\text{Lt}} \int_0^{1-h} \frac{1}{\sqrt{(1)^2 - x^2}}\, dx = \underset{h\to 0}{\text{Lt}} \left[\sin^{-1}\frac{x}{1}\right]_0^{1-h}$$

$$= \underset{h\to 0}{\text{Lt}}\ [\sin^{-1}(1-h) - \sin^{-1} 0]$$

$$= \underset{h\to 0}{\text{Lt}}\ \sin^{-1}(1-h) - 0 = \sin^{-1} 1 = \frac{\pi}{2}.$$

****Example 3.** *Find* $\int_{-1}^1 \frac{dx}{x^2}$.

Here $\frac{1}{x^2} \to \infty$ when $x \to 0$, *i.e.*, at a pt. between the limits of integration -1 and 1.

$$\therefore \quad \int_{-1}^1 \frac{1}{x^2}\, dx = \underset{h\to 0}{\text{Lt}} \int_{-1}^{0-h} \frac{1}{x^2}\, dx + \underset{h'\to 0}{\text{Lt}} \int_{0+h'}^1 \frac{1}{x^2}\, dx \qquad \text{[Def. 3, Art. 86]}$$

$$= \underset{h\to 0}{\text{Lt}} \left[-\frac{1}{x}\right]_{-1}^{-h} + \underset{h'\to 0}{\text{Lt}} \left[-\frac{1}{x}\right]_{h'}^1$$

$$= \underset{h\to 0}{\text{Lt}}\ -\left[-\frac{1}{h} - (-1)\right] + \underset{h'\to 0}{\text{Lt}}\ -\left[1 - \frac{1}{h'}\right]$$

$$= \underset{h\to 0}{\text{Lt}} \left(\frac{1}{h} - 1\right) + \underset{h'\to 0}{\text{Lt}} \left(-1 + \frac{1}{h'}\right) \qquad ...(1)$$

Now $\underset{h\to 0}{\text{Lt}} \left(\frac{1}{h} - 1\right)$ is not a definite number, also

$\underset{h'\to 0}{\text{Lt}} \left(-1 + \frac{1}{h'}\right)$ is not a definite number,

$\therefore$ the integral does not exist. [Def. Note 1, Art. 86]

****Caution.** It is a *common mistake* to work blindly thus :

$$\int_{-1}^1 \frac{1}{x^2} dx = \left[-\frac{1}{x}\right]_{-1}^1 = -[1 - (-1)] = -2,$$

which is wrong. For, the

integrand $\frac{1}{x^2}$ is + ve in the interval (– 1, 1),

$\therefore$ the value of the integral cannot be – ve. [Note 2, Art. 81]

The mistake has arisen because the theorem $\int_a^b f(x)\,dx = \phi(b) - \phi(a)$, where $\frac{d}{dx}[\phi(x)] = f(x)$ (Art. 84) holds only if $f(x)$ is *continuous* in the interval (a, b), and, therefore, does not become infinite at any point of the interval. Here $f(x) = \frac{1}{x^2} \to \infty$ when $x \to 0$, *i.e.*, at a point of the interval (– 1, 1).

MISCELLANEOUS EXAMPLES ON CHAPTER VIII

1. If m and n are positive and m is an integer, prove that

$$\int_0^1 x^{n-1}(1-x)^{m-1}\,dx = \int_0^1 x^{m-1}(1-x)^{n-1}\,dx = \frac{1.2.3\ldots(m-1)}{n(n+1)\ldots(n+m-1)}.$$

2. Evaluate $\int_0^{\frac{\pi}{2}} \frac{dx}{1+\cot x}$.

3. (*i*) Prove that $\int_0^{\frac{\pi}{4}} \log(1+\tan\theta)\,d\theta = \frac{\pi}{8}\log 2$.

(*ii*) Evaluate $\int_0^1 \frac{\log(1+x)}{1+x^2}\,dx$.

4. Prove that (*i*) $\int_0^a \frac{dx}{x+(a^2-x^2)^{\frac{1}{2}}} = \frac{\pi}{4}$.

(*ii*) $\int_0^\pi \frac{x\tan x\,dx}{\sec x+\tan x} = \pi\left(\frac{\pi}{2}-1\right)$.

5. Evaluate

(*i*) $\int_0^\pi \cos^6 x\,dx$. (*ii*) $\int_0^\pi \sin^6 x\cos^4 x\,dx$.

****6.** Prove that $\int_0^{2\pi} \frac{dx}{a+b\cos x+c\sin x} = \frac{2\pi}{\sqrt{a^2-b^2-c^2}}$,

when $a > \sqrt{b^2+c^2} > 0$.

[Put $b = r \cos \alpha$, $c = r \sin \alpha$. It will be found that

$$\int_0^{2\pi} \frac{dx}{a + b \cos x + c \sin x} = \int_0^{2\pi} \frac{dx}{a + r \cos (x - \alpha)} \qquad \text{(Put } x - \alpha = z)$$

$$= \int_{-\alpha}^{2\pi - \alpha} \frac{dz}{a + r \cos z} = \int_{-\alpha}^{0} \frac{dz}{a + r \cos z} + \int_0^{2\pi - \alpha} \frac{dz}{a + r \cos z} \qquad ...(1)$$

In the first integral on the R.H.S., *put* $u = 2\pi + z$*, *(Note this step)*

then $du = dz$; when $z = -\alpha$, $u = 2\pi - \alpha$, when $z = 0$, $u = 2\pi$.

$$\therefore \quad \int_{-\alpha}^{0} \frac{dz}{a + r \cos z} = \int_{2\pi - \alpha}^{2\pi} \frac{du}{a + r \cos u} = \int_{2\pi - \alpha}^{2\pi} \frac{dz}{a + r \cos z}.$$

$\therefore$ from (1),

$$\int_0^{2\pi} \frac{dx}{a + b \cos x + c \sin x} = \int_{2\pi - \alpha}^{2\pi} \frac{dz}{a + r \cos z} + \int_0^{2\pi - \alpha} \frac{dz}{a + r \cos z}$$

$$= \int_0^{2\pi - \alpha} \frac{dz}{a + r \cos z} + \int_{2\pi - \alpha}^{2\pi} \frac{dz}{a + r \cos z} = \int_0^{2\pi} \frac{dz}{a + r \cos z}.$$

Now evaluate $\int_0^{2\pi} \frac{dz}{a + r \cos z}$. Use theorem VI, Cor. (*i*)′, Art. 78.]

7. Evaluate $\int_0^{2\pi} \frac{\sin^2 \phi}{a - b \cos \phi} \, d\phi$. $\quad (a > b > 0)$.

8. (*i*) Evaluate $\int_0^{\pi} \theta \sin^6 \theta \cos^4 \theta \, d\theta$.

(*ii*) Apply the substitution $x = \pi - y$ to the integral $\int_0^{\pi} x \sin^6 x \cos^4 x \, dx$ and hence obtain its value.

****9.** Evaluate $\int_0^{2a} (2ax - x^2)^{\frac{n}{2}} \operatorname{vers}^{-1} \frac{x}{a} \, dx$.

Put $\frac{x}{a} = \operatorname{vers} \theta = 1 - \cos \theta$, or $x = a(1 - \cos \theta)$,

---

***Why this step.** To get the new limits of integration as $2\pi - \alpha$ and 2π.

then $dx = a \sin\theta\, d\theta$; when $x = 0$, $\theta = 0$, when $x = 2a$, $\theta = \pi$.

$$\therefore \quad \int_0^{2a} (2ax - x^2)^{\frac{n}{2}} \text{ vers}^{-1} \frac{x}{a}\, dx$$

$$= \int_0^{\pi} [2a.a(1-\cos\theta) - a^2(1-\cos\theta)^2]^{\frac{n}{2}}\, \theta.\, a \sin\theta\, d\theta$$

$$= a^{n+1} \int_0^{\pi} [2 - 2\cos\theta - (1 - 2\cos\theta + \cos^2\theta)]^{\frac{n}{2}}\, \theta \sin\theta\, d\theta$$

$$= a^{n+1} \int_0^{\pi} (1-\cos^2\theta)^{\frac{n}{2}}\, \theta \sin\theta\, d\theta$$

$$= a^{n+1} \int_0^{\pi} \theta \sin^{n+1}\theta\, d\theta \qquad ...(1)$$

[To evaluate $\int_0^{\pi} \theta \sin^{n+1}\theta\, d\theta$.]

Let $$I = \int_0^{\pi} \theta \sin^{n+1}\theta\, d\theta \qquad ...(2)$$

Then $$I = \int_0^{\pi} (\pi - \theta) \sin^{n+1}(\pi - \theta)\, d\theta$$

$$= \int_0^{\pi} (\pi - \theta) \sin^{n+1}\theta\, d\theta$$

$$= \pi \int_0^{\pi} \sin^{n+1}\theta\, d\theta - \int_0^{\pi} \theta \sin^{n+1}\theta\, d\theta \qquad ...(3)$$

Adding (2) and (3),

$$2I = \pi \int_0^{\pi} \sin^{n+1}\theta\, d\theta \qquad [\text{But } \sin^{n+1}(\pi-\theta) = \sin^{n+1}\theta]$$

$$= \pi \,.\, 2 \int_0^{\frac{\pi}{2}} \sin^{n+1}\theta\, d\theta$$

$$\therefore \quad I = \pi \int_0^{\frac{\pi}{2}} \sin^{n+1}\theta\, d\theta \qquad ...(4)$$

Case I. If n is a + ve odd integer, then $n + 1$ is a + ve even integer,

$\therefore$ from (4),

$$I = \pi \frac{n(n-2) \ldots 1}{(n+1)(n-1) \ldots 2} \cdot \frac{\pi}{2},$$

i.e., from (2), $\displaystyle\int_0^{\pi} \theta \sin^{n+1}\theta\, d\theta = \frac{n(n-2) \ldots 1}{(n+1)(n-1) \ldots 2} \cdot \frac{\pi^2}{2}$

$\therefore$ from (1), $\displaystyle\int_0^{2a} (2ax - x^2)^{\frac{n}{2}} \text{vers}^{-1} \frac{x}{a}\, dx = a^{n+1} \frac{n(n-2)\ldots 1}{(n+1)(n-1)\ldots 2} \cdot \frac{\pi^2}{2}$.

Case II. **If n is a + ve even integer, then $n + 1$ is a +ve odd integer,**

$\therefore$ from (4), $$\text{I} = \pi \frac{n(n-2)\ldots 2}{(n+1)(n-1)\ldots 3},$$

i.e., from (2), $$\int_0^{\pi} \theta \sin^{n+1} \theta\, d\theta = \frac{n(n-2)\ldots 2}{(n+1)(n-1)\ldots 3} \pi$$

$\therefore$ from (1), $$\int_0^{2a} (2ax - x^2)^{\frac{n}{2}} \text{vers}^{-1} \frac{x}{a}\, dx = a^{n+1} \frac{n(n-2)\ldots 2}{(n+1)(n-1)\ldots 3} \pi.$$

10. Evaluate $\displaystyle\int_0^{\pi} \frac{x\, dx}{a^2 \cos^2 x + b^2 \sin^2 x}$.

****11.** Evaluate $\displaystyle\int_0^{\pi} \sin^3 \theta\, (1 + 2 \cos \theta)(1 + \cos \theta)^2\, d\theta$.

[Multiply out the integrand, and use theorem VI, Cors. $(i)'$, $(ii)'$, Art. 78.]

12. Show that $\displaystyle\int_0^{\pi} \log (1 + \cos x)\, dx = \pi \log \tfrac{1}{2}$.

13. Evaluate $\displaystyle\int_0^{\infty} \frac{\log (1 + x^2)}{1 + x^2}\, dx$.

14. Find the value of $\displaystyle\int_0^1 \frac{\sin^{-1} x}{x}\, dx$.

[Put $x = \sin \theta$, and then integrate by parts.]

15. (*a*) Evaluate $\displaystyle\int_0^1 \frac{1}{x} \log \frac{1+x}{1-x}\, dx$.

(*b*) Show that $\displaystyle\int_0^{\infty} \frac{x}{1+e^x}\, dx = \frac{\pi^2}{12}$.

[(*b*) $\displaystyle\int_0^{\infty} \frac{x}{1+e^x}\, dx$ *(Divide the num. and denom. by e^x)*

$$= \int_0^{\infty} \frac{x e^{-x}}{e^{-x} + 1} dx.$$

Put $e^{-x} = z$*, then $-x \log e = \log z$, or $x = -\log z$, $\therefore$ $dx = -\frac{1}{z} dz$.]

16. (*a*) Find the limit, when $n \to \infty$, of the series

(*i*) $\frac{1}{n^3}(1 + 4 + 9 + 16 + ... + n^2)$.

(*ii*) $\frac{1}{1+n^3} + \frac{4}{8+n^3} + \frac{9}{27+n^3} + ... + \frac{r^2}{r^3+n^3} + ... + \frac{1}{2n}$.

(*b*) Show that

$\lim_{n \to \infty} [\{\sqrt{n+1} + \sqrt{n+2} + ... + \sqrt{2n}\}/n\sqrt{n}] = \frac{2}{3}(2\sqrt{2} - 1)$.

****17.** Find the limit, when n tends to infinity, of the series

(*i*) $\frac{n}{(n+1)\sqrt{(2n+1)}} + \frac{n}{(n+2)\sqrt{\{2(2n+2)\}}} + \frac{n}{(n+3)\sqrt{\{3(2n+3)\}}} + ... + \frac{n}{2n\sqrt{(n \cdot 3n)}}$.

(*ii*) $\frac{\sqrt{n}}{(3+4\sqrt{n})^2} + \frac{\sqrt{n}}{\sqrt{2}(3\sqrt{2}+4\sqrt{n})^2} + \frac{\sqrt{n}}{\sqrt{3}(3\sqrt{3}+4\sqrt{n})^2} + ... + \frac{1}{49n}$.

18. Find the limit, when $n \to \infty$, of the series

(*i*) $\frac{1}{n} + \frac{n^2}{(n+1)^3} + \frac{n^2}{(n+2)^3} + ... + \frac{1}{8n}$.

(*ii*) $\frac{n^{\frac{1}{2}}}{n^{\frac{3}{2}}} + \frac{n^{\frac{1}{2}}}{(n+3)^{\frac{3}{2}}} + \frac{n^{\frac{1}{2}}}{(n+6)^{\frac{3}{2}}} + ... + \frac{n^{\frac{1}{2}}}{[n+3(n-1)]^{\frac{3}{2}}}$.

19. Evaluate $\lim_{n \to \infty} \sum_{r=1}^{n-1} \frac{1}{n} \sqrt{\left(\frac{n+r}{n-r}\right)}$.

****20.** Find the limit, when n tends to infinity, of the sum

$$\frac{1}{n} + \frac{1}{n+1} + \frac{1}{n+2} + ... + \frac{1}{3n}.$$

[(*To find the number of terms.*) $n, n+1, n+2, ..., 3n$ is an A.P.

Let the number of terms be m. Then

$3n = n + (m-1)(1)$, or $m = 2n + 1$, $\therefore$ the number of terms is *not* $= n$

(*To find the limits of integration.*)

***Why this step.** In order to get the new limits of integration as 0 and we put $e^{-x} = z$.

It will be found that the general term $= \dfrac{1}{n+r}$.

Putting (general term, *i.e.,*) $\dfrac{1}{n+r}$ = (first term, *i.e.,*) $\dfrac{1}{n}$, $r = 0$; again putting (general term, *i.e.,*) $\dfrac{1}{n+r}$ = (last term, *i.e.,*) $\dfrac{1}{3n}$, $r = 2n$.

$\therefore$ r varies from 0 to $2n$, (multiply by h), or rh varies from 0 to $2nh$ $\left(\text{But } n = \dfrac{1}{h}\right)$

$\therefore$ rh varies from 0 to 2, (But $rh = x$)

$\therefore$ x varies from 0 to 2, which are the limits of integration.]

21. Find the limit, when n tends to infinity, of the product

$$(i)\ \left\{\left(1+\frac{1}{n}\right)\left(1+\frac{2}{n}\right)\left(1+\frac{3}{n}\right)\cdots\left(1+\frac{n}{n}\right)\right\}^{\frac{1}{n}}.$$

$$(ii)\ \left(1+\frac{1}{n}\right)\left(1+\frac{2}{n}\right)^{\frac{1}{2}}\left(1+\frac{3}{n}\right)^{\frac{1}{3}}\cdots\left(1+\frac{n}{n}\right)^{\frac{1}{n}}.$$

****22.** Show that the limit when n is infinite of

$$\left\{\phi(a).\phi\left(a+\frac{h}{n}\right).\phi\left(a+\frac{2h}{n}\right)\cdots\phi\left(a+\frac{nh}{n}\right)\right\}^{\frac{1}{n}} \text{ is } e^{\frac{1}{h}\int_a^{a+h}\log\phi(x)\,dx}.$$

Apply this result to find the limit of

$$\left\{\left(1+\frac{1}{n^2}\right)\left(1+\frac{2^2}{n^2}\right)\left(1+\frac{3^2}{n^2}\right)\cdots\left(1+\frac{n^2}{n^2}\right)\right\}^{\frac{1}{n}}.$$

[(*ii*) From part (*i*),

$$\underset{n\to\infty}{\text{Lt}}\left[\phi(a).\phi\left(a+\frac{h}{n}\right)\phi\left(a+\frac{2h}{n}\right)\cdots\phi\left(a+\frac{nh}{n}\right)\right]^{\frac{1}{n}} = e^{\frac{1}{h}\int_a^{a+h}\log\phi(x)\,dx}.$$

Put $a = 0, h = 1$, then

$$\underset{n\to\infty}{\text{Lt}}\left[\phi(0)\phi\left(\frac{1}{n}\right)\phi\left(\frac{2}{n}\right)\cdots\phi\left(\frac{n}{n}\right)\right]^{\frac{1}{n}} = e^{\int_0^1\log\phi(x)\,dx} \qquad \text{...(1)}$$

Compare $\displaystyle \operatorname*{Lt}_{n\to\infty}\left[\left(1+\frac{1}{n^2}\right)\left(1+\frac{2^2}{n^2}\right)\left(1+\frac{3^2}{n^2}\right)\cdots\left(1+\frac{n^2}{n^2}\right)\right]^{\frac{1}{n}}$

with the L.H.S. of (1),

here $\quad \phi\left(\frac{1}{n}\right)=1+\frac{1}{n^2},\ \phi\left(\frac{2}{n}\right)=1+\frac{2^2}{n^2},\ \ldots,\ \phi\left(\frac{n}{n}\right)=1+\frac{n^2}{n^2},$

$\therefore \quad \phi(x) = 1 + x^2.$

$\therefore$ required limit $= e^{\int_0^1 \log(1+x^2)\,dx}$. Integrate $\int_0^1 \log(1+x^2).\ 1\,dx$ by parts.]

CHAPTER IX

RECTIFICATION AND INTRINSIC EQUATION

SECTION I

RECTIFICATION

[☞ *Chapters IX, X and XI should be studied when the student is acquainted with Curve Tracing in Differential Calculus.*

Chapters XII and XIII may, however, be studied immediately after Chapter VIII, if so desired.]

87. Rectification. Def. The process of finding the length of the arc of a curve between two given points is called **rectification.***

I. Arc formulae for Cartesian equation.

88. To prove that the length of the arc of the curve y = f(x) between the points where x = a, x = b is

$$\int_a^b \sqrt{1+\left(\frac{dy}{dx}\right)^2}\, dx.$$

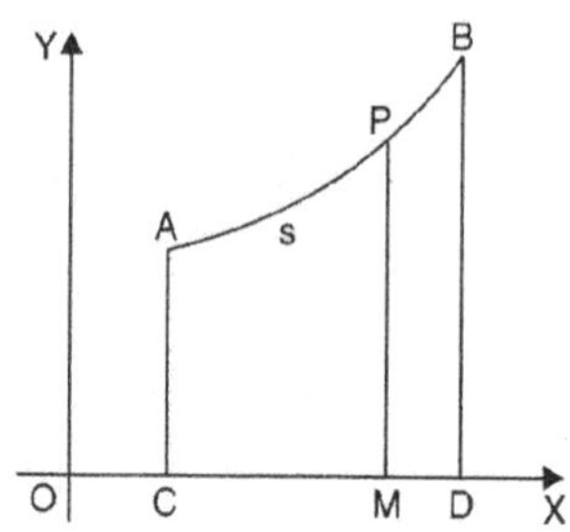

Let AB be the curve $y = f(x)$, and A, B the pts. where $x = a$, $x = b$. Let P(x, y) be any pt. on the curve. Then the arc AP is clearly a function of $x = s$ (say).

We know that $$\frac{ds}{dx} = \sqrt{1+\left(\frac{dy}{dx}\right)^2} \quad ...(1)$$

[From Diff. Calculus]

$$\therefore \quad \int_a^b \sqrt{1+\left(\frac{dy}{dx}\right)^2}\, dx = \int_a^b \frac{ds}{dx}\, dx = \Big[s \Big]_a^b$$

$$= (\text{value of } s \text{ when } x = b) - (\text{value of } s \text{ when } x = a)$$

$$= \text{arc AB} - 0,$$

i.e., $$\text{arc AB} = \int_a^b \sqrt{1+\left(\frac{dy}{dx}\right)^2}\, dx.$$

Cor. From (1), $$s = \int \sqrt{1+\left(\frac{dy}{dx}\right)^2}\, dx.$$

*Literal meaning : *finding straight line equal to* (curve).

Note. Assumptions.

(i) f(x) is continuous and single-valued in the interval (a, b).

(ii) a, b are finite, and a < b.

(iii) The integrand does not change sign in the interval (a, b).

Similarly, *the length of the arc of the curve $x = f(y)$ between the points where $y = a$, $y = b$ is*

$$\int_a^b \sqrt{1+\left(\frac{dx}{dy}\right)^2}\, dy.$$

[Proceed as in Art. 88, and instead of (1), use

$$\frac{ds}{dy} = \sqrt{1+\left(\frac{dx}{dy}\right)^2} \qquad ...(1).]$$

Example. *Find the length of the arc of the parabola $x^2 = 4ay$ extending from the vertex to one extremity of the latus rectum.*

The equation of the parabola is $x^2 = 4ay$.

Let A be the vertex and L one extremity of the latus rectum.

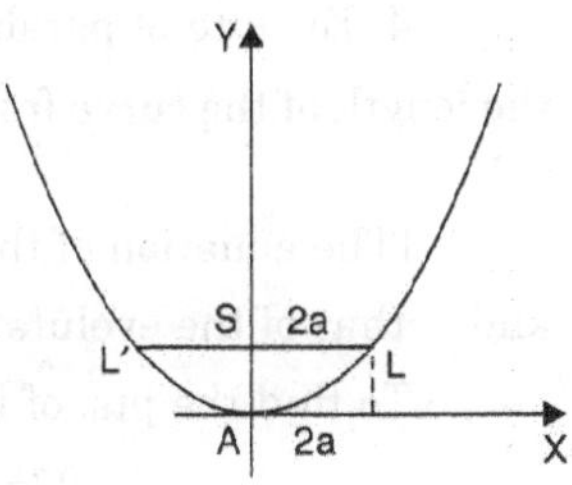

$$y = \frac{x^2}{4a}$$

$$\therefore \quad \frac{dy}{dx} = \frac{1}{4a}(2x) = \frac{x}{2a}$$

$$\therefore \quad \text{arc AL} = \int_0^{2a} \sqrt{1+\left(\frac{dy}{dx}\right)^2}\, dx$$

[$\because$ at A, $x = 0$; and at L, $x = 2a$]

$$= \int_0^{2a} \sqrt{1+\frac{x^2}{4a^2}}\, dx = \frac{1}{2a}\int_0^{2a} \sqrt{4a^2+x^2}\, dx$$

$$= \frac{1}{2a}\int_0^{2a} \sqrt{(2a)^2+x^2}\, dx$$

$$= \frac{1}{2a}\left[\frac{x\sqrt{(2a)^2+x^2}}{2} + \frac{(2a)^2}{2}\sinh^{-1}\frac{x}{2a}\right]_0^{2a}$$

$$= \frac{1}{4a}[2a.2\sqrt{2}\, a + 4a^2 \sinh^{-1} 1 - (0 + 4a^2.0)]$$

$$= \frac{4a^2}{4a}[\sqrt{2} + \sinh^{-1} 1] = a[\sqrt{2} + \log(1+\sqrt{1+(1)^2})]$$

[$\because \sinh^{-1} x = \log(x + \sqrt{1+x^2})$ [Art. 8 (*d*), (1)].

$$= a[\sqrt{2} + \log(1+\sqrt{2})].$$

EXAMPLES

1. (*a*) Show how to find the length of an arc of a curve whose equation is given in the Cartesian form.

(*b*) Find the length of the arc of the parabola $y^2 = 4ax$

(*i*) from the vertex to an extremity of the latus rectum.

(*ii*) cut off by the latus rectum.

2. *Catenary.* Show that in the catenary $y = c \cosh \dfrac{x}{c}$ the length of the arc from the vertex (where $x = 0$) to any point is given by $s = c \sinh \dfrac{x}{c}$.

[Any pt. on the curve is (x, y),

$\therefore$ the limits of integration are $x = 0$ (as given) and $x = x$.]

3. Semicubical parabola. Find the length of the arc of the semicubical parabola $ay^2 = x^3$ from the cusp (where $x = 0$) to any point.

4. Evolute of parabola. In the evolute of a parabola, viz., $4(x - 2a)^3 = 27ay^2$, show that the length of the curve from its cusp $(x = 2a)$ to the point where it meets the parabola is $2a(3\sqrt{3} - 1)$.

[The equation of the parabola is $y^2 = 4ax$ ...(1)

and that of the evolute is $27ay^2 = 4(x - 2a)^3$ ...(2)

To find the pts. of intersection, substituting the value of y^2 from (1) in (2),

$$27a(4ax) = 4(x - 2a)^3 \qquad \text{[Cancel 4]}$$

or $$27a^2x = x^3 - 6x^2a + 12xa^2 - 8a^3$$

or $$x^3 - 6x^2a - 15xa^2 - 8a^3 = 0$$

or $$(x + a)^2(x - 8a) = 0$$

or $$x = -a, -a, 8a.$$

$$\begin{aligned} & x^3 - 6x^2a - 15xa^2 - 8a^3 \\ &= (x + a)(x^2 - 7xa - 8a^2) \\ &= (x + a)(x + a)(x - 8a) \end{aligned}$$

If $x = -a$, then from (1), y is imaginary, $\therefore \; x = 8a$.

$\therefore$ the limits of integration are $x = 2a$ (as given) and $x = 8a$.]

5. Show that the length of an arc of the curve $y^n = x^{m+n}$ can be found in finite terms in the cases when $\dfrac{n}{2m}$ or $\dfrac{n}{2m} + \dfrac{1}{2}$ is an integer.

II. Arc formula for parametric equations.

89. To prove that the length of the arc of the curve x = f(t), y = φ(t) between the points where t = a, t = b is

$$\int_a^b \sqrt{\left(\frac{dx}{dt}\right)^2 + \left(\frac{dy}{dt}\right)^2}\, dt.$$

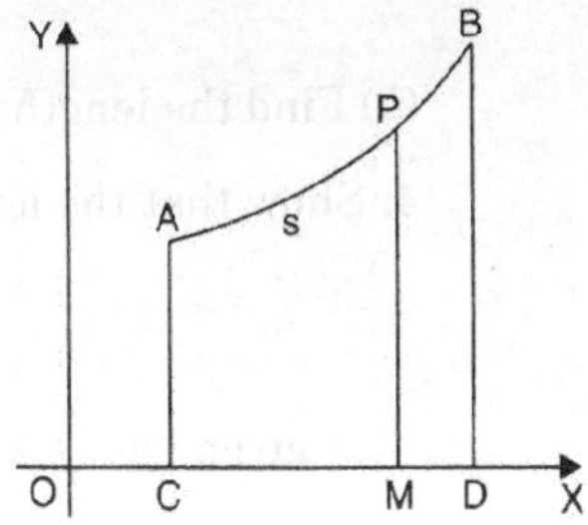

Let AB be the curve $x = f(t)$, $y = \phi(t)$, and A, B the pts. where $t = a$, $t = b$. Let P(x, y) be any pt. on the curve. Then the arc AP is clearly a function of $t = s$ (say).

We know that $$\frac{ds}{dt} = \sqrt{\left(\frac{dx}{dt}\right)^2 + \left(\frac{dy}{dt}\right)^2} \quad ...(1)$$

[From Diff. Calculus]

$$\therefore \quad \int_a^b \sqrt{\left(\frac{dx}{dt}\right)^2 + \left(\frac{dy}{dt}\right)^2}\, dt = \int_a^b \frac{ds}{dt}\, dt = \Big[s\Big]_a^b$$

= (value of s when $t = b$) – (value of s when $t = a$)

= arc AB – 0,

i.e., $$\text{arc AB} = \int_a^b \sqrt{\left(\frac{dx}{dt}\right)^2 + \left(\frac{dy}{dt}\right)^2}\, dt.$$

Cor. From (1), $$\mathbf{s} = \int \sqrt{\left(\frac{\mathbf{dx}}{\mathbf{dt}}\right)^2 + \left(\frac{\mathbf{dy}}{\mathbf{dt}}\right)^2}\, \mathbf{dt}.$$

EXAMPLES

1. Find the length of the arc of the cycloid
$$\left.\begin{aligned} x &= a(\theta + \sin\theta), \\ y &= a(1 - \cos\theta), \end{aligned}\right\}$$
between the points for which $\theta = 0$ and $\theta = 2\psi$.

2. Show that in the ellipse $x = a\cos\theta$, $y = b\sin\theta$, the perimeter may be expressed as
$$2a\pi\left\{1 - \frac{1}{2^2}e^2 - \frac{1^2.3}{2^2.4^2}e^4 - \frac{1^2.3^2.5}{2^2.4^2.6^2}e^6 - \ldots \text{ to } \infty\right\}.$$

[*The ellipse is symmetrical about both axes, and in the first quadrant, θ varies from 0 to* $\frac{\pi}{2}$.

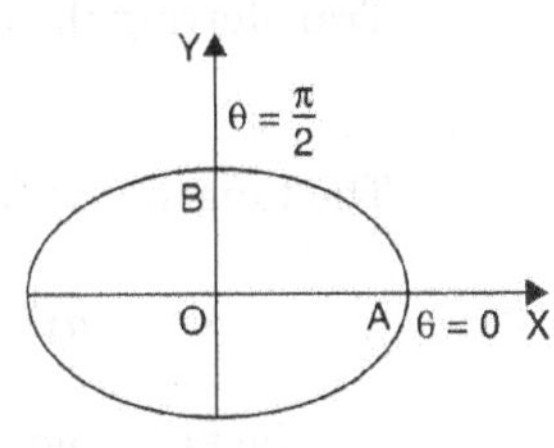

$$\therefore \quad \text{perimeter} = 4\int_0^{\frac{\pi}{2}} \sqrt{\left(\frac{dx}{d\theta}\right)^2 + \left(\frac{dy}{d\theta}\right)^2}\, d\theta.$$]

3. (*a*) In the curve $\left.\begin{aligned} x &= e^\theta \sin\theta, \\ y &= e^\theta \cos\theta, \end{aligned}\right\}$

show that $s = e^\theta\sqrt{2} + \text{C}$.

(*b*) Find the length of the arc of the curve $x = e^\theta \sin\theta$, $y = e^\theta \cos\theta$ from $\theta = 0$ to $\theta = \dfrac{\pi}{2}$.

4. Show that the length of an arc of the curve

$$x \sin\theta + y \cos\theta = f'(\theta),$$
$$x \cos\theta - y \sin\theta = f''(\theta),$$

is given by $\quad s = f(\theta) + f''(\theta) + C.$

5. Find the length of any arc of the curve $x^{\frac{2}{3}} - y^{\frac{2}{3}} = a^{\frac{2}{3}}$.

90. Tracing a curve to find the limits of integration.

In order to find the length of the arc of a curve [or its area, or the volume or the curved surface of the solid generated by the revolution of the area bounded by the curve], it is sometimes necessary to trace the curve to find the limits of integration. Hence the student is strongly urged to study from the author's *New Differential Calculus* the rules to trace a curve whose Cartesian or polar equation is given.

Example. *Find the whole length of the loop of the curve*

$$3ay^2 = x(x - a)^2.$$

To trace the curve $3ay^2 = x(x-a)^2$.

(*i*) **Symmetry.** The curve is symmetrical about the x-axis.

[$\because$ only even powers of y occur]

(*ii*) **Origin.** The curve passes through the origin.

[$\because$ putting $x = 0$, $y = 0$, we get $0 = 0$, which is true]

The tangent at the origin is

$$a^2x = 0, \quad \text{or} \quad x = 0.$$

| $3ay^2 = x(x^2 - 2ax + a^2)$ or $x^3 - 2ax^2 - 3ay^2 + a^2x = 0$

The curve meets the x-axis where, putting $y = 0$, $x(x-a)^2 = 0$, or

$$x = 0, a, a, \quad i.e., \quad \text{at } (0, 0), (a, 0).$$

Transferring the origin to $(a, 0)$, the equation becomes

$$3ay^2 = (x + a)x^2.$$

| $x^3 + ax^2 - 3ay^2 = 0$

The tangents at the *new origin* are

$$ax^2 - 3ay^2 = 0, \text{ or } y = \pm \frac{x}{\sqrt{3}}, \text{ which are real and different.}$$

$\therefore$ the new origin, *i.e.*, $(a, 0)$ is a node, and the tangents there make angles of $\pm 30°$ with the x-axis.

The curve meets the y-axis where, putting $x = 0$, $y^2 = 0$, or $y = 0, 0$, *i.e.*, at $(0, 0)$.

(*iii*) **Asymptotes.** There are no aysmptotes.

$$y^2 = \frac{x(x-a)^2}{3a}.$$

Taking + ve sign before the square root,

$$y = \frac{\sqrt{x}\,(x-a)}{\sqrt{3a}}.$$

When $x \to \infty, y \to \infty, \therefore$ the curve extends to infinity.

(*iv*) **Points where $\frac{dy}{dx} = 0$ or $\to \infty$.**

$$y = \frac{\sqrt{x}\,(x-a)}{\sqrt{3a}} = \frac{1}{\sqrt{3a}}\,(x^{\frac{3}{2}} - ax^{\frac{1}{2}}).$$

$$\therefore \quad \frac{dy}{dx} = \frac{1}{\sqrt{3a}}\left(\frac{3}{2}x^{\frac{1}{2}} - a\frac{1}{2x^{\frac{1}{2}}}\right) = \frac{3x-a}{2\sqrt{3ax}}.$$

$$\therefore \quad \frac{dy}{dx} = 0 \text{ when } x = \frac{a}{3}, \text{ and then } y = \frac{\sqrt{\frac{a}{3}}\left(-\frac{2a}{3}\right)}{\sqrt{3a}} = -\frac{2a}{9}.$$

$$\frac{dy}{dx} \to \infty \text{ when } x \to 0, \text{ and then } y \to 0,$$

i.e., at $\left(\frac{a}{3}, -\frac{2}{9}a\right)$ the tangent is $\parallel$ to the x-axis, and at (0, 0) the tangent is $\perp$ to the x-axis.

(*v*) $$y = \frac{\sqrt{x}\,(x-a)}{\sqrt{3a}}.$$

$\therefore$ y is imaginary when x is – ve.

$\therefore$ no portion of the curve lies to the left of the y-axis.

Hence the shape of the curve is that shown in the Fig.*

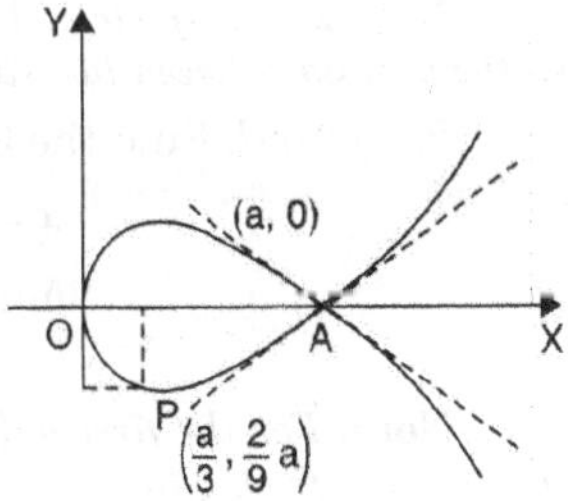

$\therefore$ for the upper half of the loop, x varies from 0 to a.

$\therefore$ length of loop

$$= 2\int_0^a \sqrt{1+\left(\frac{dy}{dx}\right)^2}\,dx$$

$$= 2\int_0^a \sqrt{1+\frac{(3x-a)^2}{12ax}}\,dx$$

***How to draw the Fig.** Draw the tangent at the origin, *viz.*, the y-axis. Plot the pt. A(a, 0) where the curve meets the x-axis, and draw the tangents there making angles of ± 30° with the x-axis. Plot the pt. P $\left(\frac{a}{3}, -\frac{2}{9}a\right)$ where the tangent is $\parallel$ to the x-axis. Now draw the curve through O, P, A as shown, and complete it from symmetry about the x-axis.

$$= 2\int_0^a \frac{\sqrt{12ax + 9x^2 - 6xa + a^2}}{2\sqrt{3}\sqrt{a}\sqrt{x}}\, dx$$

$$= \frac{1}{\sqrt{3a}}\int_0^a \frac{\sqrt{9x^2 + 6xa + a^2}}{\sqrt{x}}\, dx$$

$$= \frac{1}{\sqrt{3a}}\int_0^a \frac{3x + a}{\sqrt{x}}\, dx = \frac{1}{\sqrt{3a}}\int_0^a (3x^{\frac{1}{2}} + ax^{-\frac{1}{2}})\, dx$$

$$= \frac{1}{\sqrt{3a}}\left[3\frac{x^{\frac{3}{2}}}{\frac{3}{2}} + a\frac{x^{\frac{1}{2}}}{\frac{1}{2}}\right]_0^a = \frac{1}{\sqrt{3a}}[2a^{\frac{3}{2}} + 2a^{\frac{3}{2}} - (0+0)]$$

$$= \frac{4a^{\frac{3}{2}}}{\sqrt{3a}} = \frac{4}{\sqrt{3}}\, a.$$

EXAMPLES

1. (*i*) *Cycloid.* Find the length of the complete cycloid given by

$$\left.\begin{aligned} x &= a\,\theta + a\sin\theta, \\ y &= a - a\cos\theta. \end{aligned}\right\}$$

(*ii*) Rectify the curve $x = a(\theta + \sin\theta)$, $y = a(1 - \cos\theta)$.

Note 1. *The cycloid is symmetrical about the y-axis, and for the half of the cycloid, in the first quadrant, θ varies from 0 to π.*

Note 2. *"Complete cycloid" means "one arc of the cycloid", i.e., the portion between two successive cusps.*

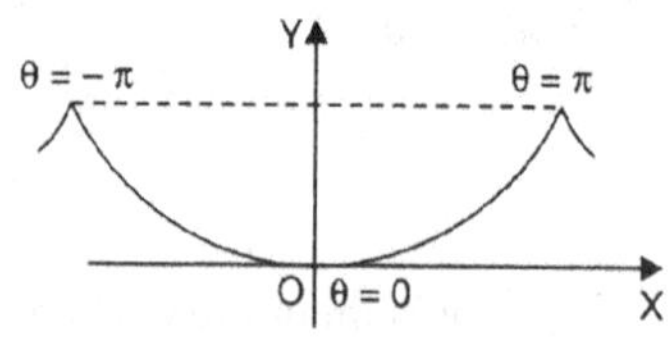

2. *Cycloid.* Find the length of one arch of the cycloid

$$x = a(\theta - \sin\theta),$$
$$y = a(1 - \cos\theta).$$

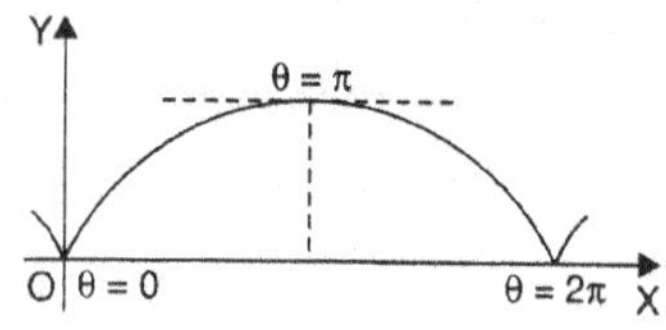

Note. *For the first half of the cycloid, in the first quadrant, θ varies from 0 to π.*

3. (*i*) Astroid or four-cusped hypocycloid. Find the whole length of the curve $x^{\frac{2}{3}} + y^{\frac{2}{3}} = a^{\frac{2}{3}}$.

(*ii*) Find the length of one quadrant of the curve

$$\left(\frac{x}{a}\right)^{\frac{2}{3}} + \left(\frac{y}{b}\right)^{\frac{2}{3}} = 1.$$

[(*i*) The equation of the curve is $x^{\frac{2}{3}} + y^{\frac{2}{3}} = a^{\frac{2}{3}}$ ...(1)

[*To find the parametric equations of the curve.*]

From (1), $\left(\frac{x}{a}\right)^{\frac{2}{3}} + \left(\frac{y}{a}\right)^{\frac{2}{3}} = 1$ ...(2) [R.H.S. = 1 (*Note this step*)]

Also $\cos^2\theta + \sin^2\theta = 1$ ...(3),

$\therefore$ comparing (2) and (3),

$$\left(\frac{x}{a}\right)^{\frac{2}{3}} = \cos^2\theta, \left(\frac{y}{a}\right)^{\frac{2}{3}} = \sin^2\theta.$$

Raising each side to the power $\frac{3}{2}$,

$$\frac{x}{a} = \cos^3\theta, \frac{y}{a} = \sin^3\theta,$$

or
$$x = a\cos^3\theta, y = a\sin^3\theta.$$

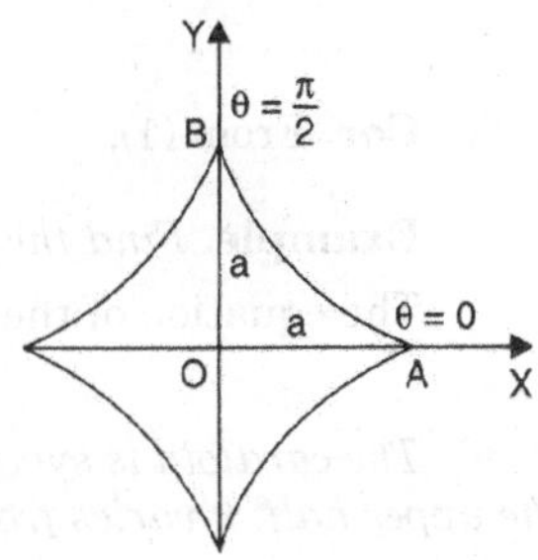

Note. *The curve is symmetrical about both axes, and in the first quadrant, θ varies from 0 to* $\frac{\pi}{2}$.]

III. Arc formulae for polar equation.

91. To prove that the length of the arc of the curve r = f(θ) between the points where θ = α, θ = β is

$$\int_\alpha^\beta \sqrt{r^2 + \left(\frac{dr}{d\theta}\right)^2}\, d\theta.$$

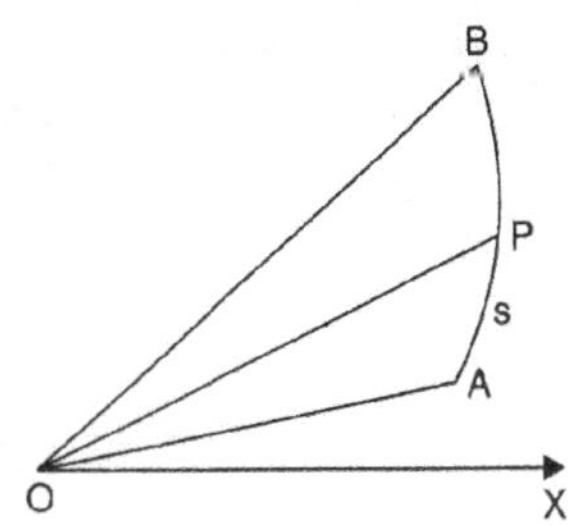

Let AB be the curve $r = f(\theta)$, and A, B the pts. where $\theta = \alpha, \theta = \beta$.

Let P(r, θ) be any pt. on the curve.

Then the arc AP is clearly a function of θ

= s (say).

We know that $$\frac{ds}{d\theta} = \sqrt{r^2 + \left(\frac{dr}{d\theta}\right)^2}$$ ...(1) [From Diff. Calculus]

$$\therefore \int_\alpha^\beta \sqrt{r^2 + \left(\frac{dr}{d\theta}\right)^2}\, d\theta = \int_\alpha^\beta \frac{ds}{d\theta}\, d\theta = \Big[s\Big]_\alpha^\beta$$

= (value of s when θ = β) – (value of s when θ = α)

= arc AB – 0,

i.e., $$\text{arc AB} = \int_\alpha^\beta \sqrt{r^2 + \left(\frac{dr}{d\theta}\right)^2}\, d\theta.$$

From (1), $$\mathbf{s} = \int \sqrt{\mathbf{r}^2 + \left(\frac{\mathbf{dr}}{\mathbf{d\theta}}\right)^2}\, \mathbf{d\theta}.$$

Similarly, *the length of the arc of the curve $\theta = f(r)$ between the points where $r = a$ and $r = b$ is*

$$\int_a^b \sqrt{1 + r^2 \left(\frac{d\theta}{dr}\right)^2}\, dr.$$

[Proceed as in Art. 94, and instead of (1), use

$$\frac{ds}{dr} = \sqrt{1 + r^2 \left(\frac{d\theta}{dr}\right)^2} \quad ...(1) \quad \text{[From Diff. Calculus]}$$

Cor. From (1), $$\mathbf{s} = \int \sqrt{1 + \mathbf{r}^2 \left(\frac{\mathbf{d\theta}}{\mathbf{dr}}\right)^2}\, \mathbf{dr}.$$

Example. *Find the perimeter of the cardioid $r = a\ (1 - \cos\theta)$.*

The equation of the cardioid is

$$r = a(1 - \cos\theta) \quad ...(1)$$

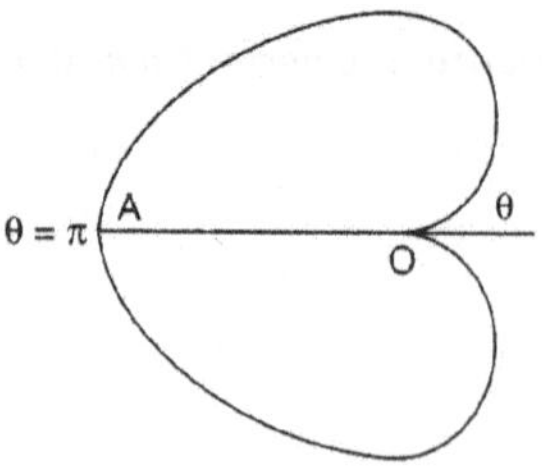

The cardioid is symmetrical about the initial line, and for the upper half, θ varies from 0 to π.

From (1), $$\frac{dr}{d\theta} = a \sin\theta.$$

∴ perimeter

$$= 2\int_0^\pi \sqrt{r^2 + \left(\frac{dr}{d\theta}\right)^2}\, d\theta = 2\int_0^\pi \sqrt{a^2\,(1 - \cos\theta)^2 + a^2 \sin^2\theta}\; d\theta$$

$$= 2a\int_0^\pi \sqrt{1 - 2\cos\theta + \cos^2\theta + \sin^2\theta}\; d\theta$$

$$= 2a\int_0^\pi \sqrt{2(1 - \cos\theta)}\; d\theta \qquad (\because \cos^2\theta + \sin^2\theta = 1)$$

$$= 2a\int_0^\pi \sqrt{2.2\sin^2\frac{\theta}{2}}\; d\theta = 4a\int_0^\pi \sin\frac{\theta}{2}\, d\theta = 4a\left[\frac{-\cos\frac{\theta}{2}}{\frac{1}{2}}\right]_0^\pi$$

$$= -8a\,[0 - 1] = 8a.$$

EXAMPLES

1. (*a*) Obtain a formula for finding the length of the arc of a curve given in polar co-ordinates.

(*b*) Find the length of any arc of the curve $r = a \cos \theta$ (*circle*).

2. Find the length of any arc of the curves

(*i*) $r = a\theta$ (Archimedean spiral). (*ii*) $r = a \sin^2 \frac{\theta}{2}$ (cardioid).

****3.** Cissoid. Find the length of any arc of the cissoid $r = a \frac{\sin^2 \theta}{\cos \theta}$.

[Length of arc $= \int_{\theta_1}^{\theta_2} \sqrt{r^2 + \left(\frac{dr}{d\theta}\right)^2}\, d\theta$ (which will be found to be)

$$= a \int_{\theta_1}^{\theta_2} \frac{\sin \theta}{\cos^2 \theta} \sqrt{1 + 3\cos^2 \theta}\, d\theta \quad ...(1)$$

Now $\int \frac{\sin \theta}{\cos^2 \theta} \sqrt{1 + 3\cos^2 \theta}\, d\theta$ (Put $\cos \theta = z$)

$$= \int \sqrt{1 + 3z^2} \left(-\frac{1}{z^2}\right) dz.$$ Integrate by parts.]

4. (*a*) Trace the curve $r = a(1 + \cos \theta)$ and find the length of the whole arc.

(*b*) Show that the arc of the upper half is bisected by $\theta = \frac{\pi}{3}$.

[**Caution.** It is a *common mistake* to work thus :

Length of whole arc

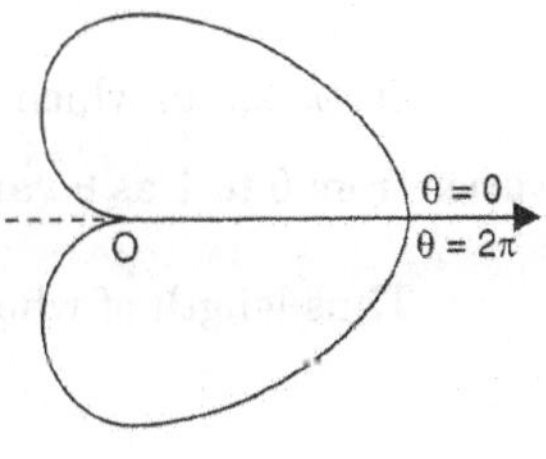

$$= \int_0^{2\pi} \sqrt{r^2 + \left(\frac{dr}{d\theta}\right)^2}\, d\theta$$

$$= \int_0^{2\pi} \sqrt{a^2 (1 + \cos \theta)^2 + a^2 \sin^2 \theta}\, d\theta$$

$$= a \int_0^{2\pi} \sqrt{1 + 2\cos \theta + \cos^2 \theta + \sin^2 \theta}\, d\theta$$

$$= a \int_0^{2\pi} \sqrt{2(1 + \cos \theta)}\, d\theta \quad (\because \cos^2 \theta + \sin^2 \theta = 1)$$

$$= a \int_0^{2\pi} \sqrt{2.2 \cos^2 \frac{\theta}{2}}\, d\theta = 2a \int_0^{2\pi} \sqrt{\cos^2 \frac{\theta}{2}}\, d\theta \quad ...(1)$$

$$= 2a \int_0^{2\pi} \cos \frac{\theta}{2}\, d\theta = 2a \left[\frac{\sin \frac{\theta}{2}}{\frac{1}{2}}\right]_0^{2\pi}$$

$= 4a\ [0 - 0] = 0$, which is obviously wrong.

The mistake has arisen, $\because$ the integrand $\cos \dfrac{\theta}{2}$ is + ve (it varies from 1 to 0) as θ varies from 0 to π, and is – ve (it varies from 0 to – 1) as θ varies from π to 2π, so that $\cos \dfrac{\theta}{2}$ changes sign at the pt. $\theta = \pi$ in the range of integration $(0, 2\pi)$.

In fact in all arc (or area or volume or surface) formulae care should be taken to see that the integrand does not change sign in the range of integration.

**{If, however, the integrand changes sign at some point c in the range of integration (a, b), the range (a, b) should be broken up into two : (a, c) and (c, b) in each of which the *positive value* of the integrand should be taken.

Thus (see (1) above) in $2a \int_0^{2\pi} \sqrt{\cos^2 \dfrac{\theta}{2}}\, d\theta$, the integrand $\cos \dfrac{\theta}{2}$ changes sign at $\theta = \pi$ in the range of integration $(0, 2\pi)$. In fact the integrand $\sqrt{\cos^2 \dfrac{\theta}{2}} = \pm \cos \dfrac{\theta}{2}$.

$\therefore$ we break up the range of integration $(0, 2\pi)$ into two :

(*i*) $(0, \pi)$ in which we take the positive value of the integrand, *viz.*, $\cos \dfrac{\theta}{2}$; and

(*ii*) $(\pi, 2\pi)$ in which again we take the *positive value* of the integrand, *viz.*, $-\cos \dfrac{\theta}{2}$ (which varies from 0 to 1 as θ varies from π to 2π).

Thus length of whole arc

$$= 2a \int_0^{\pi} \cos \frac{\theta}{2}\, d\theta + 2a \int_{\pi}^{2\pi} \left(-\cos \frac{\theta}{2}\right) d\theta$$

$$= 2a \left[\frac{\sin \frac{\theta}{2}}{\frac{1}{2}} \right]_0^{\pi} + 2a \left[-\frac{\sin \frac{\theta}{2}}{\frac{1}{2}} \right]_{\pi}^{2\pi}$$

$= 4a\ [1 - 0] - 4a\ [0 - 1] = 8a$.}

The easiest way is to find the length of the smallest part which by symmetry will give the whole required length, and see that the integrand remains positive in the range of integration.]

5. Find the length of the arc of the equiangular spiral $r = ae^{\theta \cot \alpha}$ between the points at which the radii vectors are r_1 and r_2.

The equation of the equiangular spiral is $r = ae^{\theta \cot \alpha}$.

Taking logs, $\log r = \log a + \theta \cot \alpha$. $(\because \ \log e = \log_e e = 1)$

Differentiating both sides w.r.t. θ,

$$\frac{1}{r}\frac{dr}{d\theta} = \cot \alpha.$$

Taking reciprocals of both sides,

$$r\frac{d\theta}{dr} = \frac{1}{\cot \alpha} = \tan \alpha.$$

$$\therefore \quad \text{length of arc} = \int_{r_1}^{r_2} \sqrt{1 + r^2\left(\frac{d\theta}{dr}\right)^2}\, dr = \int_{r_1}^{r_2} \sqrt{1 + \tan^2 \alpha}\, dr$$

$$= \sec \alpha . \Big[r \Big]_{r_1}^{r_2} = \sec \alpha . (r_2 - r_1).$$

6. Prove that the length of the arc of the curve $p = f(r)$ between the points where $r = a$, $r = b$ is Padal eq.

$$\int_a^b \frac{r}{\sqrt{r^2 - p^2}}\, dr.$$

$$\text{Length of arc} = \int_a^b \sqrt{1 + r^2\left(\frac{d\theta}{dr}\right)^2}\, dr \qquad \left(\text{But } r\frac{d\theta}{dr} = \tan \phi\right)$$

$$= \int_a^b \sqrt{1 + \tan^2 \phi}\, dr = \int_a^b \sec \phi\, dr$$

$$\left[\text{But } p = r \sin \phi, \text{ or } \sin \phi = \frac{p}{r}\right]$$

$$= \int_a^b \frac{r}{\sqrt{r^2 - p^2}}\, dr$$

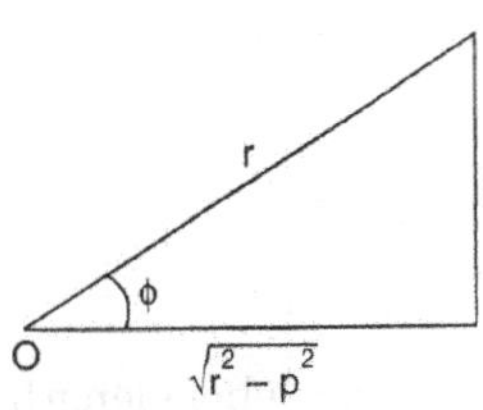

Cor. Length of arc

$$\mathbf{s} = \int \frac{\mathbf{r}}{\sqrt{\mathbf{r}^2 - \mathbf{p}^2}}\, \mathbf{dr}.$$

7. Show that the length of the arc of the equiangular spiral $p = r \sin \alpha$ between the points at which the radii vectors are r_1 and r_2, is $(r_2 - r_1) \sec \alpha$.

8.** Show that the length of the arc of the hyperbola $xy = a^2$ between the limits $x = b$ and $x = c$ is equal to the arc of the curve $p^2(a^4 + r^4) = a^4r^2$ between the limits $r = b, r = c$.

[Do not evaluate the two integrals. Show that they are equal. (Use theorem I, Art. 78.)]

IV. Length of arc of evolute.

92. In order to find the length of an arc of the evolute of a curve, it is not necessary to find the equation of the evolute. We can use the following Proposition proved in Differential Calculus :

The difference between the radii of curvature at two points of a curve is equal to the length of the arc of the evolute between the two corresponding points.

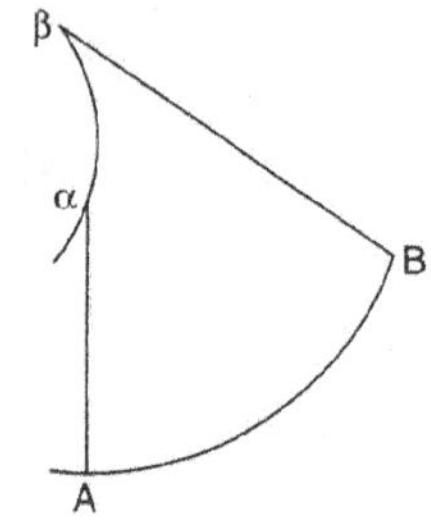

Thus* if α, β are the centres of curvature corresponding to the points A, B of a curve, then

$$\textbf{arc } \alpha\beta = B\beta - A\alpha$$
$$= (\rho \textbf{ at B}) - (\rho \textbf{ at A}).$$

Example. *Find the length of the evolute of the ellipse.*

Let α, α′, β, β′ be the centres of curvature corresponding to the extremities A, A′, B, B′, of the axes of the ellipse.

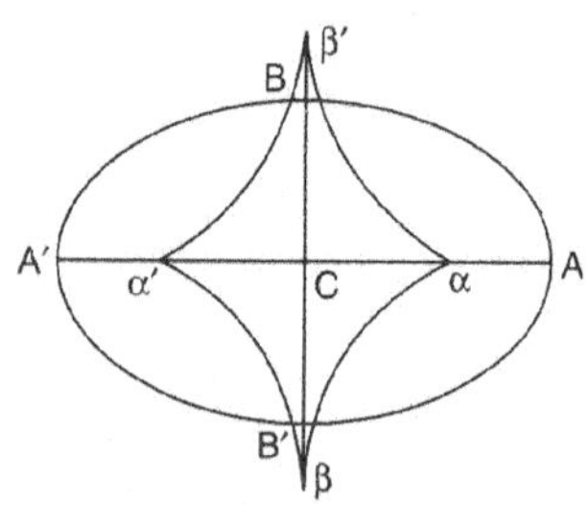

Then arc $\alpha\beta = (\rho \text{ at B}) - (\rho \text{ at A})$ (Art. 92)

[But ρ (of the ellipse) $= \dfrac{a^2b^2}{p^3}$,

where p is the central ⊥ on the tangent

(From Diff. Calculus)]

$$= \frac{a^2b^2}{b^3} - \frac{a^2b^2}{a^3} = \frac{a^2}{b} - \frac{b^2}{a}.$$

∴ whole length of evolute = 4. arc $\alpha\beta = 4\left(\dfrac{a^2}{b} - \dfrac{b^2}{a}\right)$.

***How to draw the Fig.** Draw the arc αβ of the evolute and the tangents at α, β. On these tangents, take the points A, B, and draw the curve AB such that Aα, Bβ, are the normals to it at A, B.

EXAMPLE

1. Show for the parabola $y^2 = 4ax$ that the length of the part of the evolute intercepted within the parabola is $4a(3\sqrt{3} - 1)$.

[Let α be the centre of curvature corresponding to the vertex A of the parabola, and β a pt. of intersection of the parabola and the evolute.

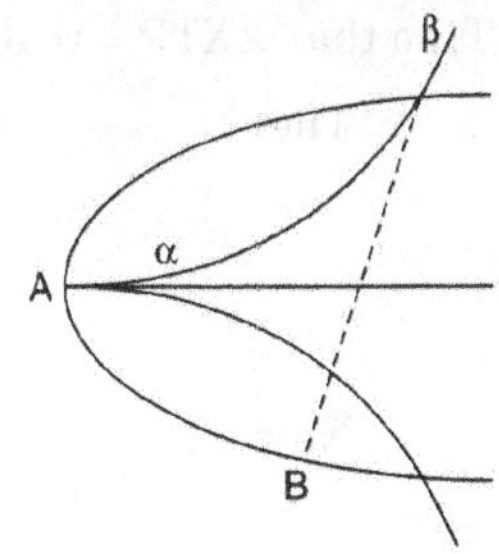

The equation of the parabola is $y^2 = 4ax$ ...(1)

and that of the evolute is $27\,ay^2 = 4(x - 2a)^3$ ...(2)

Proceed as in Ex. 4, Art. 91. It will be found that

at β, $x = 8a$, *i.e.*, $\bar{x} = 8a$ ($\because$ β is a pt. of the evolute)

i.e., $2a + 3x = 8a$, [$\because$ $\bar{x} = 2a + 3x$ (From Diff. Calculus)]

where x is the abscissa of the pt. B of the parabola corresponding to which the centre of curvature is β. (Draw the tangent at β to the evolute to meet the parabola again in B.)

$\therefore$ $x = 2a$.

Now arc $\alpha\beta = (\rho \text{ at B}) - (\rho \text{ at A})$

$$\left[\text{But } \rho \text{ (of the parabola)} = \frac{2\,\text{SP}^{\frac{3}{2}}}{\sqrt{a}} = \frac{2(a+x)^{\frac{3}{2}}}{\sqrt{a}} \text{ (From Diff. Calculus)}\right]$$

$$= \frac{2(a+2a)^{\frac{3}{2}}}{\sqrt{a}} - \frac{2(a+0)^{\frac{3}{2}}}{\sqrt{a}} \qquad [\because \text{ at B}, x = 2a\text{ ; and at A}, x = 0]$$

$$= \frac{2}{\sqrt{a}}\,[3\sqrt{3}\,a^{\frac{3}{2}} - a^{\frac{3}{2}}] = 2a(3\sqrt{3} - 1).$$

$\therefore$ length of part of the evolute intercepted within the parabola

$$= 2\,.\,\text{arc } \alpha\beta = 2.2a\,(3\sqrt{3} - 1) = 4a\,(3\sqrt{3} - 1).]$$

SECTION II

INTRINSIC EQUATION

93. Intrinsic equation. Def. If s is the length of the arc of a curve measured from a fixed point A on it to any point P, and ψ the angle which the tangent at P makes with a fixed line (generally the tangent at A), then the relation between s and ψ is called the **intrinsic equation** of the curve.

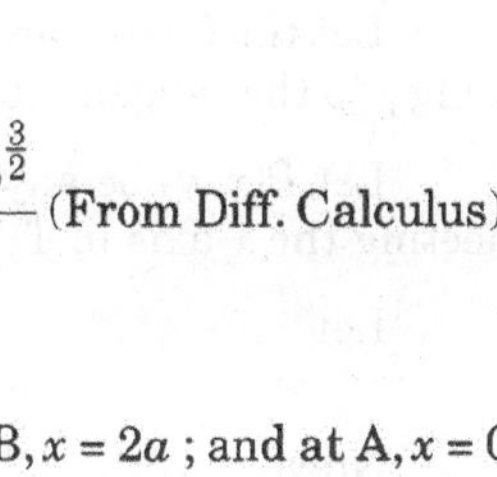

Example. *Find the intrinsic equation of a circle.*

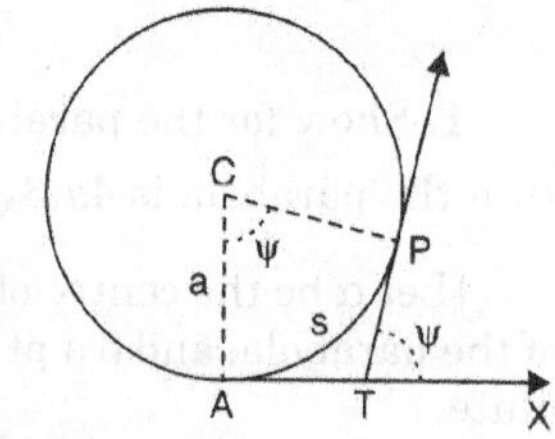

Let C be the centre and a the radius of the circle.

Let A be the fixed pt. on the circle from which the arc is measured and P any pt. on it, so that arc AP = s.

Let PT, the tangent at P, meet AX, the tangent at A, in T, so that ∠XTP = ψ. Join CA, CP.

Then ∠ACP = ∠XTP

[∵ ATPC is a cyclic quadrilateral

= ψ. since ∠CAT + ∠CPT = 2 rt. ∠s, each being = 1 rt. ∠]

Now $$\angle ACP = \frac{\text{arc AP}}{\text{radius}},\qquad \left[\theta = \frac{l}{r} \text{ (From Trigonometry)}\right]$$

i.e., $$\psi = \frac{s}{a},$$

or $s = a\psi$, which is the required intrinsic equation.

94.I-To find the intrinsic equation of a curve from the Cartesian equation.

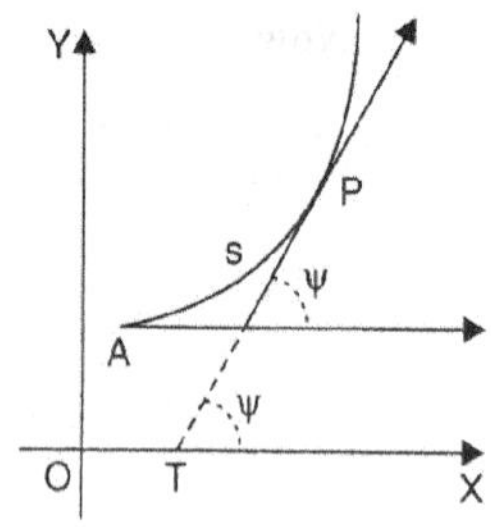

Let the Cartesian equation of the curve be $y = f(x)$, the x-axis being || to the tangent at A, the pt. from which the arc is measured.

Let P(x, y) be any pt. on the curve and PT the tangent at P meeting the x-axis in T.

Let $x = a$ at A.

Now $$\frac{dy}{dx} = f'(x)$$

∴ $$s = \int_a^x \sqrt{1+\left(\frac{dy}{dx}\right)^2}\,dx$$

$$= \int_a^x \sqrt{1+[f'(x)]^2}\,dx \qquad ...(1)$$

$$\tan\psi = \frac{dy}{dx} = f'(x) \qquad ...(2)$$

Eliminating x from (1) (after integration) and (2), we get a relation between s and ψ, which is the required intrinsic equation.

Example. *Find the intrinsic equation of the catenary*

$$y = c\cosh\frac{x}{c}.$$

The equation of the catenary is $y = c\cosh\dfrac{x}{c}$.

∴ $$\frac{dy}{dx} = c\sinh\frac{x}{c}\cdot\frac{1}{c} = \sinh\frac{x}{c}.$$

Measuring s from the point where $x = 0$,

$$s = \int_0^x \sqrt{1+\left(\frac{dy}{dx}\right)^2}\, dx = \int_0^x \sqrt{1+\sinh^2 \frac{x}{c}}\, dx$$

$$= \int_0^x \cosh \frac{x}{c}\, dx$$

$$= \left[\frac{\sinh \frac{x}{c}}{\frac{1}{c}}\right]_0^x = c\left[\sinh \frac{x}{c} - 0\right] = c \sinh \frac{x}{c} \quad \text{...(1)}$$

$$\tan \psi = \frac{dy}{dx} = \sinh \frac{x}{c} \quad \text{...(2)}$$

Eliminating x from (1) and (2) [by substituting the value of $\sinh \frac{x}{c}$ (= $\tan \psi$) from (2) in (1)],

$s = c \tan \psi$, which is the required intrinsic equation.

EXAMPLES

1. (*a*) What is meant by the intrinsic equation of a curve ?

Show that the intrinsic equation of the parabola is

$$s = a \tan \psi \sec \psi + a \log (\tan \psi + \sec \psi).$$

(*b*) Deduce that the length of the arc intercepted between the vertex and an extremity of the latus rectum is $a[\sqrt{2} + \log (1 + \sqrt{2})]$, $4a$ being the latus rectum.

[(*b*) Let the equation of the parabola be $x^2 = 4ay$.

It will be found that $\tan \psi = \frac{dy}{dx} = \frac{x}{2a}$. At an extremity of the latus rectum,

$x = 2a, \quad \therefore \quad \tan \psi = 1, \therefore \quad \psi = \frac{\pi}{4}$.]

2. Show that the intrinsic equation of the parabola $y^2 = 4ax$ is

$$s = a \cot \psi \operatorname{cosec} \psi + a \log (\cot \psi + \operatorname{cosec} \psi),$$

ψ being the angle between the x-axis and the tangent at the point whose distance from the vertex, along the parabola, is s.

3. Show that the intrinsic equation of the semicubical parabola $3ay^2 = 2x^3$ is $9s = 4a(\sec^3 \psi - 1)$.

94-II. To find the intrinsic equation of a curve from the parametric equations.

Let the parametric equations of the curve be $x = f(t)$, $y = \phi(t)$, the x-axis being || to the tangent at A, the pt. from which the arc is measured.

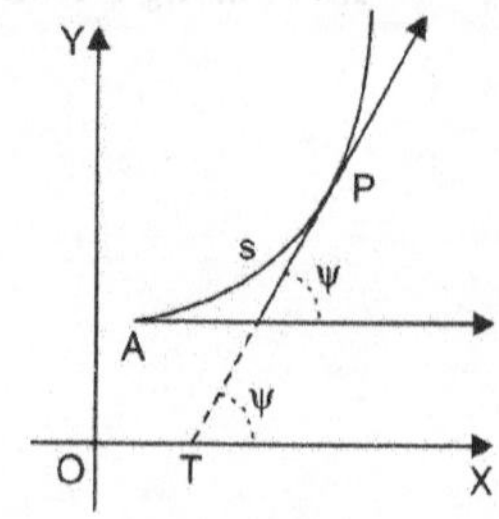

Let P(x, y) be any pt. on the curve and PT the tangent at P meeting the x-axis in T.

Let $t = a$ at A.

Now $$\frac{dx}{dt} = f'(t),\ \frac{dy}{dt} = \phi'(t).$$

$\therefore$ $$s = \int_a^t \sqrt{\left(\frac{dx}{dt}\right)^2 + \left(\frac{dy}{dt}\right)^2}\, dt$$

$$= \int_a^t \sqrt{[f'(t)]^2 + [\phi'(t)]^2}\, dt \qquad ...(1)$$

$$\tan\psi = \frac{dy}{dx} = \frac{\frac{dy}{dt}}{\frac{dx}{dt}} = \frac{\phi'(t)}{f'(t)} \qquad ...(2)$$

Eliminating t from (1) (after integration) and (2), we get a relation between s and ψ, which is the required intrinsic equation.

Example. *Find the intrinsic equation of the cycloid*

$$x = a\,(t + \sin t),$$
$$y = a(1 - \cos t).$$

The parametric equations of the cycloid are

$$x = a\,(t + \sin t), \qquad y = a\,(1 - \cos t)$$

$\therefore$ $$\frac{dx}{dt} = a\,(1 + \cos t),\ \frac{dy}{dt} = a \sin t.$$

Measuring s from the pt. where $t = 0$,

$$s = \int_0^t \sqrt{\left(\frac{dx}{dt}\right)^2 + \left(\frac{dy}{dt}\right)^2}\, dt$$

$$= \int_0^t \sqrt{a^2(1 + \cos t)^2 + a^2 \sin^2 t}\, dt$$

$$= a\int_0^t \sqrt{1 + 2\cos t + \cos^2 t + \sin^2 t}\, dt$$

$$= a\int_0^t \sqrt{2(1 + \cos t)}\, dt \qquad (\because\ \cos^2 t + \sin^2 t = 1)$$

$$= a\int_0^t \sqrt{2.2\cos^2\frac{t}{2}}\, dt = a\int_0^t 2\cos\frac{t}{2}\, dt = 2a\left[\frac{\sin\frac{t}{2}}{\frac{1}{2}}\right]_0^t$$

$$= 4a\left[\sin\frac{t}{2} - 0\right] = 4a\sin\frac{t}{2} \qquad ...(1)$$

$$\tan\psi = \frac{dy}{dx} = \frac{\frac{dy}{dt}}{\frac{dx}{dt}} = \frac{a\sin t}{a(1+\cos t)}$$

$$= \frac{2\sin\frac{t}{2}\cos\frac{t}{2}}{2\cos^2\frac{t}{2}} = \tan\frac{t}{2}$$

$$\therefore \qquad \psi = \frac{t}{2} \qquad ...(2)$$

Eliminating t from (1) and (2) [by substituting the value of $\frac{t}{2}$ (= ψ) from (2) in (1)],

$s = 4a\sin\psi$, which is the required intrinsic equation.

94-III. To find the intrinsic equation of a curve from the polar equation.

Let the polar equation of the curve be $r = f(\theta)$, the initial line being ∥ to the tangent at A, the pt. from which the arc is measured.

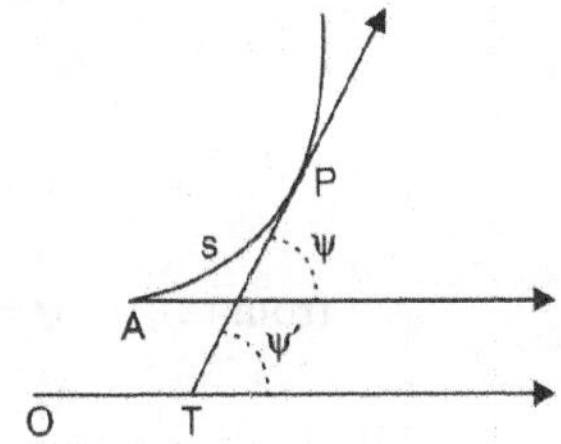

Let P(r, θ) be any pt. on the curve and PT the tangent at P meeting the initial line in T.

Let $\theta = \alpha$ at A.

Now $$\frac{dr}{d\theta} = f'(\theta).$$

$$\therefore \qquad s = \int_{\alpha}^{\theta}\sqrt{r^2 + \left(\frac{dr}{d\theta}\right)^2}\,d\theta = \int_{\alpha}^{\theta}\sqrt{[f(\theta)]^2 + [f'(\theta)]^2}\,d\theta \qquad ...(1)$$

$$\psi = \theta + \phi \qquad ...(2)$$

$$\tan\phi = r\frac{d\theta}{dr} = f(\theta).\frac{1}{f'(\theta)} = \frac{f(\theta)}{f'(\theta)} \qquad ...(3)$$

Eliminating θ and ϕ from (1) (after integration), (2), (3), we get a relation between s and ψ, which is the required intrinsic equation.

Note. *In examples it is generally useful to find from (3) the value of ϕ in terms of θ and hence from (2) the value of ψ in terms of θ ...(4), and then eliminate θ from (1) (after integration) and (4). (See the following solved Ex.)*

Example. *Find the intrinsic equation of the cardioid*

$$r = a\,(1 - \cos\theta).$$

The equation of the cardioid is $r = a\,(1 - \cos\theta)$.

$$\therefore \qquad \frac{dr}{d\theta} = a\sin\theta.$$

Measuring s from the pt. where $\theta = 0$,

$$s = \int_0^\theta \sqrt{r^2 + \left(\frac{dr}{d\theta}\right)^2}\, d\theta = \int_0^\theta \sqrt{a^2(1-\cos\theta)^2 + a^2\sin^2\theta}\, d\theta$$

$$= a\int_0^\theta \sqrt{1 - 2\cos\theta + \cos^2\theta + \sin^2\theta}\, d\theta$$

$$= a\int_0^\theta \sqrt{2(1-\cos\theta)}\, d\theta \qquad (\because \ \cos^2\theta + \sin^2\theta = 1)$$

$$= a\int_0^\theta \sqrt{2.2\sin^2\frac{\theta}{2}}\, d\theta = a\int_0^\theta 2\sin\frac{\theta}{2}\, d\theta = 2a\left[\frac{-\cos\frac{\theta}{2}}{\frac{1}{2}}\right]_0^\theta$$

$$= -4a\left[\cos\frac{\theta}{2} - 1\right] = 4a\left(1 - \cos\frac{\theta}{2}\right) \qquad \text{...(1)}$$

$$\psi = \theta + \phi \qquad \text{...(2)}$$

$$\tan\phi = r\frac{d\theta}{dr} = a(1-\cos\theta).\frac{1}{a\sin\theta} = \frac{2\sin^2\frac{\theta}{2}}{2\sin\frac{\theta}{2}\cos\frac{\theta}{2}} = \tan\frac{\theta}{2}$$

$$\therefore \qquad \phi = \frac{\theta}{2} \qquad \text{...(3)}$$

$$\therefore \text{ from (2)}, \qquad \psi = \theta + \frac{\theta}{2} = \frac{3\theta}{2} \qquad \text{...(4)}$$

Eliminating θ from (1) and (4) [by substituting the value of $\frac{\theta}{2}\left(=\frac{\psi}{3}\right)$ from (4) in (1)],

$$s = 4a\left(1 - \cos\frac{\psi}{3}\right), \text{ which is the required intrinsic equation.}$$

EXAMPLE

Find the intrinsic equation of the equiangular spiral $r = ae^{\theta \cot \alpha}$, measuring s from the point $(a, 0)$.

MISCELLANEOUS EXAMPLES ON CHAPTER IX

1. Find the length of the arc of the curve $y = \log \sec x$ from $x = 0$ to $x = \frac{\pi}{3}$.

2. Find the length of an arc of the parabola $y = x^2$, measured from the vertex.

Calculate the length of the arc to the point (1, 1), given that $\log(2 + \sqrt{5}) = 1.45$.

3. Show that the length of the arc of the parabola $y^2 = 4ax$ cut off by the line $3y = 8x$ is $a\left(\log_e 2 + \frac{15}{16}\right)$.

4. If s be the length of the arc of the catenary $y = c \cosh (x/c)$ from the vertex $(0, c)$ to the point (x, y), show that $s^2 = y^2 - c^2$.

5. Find the length of the curve $y = \log \dfrac{e^x - 1}{e^x + 1}$ from $x = 1$ to $x = 2$.

[$y = \log \dfrac{e^x - 1}{e^x + 1} = \log (e^x - 1) - \log (e^x + 1)$.

It will be found that length of curve

$$= \int_1^2 \sqrt{1 + \frac{4e^{2x}}{(e^{2x} - 1)^2}}\, dx = \int_1^2 \frac{e^{2x} + 1}{e^{2x} - 1}\, dx.$$

Divide the num. and denom. by e^x, and notice that the num. is the diff. coefficient of the denom.]

6. Show that the length of an arc of the curve $x^2 = a^2 (1 - e^{\frac{y}{a}})$ measured from $(0, 0)$ to (x, y) is $a \log \dfrac{a + x}{a - x} - x$.

[As suggested by the answer (arc in terms of x) express y in terms of x thus :

$$x^2 = a^2 (1 - e^{\frac{y}{a}}), \quad \therefore \quad a^2 e^{\frac{y}{a}} = a^2 - x^2, \text{ or } e^{\frac{y}{a}} = \frac{a^2 - x^2}{a^2}.$$

Taking logs, $\quad \dfrac{y}{a} = \log \dfrac{a^2 - x^2}{a^2}$, or $y = a \log \dfrac{a^2 - x^2}{a^2}$.]

****7. Cissoid.** Find the length of any arc of the curve $y^2(a - x) = x^3$.

[$y = \dfrac{x^{\frac{3}{2}}}{\sqrt{a - x}}$. It will be found that $\dfrac{dy}{dx} = \dfrac{\sqrt{x}\,(3a - 2x)}{2(a - x)^{\frac{3}{2}}}$.

$$\therefore \quad \text{length of arc} = \int_{x_1}^{x_2} \sqrt{1 + \left(\frac{dy}{dx}\right)^2}\, dx \text{ (which will be found to be}$$

$$= \frac{a}{2} \int_{x_1}^{x_2} \frac{\sqrt{4a - 3x}}{(a - x)^{\frac{3}{2}}}\, dx$$

$$= \frac{a}{2} \int_{x_1}^{x_2} \sqrt{4a - 3x} \cdot \frac{1}{(a - x)^{\frac{3}{2}}}\, dx. \text{ Integrate by parts.]}$$

****8.** Show that in the epicycloid

$$\left.\begin{aligned} x &= (a + b) \cos \theta - b \cos \frac{a + b}{b} \theta, \\ y &= (a + b) \sin \theta - b \sin \frac{a + b}{b} \theta, \end{aligned}\right\}$$

$$s = \frac{4b(a+b)}{a} \cos \frac{a}{2b}\,\theta,$$

s being measured from the point at which $\theta = \pi\, b/a$, *i.e.*, *a* vertex.

[**Note 1.** *Length of arc* $= \displaystyle\int_{\frac{\pi b}{a}}^{\theta} \sqrt{\left(\frac{dx}{d\theta}\right)^2 + \left(\frac{dy}{d\theta}\right)^2}\, d\theta,$

which will be found to be $\quad = -\dfrac{4b}{a}(a+b)\cos\dfrac{a\theta}{2b}$ ...(1)

If s is measured from the pt. where $\theta = \dfrac{\pi b}{a}$ *in the direction in which* θ *decreases,* then, *changing the sign in (1),* $\quad s = \dfrac{4b}{a}(a+b)\cos\dfrac{a\theta}{2b}$.

****Note 2. Explanation.**

In proving the arc formula

$$s = \int_a^b \sqrt{\left(\frac{dx}{dt}\right)^2 + \left(\frac{dy}{dt}\right)^2}\, dt$$ *(Art. 89), it was assumed that*

$$\frac{ds}{dt} = \sqrt{\left(\frac{dx}{dt}\right)^2 + \left(\frac{dy}{dt}\right)^2}\,.$$

Now, from Differential Calculus,

$$\frac{ds}{dt} = \pm\sqrt{\left(\frac{dx}{dt}\right)^2 + \left(\frac{dy}{dt}\right)^2}\,,$$

taking + ve sign before the square root if s increases as t increases and – ve sign before the square root if s increases as t decreases. Hence if s is measured from the pt. where t = a in the direction in which t decreases, then

$$\frac{ds}{dt} = -\sqrt{\left(\frac{dx}{dt}\right)^2 + \left(\frac{dy}{dt}\right)^2}\,,$$

$$\therefore \qquad s = -\int_a^b \sqrt{\left(\frac{dx}{dt}\right)^2 + \left(\frac{dy}{dt}\right)^2}\, dt. \qquad \textit{(As in Art. 89)}$$

Hence if s is measured from the pt. where t = a in the direction in which t decreases, we must change the sign in the result obtained by using the formula :

Length of arc $\displaystyle = \int_a^b \sqrt{\left(\frac{dx}{dt}\right)^2 + \left(\frac{dy}{dt}\right)^2}\, dt,$

to get s.

Similarly for the other arc formulae.]

9. Find the length of the loop of the curve $x = t^2, y = t - \frac{1}{3}t^3$.

[To trace the curve $x = t^2, y = t - \frac{1}{3}t^3$ ...(1)

(*i*) **Symmetry.** On changing t to $-t$, x remains unchanged and y is changed to $-y$,
$\therefore$ the curve is symmetrical about the x-axis.

(*ii*) **Origin.** $x = 0$ when $t^2 = 0$, or $t = 0$, and then $y = 0$,
$\therefore$ the curve passes through the origin.

It meets the x-axis where, putting $y = 0$, $t - \frac{1}{3}t^3 = 0$, or $t\left(1 - \frac{1}{3}t^2\right) = 0$,

$$\therefore \qquad t = 0, \quad \text{or} \quad t = \pm\sqrt{3}.$$

When $t = 0, x = 0$; when $t = \sqrt{3}, x = 3$; when $t = -\sqrt{3}, x = 3$.
$\therefore$ the curve meets the x-axis at (0, 0), (3, 0).
It meets the y-axis where, putting $x = 0$, $t^2 = 0$, or $t = 0$,
$\therefore$ $y = 0$, *i.e.*, at (0, 0).

(*iii*) **Asymptotes.** $x \to \infty$ when $t \to \infty$, and then $y \to -\infty$,
[From the highest degree term $-\frac{1}{3}t^3$]

$\therefore$ there are no asymptotes but the curve extends to infinity.

(*iv*) **Points where** $\frac{\mathbf{dy}}{\mathbf{dx}}$ **= 0 or** $\to \infty$. $\frac{dx}{dt} = 2t$, $\frac{dy}{dt} = 1 - \frac{1}{3}(3t^2) = 1 - t^2$.

$$\therefore \qquad \frac{dy}{dx} = \frac{\frac{dy}{dt}}{\frac{dx}{dt}} = \frac{1-t^2}{2t}.$$

$$\therefore \qquad \frac{dy}{dx} = 0 \text{ when } 1 - t^2 = 0, \text{ or } t = \pm 1.$$

When $t = 1, x = 1, y = \frac{2}{3}$; when $t = -1, x = 1, y = -1 + \frac{1}{3} = -\frac{2}{3}$,

i.e., at $\left(1, \frac{2}{3}\right), \left(1, -\frac{2}{3}\right)$ the tangent is $\parallel$ to the x-axis.

$\frac{dy}{dx} \to \infty$ when $t \to 0$, and then $x \to 0, y \to 0$,

i.e., at (0, 0) the tangent is $\perp$ to the x-axis.

When $t = \sqrt{3}$, $\frac{dy}{dx} = \frac{1-3}{2\sqrt{3}} = \frac{-2}{2\sqrt{3}} = -\frac{1}{\sqrt{3}}$;

when $t = -\sqrt{3}$, $\frac{dy}{dx} = \frac{1-3}{-2\sqrt{3}} = \frac{-2}{-2\sqrt{3}} = \frac{1}{\sqrt{3}}$,

i.e., at (3, 0), the tangents make angles of $\mp 30°$ with the x-axis.
$\therefore$ (3, 0) is a node.

(v) $t = \sqrt{x}$. ∴ t is imaginary when x is – ve, ∴ no portion of the curve lies to the left of the y-axis.

When $t \to -\infty, x \to \infty, y \to \infty$;

when $t \to +\infty, x \to \infty, y \to -\infty$.

As t increases from $-\infty$ to $+\infty$, passing through the values $-\sqrt{3}, -1, 0, 1, \sqrt{3}$ (found above), the curve DACOBAE is described as shown by the arrows in the Fig.

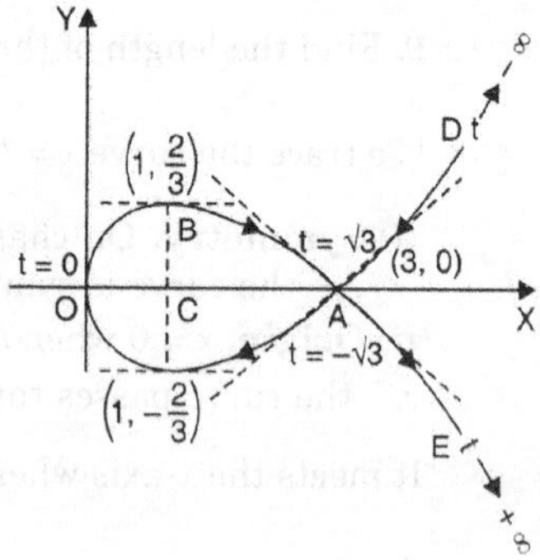

Hence the shape of the curve is that shown in the Fig.

For the upper half of the loop, t varies from 0 to $\sqrt{3}$.]

10. (i) Find the perimeter of the curve $r = a \cos \theta$.

(ii) Find the length of the arc of the parabola $\frac{2a}{r} = 1 + \cos \theta$ cut off by the latus rectum.

[(i) *The circle $r = a \cos \theta$ is symmetrical about the initial line, and for the upper half, θ varies from 0 to $\frac{\pi}{2}$.*

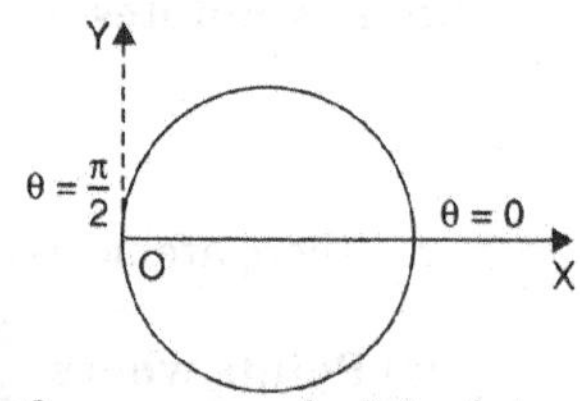

(ii) The parabola $\frac{2a}{r} = 1 + \cos \theta$ is symmetrical about the initial line. For the portion of the parabola between the vertex and the upper end of the latus rectum, θ varies from 0 to $\frac{\pi}{2}$.

$$r = \frac{2a}{1 + \cos \theta} = \frac{2a}{2\cos^2 \frac{\theta}{2}} = a \sec^2 \frac{\theta}{2}.\Bigg]$$

****11.** (a) Show that the length of the arc of that part of the cardioid $r = a(1 + \cos \theta)$, which lies on the side of the line $4r = 3a \sec \theta$ remote from the pole, is equal to $4a$.

(b) Prove that the cardioid $r = a(1 + \cos \theta)$ is divided by the line $4r \cos \theta = 3a$ into two parts such that the lengths of the arcs on either side of this line are equal.

****12.** Show that in the parabola $\frac{2a}{r} = 1 + \cos \theta$, $\frac{ds}{d\psi} = \frac{2a}{\sin^3 \psi}$, and hence show that the arc intercepted between the vertex and the extremity of the latus rectum is $a\{\sqrt{2} + \log(1 + \sqrt{2})\}$.

13. Equiangular spiral (or logarithmic spiral). Trace the curve

$$r = ae^{\theta \cot \alpha},$$

and find the length of the curve taking $s = 0$ when $\theta = 0$.

14. Limacon. Show that the perimeter of the limacon $r = a + b \cos \theta$, if $\frac{b}{a}$ is small, is approximately $2\pi a \left(1 + \frac{1}{4} \frac{b^2}{a^2}\right)$.

[The equation of the limacon is $r = a + b \cos \theta$.

$\therefore \quad \frac{b}{a}$ is small (Given) (and $\therefore \quad < 1$), $b < a$,

i.e., **a > b.**

$\therefore$ the curve consists of one closed portion.

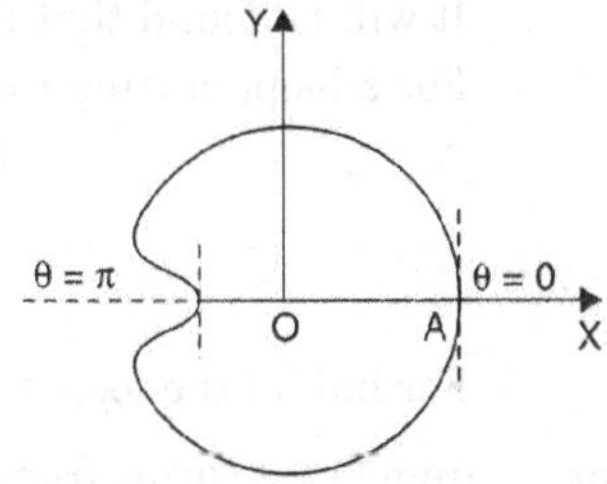

The limacon is symmetrical about the initial line, and for the upper half, θ varies from 0 to π.]

15. Show that the whole length of the limacon $r = a \cos \theta + b$ is equal to that of an ellipse whose semi-axes are equal in length to the maximum and minimum radii vectors of the limacon.

[$r = a \cos \theta + b$ ...(1)

$\therefore$ whole length of limacon $= 2 \int_0^{\pi} \sqrt{r^2 + \left(\frac{dr}{d\theta}\right)^2} \, d\theta = ?$ ...(2)

[*Assuming that a < b*]

(Do not evaluate this integral.)

From (1), max. radius vector $= a.1 + b = a + b$,

and min. radius vector $= a(-1) + b = b - a$.

$\therefore$ the parametric equations of the ellipse are

$$x = (a + b) \cos \theta, y = (b - a) \sin \theta \qquad [x = a \cos \theta, y = b \sin \theta]$$

$\therefore$ whole length of ellipse $= 4 \int_0^{\frac{\pi}{2}} \sqrt{\left(\frac{dx}{d\theta}\right)^2 + \left(\frac{dy}{d\theta}\right)^2} \, d\theta.$

(Express the integrand in terms of multiple angles) = ?

Put $2\theta = z$, and use theorem IV, Art. 78, and then theorem I, Art. 78.]

16. Reciprocal spiral (or hyperbolic spiral). Prove that the length of the arc of the hyperbolic spiral $r\theta = a$, taken from the point

$$r = a \text{ to } r = 2a, \text{ is } a \left[\sqrt{5} - \sqrt{2} + \log \frac{2 + \sqrt{8}}{1 + \sqrt{5}}\right].$$

[The equation of the hyperbolic spiral is $r\,\theta = a$, or $\theta = \frac{a}{r}$.]

****17.** Show that the length of a loop of the curve $3x^2y - y^3 = (x^2 + y^2)^3$ is $2\int_0^1 \frac{d\xi}{\sqrt{1-\xi^6}}$.

[As suggested by $(x^2 + y^2)^3$ in the equation of the curve, change the equation to the polar form (by putting $x = r\cos\theta$, $y = r\sin\theta$).

It will be found that the equation becomes $r^3 = \sin 3\theta$ ...(1)

For a loop, putting $r = 0$, $\sin 3\theta = 0$,

$$\therefore \quad 3\theta = 0, \pi \quad \textit{(two consecutive values)}$$

$$\therefore \quad \theta = 0, \frac{\pi}{3}.$$

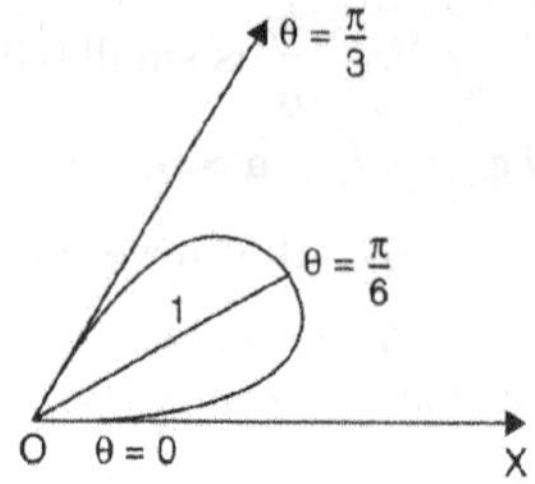

For half of the loop, θ varies from 0 to $\frac{\pi}{6}$,

or from (1), r varies from 0 to 1.

$$\therefore \quad \text{length of loop} = 2\int_0^1 \sqrt{1 + r^2\left(\frac{d\theta}{dr}\right)^2}\, dr \quad ...(2)$$

From (1), $3\theta = \sin^{-1} r^3$, $\therefore \theta = \frac{1}{3}\sin^{-1} r^3$,

$$\therefore \quad \frac{d\theta}{dr} = ? \text{ Substitute in (2).}$$

Note. Important. *To find the limits of integration for a loop, it is generally useful to put $r = 0$ and find two consecutive values of* θ.]

18. If the co-ordinates of a point on the curve $3ay^2 = x(a - x)^2$ are expressed in terms of a parameter in the form $x = 3at^2$, $y = a(t - 3t^3)$, show how the curve is traced out as t increases from $-\infty$ to $+\infty$. Show that the length of the arc of this curve from the origin to the point of the loop where the tangent to the curve makes an angle ψ with the y-axis, is $\frac{1}{9}a(3\tan\frac{1}{2}\psi + \tan^3\frac{1}{2}\psi)$.

[Proceed as in Ex. 9. It will be found that the length of the arc from the origin to the pt. t is $a(t + 3t^3)$ ...(1)

If the tangent makes an angle ψ with the y-axis, then

$$\tan\psi = \frac{dx}{dy} \text{ (which will be found to be)} = \frac{6t}{1-9t^2},$$

or

$$\frac{2\tan\frac{\psi}{2}}{1-\tan^2\frac{\psi}{2}} = \frac{2(3t)}{1-(3t)^2},$$

$$\therefore \quad \tan\frac{\psi}{2} = 3t, \quad \text{or} \quad t = \frac{1}{3}\tan\frac{\psi}{2}.$$

Substitute this value of t in (1).]

**19. In the four-cusped hypocycloid $x^{\frac{2}{3}} + y^{\frac{2}{3}} = a^{\frac{2}{3}}$ show that $s = \frac{3}{4} a \cos 2\psi$, s being measured from the vertex.

[It will be found that the parametric equations of the four-cusped hypocycloid are $x = a \cos^3 \theta$, $y = a \sin^3 \theta$. (Proceed as in Ex. 3, (*i*), Art. 90.) At the vertex $\theta = \frac{\pi}{4}$.

Proceed as in Note 1, Ex. 8. It will be found that $s = \frac{3a}{4} \cos 2\theta$. ...(1)

If ψ is the angle which the tangent makes with the x-axis,

$$\tan \psi = \frac{dy}{dx} = \frac{\frac{dy}{d\theta}}{\frac{dx}{d\theta}} = \frac{3a \sin^2 \theta \cos \theta}{3a \cos^2 \theta (-\sin \theta)} = -\tan \theta = \tan (\pi - \theta),$$

$\therefore$ $\psi = \pi - \theta$ ...(2)

Eliminate θ from (1) and (2).]

20. Find the intrinsic equation of the cardioid $r = a(1 + \cos \theta)$.

[Measuring s from the pt. where $\theta = 0$, it will be found that $s = 4a \sin \frac{\theta}{2}$ ...(1)

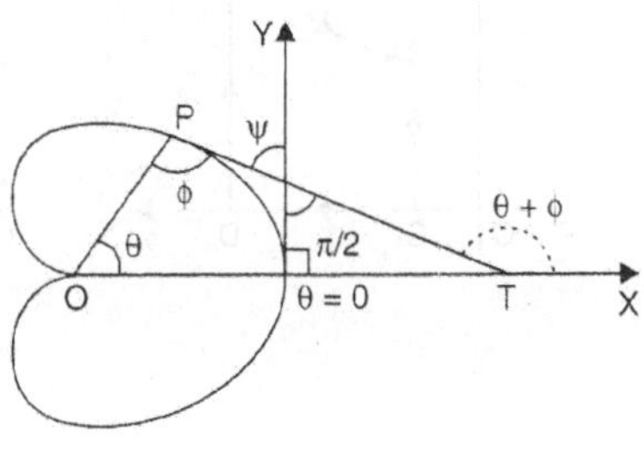

Measuring ψ *with the tangent at the pt. where* $\theta = 0$ (see the Fig.),

ψ = vertically opposite $\angle$

$= \theta + \phi - \frac{\pi}{2}$ ...(2)

It will be found that $\phi = \frac{\pi}{2} + \frac{\theta}{2}$ ...(3)

$\therefore$ from (2), $\psi = \theta + \frac{\pi}{2} + \frac{\theta}{2} - \frac{\pi}{2} = \frac{3\theta}{2}$ ...(4)

Eliminate θ from (1) and (4).]

CHAPTER X

QUADRATURE

95. Quadrature. Def. The process of finding the area bounded by a given portion of a curve is called **quadrature.***

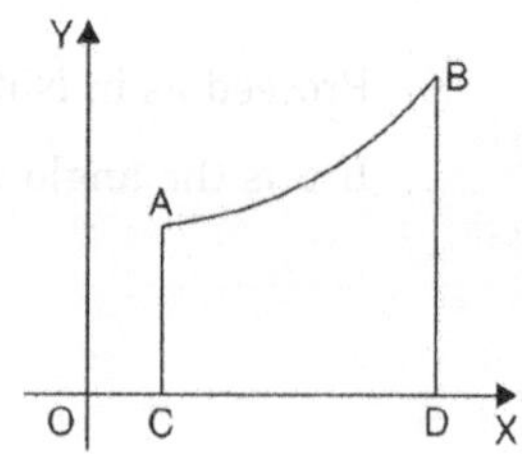

I. Area formulae for Cartesian equation.

96-I. It has been proved in Art. 77 that **the area bounded by the curve y = f(x), the x-axis and the ordinates x = a, x = b is $\int_a^b y\,dx$.**

Note. *Sign of an area.*

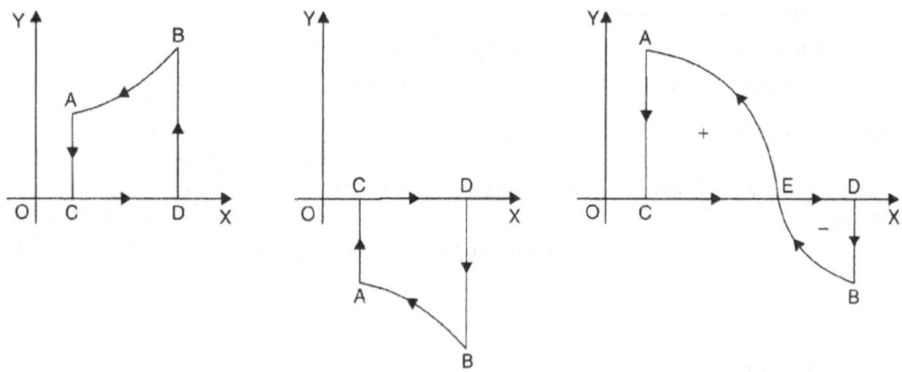

(i) (ii) (iii)

The area formula $\int_a^b y\,dx$ gives a positive result if the boundary ACDB is described in the counter-clockwise direction as in Fig. (i) above, in which the area lies above the x-axis, and it gives a negative result if the boundary ACDB is described in the clockwise direction as in Fig. (ii), in which the area lies below the x-axis.

If, however, the curve crosses the x-axis, say, at E as in Fig. (iii), the area formula $\int_a^b y\,dx$ does not give the sum of the areas of the two portions ACE and EDB but their *difference.*

Hence in order to find the sum of the numerical values of the areas ACE and EDB we must find the numerical value of each of these areas separately, and add the results.

Example 1. *Show that the area bounded by the parabola $y^2 = 4ax$, the x-axis and the ordinate $x = c$ is equal to $\frac{2}{3}$ of the rectangle of which the extreme ordinate and the abscissa of the area are adjacent sides.*

---

*Literal meaning : finding a *square* equal in area to the area bounded by a given portion of a curve.

The equation of the parabola is $y^2 = 4ax$.

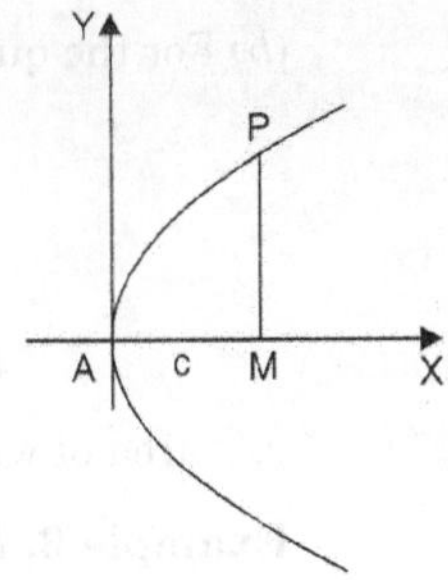

Taking + ve sign before the square root,

$$y = 2\sqrt{a}\ \sqrt{x}.$$

Let A be the vertex of the parabola and MP the ordinate $x = c$.

Then $$\text{area AMP} = \int_0^c y\,dx = \int_0^c 2\sqrt{a}\sqrt{x}\,dx$$

$$= 2\sqrt{a}\left[\frac{x^{\frac{3}{2}}}{\frac{3}{2}}\right]_0^c = \frac{4}{3}\sqrt{a}\,c^{\frac{3}{2}} \qquad ...(1)$$

Now $$MP = 2\sqrt{a}\sqrt{x} = 2\sqrt{a}\sqrt{c}.$$

$\therefore$ area of the rectangle of which MP and AM are adjacent sides

$$= 2\sqrt{a}\sqrt{c}.c = 2\sqrt{a}\,c^{\frac{3}{2}}.$$

$\therefore$ from (1), area AMP $= \frac{2}{3}.2\sqrt{a}\,c^{\frac{3}{2}} = \frac{2}{3}$ (area of the above rectangle).

Note. *The lower limit of integration is the lesser than the two values of x between the ordinates at which the required area lies.*

Thus in the above Ex. the lower limit of integration is 0.

Example 2. *Find the area bounded by the ellipse* $\frac{x^2}{a^2} + \frac{y^2}{b^2} = 1$, *the ordinates* $x = c$, $x = d$ *and the x-axis. Deduce the area of the whole ellipse.*

(*a*) The equation of the ellipse is $\frac{x^2}{a^2} + \frac{y^2}{b^2} = 1$,

or $$\frac{y^2}{b^2} = 1 - \frac{x^2}{a^2} = \frac{a^2 - x^2}{a^2}.$$

Taking + ve sign before the square root,

$$y = \frac{b}{a}\sqrt{a^2 - x^2}.$$

$\therefore$ $$\text{area} = \int_c^d y\,dx = \int_c^d \frac{b}{a}\sqrt{a^2 - x^2}\,dx$$

$$= \frac{b}{a}\left[\frac{x\sqrt{a^2 - x^2}}{2} + \frac{a^2}{2}\sin^{-1}\frac{x}{a}\right]_c^d$$

$$= \frac{b}{2a}\left[d\sqrt{a^2 - d^2} + a^2\sin^{-1}\frac{d}{a} - \left(c\sqrt{a^2 - c^2} + a^2\sin^{-1}\frac{c}{a}\right)\right]$$

$$= \frac{b}{2a}\left[d\sqrt{a^2 - d^2} - c\sqrt{a^2 - c^2} + a^2\left(\sin^{-1}\frac{d}{a} - \sin^{-1}\frac{c}{a}\right)\right]$$

(*b*) For the quadrant BCA, putting $c = 0, d = a$, area BCA

$$= \frac{b}{2a}\left[a(0) - 0 + a^2\left(\frac{\pi}{2} - 0\right)\right]$$

$$= \frac{b}{2a} \cdot a^2 \frac{\pi}{2} = \frac{\pi ab}{4}$$

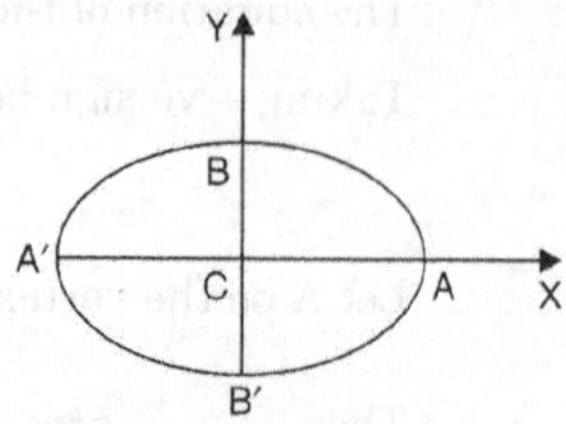

$\therefore$ area of whole ellipse = 4. area BCA = $4 \frac{\pi ab}{4} = \pi ab$.

Example 3. *Find the area*

(1) of the loop of the curve $x(x^2 + y^2) = a(x^2 - y^2)$ *;*

(2) of the portion bounded by the curve and its asymptote.

To trace the curve $x(x^2 + y^2) = a(x^2 - y^2)$ ...(1)

(*i*) **Symmetry.** The curve is symmetrical about the *x*-axis.

(*ii*) **Origin.** The curve passes through the origin.

$$x(x^2 + y^2) - a(x^2 - y^2) = 0$$

or $x^3 + xy^2 - ax^2 + ay^2 = 0$ ...(2)

The tangents at the origin are $-ax^2 + ay^2 = 0$, or $y^2 = x^2$,

or $y = \pm x$, which are real and different

$\therefore$ the origin is a node.

The curve meets the *x*-axis where, putting $y = 0$ in (2), $x^3 - ax^2 = 0$,

or $x^2(x - a) = 0$, or $x = 0, 0 ; a$, *i.e.*, at $(0, 0), (a, 0)$.

It meets the *y*-axis where, putting $x = 0$ in (2), $ay^2 = 0$, or $y = 0, 0$, *i.e.*, at $(0, 0)$.

(*iii*) **Asymptotes.** From (2), the asymptote is $x + a = 0$, or $x = -a$.

(*iv*) **Points where $\frac{dy}{dx} = 0$ or $\to \infty$.** $y^2(a + x) = x^2(a - x)$ or $y^2 = \frac{x^2(a - x)}{a + x}$.

Taking + ve sign before the square root,

$$y = x\sqrt{\frac{a - x}{a + x}} \qquad ...(3),$$

$\therefore \frac{dy}{dx} \to \infty$ * when $x \to a$, and then from (3), $y \to 0$,

i.e., at $(a, 0)$ the tangent is $\perp$ to the *x*-axis.

(*v*) From (3), $y = x\sqrt{\frac{a - x}{x - (-a)}}$,

***How to write this step.** $y = x\sqrt{\frac{a - x}{a + x}} = \sqrt{a - x} \cdot \frac{x}{\sqrt{a + x}}$ (Form *uv*)

$\therefore \frac{dy}{dx}$ will involve $\frac{1}{2\sqrt{a - x}}(-1)$ which $\to \infty$ when $x \to a$.

In general, remember : *If a factor occurs in the numerator, whose index is positive and less than 1, then $\frac{dy}{dx} \to \infty$ when that factor $\to 0$.*

$\therefore$ y is imaginary when (from the num.) $x > a$ or (from the denom.) $x < -a$,

$\therefore$ no portion of the curve lies to the right of the line $x = a$, or to the left of $x = -a$.

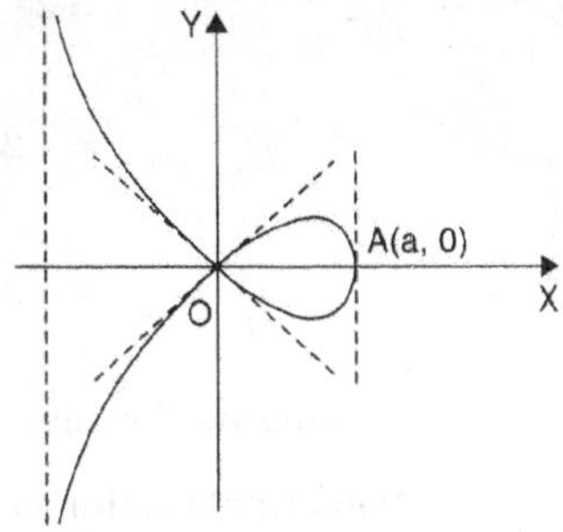

Hence the shape of the curve is that shown in the Fig.

(a) For the upper half of the loop, x varies from 0 to a.

$\therefore$ area of loop $= 2\int_0^a y\,dx = 2\int_0^a x\sqrt{\dfrac{a-x}{a+x}}\,dx$ ...(4)

Now $\int x\sqrt{\dfrac{a-x}{a+x}}\,dx$ [Rationalize the num. by multiplying the num. and denom. by $\sqrt{a-x}$]

$$= \int \frac{x(a-x)}{\sqrt{a^2-x^2}}\,dx = \int \frac{ax}{\sqrt{a^2-x^2}}\,dx - \int \frac{x^2}{\sqrt{a^2-x^2}}\,dx$$

$$= a\int (a^2-x^2)^{-\frac{1}{2}}\,x\,dx - \int \frac{-(a^2-x^2)+a^2}{\sqrt{a^2-x^2}}\,dx$$

$$= -\frac{a}{2}\int (a^2-x^2)^{-\frac{1}{2}}(-2x)\,dx + \int \sqrt{a^2-x^2}\,dx - a^2\int \frac{1}{\sqrt{a^2-x^2}}\,dx$$

$$= -\frac{a}{2}\,\frac{(a^2-x^2)^{\frac{1}{2}}}{\frac{1}{2}} + \frac{x\sqrt{a^2-x^2}}{2} + \frac{a^2}{2}\sin^{-1}\frac{x}{a} - a^2\sin^{-1}\frac{x}{a}$$

$$= \frac{(x-2a)}{2}\sqrt{a^2-x^2} - \frac{a^2}{2}\sin^{-1}\frac{x}{a} \qquad ...(5)$$

$\therefore$ from (4), area of loop $= 2\left[\dfrac{(x-2a)}{2}\sqrt{a^2-x^2} - \dfrac{a^2}{2}\sin^{-1}\dfrac{x}{a}\right]_0^a$ [Cancel 2]

$$= -a(0) - a^2\left(\frac{\pi}{2}\right) - [(-2a)a - a^2(0)]$$

$$= -a^2\frac{\pi}{2} + 2a^2 = a^2\left(2 - \frac{\pi}{2}\right).$$

(b) For half of the area bounded by the curve and its asymptote x varies from $-a$ to 0.

$\therefore$ area between curve and asymptote

$$= 2\int_{-a}^{0} y\,dx = 2\int_{-a}^{0} x\sqrt{\frac{a-x}{a+x}}\,dx \text{ which, from (5),}$$

$$= 2\left[\frac{(x-2a)}{2}\sqrt{a^2-x^2}-\frac{a^2}{2}\sin^{-1}\frac{x}{a}\right]_{-a}^{0} \qquad \text{[Cancel 2]}$$

$$= -2a(a) - a^2(0) - \left[-3a\,(0) - a^2\left(-\frac{\pi}{2}\right)\right]$$

$$= -2a^2 - a^2\,\frac{\pi}{2} = -a^2\left(2+\frac{\pi}{2}\right).$$

$\therefore$ changing the sign, area (in magnitude) $= a^2\left(2+\dfrac{\pi}{2}\right)$.

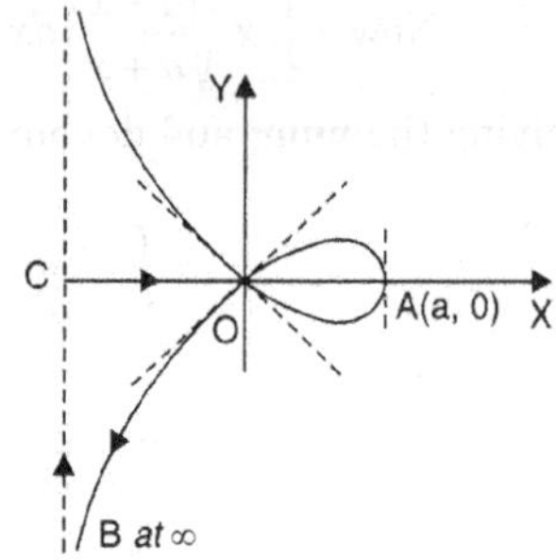

****Note. Explanation of the negative answer.** *When x varies from $-a$ to 0, x is $-$ ve, $\therefore$ $y\left(=x\sqrt{\dfrac{a-x}{a+x}}\right)$ is also $-$ ve,*

$\therefore$ $\displaystyle\int_{-a}^{0} y\,dx$ *is also $-$ ve* *(Note 2, Art. 81).*

In fact, since y is $-$ ve, the boundary BCO is described in the clockwise direction,

$\therefore$ *the area BCO is $-$ ve* *(Note, Art. 96-I).*

Example 4. *Find the area of the curve*

$x = a\cos^3 t,$

$y = b\sin^3 t.$

The parametric equations of the curve are

$$x = a\cos^3 t,\ y = b\sin^3 t \qquad ...(1)$$

The curve is symmetrical about both axes, and in the first quadrant, t varies from 0 to $\dfrac{\pi}{2}$.

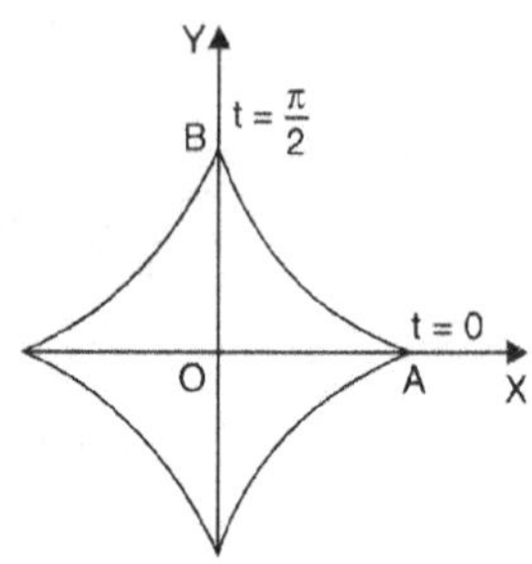

$\therefore$ area of curve $= 4\displaystyle\int y\,dx = 4\int_0^{\frac{\pi}{2}} y\,\frac{dx}{dt}\,dt$

$$= 4\int_0^{\frac{\pi}{2}} b\sin^3 t\,[a.3\cos^2 t\,(-\sin t)]\,dt$$

$$= -12ab\int_0^{\frac{\pi}{2}} \sin^4 t\cos^2 t\,dt$$

$$= -12ab\,\frac{3.1.1}{6.4.2}\,\frac{\pi}{2} = -\frac{3}{8}\,\pi ab.$$

$\therefore$ changing the sign, area (in magnitude) $= \dfrac{3}{8}\,\pi ab.$

****Note. Explanation of the negative answer.** *When t varies from 0 to $\dfrac{\pi}{2}$, x varies from a to 0, so that x decreases, $\therefore$ $\dfrac{dx}{dt}$ is $-$ ve,*

$$\therefore \quad \int_0^{\frac{\pi}{2}} y \frac{dx}{dt} \, dt \text{ is } - \text{ve} \qquad \text{[Note 2, Art. 81.]}$$

In fact, since x varies from a to 0, the boundary AOB is described in the clockwise direction, $\therefore$ the area AOB is – ve [Note, Art. 96-I].

EXAMPLES

1. Find the areas bounded by each of the following curves, the x-axis and the ordinates $x = 0, x = h$:

(*i*) $y = e^x$ (exponential curve) ; also trace the curve.

(*ii*) $y = c \cosh \dfrac{x}{c}$ (catenary).

2. Find the areas bounded by each of the following curves, the x-axis and the ordinates $x = a, x = b$:

(*i*) $xy = c^2$ (rectangular hyperbola).

(*ii*) $y = \log x$ (logarithmic curve) ; also trace the curve.

[(*ii*) **Note.** *It is assumed that* $1 < a < b$.]

3. Obtain the area bounded by a parabola and its latus rectum.

4. Find the area of the ellipse $\dfrac{x^2}{a^2} + \dfrac{y^2}{b^2} = 1$.

[*The ellipse is symmetrical about both axes, and in the first quadrant, x varies from 0 to a.*]

5. Trace the curve $x^2y^2 = a^2(y^2 - x^2)$.

Find the whole area included between the curve and its asymptotes.

6. Trace the curve $a^2y^2 = a^2x^2 - x^4$.

Find the whole area within it.

7. Trace the curve $a\, y^2 = x^2(a - x)$ and show that the area of its loop is $\dfrac{8}{15}\, a^2$.

8. (*i*) Find the whole area of the curve $y^2 = x^2 \dfrac{a^2 - x^2}{a^2 + x^2}$.

(*ii*) Find the whole area contained between the curve $x^2(x^2 + y^2) = a^2(y^2 - x^2)$ and its asymptotes.

[(*i*) It will be found that area $= 4 \displaystyle\int_0^a x \sqrt{\frac{a^2 - x^2}{a^2 + x^2}}\, dx$. Put $\sqrt{a^2 + x^2} = z$.]

9. *Cissoid.* Trace the curve $y^2 = \dfrac{x^3}{2a - x}$.

Find the area between the curve and its asymptote.

10. Trace the curve $a^4y^2 = x^5(2a - x)$, and prove that its area is to that of the circle whose radius is a as 5 to 4.

11. *Cycloid.* Find the area included between the cycloid $x = a(\theta - \sin\theta), y = a(1 - \cos\theta)$ and its base.

[**Note. Base of cycloid. Def.** The line joining the cusps of a cycloid is called the **base.**

For the first half of the cycloid in the first quadrant, θ varies from 0 to π.

$$\therefore \qquad \text{area} = 2\int y\,dx = 2\int_0^{\pi} y\,\frac{dx}{d\theta}\,d\theta.\Big]$$

12. Prove that the area between the curve $y^2(a + x) = (a - x)^3$ and its asymptote is three times the area of the circle whose radius is a.

[It will be found that area $= 2\int_{-a}^{a} (a - x)^{\frac{3}{2}} \cdot \dfrac{1}{\sqrt{a + x}}\,dx$. Integrate by parts.]

13. *Witch of Agnesi.* Trace the curve $y^2 = \dfrac{4a^2(2a - x)}{x}$ and find the area between the curve and its asymptote.

96-II. It has been proved in Art. 77 that the area bounded by the curve y = f(x), the x-axis and the ordinates x = a, x = b is $\int_a^b y\,dx$ ...(1)

Changing x to y and y to x in (1), **the area bounded by the curve x = f(y), the y-axis and the abscissae y = a, y = b is** $\int_a^b x\,dy$.

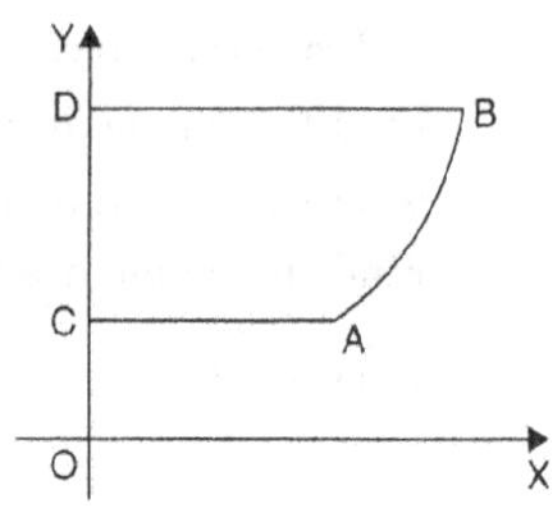

Cor. $$A = \int x\,dy.$$

[Changing x to y and y to x in Art. 77, Cor. 1.]

EXAMPLES

1. Prove that the area bounded by the curve $x = f(y)$, the y-axis and the abscissae $y = a$, $y = b$ is $\int_a^b x\,dy$.

Find the areas of the portions into which the ellipse $\dfrac{x^2}{a^2} + \dfrac{y^2}{b^2} = 1$ is divided by the line $y = c$.

2. Trace the curve $a^2x^2 = y^3(2a - y)$.

Prove that its area is equal to that of the circle whose radius is a.

3. Find the area of the cycloid $x = a\ (\theta + \sin\theta)$, $y = a(1 - \cos\theta)$.

[**Note.** "Area of the cycloid" means "area bounded by one arch of the cycloid and the line joining the cusps".]

II. Area bounded by two curves.

97. To prove that the area bounded by the curves $y = f(x)$, $y = \phi(x)$, and the ordinates $x = a$, $x = b$ is $\int_a^b [f(x) - \phi(x)]\, dx$.

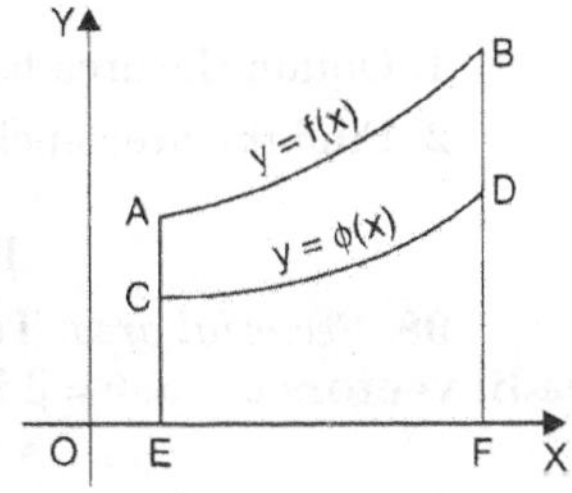

Let AB, CD be the curves $y = f(x)$, $y = \phi(x)$, and ECA, FDB the ordinates $x = a$, $x = b$.

Then area ACDB = area AEFB – area CEFD

$$= \int_a^b f(x)\, dx - \int_a^b \phi(x)\, dx$$

$$= \int_a^b [f(x) - \phi(x)]\, dx.$$

[**Aid to memory.** Area bounded by two curves = $\int (y_1 - y_2)\, dx$, where y_1 is the 'y' of the upper curve and y_2 that of the lower curve.]

Example. *Find the area above the x-axis included between the curves $y^2 = 2ax - x^2$ and $y^2 = ax$.*

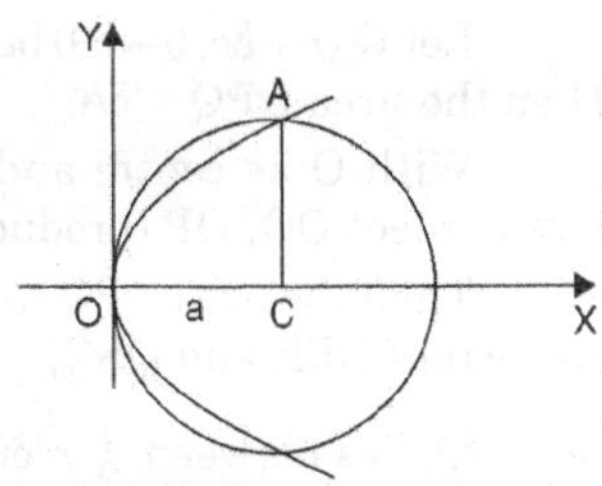

The equations of the curves are

$$y^2 = 2ax - x^2, \text{ or } x^2 + y^2 - 2ax = 0 \qquad ...(1),$$

which is a circle whose centre is $(a, 0)$ and radius a ;

$$y^2 = ax \qquad ...(2),$$

which is a parabola.

To find the pts. of intersection, substituting the value of y^2 from (2) in (1),

$$x^2 + ax - 2ax = 0, \quad \text{or} \quad x^2 - ax = 0, \quad \text{or} \quad x(x-a) = 0,$$

$\therefore\ x = 0, a$, which are the limits of integration.

$$\therefore \quad \text{required area} = \int_0^a (y_1 - y_2)\, dx = \int_0^a (\sqrt{2ax - x^2} - \sqrt{a}\sqrt{x})\, dx$$

Now $\int_0^a \sqrt{2ax - x^2}\ dx = \int_0^a \sqrt{-(x^2 - 2ax)}\ dx = \int_0^a \sqrt{-(x^2 - 2ax + a^2) + a^2}$

$$= \int_0^a \sqrt{a^2 - (x-a)^2}\ dx = \left[\frac{(x-a)\sqrt{a^2 - (x-a)^2}}{2} + \frac{a^2}{2}\sin^{-1}\frac{x-a}{2}\right]_0^a$$

$$= \frac{1}{2}\left[0 + a^2(0) - \left\{-a(0) + a^2\left(-\frac{\pi}{2}\right)\right\}\right] = \frac{\pi a^2}{4}.$$

$$\int_0^a \sqrt{a}\sqrt{x}\, dx = \sqrt{a}\left[\frac{x^{\frac{3}{2}}}{\frac{3}{2}}\right]_0^a = \frac{2}{3}\sqrt{a}\,(a^{\frac{3}{2}} - 0) = \frac{2}{3}a^2.$$

$\therefore$ from (3), required area $= \frac{\pi}{4}a^2 - \frac{2}{3}a^2 = a^2\left(\frac{\pi}{4} - \frac{2}{3}\right)$.

EXAMPLES

1. Obtain the area bounded by the curves $y^2 = 4ax$, $x^2 = 4ay$.

2. Find the area enclosed by the curves $x^2 = 4ay$ and $x^2 + 4a^2 = 8a^3/y$.

III. Area formula for polar equation.

98. *Sectorial area.* **To prove that the area bounded by the curve r = f (θ) and the radii vectors θ = α, θ = β is**

$$\int_\alpha^\beta \frac{1}{2}\,\mathbf{r}^2\, d\theta.$$

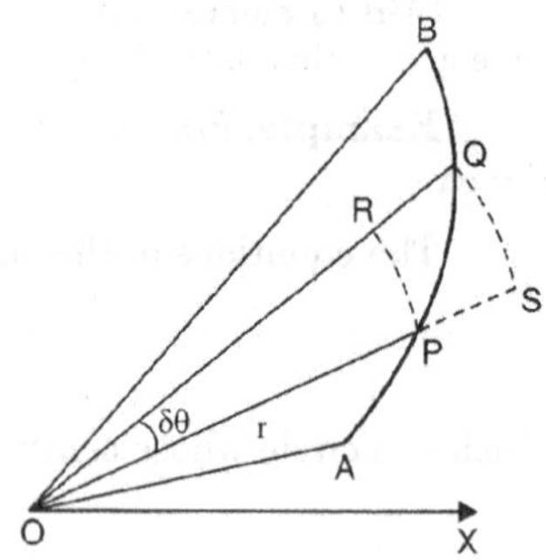

Let AB be the curve $r = f(\theta)$, and OA, OB the radii vectors $\theta = \alpha$, $\theta = \beta$.

Let P(r, θ) be any pt. on the curve. Then the area OAP is clearly a function of θ = A (say).

Let Q $(r + \delta r, \theta + \delta\theta)$ be a neighbouring pt. on the curve. Then the area OPQ = δA.

With O as centre and OP, OQ as radii draw circular arcs to meet OQ, OP (produced) respectively in R, S.

Then the area OPQ lies between the areas of the circular sectors OPR and OSQ,

i.e., δA lies between $\frac{1}{2}r^2\delta\theta$* and $\frac{1}{2}(r + \delta r)^2\,\delta\theta$

$\therefore$ $\frac{\delta A}{\delta\theta}$ lies between $\frac{1}{2}r^2$ and $\frac{1}{2}(r + \delta r)^2$

$\therefore$ taking the limits when $\delta\theta \to 0$ (and $\therefore$ $\delta r \to 0$),

$$\frac{dA}{d\theta} = \frac{1}{2}r^2 \qquad ...(1)$$

$$\therefore \quad \int_\alpha^\beta \frac{1}{2}r^2\, d\theta = \int_\alpha^\beta \frac{dA}{d\theta}\, d\theta = \Big[A\Big]_\alpha^\beta$$

= (value of A when $\theta = \beta$) – (value of A when $\theta = \alpha$)

= area OAB – 0,

i.e., area OAB $= \int_\alpha^\beta \frac{1}{2}\, r^2\, d\theta.$

Cor. From (1), $\mathbf{A} = \int \frac{1}{2}\mathbf{r}^2\, \mathbf{d\theta}.$

---

*Area of circular sector $= \frac{1}{2}$ (radius)^2 × (circular measure of angle of sector).

Note 1. Assumptions. *(i) $f(\theta)$ is continuous and single-valued in the interval (α, β).*

(ii) α, β are finite, and $\alpha < \beta$.

Note 2. *In the Fig. of Art. 98, it has been assumed that as a point moves along the curve from A to B, the radius vector constantly increases. If, however, the radius vector constantly decreases as in the adjoining Fig., the proposition is still true.*

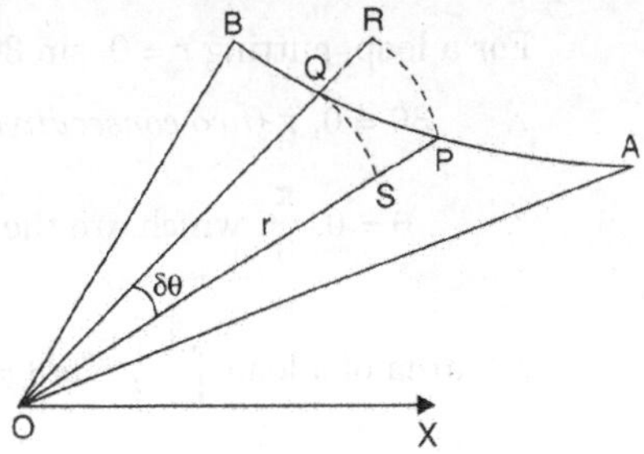

For, the area OPQ still lies between the areas of the circular sectors OSQ and OPR,

i.e., δA *lies between* $\frac{1}{2}(r + \delta r)^2\, \delta\theta$ *and* $\frac{1}{2} r^2 \delta\theta$, *and the proof holds as before.*

Note 3. *If, however, the radius vector constantly increases in some parts of the interval and constantly decreases in the others as in the adjoining Fig., the proposition is still true.*

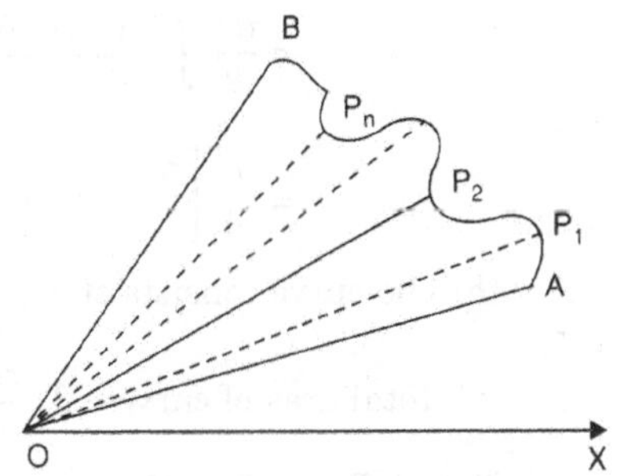

For, if the curve has a finite number of maximum and minimum radii vectors OP_1, OP_2, ..., OP_n, then area OAB

$= \text{area } OAP_1 + \text{area } OP_1P_2 + \ldots + \text{area } OP_nB,$

and the proposition is true for each of the areas on the R.H.S.

Example 1. *Obtain the area of the semicircle bounded by $r = a \cos \theta$ and the initial line.*

The equation of the circle is $r = a \cos \theta$.

For the upper semicircle, θ varies from 0 to $\frac{\pi}{2}$.

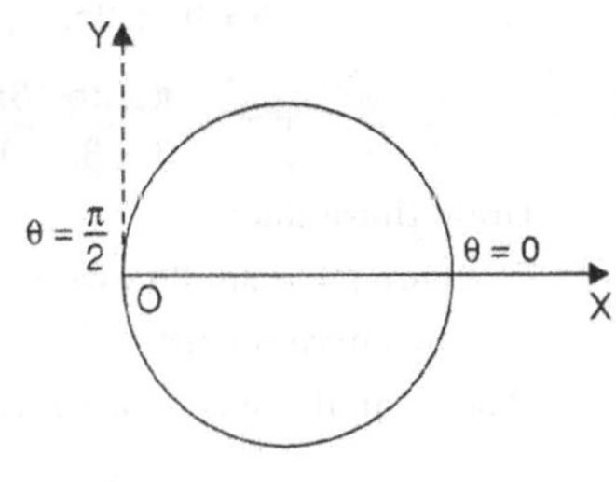

$$\therefore \quad \text{area of semicircle} = \int_0^{\frac{\pi}{2}} \frac{1}{2} r^2 d\theta$$

$$= \frac{1}{2} \int_0^{\frac{\pi}{2}} a^2 \cos^2 \theta \, d\theta$$

$$= \frac{a^2}{2} \cdot \frac{1}{2} \cdot \frac{\pi}{2} = \frac{\pi a^2}{8}.$$

Example 2. *Three-leaved rose. Obtain the area of a loop of the curve $r = a \sin 3\theta$.*

Also find the total area of the curve.

(*a*) The equation of the curve is $r = a \sin 3\theta$.

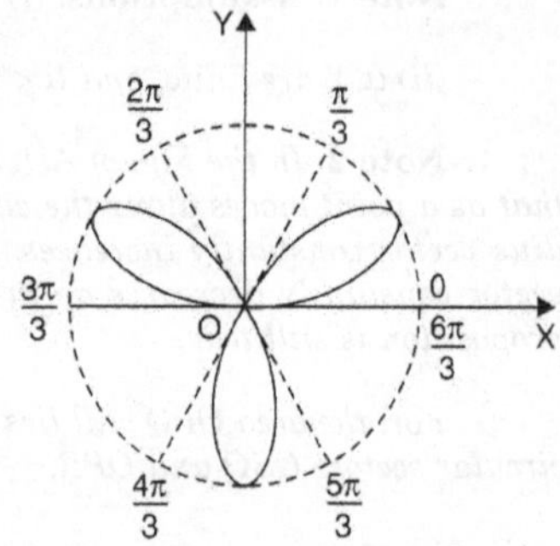

For a loop, putting $r = 0$, $\sin 3\theta = 0$,

$\therefore \quad 3\theta = 0, \pi$ *(two consecutive values)*

$\therefore \quad \theta = 0, \frac{\pi}{3}$, which are the limits of integration.

$$\therefore \quad \text{area of a loop } \int_0^{\frac{\pi}{3}} \frac{1}{2} r^2 \, d\theta = \frac{1}{2} \int_0^{\frac{\pi}{3}} a^2 \sin^2 3\theta \, d\theta$$

$$= \frac{a^2}{2} \int_0^{\frac{\pi}{3}} \frac{1 - \cos 6\theta}{2} \, d\theta = \frac{a^2}{4} \left[\theta - \frac{\sin 6\theta}{6} \right]_0^{\frac{\pi}{3}}$$

$$= \frac{a^2}{4} \left[\frac{\pi}{3} - 0 - (0 - 0) \right] = \frac{\pi a^2}{12}.$$

(*b*) The curve consists of 3 equal loops.*

$$\therefore \quad \text{total area of curve} = 3 . \frac{\pi a^2}{12} = \frac{\pi a^2}{4}.$$

Note 1. Remember : *If n is a positive odd integer, the curve $r = a \sin n\theta$ (or $r = a \cos n\theta$) consists of n equal loops, and, if n is a positive even integer, the curve consists of 2n equal loops.*

Note 2. Important. *To find the limits of integration for a loop, it is generally useful to put $r = 0$ and find two consecutive values of θ.*

---

*__How to draw the curve r = a sin 3θ.__ Putting $r = 0$, $\sin 3\theta = 0$,

$\therefore \quad 3\theta = 0, \pi, 2\pi, \ldots,$

or $\quad \theta = 0, \frac{\pi}{3}, \frac{2\pi}{3}, \frac{3\pi}{3}, \frac{4\pi}{3}, \frac{5\pi}{3}, \frac{6\pi}{3} \; (= 2\pi).$

Draw these lines.

Compare $r = a \sin 3\theta$ with $r = a \sin n\theta$. Here $n = 3$, a + ve odd integer.

$\therefore$ the curve consists of 3 equal loops. [*n* loops]

Place equal loops in *alternate* divisions thus : place first loop in the division between $\theta = 0$ and $\theta = \frac{\pi}{3}$; leave the next division between $\theta = \frac{\pi}{3}$ and $\theta = \frac{2\pi}{3}$; place loop in the next division between $\theta = \frac{2\pi}{3}$ and $\theta = \frac{3\pi}{3}$; leave the next division between $\theta = \frac{3\pi}{3}$ and $\theta = \frac{4\pi}{3}$; place loop in the next division between $\theta = \frac{4\pi}{3}$ and $\theta = \frac{5\pi}{3}$; leave the next division between $\theta = \frac{5\pi}{3}$ and $\theta = \frac{6\pi}{3}$.

Similarly for the curve $r = a \sin n\theta$, if n is a positive odd integer.

Note. If n is a positive even integer, place equal loops in *all* divisions.

EXAMPLES

1. (*a*) Show that the area enclosed between the two radii vectors $\theta = \theta_1$ and $\theta = \theta_2$ and the curve $r = f(\theta)$ is $\frac{1}{2}\int_{\theta_1}^{\theta_2} r^2\, d\theta$.

(*b*) Find the area of any sector of each of the following curves :

(*i*) $r\theta = a$ (*reciprocal spiral or hyperbolic spiral*) ($\theta = \alpha$ to $\theta = \beta$).

(*ii*) $r\theta^{\frac{1}{2}} = a$ (*lituus or trumpet*) ($\theta = \alpha$ to $\theta = \beta$).

(*iii*) $r^{\frac{1}{2}}\theta = a^{\frac{1}{2}}$ ($\theta = \alpha$ to $\theta = \beta$).

(*iv*) $r = ae^{\theta \cot \alpha}$ (*equiangular spiral*)

[$\theta = \beta$ to $\theta = \beta + \gamma$ (γ being $< 2\pi$)].

2. Find the area of one loop of each of the following curves :

(α) $r = a \sin 2\theta$ (*four-leaved rose*).

(β) $r = a \sin 4\theta$ (*eight-leaved rose*).

(γ) $r = a \sin n\theta$. Also state the total area.

3. Show that the area of a loop of the curve $r = a \cos n\theta$ is $\frac{\pi a^2}{4n}$, and state the total area in the cases n odd, n even.

4. Show that the area contained between the circle $r = a$ and the curve $r = a \cos 5\theta$ is equal to three-fourths of the area of the circle.

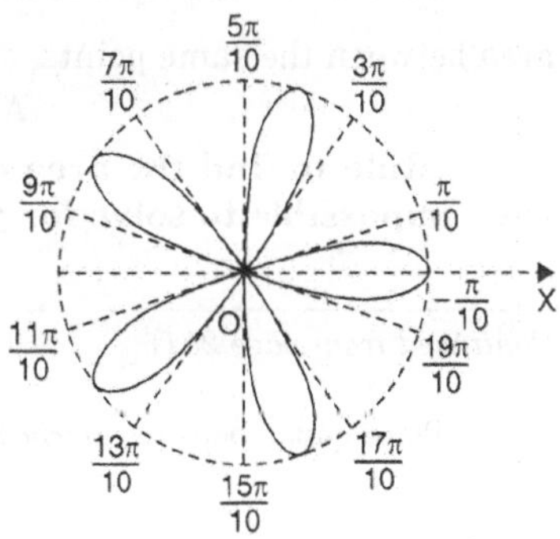

[The curve consists of 5 equal loops.* Find the area of one loop and hence the total area of the curve.

Area between circle and curve

= area of circle − area of curve.]

---

*__How to draw the curve r = a cos 5θ.__ Putting $r = 0$, $\cos 5\theta = 0$,

$\therefore \quad 5\theta = -\frac{\pi}{2}, \frac{\pi}{2}, \frac{3\pi}{2}, \ldots\ldots, \quad \left[\textit{Note the first angle} \left(-\frac{\pi}{2}\right)\right]$

or $\quad \theta = -\frac{\pi}{10}, \frac{\pi}{10}, \frac{3\pi}{10}, \frac{5\pi}{10}, \frac{7\pi}{10}, \frac{9\pi}{10}, \frac{11\pi}{10}, \frac{13\pi}{10}, \frac{15\pi}{10}, \frac{17\pi}{10}, \frac{19\pi}{10}, \left(\text{no further } \because \frac{21\pi}{10} > 2\pi\right).$

Draw these lines.

Compare $r = a \cos 5\theta$ with $r = a \cos n\theta$. Here $n = 5$, a + ve odd integer.

$\therefore$ the curve consists of 5 equal loops. [*n* loop]

[*Continued on page*]

5. Find the area of a loop of the curve $r = a \cos 3\theta + b \sin 3\theta$.

6. Find the areas bounded by

(*i*) the cardioid $r = a(1 - \cos \theta)$.

(*ii*) the cardioid $r = a(1 + \cos \theta)$.

7. Show that the whole area of the curve represented by the equation $r = a \cos \theta + b$, assuming $a < b$ is $\pi\left(\dfrac{a^2}{2} + b^2\right)$.

8. *Bernoulli's lemniscate.* Trace the curve $r^2 = a^2 \cos 2\theta$.

Find its area.

9. Find the area bounded by $r^2 = a^2 \cos^2 \theta + b^2 \sin^2 \theta$.

[The curve is symmetrical about the initial line, also about the line through the pole ⊥ to the initial line.

In the first quadrant, θ varies from 0 to $\dfrac{\pi}{2}$. $\therefore$ area $= 4\displaystyle\int_0^{\frac{\pi}{2}} \tfrac{1}{2} r^2 \, d\theta.$]

10. Find the area of a loop of the curve $r = a\,\theta \cos\theta$ between $\theta = 0$ and $\theta = \dfrac{\pi}{2}$.

11. If s be the length of the curve $r = a \tanh \dfrac{\theta}{2}$ between the origin and $\theta = 2\pi$, and A the area between the same points, show that

$$A = a\,(s - a\pi).$$

[Rule to find the area of a loop of the curve f (x, y) = 0 when it is difficult or even impossible to solve for y in terms of x :

Continued from page 261]

Place equal loops in *alternate* divisions thus : place first loop in the division between $\theta = -\dfrac{\pi}{10}$ and $\theta = \dfrac{\pi}{10}$; leave the next division between $\theta = \dfrac{\pi}{10}$ and $\theta = \dfrac{3\pi}{10}$; place loop in the next division between $\theta = \dfrac{3\pi}{10}$ and $\theta = \dfrac{5\pi}{10}$; leave the next division between $\theta = \dfrac{5\pi}{10}$ and $\theta = \dfrac{7\pi}{10}$; place loop in the next division between $\theta = \dfrac{7\pi}{10}$ and $\theta = \dfrac{9\pi}{10}$; leave the next division between $\theta = \dfrac{9\pi}{10}$ and $\theta = \dfrac{11\pi}{10}$; place loop in the next division between $\theta = \dfrac{11\pi}{10}$ and $\theta = \dfrac{13\pi}{10}$; leave the next division between $\theta = \dfrac{13\pi}{10}$ and $\theta = \dfrac{15\pi}{10}$; place loop in the next division between $\theta = \dfrac{15\pi}{10}$ and $\theta = \dfrac{17\pi}{10}$; leave the next division between $\theta = \dfrac{17\pi}{10}$ and $\theta = \dfrac{19\pi}{10}$.

Similarly for the curve $r = a \cos n\theta$, if n is a positive odd integer.

Note. If n is a positive even integer, place equal loops in *all* divisions.

(i) Change the equation to the polar form (by putting $x = r \cos\theta$, $y = r \sin\theta$).

(ii) Use the area formula : $\int \frac{1}{2} r^2 \, d\theta$.

(iii) To find the limits of integration for a loop, it is generally useful to put $r = 0$ and find two consecutive values of θ.]

12. (*a*) *Folium of Descartes.* Trace the curve $x^3 + y^3 = 3axy$ and find the area of its loop.

(*b*) Find the area of the closed portion of the Folium

$$r = \frac{3a \sin\theta \cos\theta}{\sin^3\theta + \cos^3\theta}.$$

13. Find the area of a loop of the curve $x^4 + y^4 = 2a^2xy$.

14. Prove that the area of the loop of the curve $x^5 + y^5 = 5ax^2y^2$ is $\frac{5}{2}a^2$.

****99. Area bounded by two curves whose polar equations are given. To prove that the area bounded by the curves r = f(θ), r = F(θ) and the radii vectors θ = α, θ = β is**

$$\int_\alpha^\beta \frac{1}{2} [\{\mathbf{f}(\theta)\}^2 - \{\mathbf{F}(\theta)\}^2] \, d\theta.$$

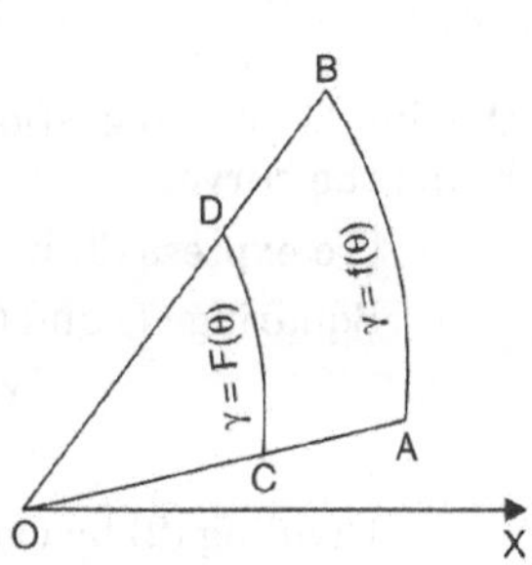

Let AB, CD be the curves $r = f(\theta)$, $r = F(\theta)$, and OCA, ODB the radii vectors $\theta = \alpha$, $\theta = \beta$. Then area CABD = area OAB – area OCD

$$= \int_\alpha^\beta \frac{1}{2} \{f(\theta)\}^2 \, d\theta - \int_\alpha^\beta \frac{1}{2} \{F(\theta)\}^2 \, d\theta$$

$$= \int_\alpha^\beta \frac{1}{2} [\{f(\theta)\}^2 - \{F(\theta)\}^2] \, d\theta.$$

[**Aid to memory.** Area between two curves = $\int \frac{1}{2} (r_1^2 - r_2^2) \, d\theta$, where r_1 is the 'r' of the outer curve and r_2 that of the inner curve.]

EXAMPLE

Trace the curve $r = a(\sec\theta + \cos\theta)$.

**Find the area between the curve and its asymptote.

[The asymptote is $r \cos\theta = a$ (*i.e.*, $x = a$, a line ∥ to the y-axis at a distance a from it).

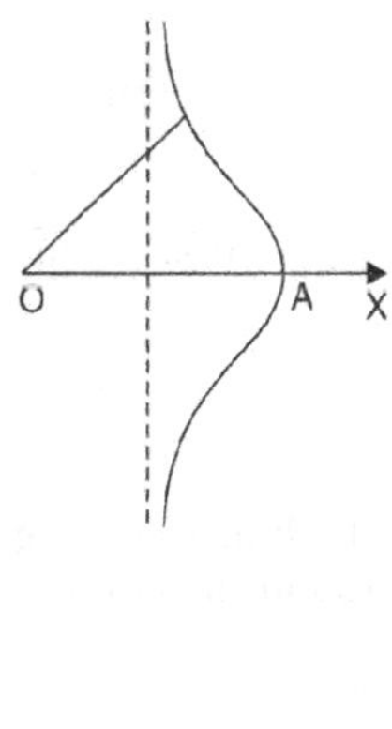

$$\text{Area} = 2 \int_0^{\frac{\pi}{2}} \frac{1}{2} (r_1^2 - r_2^2) \, d\theta$$

$$= \int_0^{\frac{\pi}{2}} [a^2 (\sec\theta + \cos\theta)^2 - a^2 \sec^2\theta] \, d\theta$$

$$= a^2 \int_0^{\frac{\pi}{2}} (2 + \cos^2 \theta)\, d\theta.]$$

IV. (Sectorial) Area formula for parametric equations.

100. *Area of a closed curve.* **To prove that the area of the closed curve x = f(t), y = φ(t) is** $\int \frac{1}{2}\left(x\frac{dy}{dt} - y\frac{dx}{dt}\right) dt$**, the limits of integration being the values of t such that the point (x, y) travels once completely round the curve.**

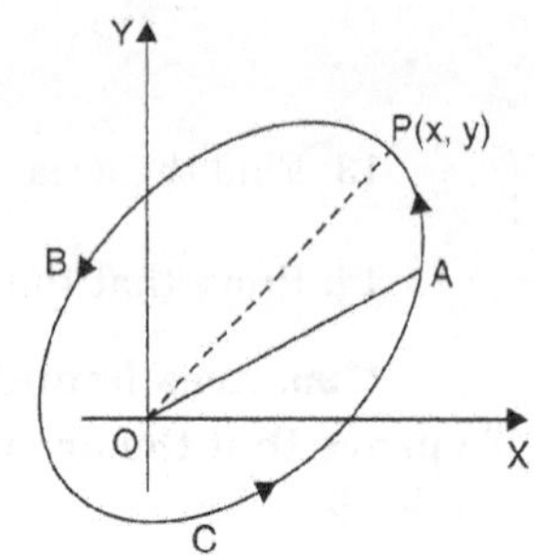

Let ABC be the curve $x = f(t)$, $y = \phi(t)$. Let P(x, y) be any pt. on the curve and (r, θ) its polar co-ordinates.

Then $x = r\cos\theta$ ...(1), $y = r\sin\theta$ ...(2)

Now area of curve = $\int \frac{1}{2} r^2\, d\theta$, the limits of integration being the values of θ such that the pt. (r, θ) travels once completely round the curve

$$= \int \frac{1}{2} r^2 \frac{d\theta}{dt}\, dt \qquad ...(3),$$

the limits of integration being the values of t such that the pt. (x, y) travels once completely round the curve.

[To express (3) in terms of x, y and their diff. coeffs. w.r.t. t.]

Squaring (1) and (2), and adding,

$$x^2 + y^2 = r^2 . 1,$$

$$\therefore \quad r^2 = x^2 + y^2 \qquad ...(4)$$

Dividing (2) by (1),

$$\frac{y}{x} = \frac{\sin\theta}{\cos\theta} = \tan\theta,$$

$$\therefore \quad \theta = \tan^{-1}\frac{y}{x}$$

$$\therefore \quad \frac{d\theta}{dt} = \frac{1}{1 + \frac{y^2}{x^2}}\left[\frac{x\frac{dy}{dt} - y\frac{dx}{dt}}{x^2}\right] = \frac{x\frac{dy}{dt} - y\frac{dx}{dt}}{x^2 + y^2} \qquad ...(5)$$

Substituting from (4) and (5) in (3),

$$\text{area of curve} = \int \frac{1}{2}(x^2 + y^2)\frac{x\frac{dy}{dt} - y\frac{dx}{dt}}{x^2 + y^2}\, dt$$

$$= \int \frac{1}{2}\left(x\frac{dy}{dt} - y\frac{dx}{dt}\right) dt \qquad ...(6)$$

the limits of integration being the values of t such that the pt. (x, y) travels once completely round the curve.

Cor. 1. From (6), **area of closed curve** = $\int \frac{1}{2}$ **(x dy – y dx)**, the limits of integration being such that the pt. (x, y) travels once completely round the curve.

Cor. 2. Sectorial area of a curve whose parametric equations are given. *The sectorial area of the curve $x = f(t)$, $y = \phi(t)$ is*

$$\int \frac{1}{2}\left(x\frac{dy}{dt} - y\frac{dx}{dt}\right) dt,$$

the limits of integration being the values of t at the ends of the bounding area.

[**Aid to memory.** Remember : Area = $\int \frac{1}{2}(x\,dy - y\,dx)$, and then

$$\text{write it} = \int \frac{1}{2}\left(x\frac{dy}{dt} - y\frac{dx}{dt}\right) dt.\ ]$$

Note 1. Assumptions. *(i) $f(t)$, $\phi(t)$ are continuous and single-valued in the interval (t_1, t_2), t_1, t_2 being the values of t such that the pt. (x, y) travels once completely round the curve.*

(ii) t_1, t_2 are finite, and $t_1 < t_2$.

(iii) The curve does not cross itself.

Note 2.

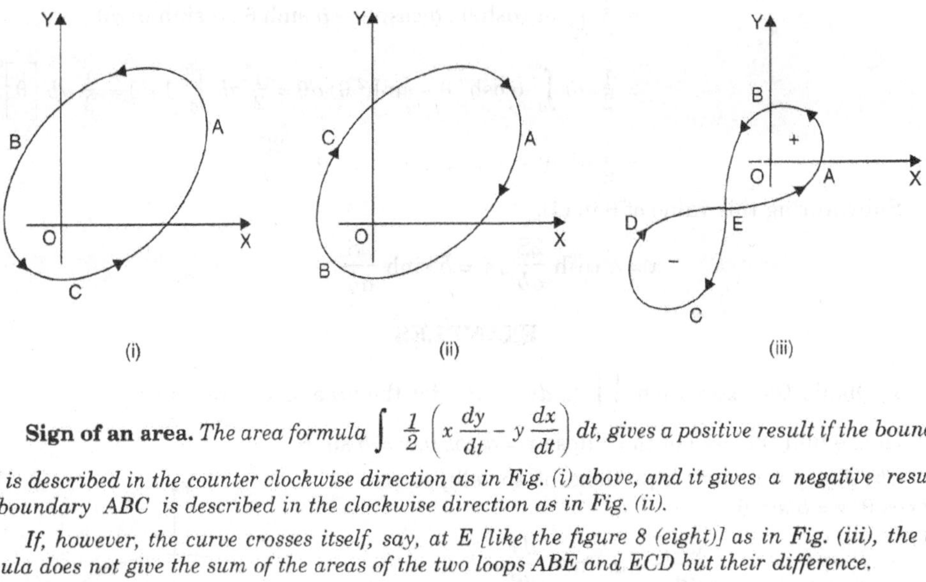

(i) (ii) (iii)

Sign of an area. *The area formula* $\int \frac{1}{2}\left(x\frac{dy}{dt} - y\frac{dx}{dt}\right) dt$, *gives a positive result if the boundary ABC is described in the counter clockwise direction as in Fig. (i) above, and it gives a negative result if the boundary ABC is described in the clockwise direction as in Fig. (ii).*

If, however, the curve crosses itself, say, at E [like the figure 8 (eight)] as in Fig. (iii), the area formula does not give the sum of the areas of the two loops ABE and ECD but their difference.

Hence in order to find the area of the curve we must find the numerical values of the areas of the loops ABE and ECD separately and add the results. Of course, if the curve consists of a number of equal loops, then in order to find the area of the curve, we find the area of one loop and multiply it by the number of loops.

Example. *Quadrature of hyperbola. If A is the vertex, O the centre, and P any point on the hyperbola* $\frac{x^2}{a^2} - \frac{y^2}{b^2} = 1$, *prove that* $x = a \cosh \frac{2S}{ab}, y = b \sinh \frac{2S}{ab}$, *where S is the sectorial area AOP.*

The equation of the hyperbola is

$$\frac{x^2}{a^2} - \frac{y^2}{b^2} = 1.$$

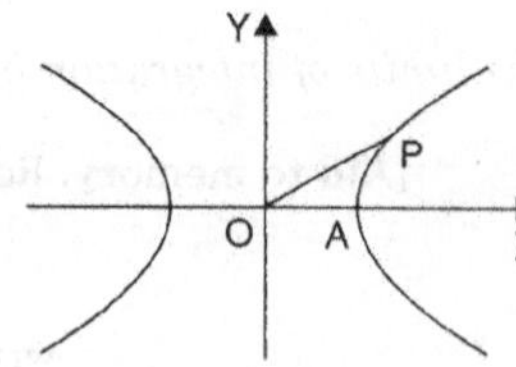

Let the co-ordinates of P be

$$x = a \cosh \theta, y = b \sinh \theta \quad ...(1)$$

[From Co-ordinate Geometry]

Then $$\frac{dx}{d\theta} = a \sinh \theta, \frac{dy}{d\theta} = b \cosh \theta.$$

$\therefore$ S = sectorial area OAP

$$= \int \frac{1}{2} (x\, dy - y\, dx)$$

$$= \int_0^\theta \frac{1}{2} \left(x \frac{dy}{d\theta} - y \frac{dx}{d\theta} \right) d\theta \quad \text{(Art. 100, Cor. 2)}$$

$$= \frac{1}{2} \int_0^\theta (a \cosh \theta \,.\, b \cosh \theta - b \sinh \theta \,.\, a \sinh \theta)\, d\theta$$

$$= \frac{1}{2} ab \int_0^\theta (\cosh^2 \theta - \sinh^2 \theta)\, d\theta = \frac{1}{2} ab \int_0^\theta 1\, d\theta = \frac{1}{2} ab \Big[\theta \Big]_0^\theta$$

$$= \frac{1}{2} ab\, [\theta - 0] = \frac{1}{2} ab\theta \quad \therefore \quad \theta = \frac{2S}{ab}.$$

Substituting this value of θ in (1),

$$x = a \cosh \frac{2S}{ab}, y = b \sinh \frac{2S}{ab}.$$

EXAMPLES

1. Obtain the expression $\frac{1}{2} \int (x\, dy - y\, dx)$ for the area of a closed curve.

Hence find the area of the ellipse $x = a \cos \theta, y = b \sin \theta$.

[(*b*) The parametric equations of the ellipse are $x = a \cos \theta, y = b \sin \theta$.

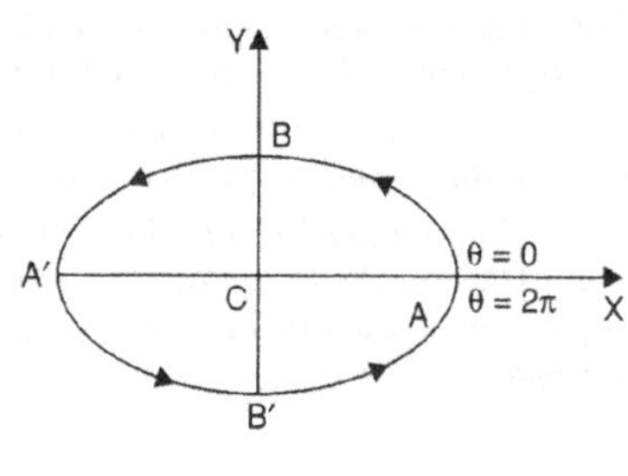

$$\therefore \quad \frac{dx}{d\theta} = -a \sin \theta, \frac{dy}{d\theta} = b \cos \theta.$$

The ellipse is a closed curve. As θ varies from 0 to 2π, the pt. (x, y) travels once completely round the curve.

$$\therefore \quad \text{area of ellipse} = \int \frac{1}{2} (x\, dy - y\, dx)$$

$$= \int_0^{2\pi} \frac{1}{2}\left(x\frac{dy}{d\theta} - y\frac{dx}{d\theta}\right) d\theta = \frac{1}{2}\int_0^{2\pi} [a \cos\theta \,.\, b\cos\theta - b\sin\theta\,(-a\sin\theta)]\, d\theta$$

$$= \frac{1}{2}\int_0^{2\pi} ab(\cos^2\theta + \sin^2\theta)\, d\theta = \frac{1}{2}\int_0^{2\pi} ab \,.\, 1\, d\theta$$

$$= \frac{ab}{2}\Big[\theta\Big]_0^{2\pi} = \frac{ab}{2}[2\pi - 0] = \pi ab.]$$

2. A closed curve is defined by the equations $x = \phi(t), y = \psi(t)$; prove that its area is given by $\frac{1}{2}\int\left(x\frac{dy}{dt} - y\frac{dx}{dt}\right)dt$, certain conventions being adopted.

Prove that the area of the curve

$$x = a\cos\theta + b\sin\theta + c,\ y = a'\cos\theta + b'\sin\theta + c',$$

is equal to $\pi(ab' - a'b)$.

[(b) Transferring the origin to (c, c'), the parametric equations of the curve become $x = a\cos\theta + b\sin\theta,\ y = a'\cos\theta + b'\sin\theta$.]

101. *Corresponding areas.* To find the area of the closed curve f (mx, ny) = 0 when the area of the closed curve f(x, y) = 0 can be found.

The equation of the curve is $f(mx, ny) = 0$ ...(1)

Put $mx = x', ny = y'$, then (1) becomes $f(x', y') = 0$ ...(2)

Now area of curve (1) $= \int y\, dx$ taken round the boundary of (1)

$$\left[\text{But } y = \frac{1}{n}y';\ x = \frac{1}{m}x',\ \therefore\ dx = \frac{1}{m}dx'\right]$$

$= \int \frac{1}{n}y' \,.\, \frac{1}{m}dx' = \frac{1}{mn}\int y'\, dx'$ taken round the boundary of (1)

$= \frac{1}{mn}$ [area of curve (2)].

****Example.** *Prove that the area of the curve* $(m^2x^2 + n^2y^2)^2 = a^2x^2 + b$ *is* $\frac{\pi}{2mn}\left(\frac{a^2}{m^2} + \frac{b^2}{n^2}\right)$.

The equation of the curve is $(m^2x^2 + n^2y^2)^2 = a^2x^2 + b^2y^2$ ...(1)

Put $mx = x', ny = y'$, then (1) becomes

$$(x'^2 + y'^2)^2 = a^2\frac{x'^2}{m^2} + b^2\frac{y'^2}{n^2} \quad ...(2) \quad \left(\because\ x = \frac{1}{m}x', y = \frac{1}{n}y'\right)$$

Now area of curve (1) $= \int y\, dx$ taken round the boundary of (1)

$$\left[\text{But } y = \frac{1}{n}y';\ x = \frac{1}{m}x',\ \therefore\ dx = \frac{1}{m}dx'\right]$$

$= \int \frac{1}{n}y' \,.\, \frac{1}{m}dx' = \frac{1}{mn}\int y'\, dx'$ taken round the boundary of (1)

$= \frac{1}{mn}$ (area of curve (2)) ...(3)

[To find the area of curve (2).]

Changing (2) to the polar form (by putting $x' = r' \cos\theta'$, $y' = r' \sin\theta'$),

$$r'^4 = a^2 \frac{r'^2 \cos^2\theta'}{m^2} + b^2 \frac{r'^2 \sin^2\theta'}{n^2}, \text{ or } r'^2 = \frac{a^2}{m^2}\cos^2\theta' + \frac{b^2}{n^2}\sin^2\theta'.$$

This curve is symmetrical about the initial line, also about the line through the pole ⊥ to the initial line ; and in the first quadrant, varies from 0 to $\frac{\pi}{2}$,

$$\therefore \quad \text{its area} = 4\int_0^{\frac{\pi}{2}} \frac{1}{2} r'^2\, d\theta' = 2\int_0^{\frac{\pi}{2}} \left(\frac{a^2}{m^2}\cos^2\theta' + \frac{b^2}{n^2}\sin^2\theta'\right) d\theta'$$

$$= 2\left(\frac{a^2}{m^2}\cdot\frac{1}{2}\cdot\frac{\pi}{2} + \frac{b^2}{n^2}\cdot\frac{1}{2}\cdot\frac{\pi}{2}\right) = \frac{\pi}{2}\left(\frac{a^2}{m^2} + \frac{b^2}{n^2}\right).$$

$\therefore$ from (3), area of curve (1)

$$= \frac{1}{mn}\cdot\frac{\pi}{2}\left(\frac{a^2}{m^2} + \frac{b^2}{n^2}\right) = \frac{\pi}{2mn}\left(\frac{a^2}{m^2} + \frac{b^2}{n^2}\right).$$

****Note. Important.** *In the equation of the curve*

$$(m^2x^2 + n^2y^2)^2 = a^2x^2 + b^2y^2,$$

The highest degree terms are $(m^2x^2 + n^2y^2)^2$,

$\therefore$ *we put* $mx = x'$, $ny = y'$.

EXAMPLE

**Find the area

(*i*) of the curve $\dfrac{x^2}{a^4} + \dfrac{y^2}{b^4} = \dfrac{1}{c^2}\left(\dfrac{x^2}{a^2} + \dfrac{y^2}{b^2}\right)^2$;

(*ii*) in the positive quadrant of

$$(m^2x^2 + n^2y^2)^{\frac{5}{2}} = ax^3 + by^3.$$

[**Note. Positive quadrant. Def.** *The* **positive quadrant** *of a curve is that in which both the x- and y-co-ordinates of points on the curve are positive.*]

V. Approximate value of a definite integral.

102. *Simpson's rule.* $\int_a^b$ **y dx is approximately**

$$\int_a^b y\,dx = \frac{h}{3}\left[(y_1 + y_{2n+1}) + 2(y_3 + y_5 + \dots + y_{2n-1}) + 4(y_2 + y_4 + \dots + y_{2n})\right]^*,$$

where $h = \dfrac{b-a}{2n}$, **and** $y_1, y_2, y_3, \dots$ **are the values of y when** $x = a, a + h, a + 2h, \dots$.

***Aid to memory.** See practical form of Simpson's rule on p. 270.

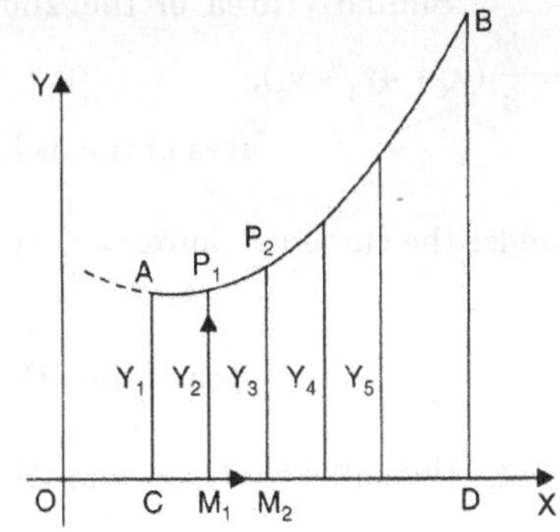

y is a function of $x = f(x)$ (say). Let AB be the curve $y = f(x)$, and CA, DB the ordinates $x = a$, $x = b$. Divide CD into $2n$ equal parts each $= h$, so that $h = \dfrac{b-a}{2n}$, and at the pts. of division $M_1, M_2, \ldots$ erect ordinates $M_1P_1, M_2P_2, \ldots$. Then $CA = y_1$, $M_1P_1 = y_2$, $M_2P_2 = y_3, \ldots$. Through A, P_1, P_2 draw a parabola (shown dotted in the Fig.) having its axis ∥ to the y-axis, and let its equation, referred to ∥ axes through M_1, be $y = A + Bx + Cx^2$ ...(1)

Then area under the parabola*

$$= \int_{-h}^{h} y\,dx = \int_{-h}^{h} (A + Bk + Cx^2)\,dx$$

$$= \left[Ax + B\frac{x^2}{2} + C\frac{x^3}{3}\right]_{-h}^{h}$$

$$= Ah + B\frac{h^2}{2} + C\frac{h^3}{3} - \left(-Ah + B\frac{h^2}{2} - C\frac{h^3}{3}\right)$$

$$= 2Ah + 2C\frac{h^3}{3} = 2h\left(A + C\frac{h^2}{3}\right) \qquad \text{...(2)}$$

[To find the values of A and C.]

∵ $A(-h, y_1)$, $P_1(0, y_2)$, $P_2(h, y_3)$ lie on the parabola (1),

$$\therefore \quad y_1 = A - Bh + Ch^2 \qquad \text{...(3)}$$

$$y_2 = A \qquad \text{...(4)}$$

$$y_3 = A + Bh + Ch^2 \qquad \text{...(5)}$$

From (4), $A = y_2$.

Adding (3) and (5),

$$y_1 + y_3 = 2A + 2Ch^2 = 2y_2 + 2Ch^2$$

$$\therefore \quad C = \frac{y_1 - 2y_2 + y_3}{2h^2}.$$

∴ from (2), area under the parabola $= 2h\left[y_2 + \dfrac{y_1 - 2y_2 + y_3}{6}\right]$

$$= \frac{h}{3}(y_1 + 4y_2 + y_3).$$

Now area of the 1st double strip (ordinates y_1, y_2, y_3) under the curve is approx.

$$= \text{area under the parabola}$$

$$= \frac{h}{3}(y_1 + 4y_2 + y_3).$$

***Area under a curve. Def.** The area bounded by a curve, the x-axis and two ordinates is called the **area under the curve.**

Similarly area of the 2nd double strip (ordinates y_3, y_4, y_5) under the curve is approx.

$= \frac{h}{3}(y_3 + 4y_4 + y_5)$,

area of the 3rd double strip (ordinates y_5, y_6, y_7)

under the curve is approx. $= \frac{h}{3}(y_5 + 4y_6 + y_7)$,

... ...

area of the nth double strip

under the curve is approx. $= \frac{h}{3}(y_{2n-1}{}^{*} + 4y_{2n} + y_{2n+1})$.

$\therefore$ adding vertically (and diagonally),

area under the curve is approx.

$$= \frac{h}{3}[(y_1 + y_{2n+1}) + 2(y_3 + y_5 + \ldots + y_{2n-1}) + 4(y_2 + y_4 + \ldots + y_{2n})],$$

i.e., $\int_a^b y\,dx$ is approx.

$$= \frac{h}{3}[(y_1 + y_{2n+1}) + 2(y_3 + y_5 + \ldots + y_{2n-1}) + 4(y_2 + y_4 + \ldots + y_{2n})].$$

Note. Practical form of Simpson's rule.

$$\int_a^b \mathbf{y\,dx \text{ is approx.}} = \frac{h}{3}[X + 2O + 4E],$$

where $h = \frac{b-a}{2n}$,

X = sum of extreme ordinates (i.e., first and last ordinates)

O = sum of other odd ordinates (i.e., omitting first and last ordinates),

E = sum of even ordinates.

[Note that X stands for 'Ex' (the first two letters of the word *Extreme*), O is the first letter of the word *Odd,* and E of *Even.*

Note further that *number of equal intervals = 2n,*

and number of ordinates (values of y) = 2n + 1.]

Example. *Given that*

$$e^0 = 1,\ e^1 = 2.72,\ e^2 = 7.39,\ e^3 = 20.09,\ e^4 = 54.60,$$

verify Simpson's rule by finding an approximate value of $\int_0^4 e^x\,dx$ *and comparing it with the exact value.*

---

*__How to write this step.__ Looking down vertically, the nth term of $y_1, y_3, y_5, \ldots$ is y_{2n-1} ($\because$ the nth term of the suffixes, 1, 3, 5, ... (A.P.) = 1 + $(n-1)2 = 2n - 1$). Similarly the nth term of $y_2, y_4, y_6, \ldots$ is y_{2n}, and that of $y_3, y_5, y_7, \ldots$ is y_{2n+1}.

Compare $\int_0^4 e^x\,dx$ with $\int_a^b y\,dx$,

here $$y = e^x,\ a = 0,\ b = 4$$

number of ordinates (given values of y) $2n + 1 = 5$, $\therefore$ $2n = 4$

$$\therefore \qquad h = \frac{b-a}{2n} = \frac{4-0}{4} = 1,$$

$\therefore$ **we find the values of y corresponding to the values of x at intervals of (h =) 1 beginning with (lower limit of given integral) 0 and ending with (upper limit) 4 thus :**

when $x = 0$, $y_1 = e^0 = 1$ (Given)

when $x = 1$, $y_2 = e^1 = 2.72$

when $x = 2$, $y_3 = e^2 = 7.39$

when $x = 3$, $y_4 = e^3 = 20.09$

when $x = 4$, $y_5 = e^4 = 54.60$

$\therefore$ X* = 1	O† = 7.39	E‡ = 2.72
54.60	2	20.09
55.60	2O = 14.78	22.81
		4
		4E = 91.24

$$\therefore \quad \int_0^4 e^x\,dx \text{ is approx.} = \frac{h}{3}\,(\text{X} + 2\text{O} + 4\text{E})$$

$= \frac{1}{3}$	55.60	= 53.873 ...
	14.78	= 53.87 (correct to two places of decimals).
	91.24	
	161.62	

Exact value $$= \int_0^4 e^x\,dx = \Big[e^x\Big]_0^4 = e^4 - 1 = 54.60 - 1 = 53.60.$$

Note. Why to calculate the answer here correct to two places of decimals. *Here the values of $e^1 = 2.72$, $e^2 = 7.39$, ..., are given correct to two places of decimals, therefore, we have calculated the answer as well correct to two places of decimals.*

EXAMPLES

1. Explain Simpson's rule for approximate integration.

---

*X = sum of extreme ordinates (*i.e.*, first and last ordinates)
$= y_1 + y_5$,

†O = sum of other odd ordinates (*i.e.*, omitting first and last ordinates)
$= y_3$,

‡E = sum of even ordinates $= y_2 + y_4$.

Calculate $\int_4^8 \frac{dx}{x}$ using 4 equal intervals.

Note 1. *In the application of Simpson's rule, when the number of places of decimals to which the answer is required, is not mentioned, it is usual to calculate the answer correct to three places of decimals.*

Note 2. How to calculate the answer correct to 3 places of decimals. *In order to calculate the answer correct to 3 places of decimals, work out all calculations upto 4 places of decimals (i.e., 1 more than the required number of places of decimals), and retain the answer correct to 3 places of decimals by the usual rule of approximation. (See the author's New Elementary Plane Trigonometry) Eighth Edition, Note 1, p. 10.)*

2. Establish Simpson's rule for approximate integration.

Compute by Simpson's Rule $\int_2^{10} \frac{dx}{1+x}$ correct to three places of decimals.

[Take Δx = 1.]

3. A curve is drawn to pass through the points given by the following table :

x	1	1.5	2	2.5	3	3.5	4
y	2	2.4	2.7	2.8	3	2.6	2.1.

Estimate the area bounded by the curve, the x-axis and the lines $x = 1$, $x = 4$.

4. A river is 80 feet wide. The depth d in feet at a distance x feet from one bank is given by the following table :

x	0	10	20	30	40	50	60	70	80
d	0	4	7	9	12	15	14	8	3.

Find approximately the area of the cross-section.

5. *Approximate value of* π. Employ Simpson's rule to calculate approximately the value of π from $\int_0^1 \frac{dx}{1+x^2}$ by dividing the interval 0 to 1 into four equal parts.

6. Obtain an approximate value of log 2 from $\int_1^2 \frac{dx}{x}$ by Simpson's rule, taking eleven ordinates.

MISCELLANEOUS EXAMPLES ON CHAPTER X

1. (*i*) If a, b are positive and $a > b$, prove that the area between the curve given by $c^{n-1}\, y = x^n$, the x-axis and the ordinates at a and b is.

$$\frac{a^{n+1} - b^{n+1}}{(n+1)c^{n-1}}.$$

(*ii*) Find the area of the figure bounded by the curve $y^n = a^{n-1}\, x$ the x-axis and the ordinate $x = h$.

2. In the catenary $y = a \cosh \dfrac{x}{a}$, prove that the area between the curve, the axis of x, and the ordinates of two points of the curve varies as the length of the intervening arc.

****3.** An ellipse of small eccentricity e has its perimeter equal to that of a circle of radius a. Show that its area is $\pi a^2 \left(1 - \dfrac{3}{32} e^4\right)$ nearly.

[Let the parametric equations of the ellipse be

$$x = a' \cos \theta,\; y = b' \sin \theta.$$

Proceed as in Ex. 2, Art. 89. It will be found that perimeter of ellipse (as far as the term in e^4*)

$$= 2\pi a' \left(1 - \frac{1}{4} e^2 - \frac{3}{64} e^4\right) = 2\pi a \text{ (Given)} \qquad \ldots(1)$$

$$\text{area of ellipse} = \pi a' b' = \pi a'^2 (1 - e^2)^{\frac{1}{2}} \qquad \ldots(2)$$

Substitute the value of a' from (1) in (2), expand by the binomial theorem, and calculate the result as far as the term in e^4.]

4. *Strophoid.* Trace the curve $y^2 (a - x) = x^2(a + x)$ and find the area of the loop.

5. Trace the curve $a^2 y = x^2(x + a)$ and show that the curve includes with the axis of x an area $\dfrac{a^2}{12}$.

6. Find the area of the loop of the curve $a^3 y^2 = x^4 (b + x)$.

7. Find the whole area of the curve $a^2 y^2 = x^3(2a - x)$.

8. Trace the curve $y^2 x + (x + a)^2(x + 2a) = 0$ and find the area of its loop.

***Why this step.** $\because$ the result to be proved is given as far as the term in e^4

$\therefore$ we also work out calculations as far as the term in e^4.

[It will be found that the shape of the curve is that shown in the Fig.

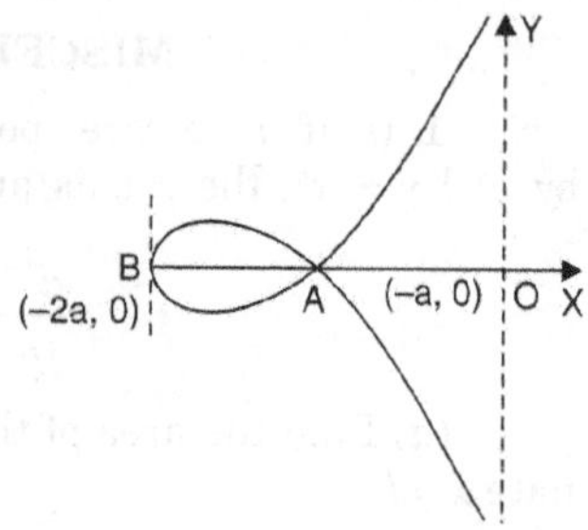

$$\text{Area of loop} = 2\int_{-2a}^{-a} y\,dx = 2\int_{-2a}^{-a} (x+a)\sqrt{\frac{x+2a}{-x}}\,dx.$$

Put $x = -2a\sin^2\theta$. It will be found that integral

$$= 8a^2\int_{\frac{\pi}{4}}^{\frac{\pi}{2}} \cos 2\theta \cos^2\theta\,d\theta. \text{ Use } \cos^2\theta = \frac{1+\cos 2\theta}{2},$$

and then use $\cos^2 2\theta = \dfrac{1+\cos 4\theta}{2}$.]

**9. Prove that the area of the curve $x^4 - 3ax^3 + a^2(2x^2 + y^2) = 0$ is $\frac{3}{8}\pi a^2$.

[It will be found that the shape of the curve is that shown in the Fig.

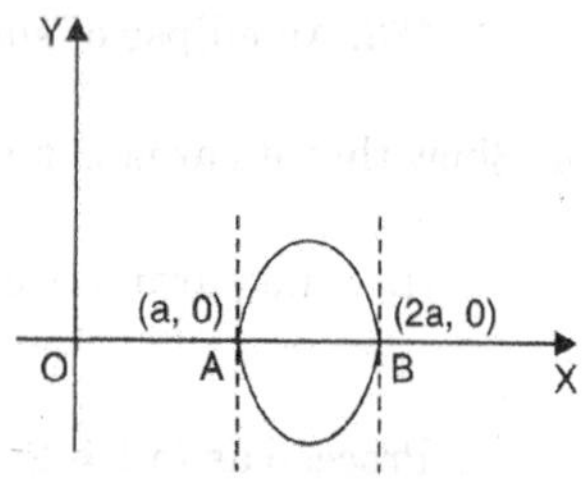

$$\text{area} = 2\int_a^{2a} y\,dx = 2\int_a^{2a} \frac{x}{a}\sqrt{-x^2+3ax-2a^2}\,dx$$

$$= \frac{2}{a}\int_a^{2a} x\sqrt{(x-a)(2a-x)}\,dx.$$

Put $x = a\cos^2\theta + 2a\sin^2\theta$.

**10. Trace roughly the curve $8a^2y^2 = x^2(a^2 - 2x^2)$ and show that the whole length of arc is πa. Show that the area enclosed by the curve is $\frac{2}{3}$ of that of the circumscribing rectangle whose sides are parallel to the axes of co-ordinates.

11. *Four-cusped hypocycloid.* Sketch the curve given by the parametric equations $x = a\cos^3 t$, $y = a\sin^3 t$. Find the area enclosed and show that the length of the curve is $6a$.

12. *Four-cusped hypocycloid.* Trace the curve $x^{\frac{2}{3}} + y^{\frac{2}{3}} = a^{\frac{2}{3}}$ and final the area enclosed by it.

13. Prove that the whole area between the four infinite branches is the tractrix

$$x = a\cos t + \frac{a}{2}\log\tan^2\frac{t}{2},\ y = a\sin t \text{ is } \pi a^2.$$

[The shape of the curve is that shown in the *Answers.*]

14. Show that the area bounded by the cissoid

$$x = a\sin^2 t,\ y = a\frac{\sin^3 t}{\cos t}$$

and its asymptote is $\dfrac{3\pi a^2}{4}$.

[The shape of the curve is that shown in the *Answers*.]

15. Show that the area of the loop of the curve $x = a(1 - t^2)$, $y = at(1 - t^2)$, $-1 \le t \le 1$, is $\dfrac{8}{15}a^2$.

[The shape of the curve is that shown in the *Answers*.]

16. Show that the area included between the parabolas

$$y^2 = 4a(x + a),\ y^2 = 4b(b - x)$$

is $\dfrac{8}{3}(a + b)\sqrt{ab}$.

[See Fig. in the *Answers*.]

****17.** Find the area common to the two curves $y^2 = ax$, $x^2 + y^2 = 4ax$.

18. Show that the larger of the two areas into which the circle $x^2 + y^2 = 64a^2$ is divided by the parabola $y^2 = 12ax$ is $\dfrac{16}{3}a^2(8\pi - \sqrt{3})$.

19. Find the area of the segment cut off from the parabola $y^2 = 4x$ by the straight line $2x - 3y + 4 = 0$.

****20.** Show that the area included between one of the branches of the curve $x^2y^2 = a^2(x^2 + y^2)$ and its asymptotes is equal to that of the square whose side is a.

[Area in the first quadrant

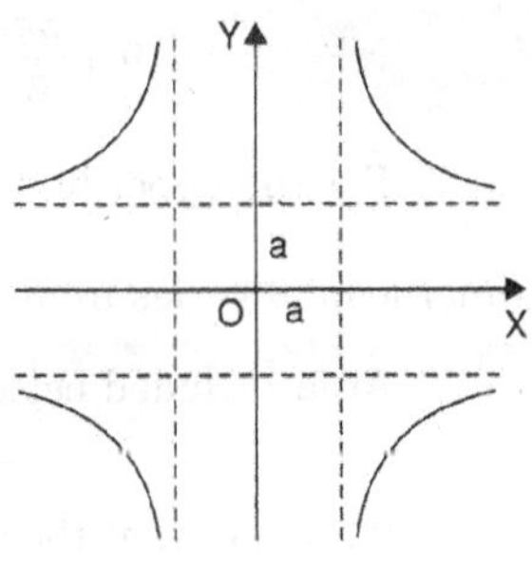

$$= \int (y_1 - y_2)\, dx = \int_a^{\infty} \left(\frac{ax}{\sqrt{x^2 - a^2}} - a \right) dx$$

$$= \int_a^{\infty} \left[\frac{a}{2}(x^2 - a^2)^{-\frac{1}{2}}(2x) - a \right] dx$$

$$= \left[\frac{a}{2} \frac{(x^2 - a^2)^{\frac{1}{2}}}{\frac{1}{2}} - ax \right]_a^{\infty}$$

$$= \underset{x \to \infty}{\text{Lt}} [a(x^2 - a^2)^{\frac{1}{2}} - ax] - [a.0 - a^2].$$

21. Prove that the area of the curve

$$r^2(2c^2 \cos^2\theta - 2ac \sin\theta \cos\theta + a^2 \sin^2\theta) = a^2c^2 \text{ is}$$

equal to πac.

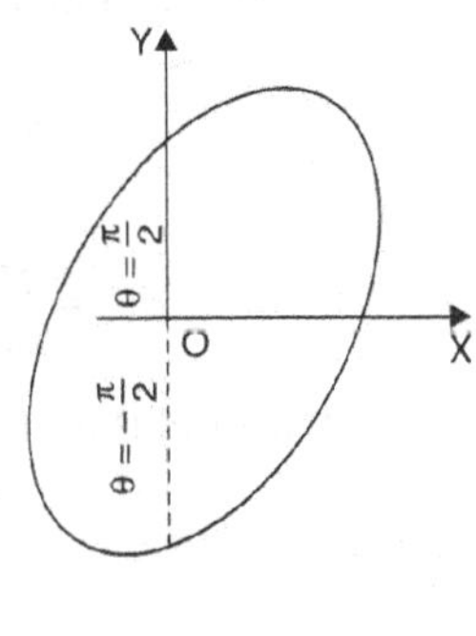

[The equation of the curve is

$$2c^2r^2 \cos^2\theta - 2ac\, r\cos\theta \,.\, r\sin\theta + a^2r^2 \sin^2\theta = a^2c^2 \qquad ...(1)$$

Changing to the Cartesian form (by putting $x = r\cos\theta$, $y = r\sin\theta$), (1) becomes

$$2c^2x^2 - 2ac\,xy + a^2y^2 = a^2c^2,$$

which is a conic whose centre is the origin.

Here "$ab - h^2$" $= 2c^2 \cdot a^2 - (-ac)^2 = 2a^2c^2 - a^2c^2 = a^2c^2$, which is + ve,

$\therefore$ the conic is an ellipse.

It is symmetrical in opposite quadrants, and for the R.H. half of the ellipse, θ varies from $-\dfrac{\pi}{2}$ to $\dfrac{\pi}{2}$.

$$\therefore \qquad \text{area} = 2\int_{-\frac{\pi}{2}}^{\frac{\pi}{2}} \frac{1}{2} r^2 \, d\theta.\Big]$$

****22.** Find the area included between the two loops of the curve

$$r = a\,(2\cos\theta + \sqrt{3}).$$

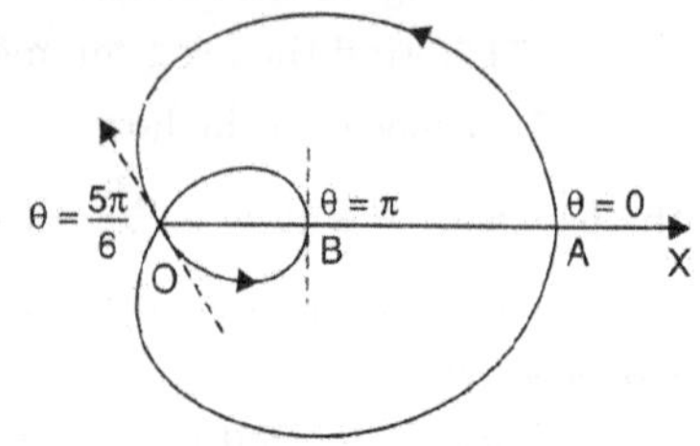

[The curve is symmetrical about the initial line. It passes through the pole where, putting $r = 0$, $2\cos\theta + \sqrt{3} = 0$,

or
$$\cos\theta = -\frac{\sqrt{3}}{2} = -\cos\frac{\pi}{6}$$
$$= \cos\left(\pi - \frac{\pi}{6}\right) = \cos\frac{5\pi}{6},$$
$$\therefore \qquad \theta = \frac{5\pi}{6}.$$

For the upper half of the outer loop, θ varies from 0 to $\dfrac{5\pi}{6}$, and for the lower half of the inner loop, θ varies from $\dfrac{5\pi}{6}$ to π.

Area included between the two loops

= area of outer loop–area of inner loop.]

23. Prove that the sum of the areas of the two loops of the limacon $r = a + b\cos\theta$, $b > a$, is equal to $\dfrac{\pi}{2}(2a^2 + b^2)$.

[It will be found that sum of the areas of the two loops

$$= 2\int_0^{\pi - \cos^{-1}\frac{a}{b}} \frac{1}{2} r^2\, d\theta + 2\int_{\pi - \cos^{-1}\frac{a}{b}}^{\pi} \frac{1}{2} r^2\, d\theta$$

$$= \int_0^{\pi - \cos^{-1}\frac{a}{b}} r^2\, d\theta + \int_{\pi - \cos^{-1}\frac{a}{b}}^{\pi} r^2\, d\theta = \int_0^{\pi} r^2\, d\theta.\Big]$$

24. Find the area common to the ellipses

$$x^2 + 2y^2 = 2c^2,\; 2x^2 + y^2 + 2c^2.$$

The equation of the first ellipse is

$$x^2 + 2y^2 = 2c^2 \qquad ...(1),$$

or $$\frac{x^2}{2c^2} + \frac{y^2}{c^2} = 1.$$

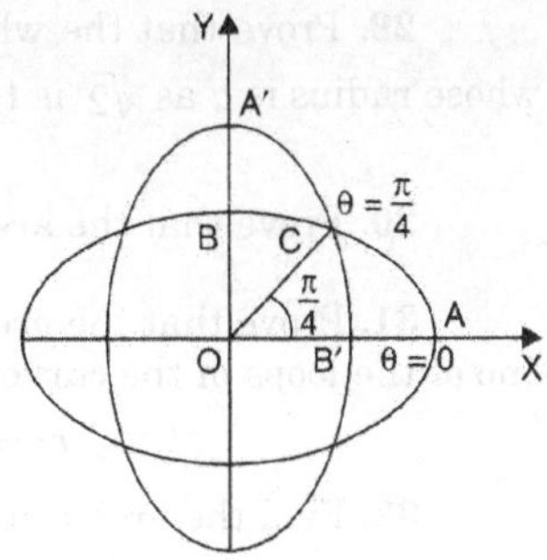

It is an ellipse whose major axis lies along the x-axis and minor axis along the y-axis. The equation of the second ellipse is

$$2x^2 + y^2 = 2c^2 \qquad ...(2),$$

or $$\frac{x^2}{c^2} + \frac{y^2}{2c^2} = 1.$$

It is an ellipse whose major axis lies along the y-axis and minor axis along the x-axis.

Changing to the polar form (by putting $x = r\cos\theta$, $y = r\sin\theta$), the equations of the ellipses are

$$r^2(\cos^2\theta + 2\sin^2\theta) = 2c^2, \text{ or } r^2 = \frac{2c^2}{\cos^2\theta + 2\sin^2\theta} \qquad ...(3)$$

$$r^2(2\cos^2\theta + \sin^2\theta) = 2c^2, \text{ or } r^2 = \frac{2c^2}{2\cos^2\theta + \sin^2\theta} \qquad ...(4)$$

To find the pts. of intersection, substituting the value of r^2 from (3) in (4), $\cos^2\theta + 2\sin^2\theta = 2\cos^2\theta + \sin^2\theta$, or $\sin^2\theta = \cos^2\theta$, or $\tan^2\theta = 1$,

or, at C, $\tan\theta = 1$, or $\theta = \frac{\pi}{4}$.

The area OB′CB is symmetrical about the line OC($y = x$).

[∵ on changing x to y and y to x, (1) becomes $y^2 + 2x^2 = 2c^2$, which is (2)]

∴ common area = 4. area OB′CB = 4.2 area OB′C

$$= 8\int_0^{\frac{\pi}{4}} \frac{1}{2} r^2\, d\theta, \text{ which from (4),}$$

$$= 4\int_0^{\frac{\pi}{4}} \frac{2c^2}{2\cos^2\theta + \sin^2\theta}\, d\theta.$$

Divide the num. and denom. by $\cos^2\theta$, and put $\tan\theta = z$.]

25. Show that the area common to the ellipses $a^2x^2 + b^2y^2 = 1$,

$$b^2x^2 + a^2y^2 = 1, \text{ where } 0 < a < b \text{ is } \frac{4}{ab}\tan^{-1}\frac{a}{b}.$$

26. Trace the curve $r = \sqrt{3}\cos 3\theta + \sin 3\theta$, and find the area of a loop.

****27.** Find the area lying between the cardioid $r = a(1 - \cos\theta)$ and its double tangent.

[**Hint.** Equation of double tangent is $x = \frac{a}{4}$]

28. Find the area of the loop of the curve $(x + y)(x^2 + y^2) = 2axy$.

29. Prove that the whole area of the curve $x^4 + y^4 = a^2(x^2 + y^2)$ is to that of the circle whose radius is a as $\sqrt{2}$ is to 1.

30. Prove that the area of the loop of the curve $x^6 + y^6 = a^2x^2y^2$ is $\dfrac{\pi a^2}{12}$.

31. Prove that the area of the loop of the curve $x^3 + y^3 = 3axy$ is three times the area of one of the loops of the curve

$$r^2 = a^2 \cos 2\theta.$$

32. Find the area common to the circles $r = a\sqrt{2}$ and $r = 2a \cos \theta$.

33. Find the ratio of the two parts into which the parabola $2a = r(1 + \cos \theta)$ divides the area of the cardioid $r = 2a(1 + \cos \theta)$.

34. Area formula for pedal equation. If $p = f(r)$ be the pedal equation of a curve, show that its area is $\dfrac{1}{2}\displaystyle\int \frac{pr\,dr}{\sqrt{r^2 - p^2}}$ taken between the proper limits.

Find the sectorial area bounded by the equiangular spiral $p = r \sin \alpha$ and the radii vectors r_1 and r_2.

$$\left[(a) \text{ Area} = \int \frac{1}{2} r^2\, d\theta = \int \frac{1}{2} r^2 \frac{d\theta}{dr} . dr\right.$$

$$= \int \frac{1}{2} r . r \frac{d\theta}{dr} . dr = \int \frac{1}{2} r \tan \phi\, dr$$

$$\left[\text{But } p = r \sin \phi, \text{ or } \sin \phi = \frac{p}{r}\right]$$

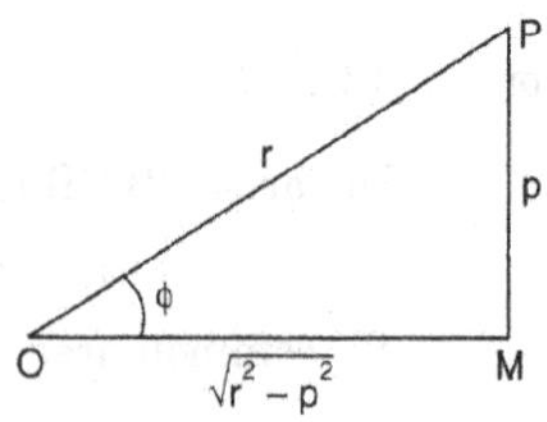

$$\left. = \int \frac{1}{2} r \frac{p}{\sqrt{r^2 - p^2}}\, dr = \frac{1}{2}\int \frac{pr\, dr}{\sqrt{r^2 - p^2}}.\ \right]$$

CHAPTER XI

VOLUME AND SURFACE OF A SOLID OF REVOLUTION

SECTION I

VOLUME OF A SOLID OF REVOLUTION

1. Volume formulae for Cartesian equation.

103-I. *Revolution about the x-axis.* **To prove that the volume of the solid generated by the revolution, about the x-axis, of the area bounded by the curve y = f(x), the x-axis and the ordinates x = a, x = b is $\int_a^b \pi y^2\,dx$.**

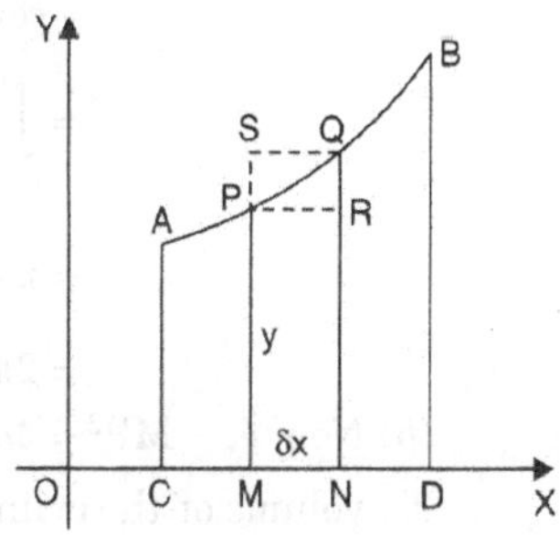

Let AB be the curve $y = f(x)$, and CA, DB the ordinates $x = a$, $x = b$.

Let P(x, y) be any pt. on the curve and MP its ordinate. Then the volume of the solid generated by the revolution, about the x-axis, of the area ACMP is clearly a function of x = V (say).

Let Q$(x + \delta x, y + \delta y)$ be a neighbouring pt. on the curve and NQ its ordinate. Then the volume of the solid generated by the revolution of the area PMNQ = δV.

From P, Q draw PR, QS ⊥ s respectively on NQ, MP produced.

Then the volume of the solid generated by the revolution of the area PMNQ lies between the volumes of the right circular cylinders generated by the rectangles. PMNR and SMNQ,

i.e., δV lies between $\pi y^2 \delta x$* and $\pi(y + \delta y)^2\,\delta x$

$\therefore$ $\dfrac{\delta V}{\delta x}$ lies between πy^2 and $\pi(y + \delta y)^2$

$\therefore$ taking the limits when $\delta x \to 0$ (and $\therefore$ $\delta y \to 0$),

$$\frac{dV}{dx} = \pi y^2 \qquad \text{...(1)}$$

$$\therefore \quad \int_a^b \pi y^2\,dx = \int_a^b \frac{dV}{dx}\,dx = \Big[V\Big]_a^b$$

= (value of V when $x = b$) – (value of V when $x = a$)

= volume of the solid generated by the revolution of the area ACDB – 0

i.e., volume of the solid generated by the revolution of the area ACDB $= \int_a^b \pi y^2\,dx.$

*Volume of right circular cylinder = $\pi r^2 h$, where r is the base-radius and h the height.

Cor. From (1), $\mathbf{V} = \int \pi \, \mathbf{y^2 \, dx}.$

Note. Assumptions. *(i) f(x) is continuous and single-valued in the interval (a, b).*

(ii) a, b are finite, and a < b.

(iii) The curve does not cross the x-axis, i.e., the axis of revolution.

Example 1. *The portion of the parabola $y^2 = 4ax$ bounded by the line $x = c$ revolves about the axis. Find the volume generated.*

Show also that it is equal to half the volume of the cylinder whose base-radius is the extreme ordinate and height the abscissa.

(*a*) The equation of the parabola is $y^2 = 4ax$.

Let A be the vertex of the parabola and MP the ordinate $x = c$.

Then required volume

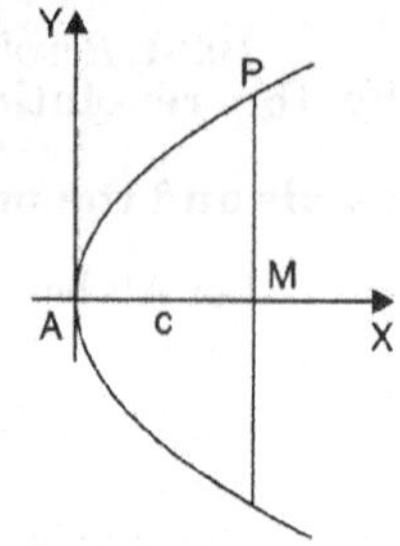

= volume of the solid generated by the revolution, about the x-axis, of the area AMP

$$= \int_0^c \pi y^2 \, dx = \pi \int_0^c 4ax \, dx$$

$$= \pi \, 4a \left[\frac{x^2}{2} \right]_0^c$$

$$= 2\pi a \, (c^2 - 0) = 2\pi \, ac^2 \qquad \text{...(1)}$$

(*b*) Now $\quad MP^2 = 4ax = 4ac.$

$\therefore$ volume of the cylinder whose base-radius is MP and height AM $\left.\right\} = \pi \, MP^2 \, . \, AM = \pi 4ac.c = 4\pi ac^2.$

$\therefore$ From (1),

volume of the solid generated by the revolution of the area AMP $\left.\right\} = \frac{1}{2}$ (volume of the above cylinder).

Example 2. *Find the volume of the prolate spheroid formed by the revolution of the ellipse $\frac{x^2}{a^2} + \frac{y^2}{b^2} = 1$ about the x-axis.*

The equation of the ellipse is $\frac{x^2}{a^2} + \frac{y^2}{b^2} = 1,$

or $$\frac{y^2}{b^2} = 1 - \frac{x^2}{a^2} = \frac{a^2 - x^2}{a^2},$$

$$\therefore \qquad y^2 = \frac{b^2}{a^2} (a^2 - x^2).$$

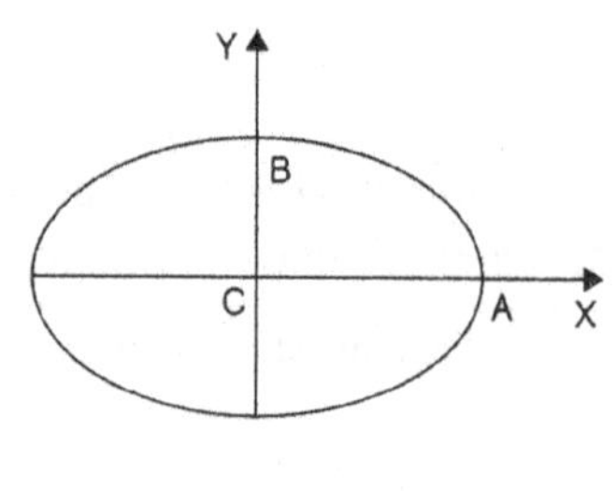

Required volume = 2. volume of the solid generated by the revolution of the area BCA

$$= 2 \int_0^a \pi y^2 \, dx = 2\pi \int_0^a \frac{b^2}{a^2} (a^2 - x^2) \, dx$$

$$= 2\pi \frac{b^2}{a^2}\left[a^2 x - \frac{x^3}{3}\right]_0^a$$

$$= 2\pi \frac{b^2}{a^2}\left[a^3 - \frac{a^3}{3} - (0 - 0)\right]$$

$$= 2\pi \frac{b^2}{a^2} \cdot \frac{2}{3} a^3 = \frac{4}{3}\pi a b^2.$$

Note. Prolate and oblate spheroids. Def. *The solid generated by the revolution of an ellipse about the major axis is called a prolate spheroid ; and that generated by the revolution about the minor axis is called an oblate spheroid.*

Example 3. *Find the volume formed by the revolution of the loop of the curve*

$y^2 = x^2 \dfrac{a - x}{a + x}$ *about the x-axis.*

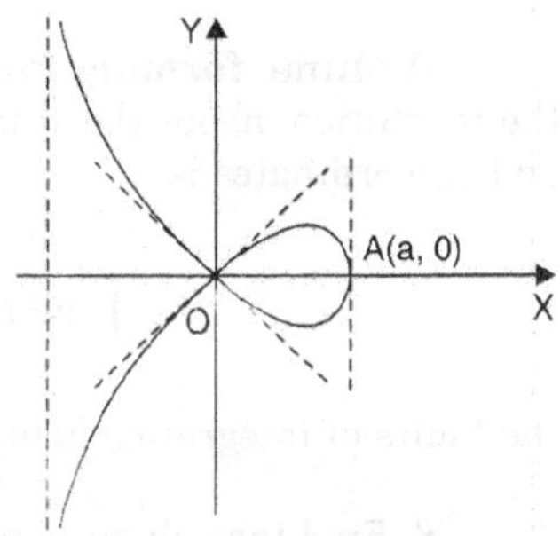

The equation of the curve is

$$y^2 = \frac{x^2(a - x)}{a + x}.$$

[Trace the curve as in solved Ex. 3, Art. 96-I.]

For the upper half of the loop, x varies from 0 to a.

$$\therefore \text{ volume} = \int_0^a \pi y^2\, dx = \pi \int_0^a \frac{x^2(a - x)}{a + x}\, dx$$

$$= \pi \int_0^a \frac{-x^3 + ax^2}{x + a}\, dx$$

$$= \pi \int_0^a \left[-x^2 + 2ax - 2a^2 + \frac{2a^3}{x + a}\right] dx$$

$$= \pi \left[-\frac{x^3}{3} + 2a\frac{x^2}{2} - 2a^2 x + 2a^3 \log(x + a)\right]_0^a$$

$$\begin{array}{r} x + a)\, -x^3 + ax^2\ (-x^2 + 2ax - 2a^2 \\ \underline{-x^3 - ax^2} \\ 2ax^2 \\ \underline{2ax^2 + 2a^2 x} \\ -2a^2 x \\ \underline{-2a^2 x - 2a^3} \\ 2a^3 \end{array}$$

$$= \pi \left[-\frac{a^3}{3} + a^3 - 2a^3 + 2a^3 \log 2a - (-0 + 0 - 0 + 2a^3 \log a)\right]$$

$$= \pi \left[-\frac{4}{3} a^3 + 2a^3 \log \frac{2a}{a}\right]$$

$$= 2\pi a^3 \left[\log 2 - \tfrac{2}{3}\right].$$

EXAMPLES

1. Find the volumes of the solids formed by the revolution, about the x-axis, of the areas bounded by each of the following curves, the x-axis and the ordinate $x = c$:

(*i*) $a^{n-1} y = x^n$. (*ii*) $y^n = a^{n-1} x$.

2. *Cone.* Find the area of the triangle formed by the line $y = x \tan \theta$, the x-axis and the line $x = a$.

If the triangle revolves round the x-axis, find the volume of the cone formed.

3. *Sphere.* Find the volume of a sphere of radius a.

4. *Ellipsoid of revolution.* Determine the entire volume of the ellipsoid which is generated by the revolution of an ellipse round its axis major.

5. Find the volume of the solid produced by the revolution of the loop of the curve $y^2 = x^2 \dfrac{a+x}{a-x}$ about the axis of x.

[**Volume formula for parametric equations.** The volume of the solid generated by the revolution, about the x-axis, of the area bounded by the curve $x = f(t)$, $y = \phi(t)$, the x-axis, and two ordinates is

$$\int \pi y^2 dx = \int \pi y^2 \frac{dx}{dt}\, dt,$$

the limits of integration being the values of t at the ends of the revolving arc.]

6. Find the volume generated by revolving one arch of the cycloid $x = a(\theta - \sin \theta)$, $y = a(1 - \cos \theta)$ about its base.

103-II. Revolution about the y-axis. It has been proved in Art. 103-I that

the volume of the solid generated by the revolution, about the x-axis, of the area bounded by the curve $y = f(x)$, the x-axis, and the ordinates $x = a$, $x = b$ is $\int_a^b \pi y^2\, dx$ ...(1)

Changing x to y and y to x in (1), we get :

the volume of the solid generated by the revolution, about the y-axis, of the area bounded by the curve x = f(y), the y-axis, and the abscissae y = a, y = b is $\int_a^b \pi x^2\, dy$.

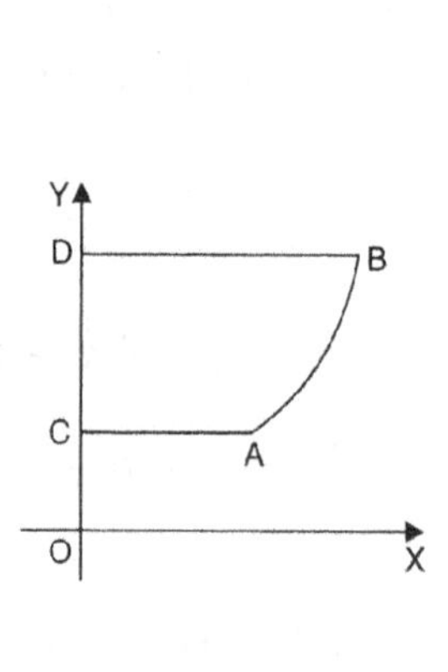

Cor. $V = \int \pi x^2\, dy.$

[Changing x to y and y to x in Art. 103-I, Cor.]

EXAMPLES

1. Find the volume of the reel-shaped solid formed by the revolution about the y-axis of the part of the parabola $y^2 = 4ax$ cut off by the latus rectum.

2. Prove that the volume of the solid generated by the revolution of an ellipse round its minor axis, is a mean proportional between those generated by the revolution of the ellipse and of the auxiliary circle round the major axis.

**103-III. Revolution about any axis.

The volume of the solid generated by the revolution, about an axis CD, of the area bounded by a curve AB, the axis CD, and the perpendiculars AC, BD on the axis, is

$$\int \pi \, \mathbf{PM}^2 \, \mathbf{d(OM)},$$

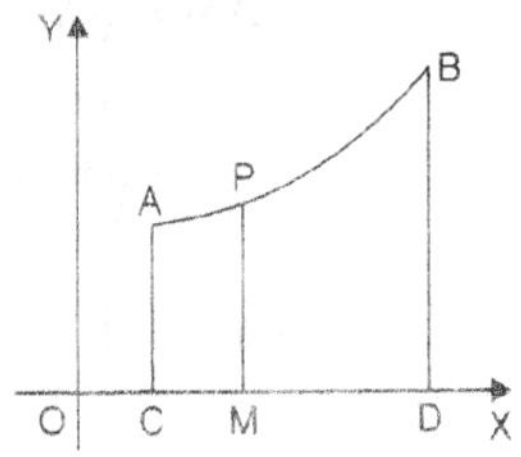

where PM is the perpendicular from any point P of the curve on the axis, and O a fixed point in the axis, the limits of integration being the values of the independent variable at the ends of the revolving arc.

Take OCD as the x-axis and OY ⊥ to it as the y-axis. Let the co-ordinates of P be (x, y).

Then volume $= \int \pi y^2 \, dx = \int \pi \, \text{PM}^2 d\,(\text{OM})$,

the limits of integration being the values of the independent variable at the ends of the revolving arc.

Note. Important. *While finding the volume (or curved surface) of the solid generated by the revolution, about* **any** *axis (which is not the x-axis or the y-axis), of an area, it is generally useful to express the integral first in terms of x and y and then in terms of a parameter by means of the parametric equations of the curve, the limits of integration being the values of the parameter at the ends of the revolving arc.*

EXAMPLES

**1. Find the volume of the spindle formed by the revolution of a parabolic arc about the line joining the vertex to one extremity of the latus rectum.

6. Find the volume of the spindle-shaped solid generated by revolving the astroid $x^{\frac{2}{3}} + y^{\frac{2}{3}} = a^{\frac{2}{3}}$ about the x-axis.

Find also the surface of the solid.

7. Find the surface of the solid generated by the revolution of the astroid $x = a\cos^3 t$, $y = a\sin^3 t$ about the axis of x.

8. Prove that the surface of the prolate spheroid formed by the revolution of an ellipse of eccentricity e about its major axis is equal to 2. area of ellipse. $\left[\sqrt{1-e^2} + \dfrac{\sin^{-1} e}{e}\right]$.

104-II. Revolution about the y-axis. It has been proved in Art. 104-I that the curved surface of the solid generated by the revolution, about the x-axis, of the area bounded by the curve $y = f(x)$, the x-axis, and the ordinates $x = a$, $x = b$ is

$$\int_{x=a}^{x=b} 2\pi y\, ds \qquad ...(1)$$

Changing x to y and y to x in (1), we get :

the curved surface of the solid generated by the revolution, about the y-axis, of the area bounded by the curve $x = f(y)$, the y-axis, and the abscissae $y = a$, $y = b$ is

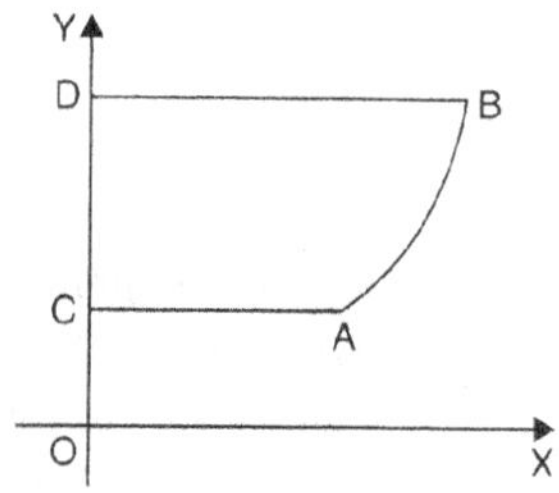

$$\int_{y=a}^{y=b} 2\pi x\, ds.$$

Cor. $\mathbf{S = \int 2\pi x\, ds.}$

[Changing x to y and y to x in Art. 104–I, Cor.]

EXAMPLE

The part of the parabola $y^2 = 4ax$ cut off by the latus rectum revolves about the tangent at the vertex. Find

(*i*) the curved surface of the reel thus formed,

(*ii*) the volume of the reel.

****104-III. Revolution about any axis.** *The curved surface of the solid generated by the revolution, about an axis CD, of the area bounded by a curve AB, the axis CD and the perpendiculars AC, BD on the axis, is*

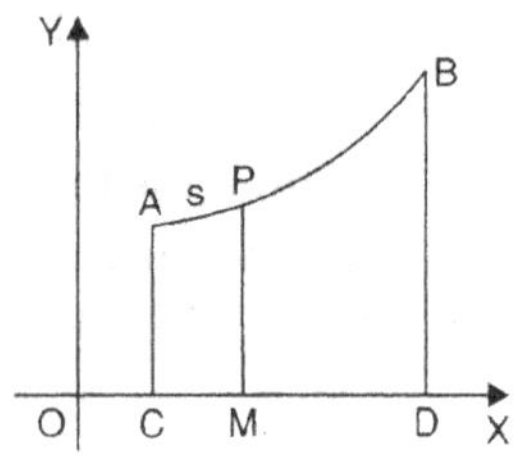

$$S = \int 2\pi\, PM\, ds,$$

where PM is the perpendicular from any point P of the curve on the axis and arc $AP = s$, the limits of integration being the values of *the independent variable at the ends of the revolving arc.*

Take a pt. O in CD as origin, OCD as the x-axis and OY ⊥ to it as the y-axis. Let the co-ordinates of P be (x, y).

Then curved surface $= \int 2\pi y \, ds = \int 2\pi \, \text{PM} \, ds,$

the limits of integration being the values of the independent variable at the ends of the revolving arc.

Note. See Note in Art. 104-III.

EXAMPLE

**A quadrant of a circle, of radius a, revolves round its chord. Show that the surface of the spindle generated $= 2\pi a^2 \sqrt{2}\left(1-\frac{\pi}{4}\right)$,

and that its volume $= \dfrac{\pi a^3}{6\sqrt{2}}(10 - 3\pi)$.

[(*i*) Surface $= \int 2\pi \, \text{PM} \, ds$ (which will be found to be)

$$= 2\pi \int \frac{x+y-a}{\sqrt{2}} \frac{ds}{d\theta} \, d\theta.$$

But the parametric equations of the circle are $x = a \cos \theta$, $y = a \sin \theta$, and in the first quadrant, θ varies from 0 to $\frac{\pi}{2}$.

(*ii*) Volume $= \int \pi \, \text{PM}^2 \, d\,(\text{AM})$ (which will be found to be)

$$= \pi \int \left[\frac{x+y-a}{\sqrt{2}}\right]^2 d\left(\frac{x-a-y}{\sqrt{2}}\right).$$

But the parametric equations of the circle are $x = a \cos \theta$, $y = a \sin \theta$ and in the first quadrant, θ varies from 0 to $\frac{\pi}{2}$.]

II. Pappus' (or Guldin's) theorems.

[☞ *These theorems (Arts. 105-I, 105-II) may be omitted by the students of the Punjab University.*]

105-I. Pappus' (or Guldin's) theorem I. If a closed curve revolves about an axis in its plane, which does not cross the curve, the volume of the solid generated is equal to the area of the curve multiplied by the length of the path of the centroid of the area.

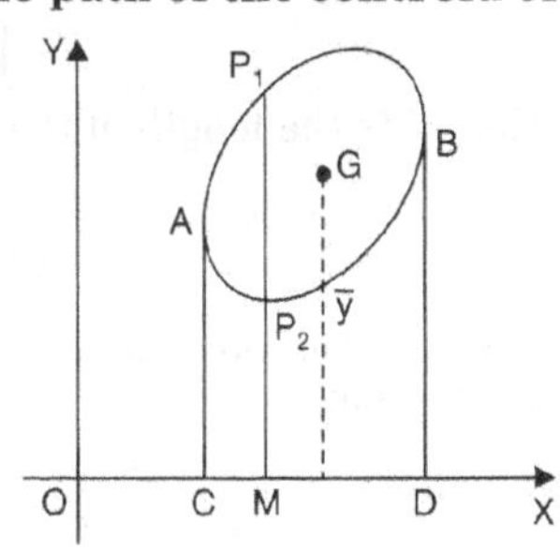

Take the axis of revolution as the x-axis and a line ⊥ to it as the y-axis.

Let AP_2BP_1 be the closed curve ; A, B the pts. where the tangent is ⊥ to the x-axis, and P_1, P_2 the pts. in which any line ⊥ to the x-axis meets the curve. Let this line meet the x-axis in M.

Let CA, DB be the ordinates of A, B, say,

$x = a, x = b.$

Now volume of the solid generated by the revolution of the area $AP_2 BP_1$

= volume of the solid generated by the revolution of the area

$ACDBP_1$ – volume of the solid generated by the revolution of the area $ACDBP_2$

$$= \int_a^b \pi y_1^2 \, dx - \int_a^b \pi y_2^2 \, dx$$

$$= \pi \int_a^b (y_1^2 - y_2^2) \, dx \quad ...(1)$$

If $\bar{y}$ is the ordinate of G, the centroid of the area of the curve, then

$$\bar{y} = \frac{\int_a^b (y_1 - y_2) \frac{1}{2} (y_1 + y_2) \, dx}{A}, \text{ where A is the area of the curve}$$

[From Statics]

$$= \frac{\frac{1}{2} \int_a^b (y_1^2 - y_2^2) \, dx}{A} \quad ...(2)$$

Substituting from (2) in (1),

volume of the solid generated $= \pi \, . \, 2A\bar{y} = A(2\pi \, \bar{y})$

= area of curve × length of path of centroid of area.

****105-II. Pappus' (or Guldin's) theorem II. If a closed curve revolves about an axis in its plane, which does not cross the curve, the curved surface of the solid generated is equal to the length of the arc of the curve multiplied by the length of the path of the centroid of the arc.**

Take the axis of revolution as the x-axis and a line ⊥ to it as the y-axis.

Let APBC be the closed curve, $P(x, y)$ any pt. on it, and arc AP = s.

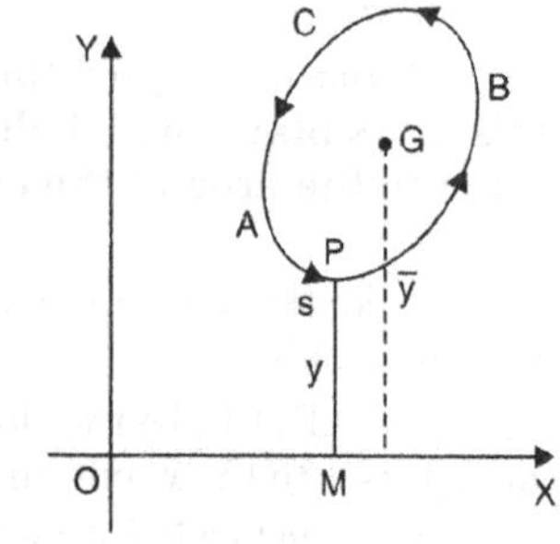

Now curved surface of the solid generated by the revolution of the area APBC

$$= \int_0^l 2\pi y \, ds,$$

where l is the length of the arc of the curve, y being a function of s

$$= 2\pi \int_0^l y \, ds \quad ...(1)$$

If $\bar{y}$ is the ordinate of G, the centroid of the arc of the curve, then

$$\bar{y} = \frac{\int_0^l y\, ds}{l} \qquad \ldots(2) \qquad \text{[From Statics]}$$

Substituting from (2) in (1),

curved surface of the solid generated $= 2\pi . l\bar{y} = l(2\pi\ \bar{y})$

= length of arc of curve × length of path of centroid of arc.

[**Note. Combined form of Pappus' (or Guldin's) theorems.** *If a closed curve revolves about an axis in its plane, which does not cross the curve, then*

(I) *volume of the solid generated*

= area of curve × length of path of centroid of area ;

(II) *curved surface of the solid generated*

= length of arc of curve × length of path of centroid of arc.

****Example. Anchor-ring.** *Find the volume and surface of an anchor-ring formed by the revolution of a circle of radius a about a line in its plane at a distance b from the centre (b > a).*

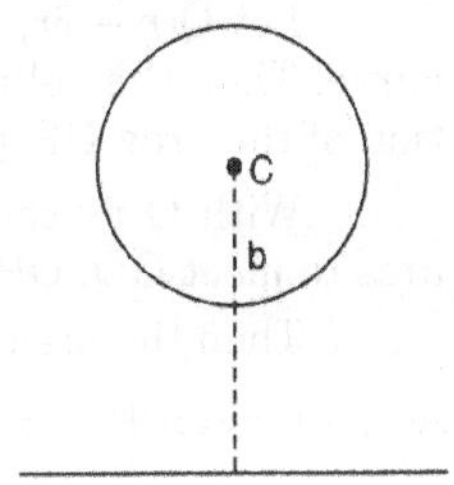

(a) The centroid of the area of the circle is at the centre.

∴ volume of anchor-ring $= \pi a^2 \times 2\pi b$

[area of curve × length of path of centroid of area (Art. 105-I)]

$= 2\pi^2 a^2 b.$

(b) The centroid of the circumference of the circle is at the centre.

∴ surface of anchor-ring

$= 2\pi a \times 2\pi b$

[length of arc of curve × length of path of centroid of area (Art. 105-II)]

$= 4\pi^2 ab.$

EXAMPLES

[☞*The following examples may be omitted by the students of the Punjab University.*]

1. Establish Pappus' theorems on surfaces and volumes of solid of revolution.

Find the volume of the surface formed by the revolution of an ellipse about the tangent at the end of the major axis.

2. The loop of the curve $2ay^2 = x(x - a)^2$ revolves about the straight line $y = a$. Find the volume of the solid generated.

[$\because$ the curve is symmetrical about the x-axis,

$\therefore$ the centroid of the area of the loop lies on the x-axis.

$\therefore$ length of path of centroid of area = $2\pi a$.]

III. Volume formula for polar equation.

[☞ *The following volume formula is generally the best for polar equation. It may be remembered without proof, if so desired.*]

106. *Revolution about the initial line.* **The volume of the solid generated by the revolution, about the initial line, of the area bounded by the curve r = f(θ) and the radii vectors θ = α, θ = β is**

$$\int_\alpha^\beta \frac{2}{3}\pi r^3 \sin\theta \, d\theta.$$

[Proof.** Let AB be the curve $r = f(\theta)$, and OA, OB the radii vectors $\theta = \alpha$, $\theta = \beta$. Let P(r, θ) be any pt. on the curve. Then the volume of the solid generated by the revolution, about the initial line, of the area OAP is clearly a function of θ = V (say).

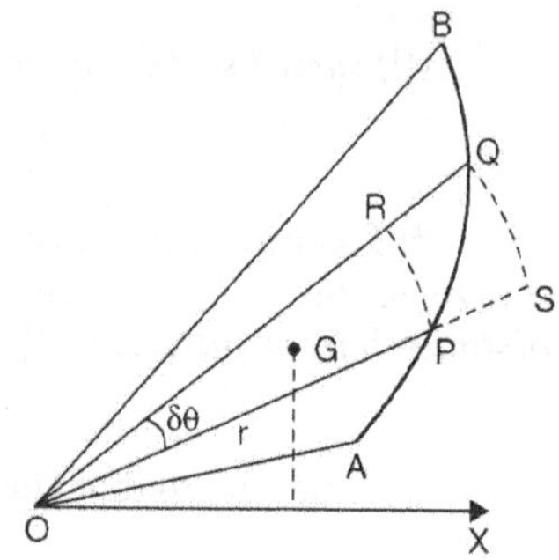

Let Q$(r + \delta r, \theta + \delta\theta)$ be a neighbouring point on the curve. Then the volume of the solid generated by the revolution of the area OPQ = δV.

With O as centre and OP, OQ as radii draw circular arcs to meet OQ, OP (produced) respectively in R, S.

Then the area OPQ lies between the areas of the circular sectors OPR and OSQ, *i.e.*, between $\frac{1}{2} r^2 \delta\theta$ and $\frac{1}{2}(r + \delta r)^2 \delta\theta$,

and the ordinate of the centroid of the area OPQ lies between $r \sin\theta$ and $\frac{2}{3}(r + \delta r)\sin(\theta + \delta\theta)$

[From Statics]

$\therefore$ the volume of the solid generated by the revolution of the area OPQ lies between

$$\frac{1}{2} r^2 \delta\theta \left[2\pi \cdot \frac{2}{3} r \sin\theta\right] \text{ and } \frac{1}{2}(r + \delta r)^2 \delta\theta \left[2\pi \cdot \frac{2}{3}(r + \delta r)\sin(\theta + \delta\theta)\right],$$

[Pappus' theorem I (Art. 105-I)]

i.e., δV lies between $\frac{2}{3}\pi r^3 \sin\theta \, \delta\theta$ and $\frac{2}{3}\pi(r + \delta r)^3 \sin(\theta + \delta\theta)\,\delta\theta$

$\therefore$ $\frac{\delta V}{\delta\theta}$ lies between $\frac{2}{3}\pi r^3 \sin\theta$ and $\frac{2}{3}\pi(r + \delta r)^3 \sin(\theta + \delta\theta)$

$\therefore$ taking the limits when $\delta\theta \to 0$ (and $\therefore$ $\delta r \to 0$),

$$\frac{dV}{d\theta} = \frac{2}{3}\pi r^3 \sin\theta \qquad \text{...(1)}$$

$$\therefore \quad \int_\alpha^\beta \frac{2}{3}\pi r^3 \sin\theta \, d\theta = \int_\alpha^\beta \frac{dV}{d\theta}\, d\theta = \Big[V \Big]_\alpha^\beta$$

= (value of V when θ = β) – (value of V when θ = α)

= volume of the solid generated by the revolution of the area OAB – 0,

i.e., $\left.\begin{array}{r}\text{volume of the solid generated}\\ \text{by the revolution of the area OAB}\end{array}\right\} = \int_\alpha^\beta \frac{2}{3}\pi r^3 \sin\theta\, d\theta.$

Cor. From (1), $\mathbf{V} = \int \frac{2}{3}\pi \mathbf{r}^3 \sin\theta\, d\theta.$]

Example. *The curve r = a (1 + cos θ) revolves about the initial line. Find the volume an surface of the figure formed.*

The equation of the curve is

$$r = a(1 + \cos\theta),$$

which is a cardioid.

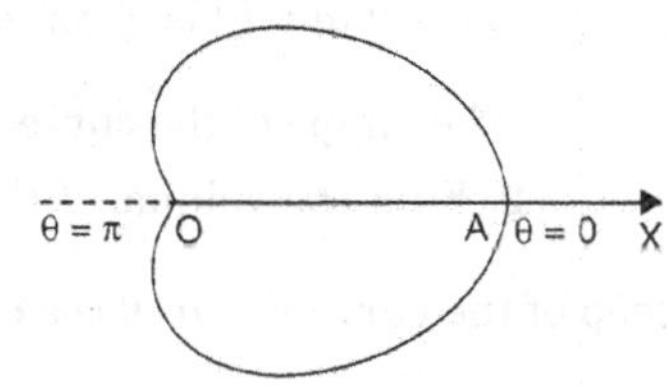

The cardioid is symmetrical about the initial line, and for the upper half, θ varies from 0 to π.

(*a*) Volume $= \int_0^\pi \frac{2}{3}\pi r^3 \sin\theta\, d\theta$

$$= \frac{2}{3}\pi \int_0^\pi a^3 (1 + \cos\theta)^3 \sin\theta\, d\theta$$

$$= -\frac{2}{3}\pi a^3 \int_0^\pi (1 + \cos\theta)^3 (-\sin\theta)\, d\theta$$

$$= -\frac{2}{3}\pi a^3 \left[\frac{(1+\cos\theta)^4}{4}\right]_0^\pi = -\frac{1}{6}\pi a^3 [0 - 16] = \frac{8}{3}\pi a^3.$$

(*b*) Surface $= \int 2\pi y\, ds = 2\pi \int_0^\pi r \sin\theta \frac{ds}{d\theta}\, d\theta$ ...(1)

Now $r = a(1 + \cos\theta), \quad \therefore \quad \frac{dr}{d\theta} = -a\sin\theta$

$$\therefore \quad \frac{ds}{d\theta} = \sqrt{r^2 + \left(\frac{dr}{d\theta}\right)^2} = \sqrt{a^2(1+\cos\theta)^2 + a^2\sin^2\theta}$$

$$= a\sqrt{1 + 2\cos\theta + \cos^2\theta + \sin^2\theta} = a\sqrt{2(1+\cos\theta)}$$

[∵ $\cos^2\theta + \sin^2\theta = 1$]

$$= a\sqrt{2.2\cos^2\frac{\theta}{2}} = 2a\cos\frac{\theta}{2}.$$

∴ from (1), surface $= 2\pi \int_0^\pi a(1 + \cos\theta)\sin\theta.\ 2a\cos\frac{\theta}{2}\, d\theta$

$$= 4\pi a^2 \int_0^\pi 2\cos^2\frac{\theta}{2}\ .\ 2\sin\frac{\theta}{2}\cos\frac{\theta}{2}\ .\ \cos\frac{\theta}{2}\, d\theta$$

$$= 16\pi a^2 \int_0^\pi \cos^4\frac{\theta}{2}\sin\frac{\theta}{2}\, d\theta$$

$$= 16\pi a^2 (-2) \int_0^{\pi} \cos^4 \frac{\theta}{2} \left(-\sin \frac{\theta}{2} . \frac{1}{2}\right) d\theta = -32\pi a^2 \left[\frac{\cos^5 \frac{\theta}{2}}{5}\right]_0^{\pi}$$

$$= -\frac{32\pi a^2}{5} [0 - 1] = \frac{32}{5} \pi a^2.$$

EXAMPLES

1. Show that the volume generated by the revolution, about the initial line, of the limacon $r = a + b \cos \theta \ (a > b)$ is $\frac{4}{3} \pi a (a^2 + b^2)$.

[The shape of the curve is that shown in the Fig. of Misc. Ex. 14, Chap. IX.]

2. Find the volume of the solid formed by the revolution about the prime radius of the loop of the curve $r^3 = a^3 \theta \cos \theta$ between $\theta = 0$ and $\theta = \frac{\pi}{2}$.

3. Show that if the area lying within the cardioid $r = 2a (1 + \cos \theta)$ and without the parabola $r(1 + \cos \theta) = 2a$, revolves about the initial line, the volume generated is $18\pi a^3$.

[The equations of the cardioid and the parabola are

$$r = 2a (1 + \cos \theta) \quad ...(1)$$

$$r = \frac{2a}{1 + \cos \theta} \quad ...(2)$$

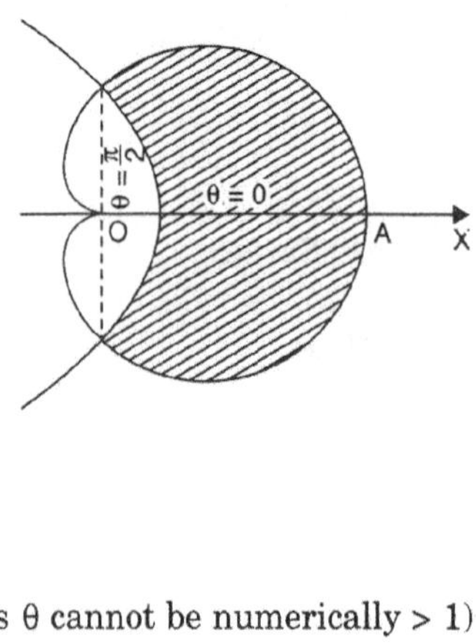

They are both symmetrical about the initial line.

To find the pts. of intersection, substituting the value of r (= $2a (1 + \cos \theta)$) from (1) in (2),

$$2a (1 + \cos \theta) = \frac{2a}{1 + \cos \theta},$$

or $$1 + 2 \cos \theta + \cos^2 \theta = 1,$$

∴ either $$\cos \theta = 0$$

or $$\cos \theta = -2, \text{ which is impossible,}$$

(∵ $\cos \theta$ cannot be numerically > 1)

∴ $$\cos \theta = 0, \text{ or } \theta = \frac{\pi}{2}.$$

∴ $$\text{volume} = \int_0^{\frac{\pi}{2}} \frac{2}{3} \pi (r_1^3 - r_2^3) \sin \theta \, d\theta$$

$$= \frac{2}{3} \pi \int_0^{\frac{\pi}{2}} \left[\{2a (1 + \cos \theta)\}^3 - \left(\frac{2a}{1 + \cos \theta}\right)^3 \right] \sin \theta \, d\theta$$

$$= \frac{16\pi a^3}{3}\left[\int_0^{\frac{\pi}{2}} (1+\cos\theta)^3 \sin\theta\, d\theta - \int_0^{\frac{\pi}{2}} (1+\cos\theta)^{-3} \sin\theta\, d\theta\right]$$

MISCELLANEOUS EXAMPLES ON CHAPTER XI

1. Find the volume of the solid generated by rotating completely about the x-axis the area enclosed between $y^2 = x^3 + 5x$ and the lines $x = 2$ and $x = 4$.

2. The part of the curve $y^2 = x^2(1 - x^2)$ between $x = 0$ and $x = 1$ rotates about the x-axis. Obtain the volume of the solid thus generated.

3. A segment is cut off from a sphere of radius a by a plane at a distance $\frac{1}{2}a$ from the centre. Show that the volume of the segment is 5/32 of the volume of the sphere.

4. Prove that the volume of the solid generated by the revolution of the curve $y = \frac{a^3}{a^2 + x^2}$ about its asymptote is $\frac{1}{2}\pi^2 a^3$.

5. A basin is formed by the revolution of the curve $x^3 = 64y$, $(y > 0)$ about the axis of y. If the depth of the basin is 8 inches, how many cubic inches of water will it hold ?

****6.** Show that the volume of the solid generated by the revolution of the cycloid $x = a(\theta + \sin\theta)$, $y = a(1 - \cos\theta)$ about the y-axis is $\pi a^3\left(\frac{3}{2}\pi^2 - \frac{8}{3}\right)$.

7. Find the volumes of the oblate and prolate spheroids generated by an ellipse whose major and minor axes are $(24\pi)^{\frac{1}{3}}$ and $(3\pi)^{\frac{1}{3}}$.

8. Trace roughly the curve $xy^2 = 4(2 - x)$. Find the area enclosed by the curve and y-axis, and also the volume of the solid formed by the revolution of the curve through four right angles about the y-axis.

****9.** The part of the ellipse $\frac{x^2}{a^2} + \frac{y^2}{b^2} = 1$ cut off by a latus rectum revolves about the tangent at the nearer vertex. Find the volume of the reel thus generated.

$$\left[\text{Volume} = 2\int \pi\,\text{PM}^2\, d(\text{AM}) = 2\pi\int_0^{\frac{b^2}{a}} (a - x)^2\, dy.\right]$$

10. The ellipse $b^2x^2 + a^2y^2 = a^2b^2$ is divided into two parts by the line $x = \frac{1}{2}a$, and the smaller part is rotated through four right angles about this line. Prove that the volume generated is $\pi a^2 b\left(\dfrac{3\sqrt{3}}{4} - \dfrac{1}{3}\pi\right)$.

11. The figure bounded by a quadrant of a circle of radius a and the tangents at its extremities revolves about one of the tangents. Prove that the volume of the solid generated is $\left(\dfrac{5}{3} - \dfrac{1}{2}\pi\right)\pi a^3$.

****12.** If the hyperbola $\dfrac{x^2}{a^2} - \dfrac{y^2}{b^2} = 1$ revolves about the x-axis, show that the volume included between the surface thus generated, the cone generated by the asymptotes and two planes perpendicular to the axis of x, at a distance h apart, is equal to that of a circular cylinder of height h and radius b.

[Let the two planes be generated by the ordinates $x = c$, $x = c + h$.

$$\text{Volume} = \int_0^{c+h} \pi(y_1^2 - y_2^2)\,dx$$

[But $y_1^2 = \dfrac{b^2}{a^2}x^2$ (from the equation of the asymptote)

and $y_2^2 = \dfrac{b^2}{a^2}(x^2 - a^2)$ (from the equation of the hyperbola)]]

****13.** A solid spheroid formed by the revolution of the ellipse $b^2x^2 + a^2y^2 = a^2b^2$ about the major axis has a cylindrical hole of circular section having the major axis as axis drilled through it. Prove that the volume of the solid which remains is $4\pi b^2 l^3/3a^2$, where $2l$ is the length of the hole.

[The required volume is generated by the revolution of the area E′FEBE′.

$$\therefore \quad \text{volume} = 2\int_0^{l} \pi(y_1^2 - y_2^2)\,dx$$

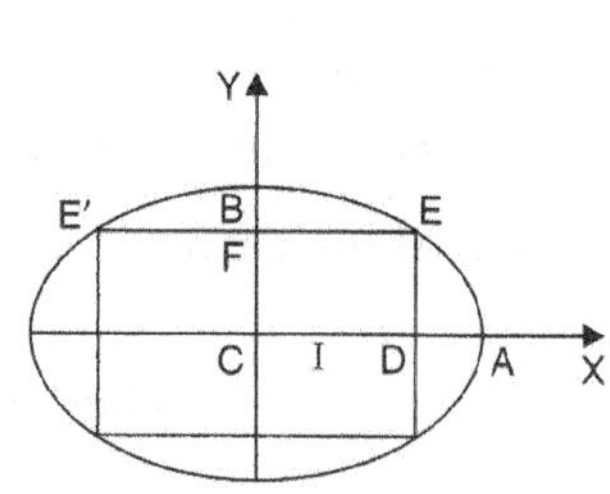

(But $y_1^2 = \dfrac{b^2(a^2 - x^2)}{a^2}$

(from the equation of the ellipse),

$y_2^2 = \dfrac{b^2(a^2 - l^2)}{a^2}$

(putting $x = l$ in the equation of the ellipse))]

14. The arc AL of a parabola, where A and L are respectively the vertex and an extremity of the latus rectum, is revolved about its axis. Find the area of the surface generated.

15. Show that the surface of the spherical zone contained between two parallel planes $= 2\pi ah$, where a is the radius of the sphere and h the distance between the planes.

16. Find the area of the surface formed by the revolution of $x^2 + 4y^2 = 16$ about its major axis.

17. Find the area of the surface swept out by the arc of the rectangular hyperbola $x^2 - y^2 = a^2$, extending from the vertex to the end of the latus rectum, when rotated through four right angles about the axis of x.

18. The co-ordinates of a point on a cycloid are

$$x = a(\theta + \sin\theta),\ y = a(1 + \cos\theta).$$

Show that the ratio of the area of the surface formed by the rotation of the arc of the cycloid between two successive cusps about the axis of x to the area enclosed by the cycloid and the axis of x is $\frac{64}{9}$.

19. *Catenoid.* Find the volume, and also the surface, generated by the revolution of the catenary $y = c\cosh\frac{x}{c}$ about the axis of x.

20. A$(0, a)$ and P(x, y) are two points on the curve whose equation is $y = a\cosh(x/a)$, and s is the length of the arc AP. If the curve makes a complete revolution about the x-axis, prove that the area S of the curved surface, bounded by planes through A and P perpendicular to the x-axis, and the corresponding volume V are connected by $a\text{S} = 2\text{V} = \pi a(ax + sy)$.

21. Prove that the surface and volume of the solid generated by the revolution, about the x-axis, of the loop of the curve $x = t^2$, $y = t - \frac{1}{3}t^3$ are respectively 3π and $\frac{3}{4}\pi$.

[The shape of the curve is that shown in the Fig. of Misc. Ex. 9, Chap. IX.]

22. Prove that the surface and volume generated by the revolution of the tractrix $x = a\cos t + \frac{1}{2}a\log\tan^2\frac{1}{2}t$, $y = a\sin t$, about its asymptote are respectively equal to the surface and half the volume of a sphere of radius a.

****23.** The arc of the cardioid $r = a(1 + \cos\theta)$ included between $-\frac{1}{2}\pi \le \theta \le \frac{1}{2}\pi$ is rotated about the line $\theta = \frac{1}{2}\pi$. Find the area of the surface generated.

[To evaluate the integral, express the integrand in a series of cosines of multiple angles.]

****24.** A circular arc revolves about its chord. Prove that the area of the surface generated is $4\pi a^2(\sin\alpha - \alpha\cos\alpha)$, where a is the radius and 2α the angle subtended by the arc at the centre.

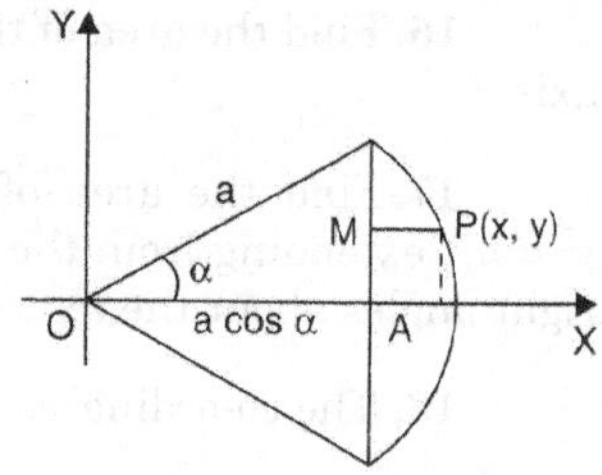

[Take O, the centre of the circle, as origin, and the join of O and the mid-pt. of the arc as the x-axis, so that the line through O ⊥ to this join is the y-axis.

Let P(x, y) be any pt. on the arc and PM ⊥ on the chord.

Now the equation of the chord is $x = a \cos \alpha$.

$$\text{Surface} = 2 \int 2\pi \text{ PM } ds$$

$$= 4\pi \int_0^{\alpha} (x - a \cos \alpha) \frac{ds}{d\theta} \, d\theta.$$

But the parametric equations of the circle are

$$x = a \cos \theta, \; y = a \sin \theta.]$$

25. Find the surface of the solid generated by the revolution of the lemniscate $r^2 = a^2 \cos 2\theta$ about the initial line.

26. The lemniscate $r^2 = a^2 \cos 2\theta$ revolves about a tangent at the pole. Show that the surface of the solid generated is $4\pi a^2$.

[Let the lemniscate revolve about the tangent OM.

Surface generated by the two loops

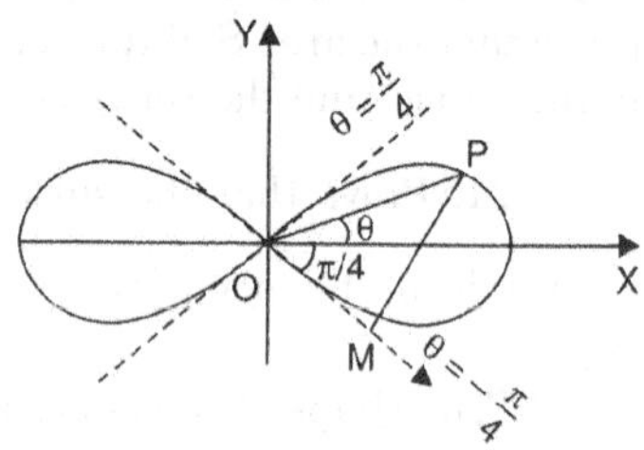

$$= 2 \int 2\pi \text{ PM } ds$$

$$= 4\pi \int_{-\frac{\pi}{4}}^{\frac{\pi}{4}} r \sin \left(\theta + \frac{\pi}{4}\right) \frac{ds}{d\theta} \, d\theta \qquad ...(1)$$

Now $\qquad r = a \sqrt{\cos 2\theta}$,

$$\therefore \qquad \frac{dr}{d\theta} = ?$$

$$\therefore \qquad \frac{ds}{d\theta} = \sqrt{r^2 + \left(\frac{dr}{d\theta}\right)^2} = ?$$

Substitute in (1) and evaluate.]

****27.** An area lies altogether on one side of an axis in its plane. Prove that the volume of the solid formed by the rotation of the area about the axis is equal to the area multiplied by the distance traversed by its centre of gravity.

Hence prove that the volume of the solid formed by the rotation about the line $\theta = 0$ of the area bounded by the curve $r = f(\theta)$ and the lines $\theta = \theta_1$, $\theta = \theta_2$, is $\frac{2}{3} \pi \int_{\theta_1}^{\theta_2} r^3 \sin \theta \, d\theta$.

Find the volume of the solid formed by the revolution of the cardioid $r = a(1 + \cos \theta)$ about the line $\theta = 0$.

28. Show that the curve $r = 1 + 2\cos\theta$ consists of an outer and an inner loop. If the area of the inner loop is rotated through two right angles about the initial line, show that the volume of the solid so formed is $\pi/12$.

****29.** Find the volume of the solid formed by the revolution of the lemniscate $r^2 = a^2 \cos 2\theta$ about the line $\theta = \dfrac{\pi}{2}$.

Note. *The volume of the solid generated by the revolution, about the line through the pole perpendicular to the initial line, of the area bounded by the curve $r = f(\theta)$ and the radii vectors $\theta = \alpha$, $\theta = \beta$ is*

$$\int_\alpha^\beta \frac{2}{3}\pi r^3 \cos\theta \, d\theta.$$

[Proceed as in Art. 109. Here the ⊥ distance of the centroid of the area OPQ from the axis of revolution (*i.e.*, line through the pole ⊥ to the initial line) lies between

$$\frac{2}{3}r\cos\theta \text{ and } \frac{2}{3}(r + \delta r)\cos(\theta + \delta\theta), \text{ etc.} \qquad \text{(From Statics)]}$$

[**Aid to memory.** If OX is the axis of revolution,

$$\text{volume} = \int_\alpha^\beta \frac{2}{3}\pi r^3 \sin\theta \, d\theta = \int_\alpha^\beta \frac{2}{3}\pi r^3 \sin XOP \, d\theta.$$

If OY (⊥ to OX) is the axis of revolution,

$$\text{volume} = \int_\alpha^\beta \frac{2}{3}\pi r^3 \sin YOP \, d\theta = \int_\alpha^\beta \frac{2}{3}\pi r^3 \sin\left(\frac{\pi}{2} - \theta\right) d\theta$$

$$= \int_\alpha^\beta \frac{2}{3}\pi r^3 \cos\theta \, d\theta.]$$

CHAPTER XII

DIFFERENTIAL EQUATIONS OF THE FIRST ORDER AND FIRST DEGREE

SECTION I

DEFINITIONS

107. (*a*) **Differential equation. Def.** A **differential equation** is an equation which involves differential coefficients or differentials.

(*b*) **Ordinary and partial differential equations. Defs.** (1) An **ordinary differential equation** is that which involves only one independent variable and differential coefficients w.r.t.* it.

Thus (*i*) $$\frac{dy}{dx} = \sin x \quad \text{or} \quad dy = \sin x\, dx,$$

(*ii*) $$y = x\frac{dy}{dx} + \frac{a}{\frac{dy}{dx}},$$

(*iii*) $$\frac{d^2y}{dx^2} + y = 0,$$ (*iv*) $$\frac{\left[1+\left(\frac{dy}{dx}\right)^2\right]^{\frac{3}{2}}}{\frac{d^2y}{dx^2}} = a,$$

are all *ordinary differential equations.*

(2) A **partial differential equation is that which involves two or more independent variables and partial differential coefficients w.r.t. any of them.

Thus $x\dfrac{\partial u}{\partial x} + y\dfrac{\partial u}{\partial y} = n\,u$ *is a partial differential equation.*

(*c*) **Order and degree of a differential equation. Defs.** The **order** of a differential equation is the order of the highest differential coefficient occurring in it.

The **degree** of a differential equation is the degree of that highest differential coefficient, when the equation has been made free from fractions and radicals as far as the differential coefficients are concerned.

Example. *Find the order and degree of the following differential equations :*

(*i*) $\dfrac{dy}{dx} = \sin x$; (*ii*) $y = x\dfrac{dy}{dx} + \dfrac{a}{\frac{dy}{dx}}$.

*The letters "w.r.t." stand, as usual, for "*with respect to*".

(iii) $\dfrac{d^2y}{dx^2} + y = 0$; (iv) $\dfrac{\left[1+\left(\dfrac{dy}{dx}\right)^2\right]^{\frac{3}{2}}}{\dfrac{d^2y}{dx^2}} = a.$

(i) The equation is $\dfrac{dy}{dx} = \sin x$, which is of the first order and first degree.

(ii) The equation is $y = x\dfrac{dy}{dx} + \dfrac{a}{\dfrac{dy}{dx}}$.

Making the equation free from fractions as far as the diff. coeffs. are concerned by multiplying throughout by $\dfrac{dy}{dx}$, it becomes

$$y\frac{dy}{dx} = x\left(\frac{dy}{dx}\right)^2 + a,$$

which is of the first order and second degree.

(iii) The equation is $\dfrac{d^2y}{dx^2} + y = 0$, which is of the second order and first degree.

(iv) The equation is $\dfrac{\left[1+\left(\dfrac{dy}{dx}\right)^2\right]^{\frac{3}{2}}}{\dfrac{d^2y}{dx^2}} = a.$

Making the equation free from fractions by multiplying throughout by $\dfrac{d^2y}{dx^2}$, it becomes

$$\left[1+\left(\frac{dy}{dx}\right)^2\right]^{\frac{3}{2}} = a\,\frac{d^2y}{dx^2}.$$

Making it free from radicals by squaring both sides, it becomes

$$\left[1+\left(\frac{dy}{dx}\right)^2\right]^3 = a^2\left(\frac{d^2y}{dx^2}\right)^2,$$

which is of the second order and second degree.

108. (*a*) **Solution (or integral). Def.** A **solution** (or **integral**) of a differential equation is a relation between the variables by means of which and the differential coefficients obtained therefrom, the equation is satisfied.

(*b*) **General and particular solutions. Defs.**

(1) The **general** (or **complete**) **solution** is that which contains a number of arbitrary constants equal to the order of the differential equation.

(2) A **particular solution** is that which can be obtained from the general solution by giving particular values to the arbitrary constants.

Example. *Show that (i)* $y = cx + \dfrac{a}{c}$, *(ii)* $y = x + a$ *are solutions of the differential equation* $y = x\dfrac{dy}{dx} + \dfrac{a}{\dfrac{dy}{dx}}$, *and indicate the kind of each solution.*

The equation is
$$y = x\frac{dy}{dx} + \frac{a}{\dfrac{dy}{dx}} \qquad ...(1)$$

(*i*) $y = cx + \dfrac{a}{c}$,

$\therefore \quad \dfrac{dy}{dx} = c.$

Substituting in (1), we get

$$cx + \frac{a}{c} = xc + \frac{a}{c}, \text{ which is true}$$

$\therefore \quad y = cx + \dfrac{a}{c}$ is a solution of (1). [Def. (Art. 108, (*a*))]

$\because$ it contains one arbitrary constant (c), and (1) is of the first order

$\therefore$ it is the general solution. [Def. (Art. 108, (*b*), (1))]

(*ii*) $y = x + a$,

$\therefore \quad \dfrac{dy}{dx} = 1.$

Substituting in (1), we get

$$x + a = x \,.\, 1 + \frac{a}{1}, \text{ which is true}$$

$\therefore \quad y = x + a$ is a solution of (1). [Def. (Art. 108, (*a*))]

$\because$ it can be obtained from the general solution $y = cx + \dfrac{a}{c}$ by putting $c = 1$

$\therefore$ it is a particular solution. [Def. (Art. 108, (*b*), (2))]

SECTION II

FIVE STANDARD CASES OF DIFFERENTIAL EQUATIONS OF THE FIRST ORDER AND FIRST DEGREE

109. Case I. Variables separable. Def. If in an equation it is possible to get all the functions of x and dx to one side, and all the functions of y and dy to the other, the **variables** are said to be **separable.**

Thus in the equation $\frac{dy}{dx} = XY$ ...(1),
where X is a function of x only and Y a function of y only, the **variables are separable.**

$$\left[\because \text{ (1) can be written as } \frac{dy}{Y} = X\,dx\right]$$

[Rule to solve an equation in which the variables are separable, *i.e.*, $\frac{\mathbf{dy}}{\mathbf{dx}} = \mathbf{XY}$, **where X is a function of x only and Y a function of y only :**

(*i*) *Multiply through out by dx.*

(*ii*) *Separate the variables x and y (i.e., get all the functions of x and dx to one side, and all the functions of y and dy to the other).*

(*iii*) *Integrate both sides, and add an arbitrary constant on one side.*]

Example. *Solve* $\frac{dy}{dx} = e^{x-y} + x^2e^{-y}$.

The equation is $\frac{dy}{dx} = e^{x-y} + x^2e^{-y} = e^{-y}(e^x + x^2)$. $\left[\text{Form } \frac{dy}{dx} = XY\right]$

(*i*) Multiplying through out by dx,

$$dy = e^{-y}(e^x + x^2)\,dx.$$

(*ii*) Separating the variables,

$$e^y\,dy = (e^x + x^2)\,dx.$$

(*iii*) Integrating both sides,

$$\int e^y\,dy = \int (e^x + x^2)\,dx + c^*$$

or $$e^y = e^x + \frac{x^3}{3} + c.$$

*[**Caution.** It is a *common mistake* not to add an arbitrary constant on one side. The resulting answer is wrong because it is not the *general* solution. {Def. (Art. 108, (*b*), (1)}]

EXAMPLES

Solve the following differential equations (Exs. 1—6) :

1. (*i*) $(a - x)\,dy - (a + y)\,dx = 0$.

(*ii*) $\frac{dy}{dx} + \sqrt{\frac{1-y^2}{1-x^2}} = 0$.

[A better form of the answer.

(*i*) It will be found that $\log(a + y) + \log(a - x) = c$ ...(1)

or $\log(a + y)(a - x) = c$, or $(a + y)(a - x) = e^c = c_1$

or changing c_1 to c, [$\because$ c being an arbitrary constant, c_1 $(= e^c)$ is also an arbitrary constant]

$$(a + y)(a - x) = c \qquad ...(2)$$

Note. *Both the solutions (1) and (2) are equally general and, therefore, equally correct. But (2), involving powers of x and y only, is simpler and neater than (1) involving logarithms. Hence (2) is preferable to (1). It is usual to express the solutions of differential equations in such forms as (2).*

(*ii*) It will be found that $\sin^{-1} y + \sin^{-1} x = c$

or $\sin^{-1} (y\sqrt{1-x^2} + x\sqrt{1-y^2}) = c$

[$\because$ $\mathbf{\sin^{-1} x + \sin^{-1} y = \sin^{-1} (x\sqrt{1-y^2} + y\sqrt{1-x^2})}$

(From Trigonometry)]

or $y\sqrt{1-x^2} + x\sqrt{1-y^2} = \sin c = c_1$

or changing c_1 to c, $y\sqrt{1-x^2} + x\sqrt{1-y^2} = c$.]

2. (*i*) $(1 + x) y\, dx + (1 - y) x\, dy = 0$.

(*ii*) $(x^2 - yx^2)\dfrac{dy}{dx} + y^2 + xy^2 = 0$.

3. $\sec^2 x \tan y\, dx + \sec^2 y \tan x\, dy = 0$.

4. (*i*) $3e^x \tan y\, dx + (1 - e^x) \sec^2 y\, dy = 0$.

5. $\left(y - x\dfrac{dy}{dx}\right) = a\left(y^2 + \dfrac{dy}{dx}\right)$.

[**Note.** *Equation reducible to variables separable. Important type.*

Rule to solve the equation $\mathbf{\dfrac{dy}{dx} = f(ax + by + c)}$ **:**

Put $ax + by + c = z$, and separate the variables z and x.]

6. (*i*) $(x + y)^2 \dfrac{dy}{dx} = a^2$.

(*ii*) $(x - y)^2 \dfrac{dy}{dx} = a^2$.

7. (*i*) $\dfrac{dy}{dx} = (4x + y + 1)^2$.

(*ii*) $(x + y + 1)\dfrac{dy}{dx} = 1$.

110. Case II. (*a*) **Homogeneous equation. Def. A homogeneous equation in x and y** is that which can be put in the form $\dfrac{dy}{dx} = \dfrac{f_1(x, y)}{f_2(x, y)}$, where $f_1(x, y)$, $f_2(x, y)$ are homogeneous functions of the same degree in x and y.

(*b*) *Homogeneous equation.* **To solve the equation $\mathbf{\dfrac{dy}{dx} = \dfrac{f_1(x, y)}{f_2(x, y)}}$, **where $\mathbf{f_1(x, y)}$, $\mathbf{f_2(x, y)}$ are homogeneous functions of the same degree in x and y.**

The equation is $\qquad \dfrac{dy}{dx} = \dfrac{f_1(x, y)}{f_2(x, y)}$ ...(1)

$\because$ $f_1(x, y), f_2(x, y)$ are homogeneous functions* of the same degree in x and y, let $f_1(x, y)$ $= x^n \phi_1\left(\dfrac{y}{x}\right), f_2(x, y) = x^n \phi_2\left(\dfrac{y}{x}\right)$.

The (1) becomes $\qquad \dfrac{dy}{dx} = \dfrac{x^n \phi_1\left(\dfrac{y}{x}\right)}{x^n \phi_2\left(\dfrac{y}{x}\right)} = \dfrac{\phi_1\left(\dfrac{y}{x}\right)}{\phi_2\left(\dfrac{y}{x}\right)}$ ...(2)

Put $y = vx$, then $\qquad \dfrac{dy}{dx} = \dfrac{dv}{dx} . x + v . 1 = x\dfrac{dv}{dx} + v.$

$\therefore$ (2) becomes $x\dfrac{dv}{dx} + v = \dfrac{\phi_1(v)}{\phi_2(v)}$.

Transposing, $\qquad x\dfrac{dv}{dx} = \dfrac{\phi_1(v)}{\phi_2(v)} - v = \dfrac{\phi_1(v) - v\phi_2(v)}{\phi_2(v)}$.

Separating the variables, $\dfrac{\phi_2(v)}{\phi_1(v) - v\phi_2(v)} dv = \dfrac{1}{x} dx,$

which can be solved. In the solution, putting $v = \dfrac{y}{x}$, we get the required solution.

[**Rule to solve a homogeneous equation in x and y,** *i.e.*, $\dfrac{dy}{dx} = \dfrac{f_1(x, y)}{f_2(x, y)}$, **where $f_1(x, y), f_2(x, y)$ are homogeneous functions of the same degree in x and y :**

Put y = vx, and separate the variables v and x.]

Example. *Solve* $(x^2 + y^2)\,dx - 2xy\,dy = 0$.

The equation is $(x^2 + y^2)\,dx - 2xy\,dy = 0$

or $\qquad \dfrac{dy}{dx} = \dfrac{x^2 + y^2}{2xy}$ ...(1)

which is homogeneous in x and y.

Put $y = vx$, then $\qquad \dfrac{dy}{dx} = \dfrac{dv}{dx} . x + v . 1 = x\dfrac{dv}{dx} + v,$

$\therefore$ (1) becomes $x\dfrac{dv}{dx} + v = \dfrac{x^2 + v^2x^2}{2x.vx} = \dfrac{1+v^2}{2v}$.

Transposing, $\qquad x\dfrac{dv}{dx} = \dfrac{1+v^2}{2v} - v = \dfrac{1-v^2}{2v}$.

---

*__Homogeneous function of the nth degree in x and y. Def.__ A homogeneous function of **the nth degree in x and y** is that which can be put in the form

$$x^n f\left(\frac{y}{x}\right).$$

Separating the variables, $\dfrac{2v}{1-v^2}\,dv = \dfrac{1}{x}\,dx.$

Integrating both sides, $\displaystyle\int \frac{2v}{1-v^2}\,dv = \int \frac{1}{x}\,dx + c$

or $\quad -\log(1-v^2) = \log x + c \;\left|\; \displaystyle\int \frac{2v}{1-v^2}\,dv = -\int \frac{-2v}{1-v^2}\,dv = -\log(1-v^2)\right.$

or $\quad \log x + \log(1-v^2) = -c,$ or $\log x(1-v^2) = -c,$ or $x(1-v^2) = e^{-c} = c_1.$

Changing c_1 to c, $x(1-v^2) = c$ $\quad\left[\text{Put } v = \dfrac{y}{x}\right]$

or $\quad x\left(1-\dfrac{y^2}{x^2}\right) = c,$ or $\dfrac{x^2-y^2}{x} = c,$ or $x^2 - y^2 = cx.$

EXAMPLES

Solve (Exs. 1—5) :

1. (*i*) $x\dfrac{dy}{dx} + \dfrac{y^2}{x} = y.$

(*ii*) $(x+y)\dfrac{dy}{dx} + (x-y) = 0.$

2. (*i*) $x\,dx + y\,dy = m(x\,dy - y\,dx).$

(*ii*) $x^2y\,dx - (x^3+y^3)\,dy = 0.$

3. (*i*) $y^2 + x^2\dfrac{dy}{dx} = xy\dfrac{dy}{dx}.$

(*ii*) $y^2\,dx + (xy + x^2)\,dy = 0.$

4. $x\dfrac{dy}{dx} - y = \sqrt{x^2+y^2}.$

5. (*i*) $(x^2-y^2)\,dx + 2xy\,dy = 0.$

**(*ii*) $\left(x\cos\dfrac{y}{x} + y\sin\dfrac{y}{x}\right)y - \left(y\sin\dfrac{y}{x} - x\cos\dfrac{y}{x}\right)x\dfrac{dy}{dx} = 0.$

**111 (*a*). Case III (*a*). Equation reducible to homogeneous.

To solve the equation $\dfrac{\mathbf{dy}}{\mathbf{dx}} = \dfrac{\mathbf{ax+by+c}}{\mathbf{a'x+b'y+c'}}$**, when** $\dfrac{\mathbf{a}}{\mathbf{a'}} \neq \dfrac{\mathbf{b}}{\mathbf{b'}}$**.**

The equation is $\dfrac{dy}{dx} = \dfrac{ax+by+c}{a'x+b'y+c'}$ ...(1)

Put $x = x' + h, y = y' + k$, where h, k are constants.

Then $dx = dx', dy = dy'$.

$\therefore$ (1) becomes $\dfrac{dy'}{dx'} = \dfrac{a(x'+h)+b(y'+k)+c}{a'(x'+h)+b'(y'+k)+c'}$

$$= \frac{ax'+by'+(ah+bk+c)}{a'x'+b'y'+(a'h+b'k+c')} \qquad ...(2)$$

Put $$ah + bk + c = 0,$$
$$a'h + b'k + c' = 0,$$

so that by cross-multiplication,

$$\frac{h}{bc' - b'c} = \frac{k}{ca' - c'a} = \frac{1}{ab' - a'b}$$

or $$h = \frac{bc' - b'c}{ab' - a'b}, \; k = \frac{ca' - c'a}{ab' - a'b} \qquad ...(3)$$

Then (2) becomes $\dfrac{dy'}{dx'} = \dfrac{ax' + by'}{a'x' + b'y'}$, which is homogeneous in x' and y', and can be solved (Art. 110, (b)). In the solution, putting $x' = x - h$, $y' = y - k$ (where h, k have the values found above), we get the required solution.

Note. *The above method fails if in (3), the denominator in the values of h and k, = 0,*

i.e., if $ab' - a'b = 0$, *or if* $\dfrac{a}{a'} = \dfrac{b}{b'}$.

111 (b). Case III (b).

To solve the equation $\dfrac{\mathbf{dy}}{\mathbf{dx}} = \dfrac{\mathbf{ax + by + c}}{\mathbf{a'x + b'y + c'}}$**, when** $\dfrac{\mathbf{a}}{\mathbf{a'}} = \dfrac{\mathbf{b}}{\mathbf{b'}}$**.**

The equation is $$\frac{dy}{dx} = \frac{ax + by + c}{a'x + b'y + c'} \qquad ...(1)$$

Now $$\frac{a}{a'} = \frac{b}{b'} = \frac{1}{m} \text{ (say)}$$

$\therefore$ $$a' = am, \; b' = bm.$$

$\therefore$ (1) becomes $$\frac{dy}{dx} = \frac{ax + by + c}{amx + bmy + c'} = \frac{ax + by + c}{m(ax + by) + c'} \qquad ...(2)$$

$$\left[\text{Form } \frac{dy}{dx} = f(ax + by)\right]$$

Put $ax + by = z$, [Note after Ex. 4, Art. 109]

then $$a + b\frac{dy}{dx} = \frac{dz}{dx},$$

or $$\frac{dy}{dx} = \frac{\dfrac{dz}{dx} - a}{b}.$$

$\therefore$ (2) becomes $$\frac{\dfrac{dz}{dx} - a}{b} = \frac{z + c}{mz + c'} \quad \text{or} \quad \frac{dz}{dx} - a = \frac{b(z + c)}{mz + c'}.$$

Transposing, $$\frac{dz}{dx} = \frac{b(z + c)}{mz + c'} + a = \frac{b(z + c) + a(mz + c')}{mz + c'}$$
$$= \frac{z(b + am) + bc + ac'}{mz + c'}.$$

Separating the variables, $\dfrac{mz + c'}{z(b + am) + bc + ac'} dz = dx$, which can be solved. In the solution, putting $z = ax + by$, we get the required solution.

[I. Rule to solve the equation $\dfrac{dy}{dx} = \dfrac{ax + by + c}{a'x + b'y + c'}$,

when $\dfrac{a}{a'} \neq \dfrac{b}{b'}$ **:**

(*i*) Put $x = x' + h$, $y = y' + k$.

(*ii*) *Put the constant terms (i.e., terms not involving x′ and y′) in the numerator and denominator of R.H.S. equal to zero, and find the values of h, k.*

(*iii*) *The equation is now homogeneous in x′ and y′.*

II. Rule to solve the equation $\dfrac{dy}{dx} = \dfrac{ax + by + c}{a'x + b'y + c'}$,

when $\dfrac{a}{a'} = \dfrac{b}{b'}$ **:**

Put $ax + by = z$, and separate the variables z and x.]

EXAMPLES

1. Solve $\dfrac{dy}{dx} = \dfrac{x + 2y - 3}{2x + y - 3}$.

The equation is $\dfrac{dy}{dx} = \dfrac{x + 2y - 3}{2x + y - 3}$ ...(1)

[Compare with $\dfrac{dy}{dx} = \dfrac{ax + by + c}{a'x + b'y + c'}$.

Here $\dfrac{a}{a'} = \dfrac{1}{2}$, $\dfrac{b}{b'} = \dfrac{2}{1}$, $\therefore \quad \dfrac{a}{a'} \neq \dfrac{b}{b'}$.]

Put $x = x' + h$, $y = y' + k$, [Rule I, Art. 111, (b), end]

then $dx = dx'$, $dy = dy'$.

$\therefore$ (1) becomes $\dfrac{dy'}{dx'} = \dfrac{x' + h + 2(y' + k) - 3}{2(x' + h) + y' + k - 3} = \dfrac{x' + 2y' + (h + 2k - 3)}{2x' + y' + (2h + k - 3)}$ ...(2)

Put $h + 2k - 3 = 0$,

$2h + k - 3 = 0$,

so that by cross-multiplication,

$$\frac{h}{-6-(-3)} = \frac{k}{-6-(-3)} = \frac{1}{1-4},$$

or $$h = \frac{-3}{-3} = 1, \; k = \frac{-3}{-3} = 1.$$

Then (2) becomes $\dfrac{dy'}{dx'} = \dfrac{x' + 2y'}{2x' + y'}$ ...(3)

which is homogeneous in x' and y'.

Put $y' = vx'$, then $\dfrac{dy'}{dx'} = \dfrac{dv}{dx'} \cdot x' + v \cdot 1 = x' \dfrac{dv}{dx'} + v.$

$\therefore$ (3) becomes $x' \dfrac{dv}{dx'} + v = \dfrac{x' + 2vx'}{2x' + vx'} = \dfrac{1+2v}{2+v}$.

Transposing, $\quad x' \dfrac{dv}{dx'} = \dfrac{1+2v}{2+v} - v = \dfrac{1-v^2}{2+v}$.

Separating the variables, $\dfrac{2+v}{1-v^2} dv = \dfrac{1}{x'} dx'$.

Integrating both sides, $\displaystyle\int \frac{2+v}{1-v^2} dv = \int \frac{1}{x'} dx' + c$ ...(4)

Now $\dfrac{2+v}{1-v^2}$ [Resolve into partial fractions (*Note this step*)]

$$= \frac{2+v}{(1+v)(1-v)} = \frac{1}{(1+v)(2)} + \frac{3}{(2)(1-v)}$$

$$\therefore \quad \int \frac{2+v}{1-v^2} dv = \frac{1}{2}\int \frac{1}{1+v} dv + \frac{3}{2}\int \frac{1}{1-v} dv$$

$$= \frac{1}{2}\log(1+v) + \frac{3}{2}\,\frac{\log(1-v)}{-1} = \frac{1}{2}\log(1+v) - \frac{3}{2}\log(1-v)$$

$\therefore$ from (4), $\frac{1}{2}\log(1+v) - \frac{3}{2}\log(1-v) = \log x' + c.$

Clearing of fractions by multiplying by 2,

$$\log(1+v) - 3\log(1-v) = 2\log x' + 2c$$

or $$2\log x' - \log(1+v) + 3\log(1-v) = -2c$$

or $$\log\frac{x'^2(1-v)^3}{1+v} = -2c, \quad \text{or} \quad \frac{x'^2(1-v)^3}{1+v} = e^{-2c} = c_1 \qquad \left[\text{Put } v = \frac{y'}{x'}\right]$$

Changing c_1 to c, $\dfrac{x'^2\left(1 - \dfrac{y'}{x'}\right)^3}{1+\dfrac{y'}{x'}} = c, \quad$ or $\quad \dfrac{(x'-y')^3}{x'+y'} = c,$

or $$(x'-y')^3 = c(x'+y') \qquad [\text{Put } x' = x - h = x - 1,\ y' = y - k = y - 1]$$

or $$[x - 1 - (y - 1)]^3 = c(x - 1 + y - 1), \text{ or } (x-y)^3 = c(x + y - 2).$$

Solve the following equations (Exs. 2—3) :

2. $(x - y)\, dy = (x + y + 1)\, dx$.

3. (*i*) $(3y + 2x + 4)\, dx - (4x + 6y + 5)\, dy = 0$.

(*ii*) $\dfrac{dy}{dx} = \dfrac{x+y+1}{2x+2y+3}$.

[(*i*) The equation is $(3y + 2x + 4)\, dx - (4x + 6y + 5)\, dy = 0$

or $$\frac{dy}{dx} = \frac{2x+3y+4}{4x+6y+5}.$$

$\left[\text{Compare with } \frac{dy}{dx} = \frac{ax+by+c}{a'x+b'y+c'}.\right.$

Here $\quad \frac{a}{a'} = \frac{2}{4} = \frac{1}{2}, \frac{b}{b'} = \frac{3}{6} = \frac{1}{2}, \quad \therefore \quad \left.\frac{a}{a'} = \frac{b}{b'}.\right]$

Put $\quad 2x + 3y = z,$ $\quad$ [$ax + by = z$ (Rule II, Art. 111, (*b*) end)]

and separate the variables z and x.]

112. Case IV. Linear equation. (*a*) **Def.** *A* **linear* differential equation** *is that in which the dependent variable and its differential coefficients occur only in the first degree and are not multiplied together.*

Thus the **standard form of a linear equation of the first order is**

$$\frac{\mathbf{dy}}{\mathbf{dx}} + \mathbf{Py} = \mathbf{Q}, \text{ where P, Q are functions of } x.$$

(*b*) **To solve the equation $\frac{dy}{dx} + Py = Q$†, where P, Q are functions of x.**

The equation is $\frac{dy}{dx} + Py = Q.$

Multiplying throughout by $e^{\int P\,dx}$,

$$\frac{dy}{dx} e^{\int P\,dx} + Pye^{\int P\,dx} = Q\,e^{\int P\,dx}$$

$$\left[\text{But } \frac{d}{dx}(y\,e^{\int P\,dx}) = \frac{dy}{dx} e^{\int P\,dx} + y\,e^{\int P\,dx} \cdot \frac{d}{dx}(\textstyle\int P dx)\right.$$
$$\left. = \frac{dy}{dx} e^{\int P\,dx} + y\,e^{\int P\,dx} \cdot P\right]$$

or $\quad \frac{d}{dx}(y\,e^{\int P\,dx}) = Q\,e^{\int P\,dx}.$

Integrating both sides w.r.t. x,

$$\boxed{y\,e^{\int P\,dx} = \int Q\,e^{\int P\,dx}\,dx + c.}$$

N.B. The solution of the linear equation, $\frac{dy}{dx} + Py = Q$, where P, Q are functions of x, is

$$y\,e^{\int P\,dx} = \int Q\,e^{\int P\,dx}\,dx + c.$$

---

*So called because it is *of the first degree* in the dependent variable and its differential coefficients.

†Called **Leibnitz's equation** after Leibnitz (1646—1716), a self-taught mathematician, who invented the Differential Calculus independently of Newton.

Note. Important. *While evaluating $e^{\int P\,dx}$, it is useful to remember that*

$$\boxed{e^{\log x} = x.}$$ (From Trigonometry)]

Example. *Solve* $x\dfrac{dy}{dx} - ay = x + 1.$

The equation is $\quad x\dfrac{dy}{dx} - ay = x + 1.$

Dividing throughout by x, $\left[\text{To make coeff. of } \dfrac{dy}{dx} = 1 \textit{ (Note this step)}\right]$

$$\frac{dy}{dx} - \frac{a}{x}\,y = \frac{x+1}{x}. \qquad \left[\text{Compare with } \frac{dy}{dx} + Py = Q\right]$$

Here $\qquad P = -\dfrac{a}{x},\ Q = \dfrac{x+1}{x}.$

$$\therefore \quad \int P\,dx = \int -\frac{a}{x}\,dx = -\,a\log x$$

$$\therefore \quad e^{\int P\,dx} = e^{-a\log x} = e^{\log x^{-a}} \qquad [\because \log m^n = n\log m]$$

$$= x^{-a} \textit{ (Note this step)} \qquad [\because e^{\log x} = x \text{ (Note, Art. 112, (b))}]$$

$$\therefore \quad y\,x^{-a} = \int \frac{x+1}{x}\,x^{-a}\,dx + c \qquad ...(1)$$

$$\left[y\,e^{\int P\,dx} = \int Q\,e^{\int P\,dx}\,dx + c \text{ (Art. 112, (b))}\right]$$

Now $\quad \displaystyle\int \frac{x+1}{x}\,x^{-a}\,dx = \int (x+1)\,x^{-a-1}\,dx$

$$= \int (x^{-a} + x^{-a-1})\,dx$$

$$= \frac{x^{-a+1}}{-a+1} + \frac{x^{-a}}{-a} = \frac{x^{1-a}}{1-a} - \frac{x^{-a}}{a}.$$

$\therefore$ from (1), $\quad y\,x^{-a} = \dfrac{x^{1-a}}{1-a} - \dfrac{x^{-a}}{a} + c.$

Dividing throughout by x^{-a},

$$y = \frac{x}{1-a} - \frac{1}{a} + c\,x^{a}.$$

Note 1. *The above "method of formula" has been advocated by eminent mathematicians like A.R. Forsyth.*

Note 2. Important. *In the standard form of a linear equation of the first order, viz.,* $\dfrac{dy}{dx} + Py = Q$, *the coefficient of* $\dfrac{dy}{dx} = 1$. *If, however, the coefficient of* $\dfrac{dy}{dx}$ *is not equal to 1, we must first divide* throughout by this coefficient (here x), reduce the equation to the form $\frac{dy}{dx}$ + Py = Q, and then use the method of formula.

EXAMPLES

Solve the following differential equations (Exs. 1—6) :

1. (i) $\frac{dy}{dx}+\frac{n}{x}y=\frac{a}{x^n}$.

(ii) $\frac{dy}{dx}+\frac{4x}{x^2+1}y=\frac{1}{(x^2+1)^3}$.

2. (i) $(x^2+1)\frac{dy}{dx}+2xy=4x^2$.

(ii) $(x+1)\frac{dy}{dx}-ny=e^x(x+1)^{n+1}$.

3. (i) $\sqrt{a^2+x^2}\frac{dy}{dx}+y=\sqrt{a^2+x^2}-x$.

(ii) $\left(\frac{e^{-2\sqrt{x}}}{\sqrt{x}}-\frac{y}{\sqrt{x}}\right)\frac{dx}{dy}=1$.

[(i) $\int\frac{1}{\sqrt{a^2+x^2}}dx=\sinh^{-1}\frac{x}{a}=\log\left[\frac{x}{a}+\sqrt{1+\frac{x^2}{a^2}}\right]$

[$\because$ $\sinh^{-1}x=\log(x+\sqrt{1+x^2})$ (Art. 8, (d), (i))]

$$=\log.\frac{x+\sqrt{a^2+x^2}}{a}.$$

Note 1. *While evaluating* $\int$ P dx, *if the answer turns out to be in the inverse hyperbolic form, it is generally useful to express it in the logarithmic form.*

Note 2. *If in an equation, only y and dy, and functions of x, and dx occur, write the equation in* $\frac{dy}{dx}$. *The equation is now linear in y.*]

4. (i) $\frac{dy}{dx}+\frac{1-2x}{x^2}y=1$.

(ii) $\cos^2 x\frac{dy}{dx}+y=\tan x$.

[(i) It will be found that $e^{\int P\,dx}=e^{-\frac{1}{x}-2\log x}=e^{-\frac{1}{x}}.e^{-2\log x}=e^{-\frac{1}{x}}.e^{\log x^{-2}}=e^{-\frac{1}{x}}.x^{-2}$.

$$\therefore \qquad y\,e^{-\frac{1}{x}}x^{-2}=\int e^{-\frac{1}{x}}x^{-2}\,dx+c \qquad ...(1)$$

Now to integrate $\int e^{-\frac{1}{x}}x^{-2}\,dx$, put $-\frac{1}{x}=z$.

Note. Important. *To evaluate the integral on the R.H.S. of (1), it is generally useful to put the function of x in the index of e* $\left(\text{here } -\frac{1}{x}\right) = z$.]

5. $x(1-x^2)\frac{dy}{dx} + (2x^2-1)\,y = ax^3$.

[**Note. Important.** *If in an equation, only x and dx, and functions of y, and dy occur, write the equation in* $\frac{dx}{dy}$. *The equation is now linear in x.*]

6. $(1+y^2)\,dx = (\tan^{-1} y - x)\,dy$.

[The equation is $(1+y^2)\,dx = (\tan^{-1} y - x)\,dy$.

Writing the equation in $\frac{dx}{dy}$,*

$$(1+y^2)\frac{dx}{dy} = \tan^{-1} y - x$$

or
$$(1+y^2)\frac{dx}{dy} + x = \tan^{-1} y.$$

Dividing throughout by $1+y^2$,

$$\frac{dx}{dy} + \frac{1}{1+y^2}\,x = \frac{1}{1+y^2}\tan^{-1} y,$$

which is linear in x.]

7. Solve $\frac{dy}{dx} + 2y\tan x = \sin x$, given that $y = 0$, when $x = \frac{\pi}{3}$.

[It will be found that $y\sec^2 x = \sec x + c$ ...(1)

When $x = \frac{\pi}{3}$, $y = 0$ (Given),

$\therefore$ substituting in (1),

$0 = 2 + c$, or $c = -2$.

Substituting this value of c in (1), $y\sec^2 x = \sec x - 2$.]

113. Case V. Equation reducible to linear. To solve the equation $\frac{dy}{dx} + Py = Qy^n$,† where P, Q are functions of x.

The equation is $\frac{dy}{dx} + Py = Qy^n$.

***Why this step.** $\because$ in the equation, only x and dx, and functions of y, and dy occur, $\therefore$ we write the equation in $\frac{dx}{dy}$.

†Called **Bernoulli's equation** after James Bernoulli (1651—1705) who first studied it. He was a fast friend of Leibnitz.

Dividing throughout by y^n, *[To make R.H.S. a function of x only]*

$$y^{-n}\frac{dy}{dx} + Py^{1-n} = Q \quad ...(1)$$

Put $y^{1-n} = z$, then $(1-n)y^{-n}\frac{dy}{dx} = \frac{dz}{dx}$, or $y^{-n}\frac{dy}{dx} = \frac{1}{1-n}\frac{dz}{dx}$.

$\therefore$ (1) becomes $\frac{1}{1-n}\frac{dz}{dx} + Pz = Q$.

Multiplying throughout by $(1-n)$, $\left[\text{To make coeff. of } \frac{dz}{dx} = 1\right]$

$$\frac{dz}{dx} + (1-n)Pz = (1-n)Q,$$

which is linear in z and can be solved (Art. 115, (*b*)).

In the solution, putting $z = y^{1-n}$, we get the required solution.

[**Rule to solve the equation $\frac{dy}{dx} + Py = Qy^n$, where P, Q are functions of x :**

(*i*) *Divide throughout by* y^n.

(*ii*) *Put* $y^{1-n} = z$.

(*iii*) *The equation is now linear in z.*]

Example. *Solve* $\frac{dy}{dx} + \frac{1}{x}y = x^2y^6$.

The equation is $\frac{dy}{dx} + \frac{1}{x}y = x^2y^6$. $\left[\text{Form } \frac{dy}{dx} + Py = Qy^n\right]$

Dividing throughout by y^6, [To make R.H.S. a function of x only]

$$y^{-6}\frac{dy}{dx} + \frac{1}{x}y^{-5} = x^2 \quad ...(1)$$

Put $y^{-5} = z$, then $-5y^{-6}\frac{dy}{dx} = \frac{dz}{dx}$, or $y^{-6}\frac{dy}{dx} = -\frac{1}{5}\frac{dz}{dx}$.

$\therefore$ (1) becomes $-\frac{1}{5}\frac{dz}{dx} + \frac{1}{x}z = x^2$.

Multiplying throughout by -5, $\left[\text{To make coeff. of } \frac{dz}{dx} = 1\right]$

$$\frac{dz}{dx} - \frac{5}{x}z = -5x^2 \quad ...(2), \text{ which is linear in } z.$$

Here $P = -\frac{5}{x}$, $Q = -5x^2$.

$\therefore$ $\int P\,dx = \int -\frac{5}{x}\,dx = -5\log x$

$\therefore$ $e^{\int P\,dx} = e^{-5\log x} = e^{\log x^{-5}} = x^{-5}$.

$\therefore$ from (2), $zx^{-5} = \int -5x^2(x^{-5})\,dx + c \quad ...(3)$

Now $\int -5x^2(x^{-5})\,dx = -5\int x^{-3}\,dx = -5\frac{x^{-2}}{-2} = \frac{5}{2}x^{-2}$.

$\therefore$ from (3), $zx^{-5} = \frac{5}{2} x^{-2} + c.$

Dividing throughout by x^{-5},

$$z = \frac{5}{2} x^3 + cx^5 \qquad \text{[Put } z = y^{-5}\text{]}$$

or $$y^{-5} = \frac{5}{2} x^3 + cx^5.$$

EXAMPLES

Solve the following differential equations (Exs. 1—5) :

1. (*i*) $\frac{dy}{dx} + \frac{xy}{1-x^2} = xy^{\frac{1}{2}}.$

(*ii*) $\frac{dy}{dx} + y \cos x = y^n \sin 2x.$

2. (*i*) $\frac{dy}{dx} = x^3y^3 - xy.$

(*ii*) $2 \frac{dy}{dx} = \frac{y}{x} + \frac{y^2}{x^2}.$

[(*i*) *Transposing the term in y to L.H.S.,*

$$\left(\text{To get the form } \frac{dy}{dx} + \text{P}y = \text{Q}y^n\right) \frac{dy}{dx} + xy = x^3y^3.\Big]$$

3. $(x^2y^3 + xy)\, dy = dx.$

[*Writing the equation in* $\frac{dx}{dy}$, $\frac{dx}{dy} = x^2y^3 + xy$,

Transposing the term in x to L.H.S., $\frac{dx}{dy} - xy = x^2y^3.$

Divide throughout by x^2, and put $x^{-1} = z$.]

[**Note. General equation reducible to linear.**

To solve the equation f ′(y) $\frac{dy}{dx}$ + Pf(y) = Q, where P, Q are functions of x.

The equation is $f'(y) \frac{dy}{dx} + \text{P}f(y) = \text{Q}$ ...(1)

Put $f(y) = z$, then $f'(y) \frac{dy}{dx} = \frac{dz}{dx}$.

$\therefore$ (1) becomes $\frac{dz}{dx} + \text{P}z = \text{Q}$,

which is linear in z and can be be solved (Art. 112, (*b*)).

In the solution, putting $z = f(y)$, we get the required solution.]

4. (*i*) $3 \frac{dy}{dx} + \frac{2}{x+1} y = \frac{x^3}{y^2}.$

(*ii*) $y\, dy + by^2\, dx = a \cos x\, dx.$

**5. (i) $3x(1-x^2)y^2\dfrac{dy}{dx}+(2x^2-1)y^3=ax^3$.

(ii) $(x+1)\dfrac{dy}{dx}+1=2e^{-y}$.

6. Integrate the equation $\dfrac{dy}{dx}+x\sin 2y=x^3\cos^2 y$.

[Dividing throughout by $\cos^2 y$, [*To make R.H.S. a function of x only*]

$$\sec^2 y\frac{dy}{dx}+2x\tan y=x^3. \qquad \left[\because \frac{\sin 2y}{\cos^2 y}=\frac{2\sin y\cos y}{\cos^2 y}=2\tan y\right]$$

Put $\tan y=z$. The equation is now linear in z.]

MISCELLANEOUS EXAMPLES ON CHAPTER XII

Solve the following differential equations (Exs. 1—10) :

1. (i) $(1-x^2)(1-y)\,dx=xy(1+y)\,dy$.

(ii) $\dfrac{dy}{dx}=\dfrac{x(2\log x+1)}{\sin y+y\cos y}$.

2. $\cos(x+y)\,dy=dx$.

3. (i) $x+y\dfrac{dy}{dx}=2y$.

(ii) $\dfrac{dy}{dx}+\dfrac{x^2+3y^2}{3x^2+y^2}=0$.

4. (i) $(x^2+2xy-y^2)\,dx+(y^2+2xy-x^2)\,dy=0$.

(ii) $(x^3-3xy^2)\,dx=(y^3-3x^2y)\,dy$.

5. (i) $x\dfrac{dy}{dx}-y=x\sqrt{x^2+y^2}$.

(ii) $(2x+4y+3)\,y'=(2y+x+1)$.

[(i) This equation though not homogeneous in x and y, can be solved by the same method as for homogeneous equation.

(ii) **Note.** *y' stands for* $\dfrac{dy}{dx}$.]

6. (i) $\dfrac{dy}{dx}+\dfrac{y}{(1-x)\sqrt{x}}=1-\sqrt{x}$.

(ii) $(1-x^2)\dfrac{dy}{dx}+2xy=x(1-x^2)^{\frac{1}{2}}$.

7. (i) $x(x-1)\dfrac{dy}{dx}-(x-2)y=x^3(2x-1)$.

(ii) $\dfrac{dy}{dx}+\dfrac{2}{x}y=\sin x$.

8. (*i*) $\dfrac{dy}{dx} + y \tan x - \sec x = 0.$

(*ii*) $\dfrac{dy}{dx} + y \sec x = \tan x.$

[(*ii*) $\int \sec x\, dx = \log \tan\left(\dfrac{\pi}{4}+\dfrac{x}{2}\right) = \log \dfrac{1+\tan\dfrac{x}{2}}{1-\tan\dfrac{x}{2}} = \log \dfrac{\cos\dfrac{x}{2}+\sin\dfrac{x}{2}}{\cos\dfrac{x}{2}-\sin\dfrac{x}{2}}$

$\left\{\text{Multiply the num. and denom. by } \left(\cos\dfrac{x}{2}+\sin\dfrac{x}{2}\right)\right\}$

$$= \log \frac{\left(\cos\dfrac{x}{2}+\sin\dfrac{x}{2}\right)^2}{\left(\cos\dfrac{x}{2}-\sin\dfrac{x}{2}\right)\left(\cos\dfrac{x}{2}+\sin\dfrac{x}{2}\right)}$$

$$= \log \frac{\cos^2\dfrac{x}{2}+\sin^2\dfrac{x}{2}+2\sin\dfrac{x}{2}\cos\dfrac{x}{2}}{\cos^2\dfrac{x}{2}-\sin^2\dfrac{x}{2}}$$

$$= \log \frac{1+\sin x}{\cos x} = \log(\sec x + \tan x).]$$

9. (*i*) $x\dfrac{dy}{dx} + 2y - x^2 \log x = 0.$

(*ii*) $x \log x . \dfrac{dy}{dx} + y = 2 \log x.$

10. (*i*) $(x + 2y^3)\dfrac{dy}{dx} = y.$

**(*ii*) $(1 + y^2) + (x - e^{-\tan^{-1} y})\dfrac{dy}{dx} = 0.$

11. (*i*) Solve $\dfrac{dy}{dx} + \dfrac{y}{x} = x^2$, given $y = 1$ when $x = 1$.

(*ii*) Integrate $(1 + x^2)\dfrac{dy}{dx} + 2yx - 4x^2 = 0$, and obtain the cubic curve satisfying this equation and passing through the origin.

Solve the following differential equations (Exs. 12—15) :

12. (*i*) $(1 - x^2)\dfrac{dy}{dx} + xy = xy^2.$

(*ii*) $\dfrac{dy}{dx} = 2y \tan x + y^2 \tan^2 x.$

13. (*i*) $xy - \dfrac{dy}{dx} = y^3 e^{-x^2}.$

(*ii*) $\cos x\, dy = y (\sin x - y)\, dx.$

14. (*i*) $x^2 y \dfrac{dy}{dx} = xy^2 - e^{-\frac{1}{x^3}}.$

(*ii*) $(x - y^2)\, dx + 2xy\, dy = 0.$

15. (*i*) $\dfrac{dy}{dx} - \dfrac{\tan y}{1+x} = (1 + x)\, e^x \sec y.$

(*ii*) $\dfrac{dy}{dx} = e^{x-y} (e^x - e^y).$

CHAPTER XIII

LINEAR EQUATIONS WITH CONSTANT COEFFICIENTS

SECTION I

LINEAR EQUATIONS WITH CONSTANT COEFFICIENTS AND RIGHT HAND SIDE ZERO

114. Linear equation. Def. A **linear differential equation** *is that in which the dependent variable and its differential coefficients occur only in the first degree and are not multiplied together.*

Thus the standard form of a linear equation of the nth order is

$$\frac{d^n y}{dx^n} + P_1 \frac{d^{n-1}y}{dx^{n-1}} + P_2 \frac{d^{n-2}y}{dx^{n-2}} + \ldots + P_n y = Q,$$

where $P_1, P_2, \ldots, P_n$ and Q are functions of x.

Linear equation with constant coefficients. Def. A **linear equation with constant coefficients** is that which can be put in the form

$$\frac{d^n y}{dx^n} + P_1 \frac{d^{n-1}y}{dx^{n-1}} + P_2 \frac{d^{n-2}y}{dx^{n-2}} + \ldots + P_n y = Q,$$

where $P_1, P_2, \ldots, P_n$ are constants, and Q is a function of x.

In this book we shall deal with linear equations with constant coefficients only.

[☞*If so desired, the following Arts. 115—117 may be omitted and the Rule of Art. 118 may be remembered without proof.*]

****115.Theorem.** *If $y = y_1, y = y_2, \ldots, y = y_n$ are n linearly independent* particular solutions of the equation*

$$\frac{d^n y}{dx^n} + P_1 \frac{d^{n-1}y}{dx^{n-1}} + P_2 \frac{d^{n-2}y}{dx^{n-2}} + \ldots + P_n y = 0,$$

where $P_1, P_2, \ldots, P_n$ are constants, then the complete solution is

$$y = c_1 y_1 + c_2 y_2 + \ldots + c_n y_n.$$

[Proof.** ∵ $y = y_1, y = y_2, \ldots, y = y_n$ are solutions of

$$\frac{d^n y}{dx^n} + P_1 \frac{d^{n-1}y}{dx^{n-1}} + P_2 \frac{d^{n-2}y}{dx^{n-2}} + \ldots + P_n y = 0 \qquad \ldots(1)$$

$$\therefore \quad \frac{d^n y_1}{dx^n} + P_1 \frac{d^{n-1}y_1}{dx^{n-1}} + P_2 \frac{d^{n-2}y_1}{dx^{n-2}} + \ldots + P_n y_1 = 0 \qquad \ldots(2)$$

**i.e.*, we cannot find constants $a_1, a_2, \ldots, a_n$ such that

$a_1y_1 + a_2y_2 + \ldots + a_ny_n = 0$ identically, *i.e.*, for *all* values of x.

Thus $x, x^2, 2x - 3x^2$ are not linearly independent.

For $2(x) - 3(x^2) - 1\,(2x - 3x^2) = 0$ identically.

Here $a_1 = 2, a_2 = -3, a_3 = -1$.

$$\frac{d^n y_2}{dx^n} + P_1 \frac{d^{n-1} y_2}{dx^{n-1}} + P_2 \frac{d^{n-2} y_2}{dx^{n-2}} + \ldots + P_n y_2 = 0 \qquad \ldots(3)$$

$$\ldots \quad \ldots \quad \ldots \quad \ldots \quad \ldots$$

$$\frac{d^n y_n}{dx^n} + P_1 \frac{d^{n-1} y_n}{dx^{n-1}} + P_2 \frac{d^{n-2} y_n}{dx^{n-2}} + \ldots + P_n y_n = 0.$$

Substituting $y = c_1 y_1 + c_2 y_2 + \ldots + c_n y_n$ in (1), we get

$$\frac{d^n}{dx^n}(c_1y_1 + c_2y_2 + \ldots + c_ny_n) + P_1 \frac{d^{n-1}}{dx^{n-1}}(c_1y_1 + c_2y_2 + \ldots + c_ny_n)$$

$$+ P_2 \frac{d^{n-2}}{dx^{n-2}}(c_1y_1 + c_2y_2 + \ldots + c_ny_n) + \ldots + P_n(c_1y_1 + c_2y_2 + \ldots + c_ny_n) = 0$$

or
$$c_1\left(\frac{d^n y_1}{dx^n} + P_1 \frac{d^{n-1} y_1}{dx^{n-1}} + P_2 \frac{d^{n-2} y_1}{dx^{n-2}} + \ldots + P_n y_1\right)$$

$$+ c_2\left(\frac{d^n y_2}{dx^n} + P_1 \frac{d^{n-1} y_2}{dx^{n-1}} + P_2 \frac{d^{n-2} y_2}{dx^{n-2}} + \ldots + P_n y_2\right)$$

$$+ \quad \ldots \quad \ldots \quad \ldots \quad \ldots \quad \ldots$$

$$+ c_n\left(\frac{d^n y_n}{dx^n} + P_1 \frac{d^{n-1} y_n}{dx^{n-1}} + P_2 \frac{d^{n-2} y_n}{dx^{n-2}} + \ldots + P_n y_n\right) = 0$$

or $\quad c_1(0) + c_2(0) + \ldots + c_n(0) = 0$ [From (2), (3), ..., above]

or $\quad 0 = 0$, which is true,

$\therefore \quad y = c_1y_1 + c_2y_2 + \ldots + c_ny_n$ is a solution of (1) [Def. (Art. 108, (*a*))]

and $\because$ it contains n arbitrary constants, and (1) is of the nth order

$\therefore$ it is the complete solution of (1). [Def. (Art. 108, (*b*), (1))]

Note. *The theorem of Art. 115 is true even if $P_1, P_2, .., P_n$ instead of being constants) are functions of x. The same proof holds.*

****116. Two theorems concerning D. Theorem I. If in the equation**

$$\frac{d^n y}{dx^n} + P_1 \frac{d^{n-1} y}{dx^{n-1}} + P_2 \frac{d^{n-2} y}{dx^{n-2}} + \ldots + P_n y = 0,$$

where $P_1, P_2, \ldots, P_n$ are constants, or in the symbolic form

$$\left(\text{by putting } \frac{d}{dx} = D, \frac{d^2}{dx^2} = D^2, \ldots, \frac{d^n}{dx^n} = D^n\right),$$

$$(D^n + P_1 D^{n-1} + P_2 D^{n-2} + \ldots + P_n) y = 0,$$

the symbolic coefficient of y, viz., $D^n + P_1 D^{n-1} + P_2 D^{n-2} + \ldots + P_n$ is factorised as if D were an ordinary algebraic quantity, then the original differential equation is obtained if D is put = $\frac{d}{dx}$, in whatever order the factors are taken.

[Proof.** Let the equation, for brevity, be of the second order

$$\frac{d^2 y}{dx^2} - (\alpha + \beta)\frac{dy}{dx} + \alpha\beta y = 0 \qquad \ldots(1)$$

or, in the symbolic form $\left[\text{by putting } \frac{d}{dx} = D, \frac{d^2}{dx^2} = D^2\right]$,

$$[D^2 - (\alpha + \beta) D + \alpha\beta] y = 0.$$

Factorising the symbolic coefficient of y,

$$(D - \alpha)(D - \beta)y = 0 \qquad \text{...(2)}$$

Putting $D = \frac{d}{dx}$, it becomes $\left(\frac{d}{dx} - \alpha\right)\left(\frac{d}{dx} - \beta\right) y = 0$

or $$\left(\frac{d}{dx} - \alpha\right)\left(\frac{dy}{dx} - \beta y\right) = 0$$

or $$\frac{d}{dx}\left(\frac{dy}{dx}\right) - \frac{d}{dx}(\beta y) - \alpha\frac{dy}{dx} + \alpha\beta y = 0$$

or $$\frac{d^2y}{dx^2} - \beta\frac{dy}{dx} - \alpha\frac{dy}{dx} + \alpha\beta y = 0 \qquad \left[\because \frac{d}{dx}(\beta y) = \beta\frac{dy}{dx}\right]$$

or $$\frac{d^2y}{dx^2} - (\alpha + \beta)\frac{dy}{dx} + \alpha\beta y = 0,$$

which is the same as (1).

Similarly if the factors in (2) had been taken in the reverse order

i.e., as $$(D - \beta)(D - \alpha) y = 0,$$

and D had been put $= \frac{d}{dx}$, it would have become the same as (1).

Similarly the theorem holds for an equation of the nth order.

Note. Important. *In simplifying $(D - \alpha)(D - \beta)y$, we begin with the factor which is next to y (viz., $D - \beta$) and perform the operation indicated by it on y; then we take the next factor (viz., $D - \alpha$) and perform the operation indicated by it on the result already obtained and so on, i.e., the symbolic factors, when used as operators, are taken in order (←) from right to left.*

Abridged notation : f (D)y = Q. The equation

$$\frac{d^ny}{dx^n} + P_1\frac{d^{n-1}y}{dx^{n-1}} + P_2\frac{d^{n-2}y}{dx^{n-2}} + \dots + P_n y = Q,$$

(where $P_1, P_2, \dots, P_n$ are constants, and Q is a function of x)

or, in the symbolic form $\left(\text{by putting } \frac{d}{dx} = D, \frac{d^2}{dx^2} = D^2, \dots, \frac{d^n}{dx^n} = D^n\right)$

$$(D^n + P_1 D^{n-1} + P_2D^{n-2} + \dots + P_n) y = Q$$

is written shortly as $f(D) y = Q$,

where $$f(D) = D^n + P_1 D^{n-1} + P_2D^{n-2} + \dots + P_n,$$

i.e., a polynomial in D. [Def. (Art. 30)]

Theorem II. If $f_1(D)$ is a factor of $f(D)$, and $y = y_1$ is a solution of $f_1(D) y = 0$, then $y = y_1$ is also a solution of $f(D) y = 0$.

Proof. Let $f(D) = f_1(D) f_2(D)$.

$\because$ $y = y_1$ is a solution of $f_1(D) y = 0$,

$\therefore \qquad f_1(D)\, y_1 = 0.$

Operating on both sides by $f_2(D)$,

$$f_2(D)\, f_1(D)\, y_1 = f_2(D)\, 0 = 0,$$

or $\qquad f_1(D)\, f_2(D)\, y_1 = 0 \qquad$ [Theorem I, Art. 116]

i.e., $\qquad f(D)\, y_1 = 0,$

$\therefore \quad y = y_1$ is a solution of $\quad f(D)\, y = 0.$

117-I. Linear equation with constant coefficients and right hand side zero.

To solve the equation

$$\frac{d^n y}{dx^n} + P_1 \frac{d^{n-1}y}{dx^{n-1}} + P_2 \frac{d^{n-2}y}{dx^{n-2}} + \ldots + P_n\, y = 0,$$

where $P_1, P_2, \ldots, P_n$ *are constants.*

Case I. Auxiliary equation having real and different roots.

The equation is

$$\frac{d^n y}{dx^n} + P_1 \frac{d^{n-1}y}{dx^{n-1}} + P_2 \frac{d^{n-2}y}{dx^{n-2}} + \ldots + P_n\, y = 0 \qquad \ldots(1)$$

or, in the symbolic form $\left(\text{by putting } \frac{d}{dx} = D, \frac{d^2}{dx^2} = D^2, \ldots, \frac{d^n}{dx^n} = D^n\right)$,

$$(D^n + P_1 D^{n-1} + P_2 D^{n-2} + \ldots + P_n)\, y = 0 \qquad \ldots(2)$$

Let the equation $D^n + P_1 D^{n-1} + P_2 D^{n-2} + \ldots + P_n = 0$ (called the **auxiliary equation**) have real and different roots $m_1, m_2, \ldots, m_n$.

Then (2) is equivalent to

$$(D - m_1)(D - m_2) \ldots (D - m_n)\, y = 0$$

$\therefore$ the solutions of the equations

$$(D - m_1)\, y = 0, (D - m_2)\, y = 0, \ldots, (D - m_n) y = 0.$$

are also the solutions of (2), *i.e.*, of (1). [Theorem II, Art. 116]

(To find the complete solution of $(D - m_1) y = 0$.)

$(D - m_1) y = 0, \quad$ or $\quad \frac{dy}{dx} - m_1 y = 0$, which is linear in y.

$\left[\text{Compare with } \frac{dy}{dx} + Py = Q\right]$

Here $P = -m_1, Q = 0$

$$\therefore \qquad \int P\, dx = \int -m_1\, dx = -m_1 x$$

$$\therefore \qquad e^{\int P\, dx} = e^{-m_1 x}$$

$$\therefore \qquad y\, e^{-m_1 x} = \int 0.e^{-m_1 x}\, dx + c_1 \qquad \text{[Art. 112, (b)]}$$

$$= c_1$$

$\therefore$ the complete solution of $(D - m_1)\, y = 0$, is $y = c_1\, e^{m_1 x}$.

Similarly the complete solution of $(D - m_2)y = 0$, is $y = c_2e^{m_2x}$,

... ...

the complete solution of $(D - m_n) y = 0$, is $y = c_n e^{m_nx}$.

$\therefore$ the complete solution of (1) is

$$y = c_1e^{m_1x} + c_2e^{m_2x} + ... + c_ne^{m_nx}. \qquad \text{[As in Art. 108]}$$

Note. *If the auxiliary equation has two real and equal roots m_1, m_1, then the solution $y = c_1 e^{m_1x} + c_2e^{m_2x} + c_3e^{m_3x} + ... + c_ne^{m_nx}$ becomes*

$$y = c_1 e^{m_1x} + c_2e^{m_1x} + c_3e^{m_3x} + ... + c_ne^{m_nx} \qquad [\because \ m_2 = m_1]$$

$$= (c_1 + c_2) e^{m_1x} + c_3e^{m_3x} + ... + c_ne^{m_nx}$$

[But $c_1 + c_2$ = one arbitrary constant C_1 (say)]

$$= C_1 e^{m_1x} + c_3e^{m_3x} + ... + c_ne^{m_nx},$$

which contains $n - 1$ (and not n) arbitrary constants

$\therefore$ *it is not the complete solution of (1).* *[Def. (Art. 108, (b), (1))]*

In order to find the complete solution in this case we proceed in the following Art. 117-II.

117-II. Case II. Auxiliary equation having real and equal roots. If the auxiliary equation has two equal roots m_1, m_1, the corresponding part of the complete solution is the complete solution of

$$(D - m_1)^2y = 0$$

or (separating one factor $D - m_1$ from the left),

$$(D - m_1)(D - m_1) y = 0 \qquad ...(1)$$

Putting (the remaining factor, *i.e.*,) $(D - m_1) y = z$ ...(2)

(1) becomes $(D - m_1) z = 0$,

or $\dfrac{dz}{dx} - m_1z = 0$, which is linear in z.

Here $P = - m_1$, $Q = 0$

$$\therefore \quad \int P\,dx = \int - m_1\,dx = - m_1x$$

$$\therefore \quad e^{\int P\,dx} = e^{-m_1x}$$

$$\therefore \quad ze^{-m_1x} = \int 0.e^{-m_1x}\,dx + c_1 \qquad \text{[Art. 112, (b)]}$$

$$= c_1$$

or $z = c_1e^{m_1x}$.

Substituting this value of z in (2), $(D - m_1)y = c_1e^{m_1x}$

or $\dfrac{dy}{dx} - m_1 y = c_1e^{m_1x}$, which is linear in y.

$$\left[\text{Compare with } \frac{dy}{dx} + \text{P}y = \text{Q}\right]$$

Here $$\text{P} = -m_1, \text{Q} = c_1 e^{m_1 x}$$

$$\therefore \quad \int \text{P}\,dx = \int -m_1\,dx = -m_1 x$$

$$\therefore \quad e^{\int \text{P}\,dx} = e^{-m_1 x}$$

$$\therefore \quad ye^{-m_1 x} = \int c_1 e^{m_1 x} \cdot e^{-m_1 x}\,dx + c_2$$

$$= \int c_1\,dx + c_2 = c_1 x + c_2$$

or $$y = e^{m_1 x}(c_1 x + c_2).$$

Changing c_1 to c_2 and c_2 to c_1 for convenience, the corresponding part of the complete solution is

$$e^{m_1 x}(c_1 + c_2 x).$$

$\therefore$ the complete solution is

$$y = e^{m_1 x}(c_1 + c_2 x) + c_3 e^{m_3 x} + \ldots + c_n e^{m_n x}.$$

Similarly *if the auxiliary equation has three real and equal roots* m_1, m_1, m_1, *the corresponding part of the complete solution is*

$$e^{m_1 x}(c_1 + c_2 x + c_3 x^2).$$

$\therefore$ the complete solution is

$$y = e^{m_1 x}(c_1 + c_2 x + c_3 x^2) + c_4 e^{m_4 x} + \ldots + c_n e^{m_n x}.$$

And so on.

Note. *In the equation* $(D^n + P_1 D^{n-1} + P_2 D^{n-2} + \ldots + P_n)\,y = 0$, *is assumed that the coefficients* P_1, $P_2, \ldots, P_n$ *are all real.* $\therefore$ *if the auxiliary equation* $D^n + P_1 D^{n-1} + P_2 D^{n-2} + \ldots + P_n = 0$, *has in imaginary roots, these roots occur in pairs (From Algebra). In order to find the complete solution in this case in the real form, we proceed in the following Art. 117-III.*

117-III. Case III. Auxiliary equation having imaginary and different roots. If the auxiliary equation has one pair of imaginary and different roots $\alpha \pm i\beta$, the corresponding part of the complete solution is

$$c_1 e^{(\alpha + i\beta)x} + c_2 e^{(\alpha - i\beta)x} \qquad \text{[To put it in the real form]}$$

$$= e^{\alpha x}[c_1 e^{i\beta x} + c_2 e^{-i\beta x}]$$

[But $e^{i\theta} = \cos\theta + i\sin\theta$* (Euler's theorem)
$e^{-i\theta} = \cos\theta - i\sin\theta$]

$$= e^{\alpha x}\,[c_1(\cos\beta x + i\sin\beta x) + c_2(\cos\beta x - i\sin\beta x)]$$

$$= e^{\alpha x}\,[\cos\beta x\,.\,(c_1 + c_2) + \sin\beta x\,.\,(ic_1 - ic_2)]$$

[But $c_1 + c_2$ = one arbitrary constant C_1 (say),
$ic_1 - ic_2$ = one arbitrary constant C_2 (say)]

$$= e^{\alpha x}\,[C_1\cos\beta x + C_2\sin\beta x].$$

Changing C_1 to c_1 and C_2 to c_2, the complete solution is

$$y = e^{\alpha x}\,(c_1\cos\beta x + c_2\sin\beta x) + c_3e^{m_3x} + \ldots + c_ne^{m_nx}.$$

117-IV. Case IV. Auxiliary equation having imaginary and equal roots.

Similarly *if the auxiliary equation has two pairs of imaginary and equal roots* $\alpha \pm i\beta$, $\alpha \pm i\beta$, *the corresponding part of the complete solution is* $e^{\alpha x}$ [$\cos\beta x.\,(c_1 + c_2x) + \sin\beta x.\,(c_3 + c_4x)$].

$\therefore$ the complete solution is

$$y = e^{\alpha x}\,[\cos\beta x.\,(c_1 + c_2x) + \sin\beta x.\,(c_3 + c_4x)] + c_5e^{m_5x} + \ldots + c_ne^{m_nx}.$$

And so on.

118. Rule to solve the equation

$$\frac{d^ny}{dx^n} + P_1\frac{d^{n-1}y}{dx^{n-1}} + P_2\frac{d^{n-2}y}{dx^{n-2}} + \ldots + P_ny = 0,$$

where $P_1, P_2, \ldots, P_n$ are constants :

(*i*) *Write the equation in the symbolic form*

$$\left[\textit{by putting } \frac{d}{dx} = D,\ \frac{d^2}{dx^2} = D^2, \ldots, \frac{d^n}{dx^n} = D^n, \textit{i.e.,}\right]$$

$$(D^n + P_1D^{n-1} + P_2D^{n-2} + \ldots + P_n)\,y = 0.$$

(*ii*) *Write down the auxiliary equation (A.E.) [by equating to zero the symbolic coefficient of y, i.e.,]*

$$D^n + P_1\,D^{n-1} + P_2D^{n-2} + \ldots + P_n = 0,$$

***Proof.** $e^{i\theta} = 1 + i\theta + \frac{(i\theta)^2}{2!} + \frac{(i\theta)^3}{3!} + \frac{(i\theta)^4}{4!} + \ldots$

[But $i^2 = -1$, $i^3 = i.i^2 = -i$, $i^4 = (i^2)^2 = (-1)^2 = 1$, ...]

$$= 1 + i\theta - \frac{\theta^2}{2!} - i\frac{\theta^3}{3!} + \frac{\theta}{4!} + \ldots$$

$$= \left(1 - \frac{\theta^2}{2!} + \frac{\theta^4}{4!} - \ldots\right) + i\left(\theta - \frac{\theta^3}{3!} + \frac{\theta^5}{5!} - \ldots\right)$$

$$= \cos\theta + i\sin\theta.$$

Changing θ to $-\theta$,

$$e^{-i\theta} = \cos(-\theta) + i\sin(-\theta) = \mathbf{\cos\theta - i\sin\theta}.$$

and solve for D as if it were an ordinary algebraic quantity.

(*iii*) *From the roots of the A.E. write down the corresponding part of the complete solution (C.S.) as follows :*

Roots of A.E.	*Corresponding part of C.S.*
(*a*) (*i*) *One real and different root* m_1.	$c_1 e^{m_1 x}$.
(*ii*) *Two real and equal roots* $m_1\, m_1$.	$e^{m_1 x}\,(c_1 + c_2 x)$.
(*iii*) *Three real and equal roots* m_1, m_1, m_1.	$e^{m_1 x}\,(c_1 + c_2 x + c_3 x^2)$.
And so on.	
(*b*) (*i*) *One pair of imaginary and different roots* $\alpha \pm i\beta$.	$e^{\alpha x}\,(c_1 \cos \beta x + c_2 \sin \beta x)$.
(*ii*) *Two pairs of imaginary and equal roots* $\alpha \pm i\beta, \alpha \pm i\beta$.	$e^{\alpha x}\,[\cos \beta x . (c_1 + c_2 x) + \sin \beta x . (c_3 + c_4 x)]$.
And so on.	

Example 1. *Solve* $\dfrac{d^2y}{dx^2} + 3\dfrac{dy}{dx} - 40y = 0.$

(*i*) The equation is $\dfrac{d^2y}{dx^2} + 3\dfrac{dy}{dx} - 40y = 0$

or, in the symbolic form, $(D^2 + 3D - 40)y = 0.$

(*ii*) The A.E. is $D^2 + 3D - 40 = 0$*

or $(D + 8)(D - 5) = 0,$

$\therefore$ $D = -8, 5.$

(*iii*) $\therefore$ the C.S. is $y = c_1 e^{-8x} + c_2 e^{5x}.$ [$c_1 e^{m_1 x}$ (Rule (Art. 118))]

Example 2. *Solve* $\dfrac{d^2y}{dx^2} + 4\dfrac{dy}{dx} + 13y = 0.$

(*i*) The equation is $\dfrac{d^2y}{dx^2} + 4\dfrac{dy}{dx} + 13y = 0$

or, in the symbolic form, $(D^2 + 4D + 13)y = 0.$

(*ii*) The A.E. is $D^2 + 4D + 13 = 0$

$$\therefore \quad D = \frac{-4 \pm \sqrt{16 - 52}}{2} = \frac{-4 \pm i6}{2} = -2 \pm i3.$$

[Compare with $\alpha \pm i\beta$, here $\alpha = -2$, $\beta = 3$]

(*iii*) $\therefore$ the C.S. is $y = e^{-2x}\,(c_1 \cos 3x + c_2 \sin 3x).$

[$e^{\alpha x}\,(c_1 \cos \beta x + c_2 \sin \beta x)$ (Rule (Art. 118))]

***How to write this step.** Equate to zero the symbolic coefficient of *y*.

Example 3. *Solve* $\frac{d^3y}{dx^3} - 3\frac{d^2y}{dx^2} + 4y = 0.$

(*i*) The equation is $\frac{d^3y}{dx^3} - 3\frac{d^2y}{dx^2} + 4y = 0$

or, in the symbolic form, $(D^3 - 3D^2 + 4)\, y = 0.$

(*ii*) The A.E. is $D^3 - 3D^2 + 4 = 0$

or $(D + 1)(D - 2)^2 = 0$

$\therefore$ $D = -1\,;\, 2, 2$*.

$D^3 - 3D^2 + 4$
$= (D + 1)(D^2 - 4D + 4)$
$= (D + 1)(D - 2)^2$

(*iii*) $\therefore$ the C.S. is $y = c_1 e^{-1.x} + e^{2x}(c_2 + c_3x)$

$[c_1e^{m_1x}\,;\, e^{m_1x}(c_1 + c_2x)$ (Rule (Art. 118))]

$= c_1 e^{-x} + e^{2x}(c_2 + c_3x).$

EXAMPLES

1. Solve (*i*) $\frac{d^2y}{dx^2} - 7\frac{dy}{dx} + 12y = 0.$

(*ii*) $\frac{d^2y}{dx^2} - 6\frac{dy}{dx} + 9y = 0.$

2. (*i*) If $\frac{d^2y}{dx^2} - a^2y = 0$, show that

$$y = c_1 e^{ax} + c_2e^{-ax} = C_1 \cosh ax + C_2 \sinh ax.$$

(*ii*) Solve $\frac{d^2y}{dx^2} \pm \mu^2 y = 0.$

[(*i*) Remember : $\mathbf{e^{\theta} = \cosh \theta + \sinh \theta}$†, $\mathbf{e^{-\theta} = \cosh \theta - \sinh \theta}$.]

3. Solve (*i*) $\frac{d^3y}{dx^3} - 6\frac{d^2y}{dx^2} + 11\frac{dy}{dx} - 6y = 0.$

(*ii*) $\frac{d^3y}{dx^3} - 4\frac{d^2y}{dx^2} + 5\frac{dy}{dx} - 2y = 0.$

(*iii*) $\frac{d^3y}{dx^3} - 2\frac{d^2y}{dx^2} + 4\frac{dy}{dx} - 8y = 0.$

4. If $\frac{d^4y}{dx^4} - a^4y = 0$, show that

$$y = c_1 \cos ax + c_2 \sin ax + c_3 \cosh ax + c_4 \sinh ax.$$

***How to write this step.** $(D - 2)^2 = 0$, or $(D - 2)(D - 2) = 0$, $\therefore$ $D = 2, 2$.

† **Proof.** $\cosh \theta + \sinh \theta = \frac{e^{\theta} + e^{-\theta}}{2} + \frac{e^{\theta} - e^{-\theta}}{2} = e^{\theta}$

or $e^{\theta} = \cosh \theta + \sinh \theta.$

Changing θ to $-\theta$, $e^{-\theta} = \cosh(-\theta) + \sinh(-\theta) = \mathbf{\cosh \theta - \sinh \theta}$.

Solve the following equations (Exs. 5—6) :

5. (*i*) $\dfrac{d^4y}{dx^4} + 13\dfrac{d^2y}{dx^2} + 36y = 0.$

(*ii*) $(D^4 + 8D^2 + 16)y = 0.$

6. $\dfrac{d^4y}{dx^4} + 4y = 0.$

7. Solve $\dfrac{d^2x}{dt^2} - 3\dfrac{dx}{dt} + 2x = 0,$

given that when $t = 0$, $x = 0$ and $\dfrac{dx}{dt} = 0.$

[It will be found that the C.S. is $x = c_1 e^t + c_2 e^{2t}$ ...(1)

But when $t = 0, x = 0$ (Given), $\therefore$ substituting in (1),

$$0 = c_1 + c_2 \quad ...(2)$$

Again from (1), $\dfrac{dx}{dt} = c_1e^t + 2c_2e^{2t}$ ...(3)

But when $t = 0$, $\dfrac{dx}{dt} = 0$ (Given), $\therefore$ substituting in (3),

$$0 = c_1 + 2c_2 \quad ...(4)$$

Find the values of c_1, c_2 from (2), (4), and substitute them in (1).]

SECTION II

LINEAR EQUATIONS WITH CONSTANT COEFFICIENTS AND RIGHT HAND SIDE A FUNCTION OF X

119. Theorem. *If $y = Y$ is the complete solution of the equation*

$$\frac{d^ny}{dx^n} + P_1\frac{d^{n-1}y}{dx^{n-1}} + P_2\frac{d^{n-2}y}{dx^{n-2}} + ... + P_n\, y = 0$$

where $P_1, P_2, ..., P_n$ are constants,

and $y = u$ is a solution, not containing arbitrary constants, of the equation

$$\frac{d^ny}{dx^n} + P_1\frac{d^{n-1}y}{dx^{n-1}} + P_2\frac{d^{n-2}y}{dx^{n-2}} + ... + P_n\, y = Q$$

where Q is a function of x,

then the complete solution of (2) is $y = Y + u$.

[Proof.** $\because$ $y = Y$ is the C.S. of

$$\frac{d^ny}{dx^n} + P_1\frac{d^{n-1}y}{dx^{n-1}} + P_2\frac{d^{n-2}y}{dx^{n-2}} + ... + P_n\, y = 0 \quad ...(1)$$

$$\therefore \quad \frac{d^nY}{dx^n} + P_1\frac{d^{n-1}Y}{dx^{n-1}} + P_2\frac{d^{n-2}Y}{dx^{n-2}} + ... + P_n\, Y = 0.$$

Again $\because$ $y = u$ is a solution, not containing arbitrary constants, of

$$\frac{d^ny}{dx^n} + P_1\frac{d^{n-1}y}{dx^{n-1}} + P_2\frac{d^{n-2}y}{dx^{n-2}} + ... + P_n\, y = Q \quad ...(2)$$

$$\therefore \qquad \frac{d^n u}{dx^n} + P_1 \frac{d^{n-1} u}{dx^{n-1}} + P_2 \frac{d^{n-2} u}{dx^{n-2}} + \dots + P_n\, u = Q.$$

Substituting $y = Y + u$ in (2), we get

$$\frac{d^n}{dx^n}(Y + u) + P_1 \frac{d^{n-1}}{dx^{n-1}}(Y + u) + P_2 \frac{d^{n-2}}{dx^{n-2}}(Y + u) + \dots + P_n (Y + u) = Q$$

or
$$\frac{d^n Y}{dx^n} + P_1 \frac{d^{n-1} Y}{dx^{n-1}} + P_2 \frac{d^{n-2} Y}{dx^{n-2}} + \dots + P_n Y$$
$$+ \frac{d^n u}{dx^n} + P_1 \frac{d^{n-1} u}{dx^{n-1}} + P_2 \frac{d^{n-2} u}{dx^{n-2}} + \dots + P_n\, u = Q$$

or $\quad 0 + Q = Q,\quad$ which is true.

$\therefore \quad y = Y + u$ is a solution of (2)

and $\because$ it contains n arbitrary constants

[Since $y = Y$, being the C.S. of (1), contains n arbitrary constants] and (2) is of the nth order

$\therefore$ it is the C.S. of (2). [Def. (Art. 108, (*b*), (1))]

Complementary function and particular integral. Defs. The part Y is called the **complementary function (C.F.),** and the part u is called the **particular integral (P.I.).**

Cor. The C.S. of (2) is y = C.F. + P.I.

Note. The theorem of Art. 119 (like that of Art. 115) is true even if $P_1, P_2, \dots, P_n$ (instead of being constants) are functions of x.

120. Inverse operator $\frac{1}{f(D)}$. Def. $\frac{1}{f(D)}$ Q is that function of x, *containing no arbitrary constants,* which, when operated on by f(D), gives Q, *i.e.,*

$$f(D)\, \frac{1}{f(D)}\, Q = Q.$$

Thus the operator $\frac{1}{f(D)}$ is the inverse of the operator f(D). It can be shown from this definition and Art. 119 that $\frac{1}{f(D)}$ *can be resolved into factors which may be taken in any order, and that* $\frac{1}{f(D)}$ *can be resolved into partial fractions.*

Cor. 1. Particular integral. $\frac{1}{f(D)}$ *Q is the particular integral of the equation $f(D)\, y = Q$.*

[Proof.** The equation is $f(D)\, y = Q$ ...(1)

Substituting $y = \frac{1}{f(D)}$ Q in (1), we get

$$f(D)\,\frac{1}{f(D)}\,Q = Q, \quad \text{or} \quad Q = Q, \qquad \text{[Def. (Art. 120)]}$$

which is true.

$$\therefore \qquad y = \frac{1}{f(D)}\,Q \text{ is a solution of (1)}$$

and $\because$ it contains no arbitrary constants [Def. (Art. 120)]

$\therefore$ it is the particular integral [Def. (Art. 119, end)].

[**Aid to memory.** The particular integral (P.I.) of (1) is the value of y obtained from (1) *as if* $f(D)$ were an ordinary algebraic quantity.]

Cor. 2. $\frac{1}{D}\,Q = \int Q\,dx$, **no arbitrary constant being added.**

** [**Proof.** Let $\frac{1}{D}\,Q = y$.

Operating on both sides by D,

$$D\,\frac{1}{D}\,Q = Dy$$

or

$$Q = Dy, \text{ i.e., } \frac{dy}{dx} = Q.$$

Integrating both sides w.r.t. x,

$$y = \int Q\,dx, \text{ no arbitrary constant being added}$$

$$\left[\because\ y\left(= \frac{1}{D}\,Q\right) \text{contains no arbitrary constant (Def. (Art. 120))}\right]$$

i.e.,

$$\frac{1}{D}\,Q = \int Q\,dx, \text{ no arbitrary constant being added.}$$

****Cor. 3.** $\frac{1}{D-a}\,Q = e^{ax}\int Qe^{-ax}\,dx$, **no arbitrary constant being added.**

** [**Proof.** Let $\frac{1}{D-a}\,Q = y$.

Operating on both sides by $(D - a)$,

$$(D-a)\,\frac{1}{D-a}\,Q = (D-a)\,y$$

$$Q = (D-a)\,y, \quad \text{i.e.,} \quad \frac{dy}{dx} - ay = Q, \text{ which is linear in } y.$$

$$\left[\text{Compare with } \frac{dy}{dx} + Py = Q\right]$$

Here $P = -a,\ Q = Q \quad \therefore \quad \int P dx = \int -a\,dx = -ax$

$$\therefore \quad e^{\int P\,dx} = e^{-ax}. \quad \therefore \quad y\,e^{-ax} = \int Q\,e^{-ax}\,dx, \text{ no arbitrary constant being added}$$

$$\left[\because\ y\left(= \frac{1}{D-a}\,Q\right) \text{contains no arbitrary constant (Def. (Art. 120))}\right]$$

or $$y = e^{ax} \int Q\, e^{-ax}\, dx,$$

i.e., $$\frac{1}{D-a}\, Q = e^{ax} \int Q\, e^{-ax}\, dx, \quad \text{no arbitrary constant being added}$$

Five standard cases of particular integrals.

121. Case I. Rule to evaluate $\dfrac{1}{f(D)}\, e^{ax}$ **:** *Put* $D = a$. $[f(a) \neq 0]$

Case II. Rule to evaluate $\dfrac{1}{f(D^2)} \cos ax$ **:** *Put* $D^2 = -a^2$ $[f(-a^2) \neq 0]$

Case III. Rule to evaluate $\dfrac{1}{f(D^2)} \sin ax$ **:** *Put* $D^2 = -a^2$ $[f(-a^2) \neq 0]$

Note 1. Cases of failure. *If in any one of the above three cases by using the usual rule, we get zero in the denominator, the rule fails. Now we proceed by the rule of Art. 123 and find the particular integral.*

** [**Case I.** To prove that $\dfrac{1}{f(D)}\, e^{ax} = \dfrac{1}{f(a)}\, e^{ax}$.] $[f(a) \neq 0]$

Proof. $D^n e^{ax} = a^n e^{ax}$ [From Differential Calculus]

$\therefore$ $f(D)\, e^{ax} = f(a)\, e^{ax}$. [$\because$ $f(D)$ is a polynomial in D]

Operating on both sides by $\dfrac{1}{f(D)}$,

$$\frac{1}{f(D)} f(D)\, e^{ax} = \frac{1}{f(D)} f(a)\, e^{ax}$$

or $$e^{ax} = f(a) \frac{1}{f(D)}\, e^{ax}$$

$\left[\because \dfrac{1}{f(D)}, f(D) \text{ are inverse operators, and } f(a) \text{ is a constant}\right]$

Dividing both sides by $f(a)$ [which is $\neq 0$],

$$\frac{1}{f(a)}\, e^{ax} = \frac{1}{f(D)}\, e^{ax}, \quad i.e., \quad \frac{1}{f(D)}\, e^{ax} = \frac{1}{f(a)}\, e^{ax}. \qquad [f(a) \neq 0]$$

Case II. To prove that $\dfrac{1}{f(D^2)} \cos ax = \dfrac{1}{f(-a^2)} \cos ax$. $[f(-a^2) \neq 0]$

Proof.

$D \cos ax = -\sin ax.\, a = -a \sin ax,$

$D^2 \cos ax = -a \cos ax.a = -a^2 \cos ax,$

$D^3 \cos ax = -a^2 (-\sin ax).a = a^3 \sin ax,$

$D^4 \cos ax = a^3 \cos ax.a = a^4 \cos ax,$

... ...

i.e., $D^2 \cos ax = (-a^2) \cos ax, \; (D^2)^2 \cos ax = (-a^2)^2 \cos ax,$

Similarly, $(D^2)^3 \cos ax = (-a^2)^3 \cos ax, \ldots,$

$\therefore \quad (D^2)^n \cos ax = (-a^2)^n \cos ax.$

$\therefore \quad f(D^2) \cos ax = f(-a^2) \cos ax.$ [$\because$ $f(D^2)$ is a polynomial in D^2]

Operating on both sides by $\dfrac{1}{f(D^2)}$,

$$\frac{1}{f(D^2)} f(D^2) \cos ax = \frac{1}{f(D^2)} f(-a^2) \cos ax$$

or
$$\cos ax = f(-a^2) \frac{1}{f(D^2)} \cos ax$$

$$\left[\because \; \frac{1}{f(D^2)}, f(D^2) \text{ are inverse operators, and } f(-a^2) \text{ is a cosntant}\right]$$

Dividing both sides by $f(-a^2)$ [which is $\neq 0$],

$$\frac{1}{f(-a^2)} \cos ax = \frac{1}{f(D^2)} \cos ax, \quad i.e., \quad \frac{1}{f(D^2)} \cos ax = \frac{1}{f(-a^2)} \cos ax.$$

Similarly $\dfrac{1}{f(D^2)} \sin ax = \dfrac{1}{f(-a^2)} \sin ax.$ *($f(-a^2) \neq 0$)]*

Note 2. *If $f(D)$ contains odd powers of D also, it can be put in the form $f(D) = f_1(D^2) + Df_2(D^2)$.*

Then
$$\frac{1}{f(D)} \cos ax = \frac{1}{f_1(D^2) + Df_2(D^2)} \cos ax$$

[Put $D^2 = -a^2$ (Rule (Art. 124, case II))]

$$= \frac{1}{f_1(-a^2) + Df_2(-a^2)} \cos ax = \frac{1}{p + qD} \cos ax \text{ (say)}$$

$$= (p - qD) \frac{1}{(p - qD)} \frac{1}{p + qD} \cos ax$$

$$\left[\because \; p - q\,D, \frac{1}{p - q\,D} \text{ are inverse operators}\right]$$

$$= (p - qD) \frac{1}{p^2 - q^2 D^2} \cos ax = (p - q\,D) \frac{1}{p^2 - q^2(-a^2)} \cos ax$$

$$= \frac{1}{p^2 + q^2 a^2} (p - q\,D) \cos ax \qquad \left[\text{But } D = \frac{d}{dx}\right]$$

$$= \frac{1}{p^2 + q^2 a^2} \left[p \cos ax - q \frac{d}{dx} (\cos ax)\right]$$

$$= \frac{1}{p^2 + q^2 a^2} [p \cos ax - q(-\sin ax) . a]$$

$$= \frac{1}{p^2 + q^2 a^2} [p \cos ax + qa \sin ax].$$

Similarly for $\dfrac{1}{f(D)} \sin ax$, if $f(D)$ contains odd powers of D also.

Case IV. Rule to evaluate $\frac{1}{f(D)} x^m$, where m is a positive integer : *Expand $\frac{1}{f(D)}$ in ascending powers of D by the Binomial theorem, taking the expansion of $[1 + \phi(D)]^{-1}$ as far as the term in D^m, and operate on x^m with each term of the expansion of $\frac{1}{f(D)}$.*

(See Note in solved Ex. 3, Art. 122.)

** **Proof.** *Consider* $\frac{1}{D-a} x^m$.

$$\frac{1}{D-a} x^m = e^{ax} \int x^m e^{-ax}\, dx \qquad \text{[Art. 120, Cor. 3]}$$

[Integrate by general rule of integration by parts]

$$= e^{ax}\left[x^m\left(\frac{e^{-ax}}{-a}\right) - m\,x^{m-1}\left(\frac{e^{-ax}}{a^2}\right) + m(m-1)\,x^{m-2}\left(\frac{e^{-ax}}{-a^3}\right) - \ldots + (-1)^n \int u^{(n)} v_n\, dx\right]$$

$$\left[\text{Compare } \int x^m e^{-ax}\, dx \text{ with } \int uv\, dx, \text{ here } u = x^m,\right.$$

$$\left. \therefore\ u^{(m)} = m\,!,\ u^{(m+1)} = 0, \ldots,\ \therefore\ \text{put } n = m\right]$$

$$= e^{ax}\left[x^m\left(\frac{e^{-ax}}{-a}\right) - m\,x^{m-1}\left(\frac{e^{-ax}}{a^2}\right) + m(m-1)\,x^{m-2}\left(\frac{e^{-ax}}{-a^3}\right) - \ldots + (-1)^m \int u^{(m)} v_m\, dx\right]$$

$$= e^{ax}\left[x^m\left(\frac{e^{-ax}}{-a}\right) - m\,x^{m-1}\left(\frac{e^{-ax}}{a^2}\right) + m(m-1)\,x^{m-2}\left(\frac{e^{-ax}}{-a^3}\right) - \ldots + (-1)^m \int m\,!\,\frac{e^{-ax}}{(-a)^m}\, dx\right]$$

$$= e^{ax}\left[-\frac{x^m}{a}e^{-ax} - \frac{m\,x^{m-1}}{a^2}e^{-ax} - \frac{m(m-1)\,x^{m-2}}{a^3}e^{-ax} - \ldots + \frac{m\,!}{a^m}\frac{e^{-ax}}{-a}\right]$$

$$= -\frac{1}{a}\left[x^m + \frac{1}{a}m\,x^{m-1} + \frac{1}{a^2}m(m-1)\,x^{m-2} + \ldots + \frac{1}{a^m}m\,!\right].$$

If we expand $\frac{1}{D-a}$ in ascending powers of D,

$$\frac{1}{D-a}x^m = \frac{1}{-a+D}x^m = \frac{1}{-a\left(1-\frac{D}{a}\right)}x^m = -\frac{1}{a}\left(1-\frac{D}{a}\right)^{-1}$$

$$= -\frac{1}{a}\left[1 + \frac{D}{a} + \frac{D^2}{a^2} + \ldots + \frac{D^m}{a^m} + \frac{D^{m+1}}{a^{m+1}} + \ldots\right]x^m$$

$$= -\frac{1}{a}\left[x^m + \frac{1}{a}m\,x^{m-1} + \frac{1}{a^2}m(m-1)\,x^{m-2} + \ldots + \frac{1}{a^m}m\,! + 0\right]$$

$$[\because\ D^m x^m = m\,!,\ \therefore\ D^{m+1} x^m = 0, \ldots.]$$

Thus to evaluate $\frac{1}{D-a} x^m$, we may expand $\frac{1}{D-a}$ in ascending powers of D as far as the term in D^m and operate on x^m by each term of the expansion.

Now to evaluate $\frac{1}{f(D)} x^m$, we may resolve $\frac{1}{f(D)}$ into partial fractions (Art. 120), and expand each partial fraction in ascending powers of D (Proved above). But the expansion of $\frac{1}{f(D)}$ in ascending powers of D is the same whether we first resolve $\frac{1}{f(D)}$ into partial fractions and then expand each partial fraction in ascending powers of D, or we expand $\frac{1}{f(D)}$, *i.e.*, $[f(D)]^{-1}$ directly in ascending powers of D. Hence the rule given above.]

Case V. Rule to evaluate $\frac{1}{f(D)}(e^{ax} V)$, where V is any function of x :

$$\frac{1}{f(D)}(e^{ax} V) = e^{ax} \frac{1}{f(D+a)} V.$$

[In words : *Take e^{ax} out from under the operator, and in the operator change D to (D + a).*]

****Proof.** By Leibnitz's theorem in Differential Calculus,

$$D^n (e^{ax} V) = a^n e^{ax} . V + {}^nC_1 a^{n-1} e^{ax} . DV + {}^nC_2 a^{n-2} e^{ax} . D^2V + ... + e^{ax} D^nV$$

$$[(uv)_n = u_n v + {}^nC_1 u_{n-1} v_1 + {}^nC_2 u_{n-2} v_2 + ... + uv_n]$$

$$= e^{ax} [a^n + {}^nC_1 a^{n-1} D + {}^nC_2 a^{n-2} D^2 + ... + D^n] V$$

$$= e^{ax} (a + D)^n V = e^{ax} (D + a)^n V$$

$$\therefore \quad f(D)(e^{ax} V) = e^{ax} f(D + a)V. \qquad [\because f(D) \text{ is a polynomial in D}]$$

Operating on both sides by $\frac{1}{f(D)}$,

$$\frac{1}{f(D)} f(D)(e^{ax} V) = \frac{1}{f(D)} [e^{ax} f(D + a) V]$$

or

$$e^{ax} V = \frac{1}{f(D)} [e^{ax} f(D + a) V]$$

$$\left[\because \frac{1}{f(D)}, f(D) \text{ are inverse operators}\right]$$

i.e.,

$$\frac{1}{f(D)} [e^{ax} f(D + a) V] = e^{ax} V \qquad ...(1)$$

Put $\qquad f(D + a)\, V = V_1$ ...(2),

so that, operating on both sides by $\dfrac{1}{f(D+a)}$,

$$\frac{1}{f(D+a)}\, f(D + a)\, V = \frac{1}{f(D+a)}\, V_1, \text{ or } V = \frac{1}{f(D+a)}\, V_1 \qquad ...(3)$$

$$\left[\because \ \frac{1}{f(D+a)},\ f(D+a) \text{ are inverse operators}\right]$$

Substituting from (2) and (3) in (1),

$$\frac{1}{f(D)}\,(e^{ax}\, V_1) = e^{ax}\, \frac{1}{f(D+a)}\, V_1.$$

Changing V_1 to V,

[$\because$ V being any function of x, V_1 {$= f(D + a)V$} is also any function of x]

$$\left.\frac{1}{f(D)}\,(e^{ax}\, V) = e^{ax}\, \frac{1}{f(D+a)}\, V.\right]$$

122. Rule to solve the equation

$$\frac{d^n y}{dx^n} + P_1 \frac{d^{n-1} y}{dx^{n-1}} + P_2 \frac{d^{n-2} y}{dx^{n-2}} + \ldots + P_n y = Q,$$

where $P_1, P_2, \ldots, P_n$ are constants, and Q is a function of x :

(i) Write the equation in the symbolic form

$$\left[\textit{by putting } \frac{d}{dx} = D, \frac{d^2}{dx^2} = D^2, \ldots, \frac{d^n}{dx^n} = D^n \textit{, i.e.,}\right]$$

$$(D^n + P_1 D^{n-1} + P_2 D^{n-2} + \ldots + P_n)\, y = Q.$$

(ii) Write down the auxiliary equation (A.E.) [by equating to zero the symbolic coefficient of y, i.e.,]

$$D^n + P_1 D^{n-1} + P_2 D^{n-2} + \ldots + P_n = 0,$$

and solve for D as if it were an ordinary algebraic quantity.

(iii) From the roots of the A.E. write down the corresponding part of the complementary function (C.F.) by the same rule as for writing down the corresponding part of the complete solution (C.S.) if the R.H.S. of the given equation were zero (instead of Q).

(iv) From step (i), find the particular integral.

$$P.I. = \frac{1}{D^n + P_1 D^{n-1} + P_2 D^{n-2} + \ldots + P_n} \cdot Q$$

$$= \frac{1}{f(D)}\, Q. \textit{ Evaluate by the usual rule (Art. 121).}$$

(v) Then the C.S. is y = C.F. + P.I.

Note. Cases of failure. *If in step (iv), by using the usual rule (Art. 121, cases I-III), we get zero in the denominator, the rule fails. Now we proceed by the rule of Art. 126 and find the particular integral.*

Example 1. *Solve* $\dfrac{d^2 y}{dx^2} - 5\dfrac{dy}{dx} + 6y = e^{4x}$.

(*i*) The equation is $\frac{d^2y}{dx^2} - 5\frac{dy}{dx} + 6y = e^{4x}$

or, in the symbolic form, $(D^2 - 5D + 6)\,y = e^{4x}$.

(*ii*) The A.E. is $D^2 - 5D + 6 = 0$, or $(D-2)(D-3) = 0$, $\therefore$ $D = 2, 3$.

(*iii*) $\therefore$ C.F. $= c_1 e^{2x} + c_2 e^{3x}$.

(*iv*) From step (*i*), P.I. $= \frac{1}{D^2 - 5D + 6} e^{4x}$

$\left[\text{Compare with } \frac{1}{f(D)} e^{ax}, \text{ here } a = 4 \,;\therefore\; \text{put } D = a = 4\right]$

$$= \frac{1}{(4)^2 - 5(4) + 6} e^{4x} = \tfrac{1}{2} e^{4x}.$$

(*v*) $\therefore$ the C.S. is $y = c_1 e^{2x} + c_2 e^{3x} + \frac{1}{2} e^{4x}$. [C.F. + P.I.]

Example 2. *Solve* $\frac{d^3y}{dx^3} + \frac{d^2y}{dx^2} - \frac{dy}{dx} - y = \sin 2x$.

(*i*) The equation is $\frac{d^3y}{dx^3} + \frac{d^2y}{dx^2} - \frac{dy}{dx} - y = \sin 2x$

or, in the symbolic form, $(D^3 + D^2 - D - 1)y = \sin 2x$.

(*ii*) The A.E. is $D^3 + D^2 - D - 1 = 0$

or $(D+1)^2(D-1) = 0$

$\therefore$ $D = -1, -1 \,;\, 1$.

$$D^3 + D^2 - D - 1 = D^2(D+1) - 1(D+1) = (D+1)(D^2-1) = (D+1)(D+1)(D-1) = (D+1)^2(D-1)$$

(*iii*) $\therefore$ C.F. $= e^{-1.x}(c_1 + c_2 x) + c_3 e^{1.x} = e^{-x}(c_1 + c_2 x) + c_3 e^x$.

(*iv*) From step (*i*),

$$\text{P.I.} = \frac{1}{D^3 + D^2 - D - 1} \sin 2x \qquad \text{(Separate D from odd powers)}$$

$$= \frac{1}{D.D^2 + D^2 - D - 1} \sin 2x$$

$\left[\text{Compare with } \frac{1}{f(D^2)} \sin ax, \text{ here } a = 2 \,;\therefore\; \text{put } D^2 = -a^2 = -2^2 = -4\right]$

$$= \frac{1}{-4D - 4 - D - 1} \sin 2x = -\frac{1}{5}\frac{1}{D+1} \sin 2x$$

$$= -\frac{1}{5}(D-1)\frac{1^*}{D-1}\frac{1}{D+1} \sin 2x \qquad \textit{[Note this step]}$$

$$= -\frac{1}{5}(D-1)\frac{1}{D^2-1} \sin 2x \qquad [\text{Put } D^2 = -a^2 = -2^2 = -4]$$

---

*__Why this step.__ To get only *even* powers of D in the denominator, we have taken $(D-1)\frac{1}{D-1}$.

Had $D - 1$ occurred in the denominator, we would have taken $(D+1)\frac{1}{D+1}$.

$$= -\frac{1}{5}(D-1)\frac{1}{-4-1}\sin 2x = \frac{1}{25}(D-1)\sin 2x \qquad \left[\text{But } D = \frac{d}{dx}\right]$$

$$= \frac{1}{25}\left[\frac{d}{dx}(\sin 2x) - \sin 2x\right] = \frac{1}{25}[\cos 2x.2 - \sin 2x].$$

(*v*) ∴ the C.S. is

$$y = e^{-x}(c_1 + c_2 x) + c_3 e^x + \frac{1}{25}[2\cos 2x - \sin 2x]. \qquad [\text{C.F.} + \text{P.I.}]$$

Example 3. *Solve* $\dfrac{d^3y}{dx^3} + 3\dfrac{d^2y}{dx^2} + 2\dfrac{dy}{dx} = x^2.$

(*i*) The equation is $\dfrac{d^3y}{dx^3} + 3\dfrac{d^2y}{dx^2} + 2\dfrac{dy}{dx} = x^2$

or, in the symbolic form, $(D^3 + 3D^2 + 2D)y = x^2$.

(*ii*) The A.E. is $D^3 + 3D^2 + 2D = 0$

or $D(D + 1)(D + 2) = 0$

∴ $D = 0, -1, -2.$

$$D^3 + 3D^2 + 2D = D(D^2 + 3D + 2) = D(D + 1)(D + 2)$$

(*iii*) ∴ C.F. $= c_1 e^{0.x} + c_2 e^{-1.x} + c_3 e^{-2x} = c_1 + c_2 e^{-x} + c_3 e^{-2x}$.

(*iv*) From step (*i*),

$$\text{P.I.} = \frac{1}{D^3 + 3D^2 + 2D}x^2 \qquad [\text{Arrange in ascending powers of D}]$$

$$= \frac{1}{2D + 3D^2 + D^3}x^2 = \frac{1}{2D\left(1 + \frac{3}{2}D + \frac{D^2}{2}\right)}x^2$$

$$= \frac{1}{2D}\left[1 + \left(\frac{3}{2}D + \frac{D^2}{2}\right)\right]^{-1} x^2$$

[Expand by the Binomial theorem as far as the term in D^2]

$$= \frac{1}{2D}\left[1 - \left(\frac{3}{2}D + \frac{D^2}{2}\right) + \left(\frac{3}{2}D + \frac{D^2}{2}\right)^2 - \ldots\right]x^2$$

[∵ $(1 + x)^{-1} = 1 - x + x^2 - \ldots.$

$$= \frac{1}{2D}\left[1 - \left(\frac{3}{2}D + \frac{D^2}{2}\right) + \frac{9}{4}D^2 \ldots\right]x^2$$

$$= \frac{1}{2D}\left[1 - \frac{3}{2}D + \frac{7}{4}D^2 \ldots\right]x^2 = \frac{1}{2}\left[\frac{1}{D} - \frac{3}{2} + \frac{7}{4}D \ldots\right]x^2 \qquad [\textit{Note this step}]$$

$$\left[\text{But } \frac{1}{D}x^2 = \int x^2\,dx = \frac{x^3}{3},\ Dx^2 = \frac{d}{dx}(x^2) = 2x\right]$$

$$= \frac{1}{2}\left[\frac{x^3}{3} - \frac{3}{2}x^2 + \frac{7}{4}(2x)\right] = \frac{1}{2}\left[\frac{x^3}{3} - \frac{3}{2}x^2 + \frac{7}{2}x\right].$$

(*v*) $\therefore$ the C.S. is

$$y = c_1 + c_2 e^{-x} + c_3 e^{-2x} + \frac{1}{2}\left[\frac{x^3}{3} - \frac{3}{2}x^2 + \frac{7}{2}x\right].$$

Note. *In order to evaluate* $\dfrac{1}{2D}\left[1+\left(\dfrac{3}{2}D+\dfrac{D^2}{2}\right)\right]^{-1} x^2,$

we have expanded $\left[1+\left(\dfrac{3}{2}D+\dfrac{D^2}{2}\right)\right]^{-1}$ *by the Binomial theorem as far as the term in* D^2. *One might think that because of the factor* $\dfrac{1}{2D}$ *in* $\dfrac{1}{2D}\left[1+\left(\dfrac{3}{2}D+\dfrac{D^2}{2}\right)\right]^{-1} x^2$, *we should have expanded* $\left[1+\left(\dfrac{3}{2}D+\dfrac{D^2}{2}\right)\right]^{-1}$ *as far as the term in* D^3 *so as to get a term* $\frac{1}{2}(...D^2)$. *But the result of operating on* x^2 *by this term would be a constant, which is already included in the complementary function.*

Hence it is only necessary to expand $\left[1+\left(\dfrac{3}{2}D+\dfrac{D^2}{2}\right)\right]^{-1}$ as far as the term in D^2. *This method is general, and usually saves much labour.*

Example 4. *Solve* $(D^2 - 5D + 6)\, y = x\, e^{4x}$.

(*i*) The equation, in the symbolic form, is

$$(D^2 - 5D + 6)\, y = x\, e^{4x}.$$

(*ii*) The A.E. is $D^2 - 5D + 6 = 0$, or $(D-2)(D-3) = 0$, $\therefore$ $D = 2, 3$.

(*iii*) $\therefore$ C.F. $= c_1 e^{2x} + c_2 e^{3x}$.

(*iv*) From step (*i*),

$$\text{P.I.} = \frac{1}{D^2 - 5D + 6}(x\, e^{4x}) = \frac{1}{D^2 - 5D + 6}(e^{4x} x)$$

[Compare with $\dfrac{1}{f(D)}(e^{ax} V)$, here $a = 4$.
Take $e^{ax} = e^{4x}$ out from under the operator,
and in the operator change D to $D + a = D + 4$]

$$= e^{4x}\frac{1}{(D+4)^2 - 5(D+4) + 6}x = e^{4x}\frac{1}{D^2 + 8D + 16 - 5D - 20 + 6}x$$

$$= e^{4x}\frac{1}{D^2 + 3D + 2}x = e^{4x}\frac{1}{2 + 3D + D^2}x$$

$$= e^{4x}\frac{1}{2\left(1+\frac{3}{2}D+\frac{D^2}{2}\right)}x = e^{4x}\frac{1}{2}\left[1+\left(\frac{3}{2}D+\frac{D^2}{2}\right)\right]^{-1}x$$

$$= e^{4x}\frac{1}{2}\left[1-\left(\frac{3}{2}D\ldots\right)\right]x = e^{4x}\frac{1}{2}\left(x - \frac{3}{2}.1\right).$$

(v) $\therefore$ the C.S. is $\quad y = c_1 e^{2x} + c_2 e^{3x} + \frac{1}{2} e^{4x}\left(x - \frac{3}{2}\right).$

EXAMPLES

Solve the following differential equations (Exs. 1—10) :

1. (i) $\frac{d^2y}{dx^2} - 3\frac{dy}{dx} + 2y = e^{5x}.$

(ii) $\frac{d^2y}{dx^2} + 2\frac{dy}{dx} + y = 2e^{2x}.$

2. (i) $(D^3 - 3D^2 + 4)y = e^{3x}.$

(ii) $\frac{d^2y}{dx^2} - 4y = 2\sin\frac{1}{2}x.$

3. (i) $\frac{d^2y}{dx^2} + \frac{dy}{dx} + y = \sin 2x.$

(ii) $\frac{d^3y}{dx^3} + y = \sin 3x - \cos^2\frac{1}{2}x.$

[(ii) P.I. $= \frac{1}{D^3+1}\left(\sin 3x - \cos^2\frac{x}{2}\right) = \frac{1}{D^3+1}\left(\sin 3x - \frac{1+\cos x}{2}\right)$ [*Note this step*]

$$= \frac{1}{D^3+1}\sin 3x - \frac{1}{2}\frac{1}{D^3+1}1 - \frac{1}{2}\frac{1}{D^3+1}\cos x.$$

Now $\frac{1}{D^3+1}1 = \frac{1}{D^3+1}e^{0.x}$ [*Note this step*]

$$= \frac{1}{0+1}e^{0.x} = 1]$$

4. (i) $\frac{d^4y}{dx^4} + 2n^2\frac{d^2y}{dx^2} + n^4y = \cos mx.$

(ii) $(D^2 - 9D + 20)y = 20x.$

5. (i) $\frac{d^2y}{dx^2} - 5\frac{dy}{dx} + 6y = x + e^{mx}.$

(ii) $\frac{d^3y}{dx^3} - 3\frac{d^2y}{dx^2} - 6\frac{dy}{dx} + 8y = x.$

6. (i) $\frac{d^2y}{dx^2} - y = 2 + 3x.$ (ii) $\frac{d^3y}{dx^3} - 13\frac{dy}{dx} + 12y = x.$

7. (i) $\frac{d^2y}{dx^2} + 4y = \sin 3x + e^x + x^2.$

(ii) $\frac{d^3y}{dx^3} + 2\frac{d^2y}{dx^2} + \frac{dy}{dx} = e^{2x} + x^2 + x.$

[(i) To evaluate D^2x^2.

$$Dx^2 = \frac{d}{dx}(x^2) = 2x.$$

Operating on both sides by D,

$$D^2x^2 = D(2x) = \frac{d}{dx}(2x) = 2.1.]$$

8. (*i*) $\dfrac{d^4y}{dx^4} - a^4y = x^4.$ (*ii*) $\dfrac{d^4y}{dx^4} - 2\dfrac{d^3y}{dx^3} + \dfrac{d^2y}{dx^2} = 6x.$

[(*ii*) To evaluate $\dfrac{1}{D^2}x.$

$$\frac{1}{D}x = \int x\,dx = \frac{x^2}{2}.$$

Operating on both sides by $\dfrac{1}{D}$,

$$\frac{1}{D^2}x = \frac{1}{D}\frac{x^2}{2} = \int \frac{x^2}{2}dx = \frac{1}{2}\cdot\frac{x^3}{3} = \frac{x^3}{6}.]$$

9. (*i*) $\dfrac{d^2y}{dx^2} + y = x\,e^{2x}.$

(*ii*) $\dfrac{d^3y}{dx^3} - 7\dfrac{dy}{dx} - 6y = e^{2x}(1 + x).$

10. (*i*) $\dfrac{d^2y}{dx^2} - 2\dfrac{dy}{dx} + 4y = e^x \cos x.$

(*ii*) $\dfrac{d^4y}{dx^4} - y = e^x \cos x.$

Cases of failure.

[☞ *All these cases of failure are of the greatest importance, and the student is strongly urged to study them very thoroughly.*]

123. Rule* to find the particular integral in cases of failure i.e., when by using the usual rule for finding the particular integral (Art. 121, cases I-III), we get zero in the denominator)

Case I′. $\dfrac{1}{f(D)}e^{ax} = x\dfrac{1}{\dfrac{d}{dD}[f(D)]}e^{ax}$, *when* $f(a) = 0.$

Case II′. $\dfrac{1}{f(D^2)}\cos ax = x\dfrac{1}{\dfrac{d}{dD}[f(D^2)]}\cos ax$, *when* $f(-a^2) = 0.$

Case III′. $\dfrac{1}{f(D^2)}\sin ax = x\dfrac{1}{\dfrac{d}{dD}[f(D^2)]}\sin ax$, *when* $f(-a^2) = 0.$

---

*This rule has been taken from the *Mathematics Student*.

$\left[\textit{i.e., } x \dfrac{1}{\textit{diff. coeff. of denom. w.r.t. } D} e^{ax} \textit{ or } \cos ax \textit{ or } \sin ax \textit{ as the case may be.} \right]$

Note. *If by using the above rule, again we get zero in the denominator, we repeat the rule, and so on. (See Ex. 2, (ii) following.)*

**[Case I′. To prove that $\dfrac{1}{f(D)} e^{ax} = x \dfrac{1}{\dfrac{d}{dD}[f(D)]} e^{ax}$, when $f(a) = 0$.

Proof. $\because$ $f(a) = 0$, $\therefore$ $D = a$ is a root of the A.E. $f(D) = 0$

$\therefore$ corresponding part of the C.F. $= c_1 e^{ax}$ ...(1)

Now $\dfrac{1}{f(D)} e^{ax} = \underset{h \to 0}{\text{Lt}} \dfrac{1}{f(D)} e^{(a+h)x*} = \underset{h \to 0}{\text{Lt}} \dfrac{1}{f(a+h)} e^{(a+h)x}$

[Rule (Art. 121, case I)

$= \underset{h \to 0}{\text{Lt}} \dfrac{1}{f(a+h)} e^{ax} \cdot e^{hx}$ [Expand e^{hx} in powers of h]

$= \underset{h \to 0}{\text{Lt}} \dfrac{1}{f(a+h)} e^{ax} \left[1 + hx + \dfrac{h^2x^2}{2!} + \ldots \right]$

[Omit the first term $\dfrac{1}{f(a+h)} e^{ax}$, which is already contained in the C.F. (From (1))]

$= \underset{h \to 0}{\text{Lt}} \dfrac{1}{f(a+h)} e^{ax} \left[hx + \dfrac{h^2x^2}{2!} + \ldots \right]$

[Expand $f(a+h)$ in powers of h by Taylor's theorem]

$= \underset{h \to 0}{\text{Lt}} \dfrac{1}{f(a) + hf'(a) + \dfrac{h^2}{2!} f''(a) + \ldots} e^{ax} \left[hx + \dfrac{h^2x^2}{2!} + \ldots \right]$

[But $f(a) = 0$ (Given)]

$= \underset{h \to 0}{\text{Lt}} \dfrac{1}{hf'(a) + \dfrac{h^2}{2!} f''(a) + \ldots} e^{ax} \left[hx + \dfrac{h^2x^2}{2!} + \ldots \right]$

[Divide the num. and denom. by h]

$= \underset{h \to 0}{\text{Lt}} \dfrac{1}{f'(a) + \dfrac{h}{2!} f''(a) + \ldots} e^{ax} \left[x + \dfrac{hx^2}{2!} + \ldots \right]$

$= \dfrac{1}{f'(a)} e^{ax} x = x \dfrac{1}{f'(a)} e^{ax}$

$= x \dfrac{1}{f'(D)} e^{ax}$ [Rule (Art. 121, case I)]

$= x \dfrac{1}{\dfrac{d}{dD}[f(D)]} e^{ax}$.]

***How to write this step.** In e^{ax}, change a to $a + h$, thus getting $e^{(a+h)x}$.

Case II′. To prove that $\dfrac{1}{f(D^2)}\cos ax = x\dfrac{1}{\dfrac{d}{dD}[f(D^2)]}\cos ax$**, when** $f(-a^2) = 0$**.**

Proof. $\because$ $f(-a^2) = 0$, $\therefore$ $D^2 = -a^2$, *i.e.*, $D = \pm ia$ are roots of the A.E. $f(D^2) = 0$.

$\therefore$ corresponding part of the C.F.

$$= e^{0.x}(c_1 \cos ax + c_2 \sin ax) = c_1 \cos ax + c_2 \sin ax \qquad ...(1)$$

Now $\dfrac{1}{f(D^2)}\cos ax = \underset{h\to 0}{\text{Lt}}\dfrac{1}{f(D^2)}\cos(a+h)x^{*} = \underset{h\to 0}{\text{Lt}}\dfrac{1}{f[-(a+h)^2]}\cos(a+h)x$

[Rule (Art. 121, case II)]

$$= \underset{h\to 0}{\text{Lt}}\frac{1}{f[-(a+h)^2]}[\cos ax\cos hx - \sin ax\sin hx]$$

[Expand $\cos hx$, $\sin hx$ in powers of h]

$$= \underset{h\to 0}{\text{Lt}}\frac{1}{f[-(a+h)^2]}\left[\cos ax.\left(1-\frac{h^2x^2}{2!}+...\right)-\sin ax.(hx-...)\right]$$

$$= \underset{h\to 0}{\text{Lt}}\frac{1}{f[-(a+h)^2]}\left[\cos ax - hx\sin ax - \frac{h^2x^2}{2!}\cos ax...\right]$$

[Omit the first term $\dfrac{1}{f[-(a+h)^2]}\cos ax$, which is already contained in the C.F. (From (1))]

$$= \underset{h\to 0}{\text{Lt}}\frac{1}{f[-(a+h)^2]}\left[-hx\sin ax - \frac{h^2x^2}{2!}\cos ax...\right]$$

$$= \underset{h\to 0}{\text{Lt}}\frac{1}{f[-a^2-(2ah+h^2)]}\left[-hx\sin ax - \frac{h^2x^2}{2!}\cos ax...\right]$$

[Expand $f[-a^2-(2ah+h^2)]$ in powers of h by Taylor's theorem]

$$= \underset{h\to 0}{\text{Lt}}\frac{1}{f(-a^2)-(2ah+h^2)f'(-a^2)+\dfrac{1}{2!}(2ah+h^2)^2 f''(-a^2)...}$$

$$\times\left[-hx\sin ax - \frac{h^2x^2}{2!}\cos ax...\right]$$

[But $f(-a^2) = 0$ (Given)]

$$= \underset{h\to 0}{\text{Lt}}\frac{1}{-2ah f'(-a^2)-...}[-hx\sin ax - ...]$$

[Divide the num. and denom. by $-h$]

***How to write this step.** In $\cos ax$, change a to $a+h$, thus getting $\cos(a+h)x$.

$$= \underset{h \to 0}{\text{Lt}} \frac{1}{2a\, f'(-a^2) + \dots} [x \sin ax + \dots]$$

$$= \frac{1}{2a\, f'(-a^2)} x \sin ax = x \frac{1}{f'(-a^2)} \frac{1}{2} \frac{\sin ax}{a}$$

$$= x \frac{1}{f'(D^2)} \frac{1}{2} \frac{1}{D} \cos ax \qquad \left[\because \frac{1}{D} \cos ax = \int \cos ax\, dx = \frac{\sin ax}{a} \right]$$

[Art. 121, case II, and Art. 120, Cor. 2]

$$= x \frac{1}{\frac{d}{dD}[f(D^2)]} \cos ax.$$

Similarly for the proof of case III′.]

Example 1. *Solve* $\dfrac{d^3y}{dx^3} + y = 3 + e^{-x} + 5e^{2x}$.

(*i*) The equation is $\dfrac{d^3y}{dx^3} + y = 3 + e^{-x} + 5e^{2x}$

or, in the symbolic form, $(D^3 + 1)\, y = 3 + e^{-x} + 5e^{2x}$.

(*ii*) The A.E. is $D^3 + 1 = 0$, or $(D + 1)(D^2 - D + 1) = 0$

$$\therefore \quad D = -1\,;\ \frac{1}{2} \pm i \frac{\sqrt{3}}{2}. \qquad \Bigg|\qquad D^2 - D + 1 = 0, \quad \therefore\ D = \frac{1 \pm \sqrt{1-4}}{2} = \frac{1 \pm i\sqrt{3}}{2}$$

(*iii*) $\therefore$

$$\text{C.F.} = c_1 e^{-1.x} + e^{\frac{1}{2}x} \left(c_2 \cos \frac{\sqrt{3}}{2} x + c_3 \sin \frac{\sqrt{3}}{2} x \right)$$

$$= c_1 e^{-x} + e^{\frac{1}{2}x} \left(c_2 \cos \frac{\sqrt{3}}{2} x + c_3 \sin \frac{\sqrt{3}}{2} x \right).$$

(*iv*) From step (*i*),

$$\text{P.I.} = \frac{1}{D^3 + 1} [3 + e^{-x} + 5e^{2x}]$$

$$= 3 \frac{1}{D^3 + 1} 1 + \frac{1}{D^3 + 1} e^{-x} + 5 \frac{1}{D^3 + 1} e^{2x} \qquad \text{...(1)}$$

Now $\dfrac{1}{D^3 + 1} 1 = \dfrac{1}{D^3 + 1} e^{0.x} = \dfrac{1}{0 + 1} e^{0.x} = 1,$

$$\frac{1}{D^3 + 1} e^{-x} = \frac{1}{-1 + 1} e^{-x} = \frac{1}{0} e^{-x}, \ \therefore \ \text{the rule fails}\,;$$

$$\therefore \quad \frac{1}{D^3 + 1} e^{-x} = x \frac{1}{3D^2} e^{-x} \qquad \left[x \frac{1}{\text{diff. coeff. of denom. w.r.t. D}} e^{-x} \text{ (Art. 123, case I′)} \right]$$

$$= \frac{x}{3} \frac{1}{(1)} e^{-x} = \frac{x}{3} e^{-x},$$

$$\frac{1}{D^3+1}e^{2x} = \frac{1}{8+1}e^{2x} = \frac{1}{9}e^{2x},$$

$\therefore$ from (1), $\qquad$ P.I. $= 3 + \frac{x}{3}e^{-x} + \frac{5}{9}e^{2x}$.

(*v*) $\therefore$ the C.S. is $\quad y = c_1 e^{-x} + e^{\frac{1}{2}x}\left(c_2 \cos\frac{\sqrt{3}}{2}x + c_3 \sin\frac{\sqrt{3}}{2}x\right) + 3 + \frac{x}{3}e^{-x} + \frac{5}{9}e^{2x}$.

Example 2. *Solve* $\frac{d^2y}{dx^2} + a^2y = \cos ax$.

(*i*) The equation is $\frac{d^2y}{dx^2} + a^2y = \cos ax$

or, in the symbolic form, $(D^2 + a^2)y = \cos ax$.

(*ii*) The A.E. is $D^2 + a^2 = 0$, or $D^2 = -a^2$, $\therefore$ $D = \pm ia$.

(*iii*) $\therefore$ C.F. $= e^{0.x}(c_1 \cos ax + c_2 \sin ax) = c_1 \cos ax + c_2 \sin ax$.

(*iv*) From step (*i*),

$$\text{P.I.} = \frac{1}{D^2+a^2}\cos ax = \frac{1}{-a^2+a^2}\cos ax$$

$$= \frac{1}{0}\cos ax, \therefore \text{ the rule fails.}$$

$$\therefore \frac{1}{D^2+a^2}\cos ax = x\frac{1}{2D}\cos ax$$

$$\left[x\frac{1}{\text{diff. coeff. of denom. w.r.t. D}}\cos ax \text{ (Art. 123, case II}')\right]$$

$$= x\,\frac{1}{2}\,\frac{\sin ax}{a} \qquad \left[\because \frac{1}{D}\cos ax = \int \cos ax\,dx = \frac{\sin ax}{a}\right]$$

$$= \frac{x}{2a}\sin ax.$$

(*v*) $\therefore$ the C.S. is $y = c_1 \cos ax + c_2 \sin ax + \frac{x}{2a}\sin ax$.

EXAMPLES

Solve the following differential equations (Exs. 1—3) :

1. (*i*) $(D^2 - a^2)\,y = e^{ax} + e^{nx}$.

(*ii*) $\frac{d^2y}{dx^2} - y = \cosh x$.

$\left[(ii)\ \text{P.I.} = \frac{1}{D^2-1}\cosh x = \frac{1}{D^2-1}\left[\frac{e^x+e^{-x}}{2}\right] = \frac{1}{2}\left[\frac{1}{D^2-1}e^x + \frac{1}{D^2-1}e^{-x}\right].\right]$

2. (*i*) $\frac{d^3y}{dx^3} - y = (e^x + 1)^2$.

(*ii*) $\dfrac{d^3y}{dx^3}+3\dfrac{d^2y}{dx^2}+3\dfrac{dy}{dx}+y=e^{-x}$.

[(*ii*) P.I. $=\dfrac{1}{(D+1)^3}e^{-x}=\dfrac{1}{0}e^{-x}$, ∴ the rule fails.

$$\therefore \quad \frac{1}{(D+1)^3}e^{-x}=x\cdot\frac{1}{3(D+1)^2}e^{-x}=\frac{x}{3}\frac{1}{(D+1)^2}e^{-x} \quad ...(1)$$

$$=\frac{x}{3}\frac{1}{0}e^{-x}, \quad \therefore \text{ the rule fails.}$$

$$\therefore \text{ from (1), } \frac{1}{(D+1)^3}e^{-x}=\frac{x}{3}\left[x\cdot\frac{1}{2(D+1)}e^{-x}\right]=\frac{x^2}{6}\frac{1}{D+1}e^{-x} \quad ...(2)$$

$$=\frac{x^2}{6}\frac{1}{0}e^{-x}, \quad \therefore \text{ the rule fails.}$$

$$\therefore \text{ from (2), } \frac{1}{(D+1)^3}e^{x}=\frac{x^2}{6}\left[x\cdot\frac{1}{1}e^{-x}\right]=\frac{x^3}{6}e^{-x}.\Bigg]$$

3. (*i*) $(D^3-3D^2+4D-2)y=e^x+\cos x$.

(*ii*) $\dfrac{d^2y}{dx^2}+4y=e^x+\sin 2x$.

MISCELLANEOUS EXAMPLES ON CHAPTER XIII

Solve the following differential equations (Exs. 1—9) :

1. (*i*) $\dfrac{d^2y}{dx^2}+(a+b)\dfrac{dy}{dx}+aby=0$.

(*ii*) $\dfrac{d^2y}{dx^2}-4\dfrac{dy}{dx}+4y=0$.

(*iii*) $\dfrac{d^2y}{dx^2}+y=0$, given $y=2$ for $x=0$, $y=-2$ for $x=\dfrac{1}{2}\pi$.

2. (*i*) $\dfrac{d^4y}{dx^4}+m^4y=0$.

(*ii*) $(D^4+2D^3+3D^2+2D+1)y=0$.

[(*ii*) (To factorise) $\overline{D^4+2D^3}+3D^2+2D+1$

(Complete the square in the first two terms)

$$=(D^4+2D^3+D^2)+2D^2+2D+1$$

$$=(D^2+D)^2+2(D^2+D)+1=(\overline{D^2+D}+1)^2.]$$

3. (*i*) $\dfrac{d^2y}{dx^2}+2p\dfrac{dy}{dx}+(p^2+q^2)y=e^{ax}$.

(*ii*) $\dfrac{d^2y}{dx^2}-4y=e^x+\sin 2x$.

4. (*i*) $\dfrac{d^2y}{dx^2} - \dfrac{dy}{dx} - 2y = \sin 2x$

(*ii*) $\left(\dfrac{d^2y}{dx^2} - 8\dfrac{dy}{dx} + 9y = 40 \sin 5x\right)$

(*iii*) $\dfrac{d^2y}{dx^2} - 3\dfrac{dy}{dx} + 2y = \cos 3x.$

(*iv*) $\dfrac{d^2y}{dx^2} + 2k\dfrac{dy}{dx} + k^2y = A\cos px.$

5. (*i*) $\dfrac{d^2y}{dx^2} - 13\dfrac{dy}{dx} + 12y = x.$

(*ii*) $\dfrac{d^2y}{dx^2} - 2\dfrac{dy}{dx} + 3y = \cos x + x^2.$

6. (*i*) $3\dfrac{d^2y}{dx^2} + \dfrac{dy}{dx} - 14y = 13e^{2x}.$

(*ii*) $4\dfrac{d^2y}{dx^2} + 12\dfrac{dy}{dx} + 9y = 144\, e^{-3/2x}.$

7. (*i*) $D^2y - 3Dy + 2y = \cosh x.$

(*ii*) $\dfrac{d^3y}{dx^3} - 4\dfrac{d^2y}{dx^2} + 5\dfrac{dy}{dx} - 2 = 0.$

8. (*i*) $\dfrac{d^2y}{dx^2} + y = \sin x \sin 2x.$ (*ii*) $\dfrac{d^4y}{dx^4} + 2\dfrac{d^2y}{dx^2} + y = \cos x.$

[(*ii*) P.I. $= \dfrac{1}{(D^2+1)^2}\cos x$ [Put $D^2 = -a^2 = -1$]

$= \dfrac{1}{0}\cos x$, $\therefore$ the rule fails. $\therefore$ $\dfrac{1}{(D^2+1)^2}\cos x = x\dfrac{1}{2(D^2+1)\,2D}\cos x$

$\left[x\dfrac{1}{\text{diff. coeff. of denom. w.r.t. D}}\cos x\right]$

$$= \frac{x}{4}\left[\frac{1}{D^2+1}\sin x\right] \qquad ...(1)$$

$\left(\because \dfrac{1}{D}\cos x = \int \cos x\, dx = \sin x\right)$ (Put $D^2 = -a^2 = -1$)

$$= \frac{x}{4}\left[\frac{1}{0}\sin x\right], \quad \therefore \text{ the rule fails.}$$

$\therefore$ from (1),

$$\frac{1}{(D^2+1)^2}\cos x = \frac{x}{4}\left[x\frac{1}{2D}\sin x\right] = \frac{x^2}{8}(-\cos x).]$$

****9.** (*i*) $(D^5 - D)y = 12e^x + 8\sin x - 2x.$

(*ii*) $(D-1)^2(D^2+1)^2\, y = \sin^2 \frac{1}{2}x + e^x.$

$$\left[(ii)\ \text{P. I.} = \frac{1}{(D-1)^2(D^2+1)^2}\left[\sin^2\frac{x}{2}+e^x\right]\right.$$

$$= \frac{1}{(D-1)^2(D^2+1)^2}\left[\frac{1-\cos x}{2}+e^x\right]$$

$$= \frac{1}{2}\frac{1}{(D-1)^2(D^2+1)^2}\,1 - \frac{1}{2}\frac{1}{(D-1)^2(D^2+1)^2}\cos x$$

$$+ \frac{1}{(D-1)^2(D^2+1)^2}\,e^x \qquad ...(1)$$

Now $\dfrac{1}{(D-1)^2(D^2+1)^2}\,1 = \dfrac{1}{(D-1)^2(D^2+1)^2}\,e^{0.x} = \dfrac{1}{2}$ $\qquad D = a = 0$

$$\frac{1}{(D-1)^2(D^2+1)^2}\cos x = \frac{1}{(D^2+1)^2}\frac{1}{D^2-2D+1}\cos x \qquad (\text{Put } D^2 = -a^2 = -1)$$

$$= \frac{1}{(D^2+1)^2}\frac{1}{-2D}\cos x = -\frac{1}{2}\frac{1}{(D^2+1)^2}\sin x = ?$$

$$\frac{1}{(D-1)^2(D^2+1)^2}\,e^x = \frac{1}{(D-1)^2}\frac{1}{(D^2+1)^2}\,e^x \qquad (\text{Put } D = a = 1)$$

$$\left.= \frac{1}{(D-1)^2}\frac{1}{4}\,e^x = ?\ \right]$$

10. (*i*) Find the solution of the equation $\dfrac{d^2y}{dx^2} - y = 1$, which vanishes when $x = 0$ and tends to a finite limit as $x \to -\infty$.

(*ii*) Solve $(D^2 - 3D + 2)y = e^x$, if $y = 3$ and $Dy = 3$ when $x = 0$.

[(*i*) It will be found that the C.S. is

$$y = c_1 e^x + c_2 e^{-x} - 1 \qquad ...(1)$$

But when $x = 0, y = 0$ (Given), $\therefore$ substituting in (1),

$$0 = c_1 + c_2 - 1 \qquad (2)$$

When $x \to -\infty$, y is finite (Given), $\therefore$ from (1),

$$\text{finite} = \underset{x \to -\infty}{\text{Lt}} (c_1 e^x + c_2 e^{-x} - 1)$$

But when $x \to -\infty, e^x \to 0$ $\qquad \left(\because\ e^{-\infty} = \dfrac{1}{e^\infty} = \dfrac{1}{\infty} = 0\right)$

and $\qquad e^{-x} \to \infty \qquad (\because\ e^\infty = \infty)$

$\therefore$ coeff. of $e^{-x} = 0$, *i.e.*, $c_2 = 0$ $\qquad ...(3)$

Find the values of c_1, c_2 from (2), (3), and substitute them in (1)].

11. Solve $\dfrac{d^2y}{dx^2} + 2\dfrac{dy}{dx} + 10y + 37\sin 3x = 0$; and find the value of y when $x = \dfrac{\pi}{2}$, if it is given that $y = 3$ and $\dfrac{dy}{dx} = 0$ when $x = 0$.

12. Solve $(D^4 - 2D^3 - 3D^2 + 4D + 4)\, y = x^2 e^x$

13. Solve (*i*) $\dfrac{d^2y}{dx^2} + 4\dfrac{dy}{dx} - 12y = (x-1)\, e^{2x}$.

(*ii*) $\dfrac{d^3y}{dx^3} - 3\dfrac{d^2y}{dx^2} + 3\dfrac{dy}{dx} - y = x\,e^x + e^x$.

14. Solve $\dfrac{d^2y}{dx^2} + y = e^{-x} + \cos x + x^3 + e^x \sin x$

15. Solve $\dfrac{d^4y}{dx^4} - y = x \sin x$

16. (*i*) $\dfrac{d^2y}{dx^2} - 2\dfrac{dy}{dx} + y = xe^x \sin x$

**(*ii*) $\dfrac{d^2y}{dx^2} - 4\dfrac{dy}{dx} + 4y = 8x^2 e^{2x} \sin 2x$

[P.I. $= \dfrac{1}{(D-1)^2}\, [e^x (x \sin x)]$

$= e^x \dfrac{1}{D^2}\, (x \sin x)]$.

ANSWERS

CHAPTER I

Page 4.

1. $\dfrac{x^{a+1}}{a+1}, \dfrac{x^{10}}{10}, \dfrac{x^2}{2}.$ **2.** $-\dfrac{x^{-8}}{8}, -\dfrac{1}{10x^{10}}, -\dfrac{1}{(n-1)x^{n-1}}.$

3. $\dfrac{3}{5}x^{\frac{5}{3}}, \dfrac{n}{n+1}x^{\frac{n+1}{n}}, \dfrac{2}{3}x^{\frac{3}{2}}.$ **4.** $3x^{\frac{1}{3}}, 4x^{\frac{1}{4}}.$

Page 7.

Ex. $a\dfrac{x^2}{2}, a\log x.$

Pages 8—9.

1. (*i*) $a\log x + x.$ (*ii*) $ax + b\dfrac{x^2}{2} + \dfrac{c}{2x^2}.$ (*iii*) $a\dfrac{x^3}{3} + b\dfrac{x^2}{2} + cx.$

2. (*i*) $\dfrac{x^4}{4} - \dfrac{3^x}{\log 3}.$ (*ii*) $a\dfrac{x^2}{2} + \dfrac{a^x}{\log a} - \dfrac{x^{a+1}}{a+1}.$

Page 15.

1. (*i*) $\dfrac{1}{100}.$ (*ii*) $\dfrac{1}{n+1}[2^{n+1} - 1].$ (*iii*) $\log\frac{3}{2}.$

2. (*i*) 1. (*ii*) 1. (*iii*) 1.

3. $\frac{1}{2}(b^2 - a^2) + \sin b - \sin a.$

MISCELLANEOUS EXAMPLES ON CHAPTER I

Pages 16—17.

1. (*i*) $\frac{2}{7}x^{\frac{7}{2}} + \frac{4}{5}x^{\frac{5}{2}} + 2x^{\frac{3}{2}}.$ (*ii*) $4\log x - \dfrac{3}{x} - \dfrac{1}{x^2}.$

(*iii*) $\dfrac{x^4}{4} + 3\dfrac{x^2}{2} + 3\log x - \dfrac{1}{2x^2}.$ **2.** $\dfrac{\left(\frac{a}{b}\right)^x}{\log\frac{a}{b}} + 2x + \dfrac{\left(\frac{b}{a}\right)^x}{\log\frac{b}{a}}.$

3. (*i*) $-(\cos x + \sin x).$ (*ii*) $\sec x + 5\cot x$

4. (*i*) $\sec x.$ (*ii*) $\tan x + \cot x.$

CHAPTER II

Page 20.

1. (*i*) $\sin^{-1} x.$ (*ii*) $\frac{1}{3}\tan^{-1}\dfrac{x}{3}.$ (*iii*) $\frac{1}{2}\sec^{-1}\dfrac{x}{2}.$

2. (*i*) $\frac{1}{6}\tan^{-1}\frac{2x}{3}$. (*ii*) $\frac{1}{2\sqrt{3}}\sec^{-1}\frac{x}{2}$. **3.** $x - 2\tan^{-1}\frac{x}{2}$.

Page 21.

Ex. (*i*) $\sinh^{-1} x$. (*ii*) $\cosh^{-1} x$.

Pages 22—23.

1. $-\frac{1}{(n-1)(a+x)^{n-1}}, \frac{1}{(n-1)(a-x)^{n-1}}, -\frac{2a}{a^2-x^2}$.

2. (*i*) $\log(x+1), \log(a+x)$. (*ii*) $-\frac{1}{b}\log(a-bx), \log\frac{x-a}{x+a}, \log(a^2-x^2)$.

3. $\frac{(2x+5)^{\frac{3}{2}}}{3}, \frac{\log(2x+1)}{2}, -\frac{\cos(3x+2)}{3}, \frac{\cot(a-bx)}{b}$.

4. $\frac{2}{3(b-a)}[(x-a)^{\frac{3}{2}} - (x-b)^{\frac{3}{2}}]$. **5.** $ax + \frac{b^x}{\log b} - \frac{c^{2x}}{2\log c}$.

6. $x - \log(1+x),\ x + a\log(x-a)$. **7.** $x^2 + 3x + 3\log(x-1)$.

8. $\frac{1}{2}\left[-\frac{\cos(m+n)x}{m+n} - \frac{\cos(m-n)x}{m-n}\right]$.

9. $\frac{1}{2}\left[\frac{\sin(m-n)x}{m-n} - \frac{\sin(m+n)x}{m+n}\right], \frac{1}{2}\left[x - \frac{\sin 2mx}{2m}\right]$.

10. $\frac{1}{2}\left[\frac{\sin(m+n)x}{m+n} + \frac{\sin(m-n)x}{m-n}\right], \frac{1}{2}\left[\frac{\sin 2mx}{2m} + x\right]$.

11. (*i*) $\frac{1}{4}\left[-\frac{\cos 2x}{2} - \frac{\cos 4x}{4} + \frac{\cos 6x}{6}\right]$.

(*ii*) $\frac{1}{4}\left[-2\frac{\cos mx}{m} - \frac{\cos(m+2)x}{m+2} - \frac{\cos(m-2)x}{m-2}\right]$.

Pages 24—25.

1. $\frac{(ax^2+bx+c)^{n+1}}{n+1}, -\frac{1}{(n-1)(ax^2+bx+c)^{n-1}}$.

2. (*a*) $\frac{(x^2+1)^{\frac{3}{2}}}{3}, -\frac{(a^2-x^2)^{\frac{n+2}{2}}}{n+2}$. (*b*) $\sqrt{x^2+2x+3}$.

3. $-\sqrt{1-x^2}, \sqrt{x^2-1}, \sqrt{a^2+x^2}$.

4. $\sqrt{x^2+a^2} + b\sinh^{-1}\frac{x}{a}, \sqrt{x^2-a^2} - b\cosh^{-1}\frac{x}{a}, -2\sqrt{1-x^2} + 3\sin^{-1}x$.

5. (*i*) $\frac{\sin^6\theta}{6}$. (*ii*) $-\frac{\cos^4 x}{4}$.

6. (*i*) $-\dfrac{\cos^4 x}{2}$. (*ii*) $\dfrac{\tan^{n+1} x}{n+1}$.

7. (*i*) $\dfrac{(e^x - a)^{n+1}}{n+1}$. (*ii*) $-\dfrac{1}{\log x}$.

8. $\dfrac{(\sin^{-1} x)^2}{2}, \dfrac{(\tan^{-1} x)^2}{2}, \dfrac{(\sec^{-1} x)^2}{2}$. 9. $\sqrt{x^2 - 1} + \cosh^{-1} x$.

10. $\sin^{-1} x - \sqrt{1 - x^2}$.

Page 26.

1. $\log (e^x - a), \log (e^x + e^{-x}), \log \tan x$.
2. $\log \sin^{-1} x, \log \tan^{-1} x, \log \sec^{-1} x$. 3. $\log (\log x)$.
4. $\log \{\log (\log x)\}, \log [\log \{\log (\log x)\}], l^{n+1} (x)$.
5. (*i*) $\log (\log \sin x)$. (*ii*) $\log (\log \tan x)$.
6. $\frac{1}{3} \log (x^3 + a^3), \dfrac{1}{n} \log (x^n + a^n)$. 7. $\dfrac{1}{e} \log (x^e - e^x)$.

Page 27.

1. $\log \tan x$. 2. $\log \sin x - \operatorname{cosec} x$.

Pages 28—29. Art. 20.

1. $\frac{1}{2} \log \tan \left(\dfrac{\pi}{4} + \dfrac{x}{2}\right)$.
2. (*i*) $\dfrac{1}{a} \log \tan \dfrac{ax + b}{2}$. (*ii*) $\frac{1}{2} \log \tan x$.
3. $\frac{1}{2} \log \tan \left(\dfrac{\pi}{4} + x\right), \frac{1}{2} \log \tan \left(\dfrac{\pi}{4} + x\right), \frac{1}{3} \log \tan \dfrac{3x}{2}$.
4. (*i*) $\dfrac{1}{\sqrt{2}} \log \tan \left(\dfrac{x}{2} + \dfrac{\pi}{8}\right)$. (*ii*) $\dfrac{1}{\sqrt{a^2 + b^2}} \log \tan \left(\dfrac{x}{2} + \dfrac{1}{2} \tan^{-1} \dfrac{a}{b}\right)$.
5. $-\dfrac{1}{a^2 + b^2} \cot \left(x + \tan^{-1} \dfrac{b}{a}\right)$.
6. (*i*) $2 \left(\sin \dfrac{x}{2} - \cos \dfrac{x}{2}\right)$. (*ii*) $2 \left(\sin \dfrac{x}{2} - \cos \dfrac{x}{2}\right) - \sqrt{2} \log \tan \left(\dfrac{x}{4} + \dfrac{\pi}{8}\right)$.
7. $\log \tanh \dfrac{x}{2}$.

Pages 29—30. Art. 21.

1. $\dfrac{e^{m \tan^{-1} x}}{m}$. 2. (*i*) $-\cos x^n$. (*ii*) $\frac{1}{3} \tan x^3$.

3. (*i*) $-\cos e^x$. (*ii*) $a \sin e^x + b \log \cosh x$. **4.** $\sin (\log x)$.

5. (*i*) $\tan^{-1} x^2$. (*ii*) $\frac{1}{3} \tan^{-1} x^3$.

6. (*i*) $\frac{1}{2} \sin^{-1} \frac{x^2}{a^2}$. (*ii*) $\frac{1}{4} \sinh^{-1} x^4$.

7. (*i*) $\frac{1}{ab} \tan^{-1} \left(\frac{a}{b} \tan \theta\right)$. (*ii*) $\frac{1}{a\sqrt{a^2-b^2}} \tan^{-1} \left(\frac{a \tan x}{\sqrt{a^2-b^2}}\right)$.

8. $(a > b) - \frac{1}{\sqrt{a-b}} \sinh^{-1} \left[\sqrt{\frac{a-b}{b}} \cos x\right]$; $(a < b) - \frac{1}{\sqrt{b-a}} \sin^{-1} \left[\sqrt{\frac{b-a}{b}} \cos x\right]$.

9. $-\frac{2}{b^2}\left[\log (a + b \cos x) + \frac{a}{a + b \cos x}\right]$. **10.** $-\cosh^{-1} (\cos \theta + \sin \theta)$.

Pages 32—33. Art. 22.

1. (*i*) $\frac{\pi}{2}$. (*ii*) $\frac{\pi}{4}$.

2. (*i*) $\tan^{-1} e^x$. (*ii*) $2 \tan^{-1} \frac{e-1}{e+1}$.

3. $\frac{\pi}{2ab}$. **4.** $\frac{\pi}{4}$.

MISCELLANEOUS EXAMPLES ON CHAPTER II

Pages 33—35.

1. (*i*) $2 \log 2 - 1$. (*ii*) $\frac{2}{3}[(1+x)^{\frac{3}{2}} + x^{\frac{3}{2}}]$. **2.** $\frac{2}{\alpha}$.

5. $2 - \frac{\pi}{2}$. **6.** (*i*) $\log (1 + \log x)$. (*ii*) $\frac{1}{4} \log \left(1 + \frac{2}{\sqrt{3}}\right)$.

(*iii*) $\frac{1}{2(a^2 - b^2)} \log (a^2 \sin^2 x + b^2 \cos^2 x)$.

7. (*i*) $2 \sin x - \log \tan \left(\frac{\pi}{4} + \frac{x}{2}\right)$. (*ii*) $-\frac{1}{5} \cot (x + \tan^{-1} \frac{1}{2})$.

8. (*i*) $\frac{1}{2} [\log (\sec x + \tan x)]^2$. (*ii*) $\frac{2}{3} \sin^{-1} \left(\frac{x}{a}\right)^{\frac{3}{2}}$. **9.** $\frac{1}{2} \tan^{-1} (2 \tan x)$.

11. (*i*) $x \cos a + \sin a \log \sin (x - a)$. (*ii*) $\frac{1}{\sin (a-b)} \log \frac{\sin (x-a)}{\sin (x-b)}$.

12. $-\cos \alpha \,.\, \sin^{-1} \frac{\cos x}{\cos \alpha} - \sin \alpha \,.\, \cosh^{-1} \frac{\sin x}{\sin \alpha}$.

13. $-\frac{2}{\sin \alpha} \sqrt{\frac{\sin (x + \alpha)}{\sin x}}$.

CHAPTER III

Pages 38—39.

1. $e^x(x-1)$, $x \sin x + \cos x$, $x \tan x + \log \cos x$.

2. $x \cosh x - \sinh x$, $\dfrac{x \sin 2x}{2} + \dfrac{\cos 2x}{4}$.

3. (*i*) $\dfrac{1}{2}\left[-\dfrac{x \cos 2x}{2} + \dfrac{\sin 2x}{4}\right]$. (*ii*) $\dfrac{1}{2}\left[\dfrac{x^2}{2} + \dfrac{x \sin 2x}{2} + \dfrac{\cos 2x}{4}\right]$

(*iii*) $\dfrac{1}{4}\left[3(x \sin x + \cos x) + \dfrac{x \sin 3x}{3} + \dfrac{\cos 3x}{9}\right]$.

4. $\dfrac{1}{4}\left[x\left(-\dfrac{\cos 2x}{2} - \dfrac{\cos 4x}{4} + \dfrac{\cos 6x}{6}\right) + \dfrac{\sin 2x}{4} + \dfrac{\sin 4x}{16} - \dfrac{\sin 6x}{36}\right]$

5. (*i*) $\dfrac{x^3}{3} \log x - \dfrac{x^3}{9}$. (*ii*) $\dfrac{x^4}{4} \tan^{-1} x - \dfrac{1}{4}\left(\dfrac{x^3}{3} - x + \tan^{-1} x\right)$.

(*iii*) $\dfrac{x^3}{3} \sin^{-1} x - \dfrac{1}{9}(1-x^2)^{\frac{3}{2}} + \dfrac{1}{3}\sqrt{1-x^2}$.

6. (*i*) $x \sin^{-1} x + \sqrt{1-x^2}$. (*ii*) $x \tan^{-1} x - \dfrac{1}{2} \log(1+x^2)$.

(*iii*) $x \sec^{-1} x - \cosh^{-1} x$. **7.** $-\sqrt{1-x^2} \sin^{-1} x + x$.

8. $(a+x) \tan^{-1} \sqrt{\dfrac{x}{a}} - \sqrt{ax}$. **9.** $\dfrac{\pi^2}{4}$.

10. $2\left[x \tan^{-1} x - \dfrac{1}{2} \log(1+x^2)\right]$. **11.** $\dfrac{\sin 2\theta}{2} \log \dfrac{\cos\theta + \sin\theta}{\cos\theta - \sin\theta} + \dfrac{\log \cos 2\theta}{2}$.

12. $\dfrac{\sin x - x \cos x}{x \sin x + \cos x}$. **13.** $\cosh x \tan \dfrac{x}{2}$.

Page 42.

1. $\dfrac{x\sqrt{1-x^2}}{2} + \dfrac{1}{2} \sin^{-1} x$, $\dfrac{x\sqrt{x^2-1}}{2} - \dfrac{1}{2} \cosh^{-1} x$, $\dfrac{x\sqrt{1+x^2}}{2} + \dfrac{1}{2} \sinh^{-1} x$

3. $-\left[\dfrac{\cot x \operatorname{cosec} x}{2} + \dfrac{1}{2} \sinh^{-1}(\cot x)\right]$.

4. (*i*) $-\dfrac{x\sqrt{1-x^2}}{2} + \dfrac{1}{2} \sin^{-1} x$. (*ii*) $\dfrac{x\sqrt{1+x^2}}{2} - \dfrac{1}{2} \sinh^{-1} x$.

5. (*i*) $\dfrac{x\sqrt{x^2+4}}{2} - \sinh^{-1} \dfrac{x}{2}$. (*ii*) $\dfrac{1}{3}(x^2+1)^{\frac{3}{2}} + \dfrac{x\sqrt{x^2+1}}{2} + \dfrac{1}{2} \sinh^{-1} x$.

6. (*i*) $-\dfrac{x\sqrt{1-x^2}}{2} + 2\sqrt{1-x^2} + \dfrac{7}{2} \sin^{-1} x$. (*ii*) $\dfrac{x\sqrt{x^2+1}}{2} + 2\sqrt{x^2+1} + \dfrac{5}{2} \sinh^{-1} x$

7. (i) $\dfrac{x\sqrt{1-x^2}}{2}-\sqrt{1-x^2}-\dfrac{1}{2}\sin^{-1}x$. (ii) $\dfrac{x\sqrt{x^2-4}}{2}+3\sqrt{x^2-4}+4\cosh^{-1}\dfrac{x}{2}$.

8. $\dfrac{x^2}{2}\sin^{-1}x+\dfrac{1}{4}x\sqrt{1-x^2}-\dfrac{1}{4}\sin^{-1}x$.

Page 43.

1. $e^x\sin x$. 2. (i) $e^x\tan\dfrac{x}{2}$. (ii) $-e^x\cot\dfrac{x}{2}$.

Page 44.

1. $e^x(x^2-2x+2)$.
2. (i) $x^2\sin x+2x\cos x-2\sin x$. (ii) $-x^2\cos x+2x\sin x+2\cos x$.
(iii) $x^2\sinh x-2x\cosh x+2\sinh x$. 3. π^2-4.
4. $\dfrac{x^{n+1}}{n+1}(\log x)^2-\dfrac{2}{(n+1)^2}x^{n+1}\log x+\dfrac{2}{(n+1)^3}x^{n+1}$.

Pages 47—49.

1. (i) $\dfrac{1}{2\sqrt{m^2+4}}e^{mx}\sin\left(2x-\tan^{-1}\dfrac{2}{m}\right)$.

(ii) $\dfrac{a+c}{2}e^x+\dfrac{b}{\sqrt{5}}e^x\sin(2x-\tan^{-1}2)+\dfrac{c-a}{2\sqrt{5}}e^x\cos(2x-\tan^{-1}2)$.

2. $\dfrac{1}{4\sqrt{m^2+16}}e^{mx}\sin\left(4x-\tan^{-1}\dfrac{4}{m}\right)$.

3. $\dfrac{1}{\sqrt{m^2+1}}e^{m\theta}\cos\left(\theta-\tan^{-1}\dfrac{1}{m}\right)$, where $x=\sin\theta$.

4. (i) $\dfrac{1}{\sqrt{m^2+1}}e^{m\theta}\cos\left(\theta-\tan^{-1}\dfrac{1}{m}\right)$, where $x=\tan\theta$.

(ii) $\dfrac{1}{2}\left[\dfrac{e^{m\theta}}{m}+\dfrac{1}{\sqrt{m^2+4}}e^{m\theta}\cos\left(2\theta-\tan^{-1}\dfrac{2}{m}\right)\right]$, where $x=\tan\theta$.

(iii) $\dfrac{1}{4}\left[\dfrac{3}{\sqrt{m^2+1}}e^{m\theta}\cos\left(\theta-\tan^{-1}\dfrac{1}{m}\right)+\dfrac{1}{\sqrt{m^2+9}}e^{m\theta}\cos\left(3\theta-\tan^{-1}\dfrac{3}{m}\right)\right]$,
where $x=\tan\theta$.

5. $\dfrac{x}{\sqrt{1+b^2}}\cos\left(b\log\dfrac{x}{a}-\tan^{-1}b\right)$. 9. $\dfrac{1}{a^2+b^2}[a\sinh ax\sin bx-b\cosh ax\cos bx]$.

Page 50.

1. (i) $e^x(x^3-3x^2+6x-6)$. (ii) $e^x(x^4-4x^3+12x^2-24x+24)$.

2. $265\,e - 720$.

3. (*i*) $-x^3 \cos x + 3x^2 \sin x + 6x \cos x - 6 \sin x$.

(*ii*) $\frac{1}{2}\left[\frac{x^4}{4} - \frac{x^3}{2}\sin 2x - \frac{3}{4}x^2\cos 2x + \frac{3}{4}x\sin 2x + \frac{3}{8}\cos 2x\right]$.

4. (*i*) $\pi^5 - 20\pi^3 + 120\pi$. (*ii*) $-5\pi^4 + 60\pi^2 - 240$.

5. (*i*) $x^3 \cosh x - 3x^2 \sinh x + 6x \cosh x - 6 \sinh x$.

(*ii*) $x^5 \sinh x - 5x^4 \cosh x + 20x^3 \sinh x - 60x^2 \cosh x + 120\,x \sinh x - 120 \cosh x$

Page 51. Art. 29.

1. $\frac{1}{2}e^x(x-1) - \frac{1}{2}e^x\left[x\frac{1}{\sqrt{5}}\cos(2x - \tan^{-1}2) - \frac{1}{5}\cos(2x - 2\tan^{-1}2)\right]$

2. $2^x\left[\frac{x^2}{r}\sin(2x-\alpha) - \frac{2x}{r^2}\sin(2x-2\alpha) + \frac{2}{r^3}\sin(2x-3\alpha)\right]$,

where $r = \sqrt{(\log 2)^2 + 4}$, and $\alpha = \tan^{-1}\frac{2}{\log 2}$.

MISCELLANEOUS EXAMPLES ON CHAPTER III

Pages 51—52.

1. (*i*) $e^{ax}\left(\frac{x}{a} - \frac{1}{a^2}\right)$. (*ii*) $x\tan x + \log\cos x - \frac{x^2}{2}$. (*iii*) $x\tan\frac{x}{2}$.

2. (*i*) $\frac{x^2}{2}\log x - \frac{x^2}{4}$. (*ii*) $\frac{1}{2}[(x^2+1)\tan^{-1}x - x]$.

3. (*i*) $\frac{1}{3}\left[x^3\tan^{-1}x - \frac{x^2}{2} + \frac{1}{2}\log(1+x^2)\right]$. (*ii*) $x\tan^{-1}x - \frac{1}{2}\log(1+x^2) - \frac{(\tan^{-1}x)^2}{2}$.

4. (*i*) $x\log(x + \sqrt{a^2+x^2}) - \sqrt{a^2+x^2}$. (*ii*) $x\log(1+x^2) - 2x + 2\tan^{-1}x$.

5. (*i*) $\frac{1}{\sqrt{1+x^2}}(x - \tan^{-1}x)$. (*ii*) $\frac{x}{\sqrt{1-x^2}}\sin^{-1}x + \frac{1}{2}\log(1-x^2)$.

6. (*b*) $a^2\left(\theta\cos 4\theta - \frac{\sin 4\theta}{4}\right)$, where $x = 2a\cos 2\theta$.

7. (*i*) $e^x \log x$. (*ii*) $e^x\frac{x-1}{x+1}$.

8. (*i*) $\frac{1}{a^2+b^2}e^{ax}[a\cos(bx+c) + b\sin(bx+c)]$

$$= \frac{1}{\sqrt{a^2+b^2}}e^{ax}\cos\left(bx + c - \tan^{-1}\frac{b}{a}\right).$$

(ii) $\frac{e^{2x}}{4}\left[\frac{3}{\sqrt{5}}\cos\left(x-\tan^{-1}\frac{1}{2}\right)+\frac{1}{\sqrt{13}}\cos\left(3x-\tan^{-1}\frac{3}{2}\right)\right]$.

(iii) $\frac{e^{ax}}{4}\left[\frac{3}{\sqrt{a^2+1}}\sin\left(x-\tan^{-1}\frac{1}{a}\right)-\frac{1}{\sqrt{a^2+9}}\sin\left(3x-\tan^{-1}\frac{3}{a}\right)\right]$.

9. $\frac{\alpha(\alpha^2+\beta^2+\gamma^2)}{(\alpha^2+\beta^2+\gamma^2)^2-4\beta^2\gamma^2}$.

10. (i) $e^{mx}\left[\frac{x^2}{m}-\frac{2x}{m^2}+\frac{2}{m^3}\right]$. (ii) $\frac{x^3}{3}\left[(\log x)^2-\frac{2}{3}\log x+\frac{2}{9}\right]$.

(iii) $x\,[(\log x)^2-2\log x+2]$. **11.** $\frac{3}{25}$.

12. (a) $\frac{2}{27}-\frac{\pi^2}{12}$. (b) $e^x\left[\frac{x}{\sqrt{2}}\cos\left(x-\frac{\pi}{4}\right)-\frac{1}{2}\sin x+\frac{1}{\sqrt{2}}\sin\left(x-\frac{\pi}{4}\right)\right]$.

CHAPTER IV

Page 56.

1. (i) $\frac{1}{4}\log\frac{x-2}{x+2}$. (ii) $x+\log\frac{x-2}{x+2}$.

2. $\frac{1}{2\sqrt{3}}\log\frac{\sqrt{3}+\tan x}{\sqrt{3}-\tan x}$.

Page 57.

1. (i) $\frac{1}{3}\log\frac{x-1}{2x+1}$. [**Note.** The constant $\frac{1}{3}\log 2$ has been omitted.]

(ii) $\tan^{-1}(2x-1)$. **2.** $\frac{1}{\sqrt{2}}\tan^{-1}\frac{1}{2\sqrt{2}}$.

Page 58.

1. $\frac{1}{2}\log(x^2+4x+5)-\tan^{-1}(x+2)$. **2.** $\log(x+1)+\frac{1}{x+1}$.

3. $\frac{1}{2}\log(2x^2+2x+1)-2\tan^{-1}(2x+1)$. **4.** $-\log(3-x)$.

5. (i) $2x-\frac{5}{2}\log(x^2+4x+5)+4\tan^{-1}(x+2)$.

(ii) $x-2\log(x^2+2x+10)-\frac{5}{3}\tan^{-1}\frac{x+1}{3}$.

Pages 59—62.

1. (i) $\frac{1}{a-b}[a\log(x-a)-b\log(x-b)]$.

(*ii*) $\dfrac{a^2}{(b-a)(c-a)}\log(x+a)+\dfrac{b^2}{(a-b)(c-b)}\log(x+b)+\dfrac{c^2}{(a-c)(b-c)}\log(x+c)$.

2. (*i*) $-\frac{1}{6}\log(x+1)+\frac{4}{15}\log(x-2)+\frac{9}{10}\log(x+3)$.

(*ii*) $\dfrac{a^2+la+m}{(a-b)(a-c)}\log(x-a)+\dfrac{b^2+lb+m}{(b-a)(b-c)}\log(x-b)+\dfrac{c^2+lc+m}{(c-a)(c-b)}\log(x-c)$.

3. (*i*) $-\frac{1}{4}\log x+\frac{1}{8}\log(x^2-4)$. (*ii*) $\log\dfrac{x^2-1}{x}$. (*iii*) $\frac{1}{2}\log x+\frac{3}{4}\log(2-x^2)$.

4. $x+\dfrac{a^2}{a-b}\log(x-a)+\dfrac{b^2}{b-a}\log(x-b)$.

5. (*i*) $x+\dfrac{(c-a)(c-b)}{c-d}\log(x-c)+\dfrac{(d-a)(d-b)}{d-c}\log(x-d)$.

(*ii*) $x+\dfrac{(a'-a)(a'-b)(a'-c)}{(a'-b')(a'-c')}\log(x-a')$

$+\dfrac{(b'-a)(b'-b)(b'-c)}{(b'-a')(b'-c')}\log(x-b')+\dfrac{(c'-a)(c'-b)(c'-c)}{(c'-a')(c'-b')}\log(x-c')$.

6. $\dfrac{x^3}{3}+x^2+9x-\frac{1}{6}\log(x-1)-\frac{32}{15}\log(x+2)+\frac{243}{10}\log(x-3)$.

7. (*i*) $\log\frac{4}{3}$.

(*ii*) $\frac{1}{2}\log(1+\cos x)+\frac{1}{6}\log(1-\cos x)-\frac{2}{3}\log(1+2\cos x)$.

8. $\frac{1}{3}[\log x^3+\frac{1}{2}\log(1+x^3)-\frac{3}{2}\log(1+3x^3)]$.

9. $\frac{1}{5}\log\dfrac{x^5}{x^5+1}$.

10. (*i*) $\log\dfrac{e^x-1}{e^x}$. (*ii*) $x+\log(1+e^x)-2\log(1+2e^x)$.

11. $\dfrac{1+x^3}{3}\log(1+x)-\dfrac{1-x^3}{3}\log(1-x)-\dfrac{2}{3}\left(\dfrac{x^3}{3}+x\right)$.

Page 64.

1. (*i*) $-\dfrac{1}{2}\dfrac{1}{x-1}+\dfrac{1}{4}\log\dfrac{x+1}{x-1}$.

(*ii*) $\dfrac{1}{6}\dfrac{1}{(x+1)^3}+\dfrac{1}{8}\dfrac{1}{(x+1)^2}+\dfrac{1}{8}\dfrac{1}{x+1}+\dfrac{1}{16}\log\dfrac{x-1}{x+1}$.

2. (*i*) $-\dfrac{1}{ax}+\dfrac{b}{a^2}\log\dfrac{a+bx}{x}$.

(*ii*) $-\dfrac{1}{2}\dfrac{1}{x+1}-\dfrac{3}{4}\log(x+1)+\dfrac{2}{3}\log(x+2)+\dfrac{1}{12}\log(x-1)$.

3. (*i*) $-\frac{2}{9}\frac{1}{x-1}+\frac{1}{9}\frac{1}{x+2}+\frac{1}{27}\log\frac{x+2}{x-1}$. (*ii*) $\frac{1}{3}+\frac{1}{4}\log 3$.

4. $\frac{1}{2}\frac{1}{1+\sin x}+\frac{1}{4}\log\frac{1+\sin x}{1-\sin x}$. **5.** $-\frac{1}{e^x-1}+\log\frac{e^x}{e^x-1}$.

Pages 66—69.

1. (*i*) $\frac{1}{18}\log(x-3)-\frac{1}{36}\log(x^2+9)-\frac{1}{18}\tan^{-1}\frac{x}{3}$.

(*ii*) $-\frac{1}{2}\log(x+2)+\frac{1}{4}\log(x^2+4)$.

2. (*i*) $\frac{1}{3}\log(x-1)-\frac{1}{6}\log(x^2+x+1)+\frac{1}{\sqrt{3}}\tan^{-1}\frac{2x+1}{\sqrt{3}}$.

(*ii*) $\frac{1}{2}\log(1+x)-\frac{1}{4}\log(1+x^2)+\frac{1}{2}\tan^{-1}x$.

3. $-\frac{1}{2}\log(x+1)+\frac{1}{6}\log(x^3-1)-\frac{1}{\sqrt{3}}\tan^{-1}\frac{2x+1}{\sqrt{3}}$.

4. (*i*) $\frac{1}{4}(1+\log 2)$. (*ii*) $\frac{1}{2}\frac{1}{1-x}-\frac{1}{2}\log(1-x)+\frac{1}{4}\log(1+x^2)$.

5. (*i*) $-\frac{1}{3}\frac{1}{2+x}+\frac{1}{3}\log(2+x)-\frac{1}{6}\log(1+x+x^2)+\frac{1}{3\sqrt{3}}\tan^{-1}\frac{1+2x}{\sqrt{3}}$.

(*ii*) $-\frac{1}{x-1}+\log(x-1)-\frac{1}{2}\log(x^2-x+1)-\frac{1}{\sqrt{3}}\tan^{-1}\frac{2x-1}{\sqrt{3}}$.

6. (*i*) $\frac{1}{a^2}\frac{1}{x}-\frac{2}{a^3}\log x+\frac{1}{a^3}\log(x-a)+\frac{1}{2a^3}\log(x^2+a^2)$.

(*ii*) $\frac{1}{2}\frac{1}{x+1}+\log\frac{x}{x+1}-\frac{1}{2}\tan^{-1}x$.

7. $\frac{1}{2}\log\frac{x^2+1}{x^2-x+1}+\frac{1}{\sqrt{3}}\tan^{-1}\frac{2x-1}{\sqrt{3}}$. **9.** $\frac{1}{2a}$.

10. $-\frac{1}{2}\frac{1+x}{1+x^2}-\frac{1}{2}\log(1-x)+\frac{1}{4}\log(1+x^2)$.

Page 70.

1. (*i*) $\frac{1}{3\sqrt{2}}\tan^{-1}\left[\frac{3x}{(1-x^2)\sqrt{2}}\right]$. (*ii*) $\frac{1}{6}\log\frac{x-1}{x+1}+\frac{\sqrt{2}}{3}\tan^{-1}\frac{x}{\sqrt{2}}$.

(*iii*) $\frac{1}{10}\log\frac{x-1}{x+1}-\frac{1}{10}\tan^{-1}\frac{x}{2}$. **2.** $\frac{1}{a(b^2-a^2)}\tan^{-1}\frac{x}{a}+\frac{1}{b(a^2-b^2)}\tan^{-1}\frac{x}{b}$.

3. $x+\frac{(a^2-c^2)(b^2-c^2)}{c(d^2-c^2)}\tan^{-1}\frac{x}{c}+\frac{(a^2-d^2)(b^2-d^2)}{d(c^2-d^2)}\tan^{-1}\frac{x}{d}$.

5. $\frac{1}{4}\left[-\frac{1}{2}\log\frac{1+\sin x}{1-\sin x}+\frac{1}{\sqrt{2}}\log\frac{1+\sqrt{2}\sin x}{1-\sqrt{2}\sin x}\right]$.

6. $-\frac{1}{\sqrt{3}}\log\frac{\sqrt{3}+\tan\theta}{\sqrt{3}-\tan\theta}+\theta$.

Page 71.

1. (*i*) $\frac{1}{2}\left[-2\log x(x^2+1)+3\log(x^2-1)\right]$.

(*ii*) $\frac{1}{2}\left[\frac{1}{2}\frac{1}{(x^2+1)^2}+\frac{1}{x^2+1}+\log\frac{x^2}{x^2+1}\right]$.

2. $x-\frac{1}{4}\log(x-1)-\frac{3}{4}\log(x+1)-\frac{1}{2}\tan^{-1}x+2\log x-\frac{1}{2}\log(x^2+1)$.

Pages 72—73. Art. 40-II.

1. (*i*) $\frac{1}{2}\log\frac{x^2-x+1}{x^2+x+1}$. (*ii*) $\frac{1}{\sqrt{2}}\tan^{-1}\frac{x^2-1}{x\sqrt{2}}$.

2. (*i*) $\tan^{-1}\frac{x^2-1}{x}-\sqrt{3}\tan^{-1}\frac{2x^2-1}{\sqrt{3}}$. (*ii*) $\frac{1}{\sqrt{3}}\tan^{-1}\frac{x^2-1}{x\sqrt{3}}-\frac{2}{\sqrt{3}}\tan^{-1}\frac{2x^2+1}{\sqrt{3}}$.

3. (*i*) $\frac{1}{2\sqrt{2}}\tan^{-1}\frac{x^2-1}{x\sqrt{2}}-\frac{1}{4\sqrt{2}}\log\frac{x^2-x\sqrt{2}+1}{x^2+x\sqrt{2}+1}$

(*ii*) $\frac{1}{2\sqrt{2}}\tan^{-1}\frac{x^2-1}{x\sqrt{2}}+\frac{1}{4\sqrt{2}}\log\frac{x^2-x\sqrt{2}+1}{x^2+x\sqrt{2}+1}$.

4. (*i*) $\frac{1}{a}\tan^{-1}\frac{x^2-a^2}{ax}$ (*ii*) $\frac{1}{2a}\log\frac{x^2-ax+a^2}{x^2+ax+a^2}$

5. (*i*) $\frac{\pi}{2\sqrt{2}}+\frac{1}{\sqrt{2}}\log(\sqrt{2}-1)$ (*ii*) $\frac{\pi}{2\sqrt{2}}-\frac{1}{\sqrt{2}}\log(\sqrt{2}-1)$ **6.** $\frac{\pi}{2}$

MISCELLANEOUS EXAMPLES ON CHAPTER IV

Pages 73—75.

1. (*i*) $\frac{1}{5}\log\frac{x-3}{x+2}$. (*ii*) $\frac{\alpha}{2}$ cosech α.

2. (*i*) $\frac{35}{12}\log(x+10)+\frac{1}{12}\log(x-2)$. (*ii*) $\frac{1}{n}\log\frac{x^n}{x^n+1}$.

3. (*i*) $-\frac{2}{15}+\log\frac{6}{5}$. (*ii*) $-\frac{1}{3}\frac{1}{x-1}+\frac{2}{9}\log\frac{x-1}{x+2}$.

4. (*i*) $-\frac{5}{(x-2)^2}-\frac{2}{x-2}+4\log(x-2)-3\log(x-1)$. (*ii*) $-\frac{x}{x^2-1}$.

5. (*i*) $\frac{1}{2}\log 2$.

(*ii*) $\frac{1}{3}\log(x-1)-\frac{1}{6}\log(x^2+x+1)-\frac{1}{\sqrt{3}}\tan^{-1}\frac{2x+1}{\sqrt{3}}$.

6. $-\frac{1}{4}\frac{1}{2x^2+4x+3}+\frac{1}{2\sqrt{2}}\tan^{-1}[(x+1)\sqrt{2}]+\frac{1}{2}\frac{x+1}{2x^2+4x+3}$.

7. (i) $\frac{1}{3\sqrt{2}}\tan^{-1}\frac{x}{\sqrt{2}}+\frac{1}{6}\log\frac{1+x}{1-x}$. (ii) $-\frac{1}{2}\tan^{-1}x+\frac{1}{4}\log\frac{x-1}{x+1}$.

9. (i) $x-\frac{3}{2}\tan^{-1}x+\frac{1}{2}\frac{x}{1+x^2}$. (ii) $\frac{3}{2}-\frac{\pi}{4}$.

10. (a) $\frac{x^4}{4(x^2+1)^2}$. (b) $-\frac{1}{2(x^2+1)}+\frac{1}{4(x^2+1)^2}$.

11. $\frac{1}{2}\log\frac{x^2+1}{x^2+3}$. 12. (i) $\frac{1}{2\sqrt{3}}\tan^{-1}\frac{x^2-1}{x\sqrt{3}}+\frac{1}{4}\log\frac{x^2-x+1}{x^2+x+1}$.

13. (i) $\frac{1}{2a\sqrt{2}}\tan^{-1}\frac{x^2-a^2}{ax\sqrt{2}}+\frac{1}{4a\sqrt{2}}\log\frac{x^2-ax\sqrt{2}+a^2}{x^2+ax\sqrt{2}+a^2}$.

(ii) $\frac{1}{6\sqrt{14}}\tan^{-1}\frac{x^2-3}{x\sqrt{14}}-\frac{1}{6\sqrt{2}}\tan^{-1}\frac{x^2+3}{x\sqrt{2}}$.

CHAPTER V

Page 78.

1. (i) $\sinh^{-1}\frac{x+1}{\sqrt{3}}$. (ii) $\cosh^{-1}\frac{2x+1}{3}$. (iii) $\frac{1}{\sqrt{2}}\sinh^{-1}\frac{2x-1}{\sqrt{5}}$.

2. $\sin^{-1}\frac{2}{\sqrt{3}}-\sin^{-1}\frac{1}{\sqrt{3}}, \frac{1}{\sqrt{3}}\sin^{-1}\frac{3x-4}{5}$.

3. (i) $\sin^{-1}\frac{x-a}{a}$. (ii) $\sin^{-1}(2x-1)$. (iii) $\frac{1}{2}\sin^{-1}(2x-1)$.

4. (i) $\frac{(x-1)\sqrt{x^2-2x}}{2}-\frac{1}{2}\cosh^{-1}(x-1)$. (ii) $(b^2>4ac)$

$$\frac{1}{8c^{\frac{3}{2}}}\left[2(b+2cx)\sqrt{c(a+bx+cx^2)}-(b^2-4ac)\cosh^{-1}\frac{b+2cx}{\sqrt{b^2-4ac}}\right].\ (b^2<4ac)$$

$$\frac{1}{8c^{\frac{3}{2}}}\left[2(b+2cx)\sqrt{c(a+bx+cx^2)}+(4ac-b^2)\sinh^{-1}\frac{b+2cx}{\sqrt{4ac-b^2}}\right].$$

(iii) $\frac{1}{8c^{\frac{3}{2}}}\left[2(2cx-b)\sqrt{c(a+bx-cx^2)}+(b^2+4ac)\sin^{-1}\frac{2cx-b}{\sqrt{b^2+4ac}}\right]$.

5. $\frac{\pi}{8}(\beta-\alpha)^2$.

Pages 79—80.

1. (i) $2\sqrt{x^2+x+1}+2\sinh^{-1}\frac{2x+1}{\sqrt{3}}$. (ii) $-\sqrt{9+8x-x^2}+4\sin^{-1}\frac{x-4}{5}$.

2. $\frac{1}{2}(x-1)\sqrt{x^2+2x+3}$.

3. $\frac{1}{3}(x^2-x+1)^{\frac{3}{2}}+\frac{3}{8}(2x-1)\sqrt{x^2-x+1}+\frac{9}{16}\sinh^{-1}\frac{2x-1}{\sqrt{3}}$.

4. $\frac{1}{3}(x^2+2x+3)^{\frac{3}{2}}-(x+1)\sqrt{x^2+2x+3}+2\sinh^{-1}\frac{x+1}{\sqrt{2}}$.

Page 82.

1. (*i*) $\tan^{-1}\sqrt{x}$. (*ii*) $2\tan^{-1}\sqrt{x-1}$.

2. (*i*) $\log\frac{\sqrt{x+1}-1}{\sqrt{x+1}+1}$. (*ii*) $2\tan^{-1}\sqrt{x}$.

3. (*i*) $\frac{2}{\sqrt{3}}\tan^{-1}\sqrt{\frac{1+2x}{3}}$. (*ii*) $\frac{1}{2}\log\frac{5}{3}$.

4. (*i*) $2\sqrt{x+3}-\sqrt{2}\log\frac{\sqrt{x+3}-\sqrt{2}}{\sqrt{x+3}+\sqrt{2}}$. (*ii*) $\frac{2}{3}(x-2)^{\frac{3}{2}}+2\sqrt{3}\tan^{-1}\sqrt{\frac{x-2}{3}}$.

Pages 85—86.

1. (*i*) $-\text{cosech}^{-1}x$. (*ii*) $-\frac{1}{\sqrt{2}}\sinh^{-1}\frac{x+1}{x-1}$.

2. (*i*) $-\frac{1}{\sqrt{3}}\cosh^{-1}\frac{3-2x}{x}$. (*ii*) $\sqrt{\frac{x-1}{x+1}}$.

3. (*i*) $\frac{\pi}{3}$. (*ii*) $-\sinh^{-1}\frac{1-x}{(1+x)\sqrt{3}}$.

4. (*i*) $\sinh^{-1}x-\frac{1}{\sqrt{2}}\sinh^{-1}\frac{x+1}{x-1}$.

(*ii*) $\sqrt{x^2+4x+6}-\sinh^{-1}\frac{x+2}{\sqrt{2}}-\frac{1}{\sqrt{2}}\text{cosech}^{-1}\frac{x+2}{\sqrt{2}}$.

5. (*i*) $x\sqrt{\frac{x+1}{x-1}}$.

(*ii*) $\sqrt{1+x+x^2}-\frac{1}{2}\sinh^{-1}\frac{1+2x}{\sqrt{3}}-\sinh^{-1}\frac{1-x}{(1+x)\sqrt{3}}$.

6. $x\sin^{-1}x\,.\left(1+\log\frac{a}{x}\right)+\sqrt{1-x^2}\left(2+\log\frac{a}{x}\right)-\text{sech}^{-1}x$.

7. $\frac{\sqrt{x^2-a^2}}{a^2x}$. 8. $-\frac{1}{\sqrt{a+c}}\sinh^{-1}\left[\sqrt{\frac{a+c}{b+c}}\cot\theta\right]$.

9. $-2\sec^{-1}\frac{1+x}{\sqrt{x}}$.

Page 88.

1. (*i*) $\frac{1}{\sqrt{2}}\tan^{-1}\frac{x-1}{\sqrt{2x}}-\frac{1}{2\sqrt{2}}\log\frac{x-\sqrt{2x}+1}{x+\sqrt{2x}+1}$.

(*ii*) $\frac{1}{\sqrt{2}}\log\frac{x-\sqrt{2(x+2)}+3}{x+\sqrt{2(x+2)}+3}$.

2. $2\sqrt{x-1}+\frac{3}{\sqrt{2}}\tan^{-1}\frac{x-2}{\sqrt{2(x-1)}}+\frac{1}{2\sqrt{2}}\log\frac{x-\sqrt{2(x-1)}}{x+\sqrt{2(x-1)}}$.

Pages 90—91.

1. (*i*) $-\frac{1}{2\sqrt{2}}\log\frac{x\sqrt{2}+\sqrt{x^2+1}}{x\sqrt{2}-\sqrt{x^2+1}}$.

2. $\frac{1}{8\sqrt{2}}\log\frac{x\sqrt{2}+\sqrt{x^2-4}}{x\sqrt{2}-\sqrt{x^2-4}}$.

3. $(a^2>b^2)\ \frac{1}{\sqrt{a^2-b^2}}\tan^{-1}\sqrt{\frac{x^2+b^2}{a^2-b^2}}$. $(a^2<b^2)\ \frac{1}{2\sqrt{b^2-a^2}}\log\frac{\sqrt{x^2+b^2}-\sqrt{b^2-a^2}}{\sqrt{x^2+b^2}+\sqrt{b^2-a^2}}$.

4. $(b^2>c^2)\ \frac{1}{\sqrt{b^2-c^2}}\tan^{-1}\sqrt{\frac{x^2+c^2}{b^2-c^2}}-\frac{a}{2b\sqrt{b^2-c^2}}\log\frac{b\sqrt{x^2+c^2}-x\sqrt{b^2-c^2}}{b\sqrt{x^2+c^2}+x\sqrt{b^2-c^2}}$. $(b^2<c^2)$

$\frac{1}{2\sqrt{c^2-b^2}}\log\frac{\sqrt{x^2+c^2}-\sqrt{c^2-b^2}}{\sqrt{x^2+c^2}+\sqrt{c^2-b^2}}-\frac{a}{b\sqrt{c^2-b^2}}\tan^{-1}\frac{b\sqrt{x^2+c^2}}{x\sqrt{c^2-b^2}}$.

5. $-\sinh^{-1}x-\frac{1}{\sqrt{2}}\log\frac{\sqrt{1+x^2}-x\sqrt{2}}{\sqrt{1+x^2}+x\sqrt{2}}$.

6. $-\frac{1}{2}\sinh^{-1}\frac{1-x}{x}-\frac{1}{2\sqrt{5}}\sinh^{-1}\frac{3x-1}{x-2}$.

Page 94.

2. $\frac{x^4}{4}+\frac{x^9}{3}+\frac{3x^{14}}{14}+\frac{x^{19}}{19}$.

4. (*i*) $\frac{x}{c\sqrt{ax^2+c}}$. (*ii*) $\frac{1}{na^{\frac{n}{2}}}\log\frac{\sqrt{a^n+x^n}-a^{\frac{n}{2}}}{\sqrt{a^n+x^n}+a^{\frac{n}{2}}}$.

Page 95. Art. 48.

1. $(2\sqrt{2}-\sqrt{3}-1)\frac{\pi}{2}$.

2. (*i*) $\frac{3}{2}\log(x^{\frac{2}{3}}+1)$.

(*ii*) $\frac{2}{a-b}\left[\sqrt{a(x-b)}-\sqrt{b(x-a)}-\sqrt{ab}\left(\tan^{-1}\sqrt{\frac{x-b}{b}}-\tan^{-1}\sqrt{\frac{x-a}{a}}\right)\right]$.

Pages 95—96. Art. 49.

1. $6\left[\frac{1}{3}(1+x)^{\frac{1}{2}}-\frac{1}{2}(1+x)^{\frac{1}{3}}+(1+x)^{\frac{1}{6}}-\log\{1+(1+x)^{\frac{1}{6}}\}\right]$.

2. $4\left[\frac{2}{3}+\tan^{-1}2\right]$.

Pages 97—98. Art. 50.

1. $\frac{1}{2}\left[\frac{(x+\sqrt{a^2+x^2})^{n+1}}{n+1}+a^2\frac{(x+\sqrt{a^2+x^2})^{n-1}}{n-1}\right]$.

Page 98. Art. 51.

1. (*i*) $\frac{x}{a^2\sqrt{a^2-x^2}}$. (*ii*) $\frac{x}{a^2\sqrt{a^2+x^2}}$. 3. $-\frac{a+x}{a^2\sqrt{2ax+x^2}}$.

MISCELLANEOUS EXAMPLES ON CHAPTER V

Pages 98—102.

1. π.

2. (*i*) $(a>b)\ 2\sinh^{-1}\sqrt{\frac{x-a}{a-b}}$; $(a<b)\ 2\cosh^{-1}\sqrt{\frac{x-a}{b-a}}$. (*ii*) π

3. (*i*) $(a>b)\ \sqrt{(x-a)(x-b)}-(a-b)\cosh^{-1}\sqrt{\frac{x-b}{a-b}}$.

$(a<b)\ \sqrt{(x-a)(x-b)}+(b-a)\sinh^{-1}\sqrt{\frac{x-b}{b-a}}$.

(*ii*) $\sqrt{x(a+x)}+a\sinh^{-1}\sqrt{\frac{x}{a}}$. (*iii*) $a^2\left(\frac{\pi}{4}-\frac{1}{2}\right)$.

4. $\sqrt{e^{2x}+ae^x}+a\sinh^{-1}\sqrt{\frac{e^x}{a}}$.

5. (*i*) $\frac{1}{4}(2x+5)\sqrt{x^2+x+1}+\frac{15}{8}\sinh^{-1}\frac{2x+1}{\sqrt{3}}$.

(*ii*) $(3-x)\sqrt{3-2x-x^2}+9\sin^{-1}\frac{x+1}{2}$.

6. 0.

7. (*i*) $(a>b)\ \frac{1}{\sqrt{a-b}}\log\frac{\sqrt{x+a}-\sqrt{a-b}}{\sqrt{x+a}+\sqrt{a-b}}$. $(a<b)\ \frac{2}{\sqrt{b-a}}\tan^{-1}\sqrt{\frac{x+a}{b-a}}$.

(ii) $-6\sqrt{1+x}+2\log\dfrac{\sqrt{1+x}-1}{\sqrt{1+x}+1}$.

8. (a) (i) $-\sqrt{\dfrac{1-x}{1+x}}$. (ii) $-\dfrac{1}{a}\sqrt{\dfrac{x+a}{x-a}}$.

9. (i) $-\sin^{-1}\dfrac{-1-3x}{(1+x)\sqrt{5}}$. (ii) $-\dfrac{1}{\sqrt{2}}\sin^{-1}\dfrac{-x\sqrt{2}}{x+1}$.

(iii) $-\dfrac{2}{b-a}\sqrt{\dfrac{b-x}{x-a}}$.

10. (i) $\sqrt{x^2-a^2}-a\sec^{-1}\dfrac{x}{a}$. (ii) $-\dfrac{\sqrt{1+x^2}}{x}$.

11. (i) $-\dfrac{1}{\sqrt{2}}\sec^{-1}\dfrac{x^2+1}{x\sqrt{2}}$. (ii) $-\dfrac{1}{2x^2}\cos^{-1}x+\frac{1}{2}\dfrac{\sqrt{1-x^2}}{x}$.

12. $\sec^{-1}(\cos x+\sec x)$. **13.** $\dfrac{1}{ab}\tan^{-1}\left[\dfrac{b}{a}\sqrt{\dfrac{x^2-b^2}{a^2-x^2}}\right]$.

14. $\dfrac{1}{\sqrt{b^2+c^2}}\sinh^{-1}\left[\sqrt{\dfrac{b^2+c^2}{a^2+c^2}}\tan\theta\right]$.

15. $-\dfrac{1}{\sqrt{(b^2-a^2)(c^2-b^2)}}\tan^{-1}\dfrac{\sqrt{(b^2-a^2)(x^2+2ax+c^2)}}{(x+a)\sqrt{c^2-b^2}}$. $(a<b<c)$

Similarly for the other cases.

16. $\dfrac{1}{2\sqrt{5}}\log\dfrac{\sqrt{x^2+9}-\sqrt{5}}{\sqrt{x^2+9}+\sqrt{5}}-\dfrac{1}{2\sqrt{5}}\tan^{-1}\left[\dfrac{2\sqrt{x^2+9}}{x\sqrt{5}}\right]$.

17. (i) $\dfrac{3}{56}(1+x^2)^{\frac{4}{3}}(4x^2-3)$. (ii) $\frac{1}{3}\log\dfrac{\sqrt{x^3+1}-1}{\sqrt{x^3+1}+1}$.

18. (i) $2[\sqrt{x}-\tan^{-1}\sqrt{x}]$. (ii) $\frac{3}{5}(x-3)^{\frac{2}{3}}(x+12)$.

(iii) $\log(x+\sqrt{x-1})-\dfrac{2}{\sqrt{3}}\tan^{-1}\left[\dfrac{2\sqrt{x-1}+1}{\sqrt{3}}\right]$.

19. $-\cot\theta\log(\cos\theta+\sqrt{\cos 2\theta})-\cot\theta-\theta+\dfrac{\sqrt{\cos 2\theta}}{\sin\theta}$.

CHAPTER VI

Pages 103—104.

1. (i) $-\left[\cos x-\dfrac{\cos^3 x}{3}\right]$. (ii) $-\left[\cos x-2\dfrac{\cos^3 x}{3}+\dfrac{\cos^5 x}{5}\right]$. (iii) $\dfrac{16}{35}$.

2. (i) $\dfrac{43}{60\sqrt{2}}$. (ii) $\sin x-\sin^3 x+3\dfrac{\sin^5 x}{5}-\dfrac{\sin^7 x}{7}$.

Page 106.

1. (*i*) $\frac{1}{2}\left[x - \frac{\sin 2x}{2}\right]$. (*ii*) $\frac{1}{8}\left[3x - 2\sin 2x + \frac{\sin 4x}{4}\right]$.

(*iii*) $\frac{1}{8}\left[\frac{3\pi}{4} - 2\right]$.

2. $\frac{1}{2}\left[\frac{\pi}{4} + \frac{1}{2}\right]$. 3. (*i*) $\frac{1}{32}\left[\frac{5\pi}{2} + \frac{22}{3}\right]$.

(*ii*) $\frac{1}{2^7}\left[\frac{\sin 8x}{8} - 4\frac{\sin 6x}{3} + 7\sin 4x - 28\sin 2x + 35x\right]$.

Pages 107—108.

1. (*i*) $\frac{\sin^3 x}{3} - \frac{\sin^5 x}{5}$. (*ii*) $-\left[\frac{\cos^4 x}{4} - \frac{\cos^6 x}{6}\right]$.

2. (*i*) $-\left[\frac{\cos^5 x}{5} - \frac{\cos^7 x}{7}\right]$. (*ii*) $\frac{\sin^3\theta}{3} - 3\frac{\sin^5\theta}{5} + 3\frac{\sin^7\theta}{7} - \frac{\sin^9\theta}{9}$.

3. (*i*) $\frac{\sin^6\theta}{6} - \frac{\sin^8\theta}{8}$. (*ii*) $-\left[\frac{\cos^8\theta}{8} - \frac{\cos^{10}\theta}{5} + \frac{\cos^{12}\theta}{12}\right]$.

4. (*i*) $\frac{\sin^9\theta}{9} - \frac{\sin^{11}\theta}{11}, \frac{2}{99}$. (*ii*) $\frac{\sec^3 x}{3} - \sec x$.

5. (*i*) $\frac{8}{105} - \frac{71}{840\sqrt{2}}$. (*ii*) $\frac{\sin^4 x}{4} - 2\frac{\sin^6 x}{3}$.

6. $\theta(\sec\theta + \cos\theta) - \left[\log\tan\left(\frac{\pi}{4} + \frac{\theta}{2}\right) + \sin\theta\right]$, where $\pi = \sin\theta$

Page 110.

1. $\frac{1}{4}\left[2\frac{\sin nx}{n} - \frac{\sin(n+2)x}{n+2} - \frac{\sin(n-2)x}{n-2}\right]$.

2. (*i*) $\frac{1}{32}\left[\frac{\pi}{2} + \frac{2}{3}\right]$. (*ii*) $\frac{1}{128}\left[3x - \sin 4x + \frac{\sin 8x}{8}\right]$

3. $\frac{1}{2^9}\left[\frac{\sin 10x}{10} + \frac{\sin 8x}{4} - \frac{\sin 6x}{2} - 2\sin 4x + \sin 2x + 6x\right]$.

Page 112.

1. (*i*) $\frac{\tan^3 x}{3}$. (*ii*) $-\frac{\cot^3 x}{3}$.

2. (*i*) $-\cot x + \tan x$. (*ii*) $-\frac{\cot^3 x}{3} - 2\cot x + \tan x$.

(*iii*) $-\frac{\cot^3 x}{3} - 3\cot x + 3\tan x + \frac{\tan^3 x}{3}$.

3. $2\sqrt{\tan x}$.

Page 113.

Ex. (*ii*) $-2\left[\cot\frac{\theta}{2}+\frac{1}{3}\cot^3\frac{\theta}{2}\right]$.

Pages 116—117.

1. (*i*) $\frac{\pi}{3\sqrt{3}}$. (*ii*) $\frac{1}{3}\cosh^{-1}\frac{5+4\cos x}{4+5\cos x}$.

2. (*b*) $(a^2>1)\ \frac{\pi}{a^2-1}$, $(a^2<1)\ \frac{\pi}{1-a^2}$.

3. (*i*) $\operatorname{cosec}\alpha\cosh^{-1}\frac{1+\cos\alpha\cos x}{\cos\alpha+\cos x}$. (*ii*) $x\cos\alpha+\sin\alpha\cosh^{-1}\frac{1+\cos\alpha\cos x}{\cos\alpha+\cos x}$.

4. $\frac{1}{5\sqrt{3}}\cos^{-1}\frac{1+2\cos\theta}{2+\cos\theta}-\frac{2}{5\sqrt{3}}\cosh^{-1}\frac{2-\cos\theta}{-1+2\cos\theta}$.

Page 118.

Ex. (*i*) $\frac{1}{3}\cos^{-1}\frac{4}{5}$. (*ii*) $\frac{1}{3}\log 2$.

Pages 119—121. Art. 60.

1. (*i*) $\frac{1}{\sqrt{3}}\cos^{-1}\frac{1+2\cos\left(x-\frac{\pi}{4}\right)}{2+\cos\left(x-\frac{\pi}{4}\right)}$. (*ii*) $-\cosh^{-1}\frac{\sqrt{10}-3\cos(x-\tan^{-1}3)}{-3+\sqrt{10}\cos(x-\tan^{-1}3)}$.

3. $\frac{1}{2}\left[\log\tan\frac{x}{2}+2\tan\frac{x}{2}+\frac{1}{2}\tan^2\frac{x}{2}\right]$.

Page 121. Art. 61.

1. (*i*) $\frac{ac+be}{c^2+e^2}\theta+\frac{bc-ae}{c^2+e^2}\log(c\sin\theta+e\cos\theta)$.

(*ii*) $\frac{a}{a^2+b^2}\theta+\frac{b}{a^2+b^2}\log(a\cos\theta+b\sin\theta)$.

2. $\frac{\sin 2\theta}{2}\log(1+\tan q)-\frac{1}{2}\theta+\frac{1}{2}\log(\cos\theta+\sin\theta)$.

Page 122.

Ex. $\frac{7}{25}x-\frac{1}{25}\log(3\sin x+4\cos x+1)$

$$-\frac{7}{50\sqrt{6}}\cosh^{-1}\frac{5+\cos(x-\tan^{-1}\frac{3}{4})}{1+5\cos(x-\tan^{-1}\frac{3}{4})}.$$

Pages 123—124. Art. 63.

1. $\dfrac{2}{\sqrt{7}}\tan^{-1}\left(\dfrac{1}{\sqrt{7}}\tanh\dfrac{x}{2}\right)$. 2. $\dfrac{1}{\sqrt{a^2+b^2}}\log\dfrac{\sqrt{a^2+b^2}+a\tanh\dfrac{x}{2}-b}{\sqrt{a^2+b^2}-a\tanh\dfrac{x}{2}+b}$.

MISCELLANEOUS EXAMPLES ON CHAPTER VI

Page 124.

1. (*a*) $-\left[\dfrac{2}{3}\cos^{\frac{3}{2}}\theta-\dfrac{4}{7}\cos^{\frac{7}{2}}\theta+\dfrac{2}{11}\cos^{\frac{11}{2}}\theta\right]$. (*b*) $\dfrac{8}{45}$.

2. (*i*) $\dfrac{\pi}{4}$. (*ii*) $\dfrac{1}{32a^3}\left[\dfrac{\pi}{2}-\dfrac{2}{3}\right]$.

5. $\dfrac{1}{2}\log 3, \tan^{-1}\dfrac{1}{3}$.

6. (*i*) $\dfrac{12}{13}x-\dfrac{5}{13}\log(3\cos x+2\sin x)$. (*ii*) $c\,\theta+b\log(1+\sin\theta)+(a-c)(\tan\theta-\sec\theta)$.

7. (*i*) $(a^2>b^2)\ \dfrac{1}{b}x-\dfrac{a}{b\sqrt{a^2-b^2}}\cos^{-1}\dfrac{b+a\cos x}{a+b\cos x}$.

$(a^2<b^2)\ \dfrac{1}{b}x-\dfrac{a}{b\sqrt{b^2-a^2}}\cosh^{-1}\dfrac{b+a\cos x}{a+b\cos x}$.

(*ii*) $\pi(a-\sqrt{a^2-1})$.

CHAPTER VII

Pages 126—128.

1. $\dfrac{x^{m+1}}{m+1}\left[\log x-\dfrac{1}{m+1}\right], \dfrac{x^{m+1}}{m+1}\left[(\log x)^2-\dfrac{2}{m+1}\log x+\dfrac{2}{(m+1)^2}\right]$.

4. $\displaystyle\int\frac{x^m}{(\log x)^n}dx=-\frac{x^{m+1}}{(n-1)(\log x)^{n-1}}+\frac{m+1}{n-1}\int\frac{x^m}{(\log x)^{n-1}}dx$.

5. $n\,!$.

Pages 129—130.

1. $\displaystyle\int x^n\sin x\,dx=-x^n\cos x+n\,x^{n-1}\sin x-n(n-1)\int x^{n-2}\sin x\,dx$.

2. $\dfrac{5}{16}\pi^4-15\pi^2+120$.

3. $\displaystyle\int x^m\cos nx\,dx=\frac{x^m\sin nx}{n}+\frac{m}{n^2}x^{m-1}\cos nx-\frac{m(m-1)}{n^2}\int x^{m-2}\cos nx\,dx$.

Case I. If m is a positive odd integer, then ultimately we are left with $\int x\cos nx\,dx$ $=\dfrac{x\sin nx}{n}+\dfrac{\cos nx}{n^2}$.

Case II. If m is a positive even integer, then ultimately we are left with $\int \cos nx\, dx = \frac{\sin nx}{n}$.

Pages 131—132.

1. $\int_0^{\pi} e^{-x} \sin^n x\, dx = \frac{n(n-1)}{n^2+1} \int_0^{\pi} e^{-x} \sin^{n-2} x\, dx \; (n > 1).$

2. (*i*) $\int e^{ax} \sin^n x\, dx$

$$= \frac{e^{ax} \sin^{n-1} x\,(a \sin x - n \cos x)}{a^2+n^2} + \frac{n(n-1)}{a^2+n^2} \int e^{ax} \sin^{n-2} x\, dx.$$

(*ii*) $e^{ax} \left[\frac{\sin^2 x(a \sin x - 3 \cos x)}{a^2+3^2} + \frac{3.2}{(a^2+3^2)(a^2+1^2)} (a \sin x - \cos x) \right].$

4. (*i*) $\int e^{ax} \cos^n x\, dx = \frac{e^{ax} \cos^{n-1} x\,(a \cos x + n \sin x)}{a^2+n^2} + \frac{n(n-1)}{a^2+n^2} \int e^{ax} \cos^{n-2} x\, dx.$

(*ii*) $e^{ax} \left[\frac{\cos^3 x\,(a \cos x + 4 \sin x)}{a^2+4^2} + \frac{4.3 \cos x\,(a \cos x + 2 \sin x)}{(a^2+4^2)(a^2+2^2)} + \frac{4.3.2.1}{(a^2+4^2)(a^2+2^2)a} \right].$

Pages 138—141.

1. $\int x^{m-1} X^p\, dx = \frac{x^m X^p}{m+pn} + \frac{pna}{m+pn} \int x^{m-1} X^{p-1}\, dx.$

3. $\int x^{m-1} (a+bx^n)^p\, dx$

$$= \frac{x^{m-n}(a+bx^n)^{p+1}}{b(m+pn)} - \frac{a(m-n)}{b(m+pn)} \int x^{m-n-1} (a+bx^n)^p\, dx \;;$$

$$-(1-x^3)^{\frac{2}{3}} \left[\frac{x^6}{8} + \frac{6}{8.5} x^3 + \frac{6.3}{8.5.2} \right].$$

8. (*i*) $\int x^m \sqrt{2ax-x^2}\, dx = -\frac{x^{m-1}(2ax-x^2)^{\frac{3}{2}}}{m+2} + \frac{(2m+1)a}{m+2} \int x^{m-1} \sqrt{2ax-x^2}\, dx.$

(*ii*) $\frac{7}{8} \pi a^5.$

9. $\int x^m (1+x^2)^{\frac{n}{2}}\, dx = \frac{x^{m-1}(1+x^2)^{\frac{n}{2}+1}}{m+n+1} - \frac{m-1}{m+n+1} \int x^{m-2} (1+x^2)^{\frac{n}{2}}\, dx \;;$

$$(1+x^2)^{\frac{9}{2}} \left[\frac{x^4}{13} - \frac{4}{13.11} x^2 + \frac{4.2}{13.11.9} \right].$$

10. (a) $\displaystyle\int \frac{x^m}{(a^3+x^3)^{\frac{n}{3}}}\,dx = \frac{x^{m-2}}{(m-n+1)(a^3+x^3)^{\frac{n}{3}-1}} - \frac{(m-2)a^3}{m-n+1}\int \frac{x^{m-3}}{(a^3+x^3)^{\frac{n}{3}}}\,dx.$

(b) $\displaystyle\int \frac{x^m}{(x^3-1)^{\frac{1}{3}}}\,dx = \frac{x^{m-2}(x^3-1)^{\frac{2}{3}}}{m} + \frac{m-2}{m}\int \frac{x^{m-3}}{(x^3-1)^{\frac{1}{3}}}\,dx\,;$

$$(x^3-1)^{\frac{2}{3}}\left[\frac{x^6}{8}+\frac{6}{8.5}x^3+\frac{6.3}{8.5.2}\right].$$

11. $\displaystyle\int \frac{x^m}{\sqrt{ax^2+2bx+c}}\,dx = \frac{x^{m-1}\sqrt{ax^2+2bx+c}}{ma}$

$$-\frac{(2m-1)b}{ma}\int \frac{x^{m-1}}{\sqrt{ax^2+2bx+c}}\,dx - \frac{(m-1)c}{ma}\int \frac{x^{m-2}}{\sqrt{ax^2+2bx+c}}\,dx.$$

Page 143.

1. (a) $\displaystyle\int \sin^m x\cos^n x\,dx = -\frac{\sin^{m-1}x\cos^{n+1}x}{m+n} + \frac{m-1}{m+n}\int \sin^{m-2}x\cos^n x\,dx$

$$-\frac{\sin^3 x\cos^3 x}{6} - \frac{1}{8}\sin x\cos^3 x + \frac{1}{16}\left(x+\frac{\sin 2x}{2}\right).$$

(b) $-\dfrac{\sin^3 x}{2\cos x} + \frac{3}{2}(\tan x - x).$

[**Hint.** Connect $\int \sin^p x\cos^q x\,dx$ with $\int \sin^{p-2}x\cos^q x\,dx$, and put $p = 4$, $q = -2$.]

2. $\displaystyle\int \sin^p x\cos^q x\,dx = \frac{\sin^{p+1}x\cos^{q+1}x}{p+1} + \frac{p+q+2}{p+1}\int \sin^{p+2}x\cos^q x\,dx.$

3. $\displaystyle\int \sin^p x\cos^q x\,dx = \int \frac{\sin^{p+1}x\cos^{q-1}x}{p+q} + \frac{q-1}{p+q}\int \sin^p x\cos^{q-2}x\,dx.$

4. $\displaystyle\int \sin^p x\cos^q x\,dx = -\frac{\sin^{p+1}x\cos^{q+1}x}{q+1} + \frac{p+q+2}{q+1}\int \sin^p x\cos^{q+2}x\,dx.$

5. $\displaystyle\int \sin^p x\cos^q x\,dx = -\frac{\sin^{p-1}x\cos^{q+1}x}{q+1} + \frac{p-1}{q+1}\int \sin^{p-2}x\cos^{q+2}x\,dx.$

Page 145.

1. (i) $-\dfrac{\sin^3 x\cos x}{4} - \dfrac{3}{8}\sin x\cos x + \dfrac{3}{8}x.$

(ii) $-\dfrac{\sin^5 x\cos x}{6} - \dfrac{5}{24}\sin^3 x\cos x - \dfrac{5}{16}\sin x\cos x + \dfrac{5}{16}x.$

2. $-\dfrac{\sin^7 x\cos x}{8}-\dfrac{7}{48}\sin^5 x\cos x-\dfrac{35}{192}\sin^3 x\cos x-\dfrac{35}{128}\sin x\cos x+\dfrac{35}{128}x.$

3. $\int \cos^n x\,dx=\dfrac{\sin x\cos^{n-1}x}{n}+\dfrac{n-1}{n}\int \cos^{n-2}x\,dx.$

Pages 148—149.

1. (i) $\dfrac{\pi}{4}$. (ii) $\dfrac{2}{3}$. (iii) $\dfrac{3\pi}{16}$.

2. (i) $\dfrac{8}{15}$. (ii) $\dfrac{63\pi}{512}$.

4. $\int \dfrac{x^{2n+1}\,dx}{(1-x^2)^{\frac{1}{2}}}=-\dfrac{x^{2n}(1-x^2)^{\frac{1}{2}}}{2n+1}+\dfrac{2n}{2n+1}\int\dfrac{x^{2n-1}\,dx}{(1-x^2)^{\frac{1}{2}}}$; $\dfrac{1.3.5\ldots(2n-1)}{2.4.6\ldots 2n}\cdot\dfrac{\pi}{2}$.

Pages 151—152.

1. (i) $\dfrac{5\pi}{32}$. (ii) $\dfrac{128}{315}$. 2. (i) $\dfrac{\pi}{8}$. (ii) $\dfrac{2}{9}$.

3. (i) $\dfrac{\pi}{4}$. (ii) $\dfrac{(2n-2)(2n-4)\ldots 2}{(2n-1)(2n-3)\ldots 3}$.

Pages 157—158.

1. (i) $\dfrac{3\pi}{512}$. (ii) $\dfrac{8}{693}$. 2. (i) $\dfrac{8}{315}$. (ii) $\dfrac{1}{60}$.

3. $\dfrac{7\pi}{384}$. 5. (b) (i) $\dfrac{\pi}{32}$. (ii) $\dfrac{\pi}{32}$.

6. (i) $\dfrac{3\pi}{128}$. (ii) $\dfrac{\pi a^3}{2}$. 7. (i) $\dfrac{5}{8}\pi a^4$. (ii) $\dfrac{7}{8}\pi a^5$.

8. $\dfrac{(2m+1)(2m-1)\ldots 1}{(m+2)!}\pi a^{m+2}$.

Page 159.

1. (i) $\dfrac{\tan^2 x}{2}+\log\cos x$. (ii) $\dfrac{\tan^3 x}{3}-\tan x+x$.

2. $\int \tanh^n x\,dx=\int \tanh^{n-2}x\,dx-\dfrac{\tanh^{n-1}x}{n-1}$.

Page 160.

1. (i) $-\dfrac{\cot^2 x}{2}-\log\sin x$. (ii) $-\dfrac{\cot^3 x}{3}+\cot x+x$.

2. $-\dfrac{\cot^4 x}{4}+\dfrac{\cot^2 x}{2}+\log\sin x$.

Page 161.

1. $\int \sec^n x\,dx=\dfrac{\sec^{n-2}x\tan x}{n-1}+\dfrac{n-2}{n-1}\int \sec^{n-2}x\,dx.$

(i) $\dfrac{\sec x\tan x}{2}+\dfrac{1}{2}\log\tan\left(\dfrac{\pi}{4}+\dfrac{x}{2}\right)$.

(ii) $\dfrac{\sec^3 x \tan x}{4} + \dfrac{3}{8} \sec x \tan x + \dfrac{3}{8} \log \tan\left(\dfrac{\pi}{4} + \dfrac{x}{2}\right)$.

2. $\dfrac{\sec^5 x \tan x}{6} + \dfrac{5}{24} \sec^3 x \tan x + \dfrac{5}{16} \sec x \tan x + \dfrac{5}{16} \log \tan\left(\dfrac{\pi}{4} + \dfrac{x}{2}\right)$.

Page 162.

Ex. $\displaystyle\int \operatorname{cosec}^n x \, dx = -\frac{\operatorname{cosec}^{n-2} x \cot x}{n-1} + \frac{n-2}{n-1} \int \operatorname{cosec}^{n-2} x \, dx.$

(i) $-\dfrac{\operatorname{cosec} x \cot x}{2} + \dfrac{1}{2} \log \tan \dfrac{x}{2}$.

(ii) $-\dfrac{\operatorname{cosec}^3 x \cot x}{4} - \dfrac{3}{8} \operatorname{cosec} x \cot x + \dfrac{3}{8} \log \tan \dfrac{x}{2}$.

Page 164.

2. (i) $\dfrac{2}{5}$.

Pages 165—166.

Ex. $-\dfrac{181}{105}$.

Pages 168—169. Art. 76-III.

1. $\dfrac{5\pi}{27}$.

2. $\dfrac{1}{3} \dfrac{\sin x}{(1 + 2\cos x)^2} - \dfrac{1}{3} \dfrac{\sin x}{1 + 2\cos x} + \dfrac{1}{3\sqrt{3}} \cosh^{-1} \dfrac{2 + \cos x}{1 + 2\cos x}$.

MISCELLANEOUS EXAMPLES ON CHAPTER VII

Pages 169—177.

3. $\dfrac{24}{85}$. **4.** $\dfrac{16}{35} a^7$.

5. $\dfrac{q^n (n!)}{(qn + p + 1)[q(n-1) + p + 1] \dots (q + p + 1)(p + 1)}$.

8. $\dfrac{5\pi a^4}{128}$. **9.** $\dfrac{\pi a^6}{32}$.

10. (i) $\displaystyle\int x^m (a + bx)^{\frac{2n+1}{2}} dx = \frac{2x^m (a + bx)^{\frac{2n+3}{2}}}{b(2m + 2n + 3)} - \frac{2am}{b(2m + 2n + 3)} \int x^{m-1} (a + bx)^{\frac{2n+1}{2}} dx.$

(ii) $\displaystyle\int x^{2m} (a^2 + x^2)^{n + \frac{1}{2}} dx = \frac{x^{2m-1} (a^2 + x^2)^{n + \frac{3}{2}}}{2m + 2n + 2} - \frac{(2m-1) a^2}{2m + 2n + 2} \int x^{2m-2} (a^2 + x^2)^{n + \frac{1}{2}} dx.$

11. $\displaystyle\int \frac{dx}{(a^2 + x^2)^n} = \frac{x}{(2n-2)a^2 (a^2 + x^2)^{n-1}} + \frac{2n-3}{(2n-2)a^2} \int \frac{dx}{(a^2 + x^2)^{n-1}}.$

13. $\dfrac{x}{8(x^2 + 2)^2} + \dfrac{3}{32} \dfrac{x}{x^2 + 2} + \dfrac{3}{32\sqrt{2}} \tan^{-1} \dfrac{x}{\sqrt{2}}$.

17. (*i*) $\frac{\pi}{32}$. (*ii*) $\frac{8}{1287}$. **18.** (*i*) $\frac{\pi a^3}{2}$. (*ii*) $\frac{5}{8}\pi a^4$.

19. (*i*) $\frac{7}{8}\pi a^5$. (*ii*) $\frac{63}{8}\pi a^5$. **21.** (*i*) $\frac{1}{12}$. (*ii*) $\frac{\pi}{4a^3b^3}(a^2+b^2)$.

22. $\frac{1}{2}\log 2 - \frac{1}{4}$. **23.** (*i*) $\frac{1}{2}\log 2 - \frac{1}{4}$.

25. $\displaystyle\int \frac{x}{\sin^n x}\,dx = -\frac{x\cos x}{(n-1)\sin^{n-1}x} - \frac{1}{(n-1)(n-2)\sin^{n-2}x} + \frac{n-2}{n-1}\int \frac{x}{\sin^{n-2}x}\,dx.$

26. (*a*) $\frac{\tanh x \operatorname{sech} x}{2} + \frac{1}{2}\sin^{-1}(\tanh x)$. (*b*) $\frac{3\pi}{16}$.

[**Hint.** $\displaystyle\underset{x\to\infty}{\text{Lt}} \operatorname{sech} x = \underset{x\to\infty}{\text{Lt}} \frac{1}{\cosh x} = \underset{x\to\infty}{\text{Lt}} \frac{1}{\frac{e^x+e^{-x}}{2}}$

$\displaystyle = \underset{x\to\infty}{\text{Lt}} \frac{2}{e^x+e^{-x}}$ [Divide the num. and denom. by e^x]

$\displaystyle = \underset{x\to\infty}{\text{Lt}} \frac{2e^{-x}}{1+e^{-2x}} = \frac{2(0)}{1+0} = 0.$ Similarly $\underset{x\to\infty}{\text{Lt}} \tanh x = 1.$]

32. (*a*) $\frac{1}{2}$.

(*b*) $\displaystyle\frac{1}{(1-e^2)^{\frac{3}{2}}}\left[2\tan^{-1}\left(\sqrt{\frac{1-e}{1+e}}\tan\frac{\theta}{2}\right) - e\sqrt{1-e^2}\,\frac{\sin\theta}{1+e\cos\theta}\right].$

CHAPTER VIII

Pages 187—188.

1. (*c*) 0. **4.** π.

Pages 194—195.

1. $\frac{4}{15}$. **2.** (*a*) $\frac{3\pi}{8}$. **6.** $-\frac{\pi}{2}\log 2$. **7.** (*i*) $\frac{\pi}{2}\log 2$.

Page 196.

Ex. In the Fig. of theorem VII, Art. 78,

the theorem $\displaystyle\int_a^{na} f(x)\,dx = (n-1)\int_0^a f(x)\,dx$

expresses that area $P_1M_1M_nP_n = (n-1)$. area $P_0OM_1P_1$,

which is obvious.

Pages 197—198.

2. $-\frac{\pi^2}{6}$. **3.** (*i*) $-\frac{\pi^2}{24}$. (*ii*) $\log 2 - 1$.

Page 203.

2. $\dfrac{b^4 - a^4}{4}$. 5. (b) 1.

Pages 211—213.

1. (i) $\dfrac{\pi}{4}$. (ii) $\dfrac{\pi}{2}$. (iii) $\dfrac{(2k-1)(2k-3)\ldots 1}{2k(2k-2)\ldots 2}$.

3. (i) log 2. (ii) $\dfrac{\pi}{2}$. 4. (i) $\dfrac{\pi}{2}$. 5. (a) $\dfrac{1}{4}\log 2$.

Page 214.

3. $\dfrac{3}{2}$.

MISCELLANEOUS EXAMPLES ON CHAPTER VIII

Pages 217—223.

2. $\dfrac{\pi}{4}$. 3. (ii) $\dfrac{\pi}{8}\log 2$. 5. (i) $\dfrac{5\pi}{16}$. (ii) $\dfrac{3\pi}{256}$.

7. $\dfrac{2\pi}{b^2}(a - \sqrt{a^2 - b^2})$. 8. (i) $\dfrac{3\pi^2}{512}$. (ii) $\dfrac{3\pi^2}{512}$. 10. $\dfrac{\pi^2}{2ab}$.

11. $\dfrac{8}{3}$. 13. $\pi \log 2$. 14. $\dfrac{\pi}{2}\log 2$. 15. (a) $\dfrac{\pi^2}{4}$.

16. (a) (i) $\dfrac{1}{3}$. (ii) $\dfrac{1}{3}\log 2$. 17. (i) $\dfrac{\pi}{3}$. (ii) $\dfrac{1}{14}$.

18. (i) $\dfrac{3}{8}$. (ii) $\dfrac{1}{3}$. 19. $\dfrac{\pi}{2} + 1$. 20. log 3.

21. (i) $\dfrac{4}{e}$. (ii) $e^{\frac{\pi^2}{12}}$. 22. $2e^{\frac{\pi-4}{2}}$.

CHAPTER IX

Page 226.

1. (a) The length of the arc of the curve $y = f(x)$ between the pts. where $x = a, x = b$ is

$$\int_a^b \sqrt{1+\left(\frac{dy}{dx}\right)^2}\, dx.$$

(b) (i) $a\,[\sqrt{2} + \log(1 + \sqrt{2})]$. (ii) $2a\,[\sqrt{2} + \log(1 + \sqrt{2})]$

3. $\dfrac{1}{27\sqrt{a}}\,[(4a + 9x)^{\frac{3}{2}} - (4a)^{\frac{3}{2}}]$.

Pages 227—228.

1. $4a \sin \psi$. 3. (b) $\sqrt{2}\,(e^{\frac{\pi}{2}} - 1)$.

5. $\dfrac{1}{2}\,[(x_2^{\frac{2}{3}} + y_2^{\frac{2}{3}})^{\frac{3}{2}} - (x_1^{\frac{2}{3}} + y_1^{\frac{2}{3}})^{\frac{3}{2}}]$.

Pages 230—231.

1. (*a*) $8a$. (*b*) $8a$. **2.** $8a$. **3.** (*i*) $6a$. (*ii*) $\dfrac{a^2+ab+b^2}{a+b}$.

Pages 232—236.

1. (*a*) The length of the arc of the curve $r = f(\theta)$ between the pts. where $\theta = \alpha$, $\theta = \beta$ is $\displaystyle\int_\alpha^\beta \sqrt{r^2+\left(\frac{dr}{d\theta}\right)^2}\,d\theta$. (*b*) $a(\theta_2-\theta_1)$.

2. (*i*) $\dfrac{a}{2}\left[\theta\sqrt{\theta^2+1}+\log(\theta+\sqrt{\theta^2+1})\right]_{\theta_1}^{\theta_2}$. (*ii*) $2a\left(\cos\dfrac{\theta_1}{2}-\cos\dfrac{\theta_2}{2}\right)$.

3. $a\left[\dfrac{\sqrt{1+3\cos^2\theta}}{\cos\theta}-\sqrt{3}\log(\sqrt{3}\,.\cos\theta+\sqrt{1+3\cos^2\theta})\right]_{\theta_1}^{\theta_2}$.

4. (*a*) 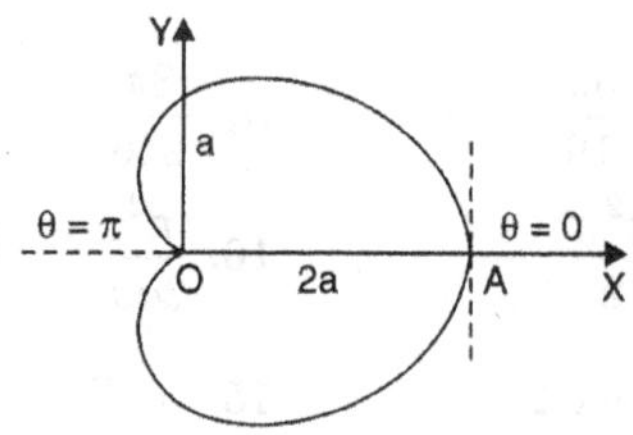

; $8a$

Page 239.

1. (*a*) See Def., Art. 93.

Page 242. Art. 94-III.

Ex. $s = a\sec\alpha.\,[e^{(\psi-\alpha)\cot\alpha}-1]$.

MISCELLANEOUS EXAMPLES ON CHAPTER IX

Pages 242—249.

1. $\log(2+\sqrt{3})$. **2.** $\dfrac{x\sqrt{1+4x^2}}{2}+\dfrac{1}{4}\log(2x+\sqrt{1+4x^2})$; 1.48.

5. $\log(e+e^{-1})$.

7. $\dfrac{a}{2}\left[2\sqrt{\dfrac{4a-3x}{a-x}}+\sqrt{3}\log\dfrac{6x-7a+2\sqrt{3}\sqrt{3x^2-7xa+4a^2}}{a}\right]_{x_1}^{x_2}$.

9. $4\sqrt{3}$. **10.** (*i*) πa. (*ii*) $2a[\sqrt{2}+\log(1+\sqrt{2})]$.

13. 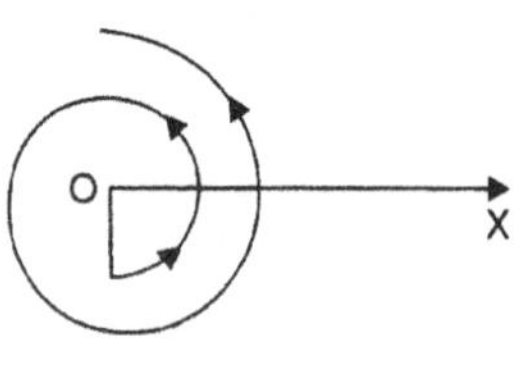

; $a\sec\alpha.\,(e^{\theta\cot\alpha}-1)$.

18. 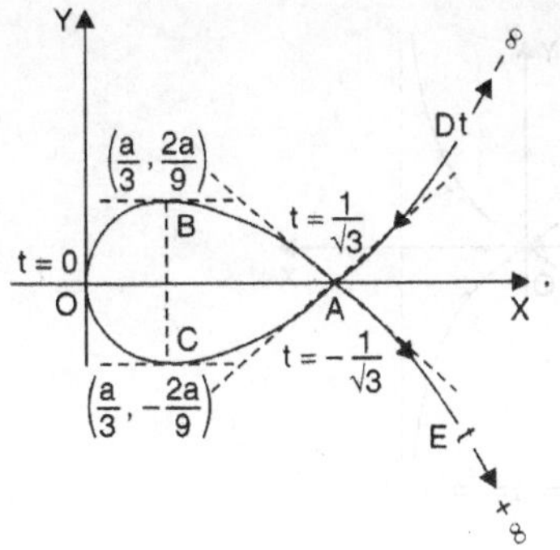

20. $s = 4a \sin \dfrac{\psi}{3}$.

CHAPTER X

Pages 255—256. Art. 96-I.

1. (*i*) $e^h - 1$. 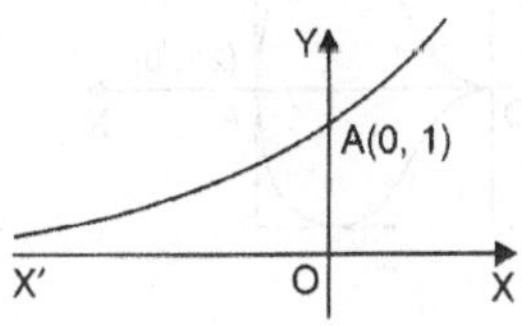

(*ii*) $c^2 \sinh \dfrac{h}{c}$.

2. (*i*) $c^2 \log \dfrac{b}{a}$. (*ii*) $b \log b - a \log a - (b - a)$. 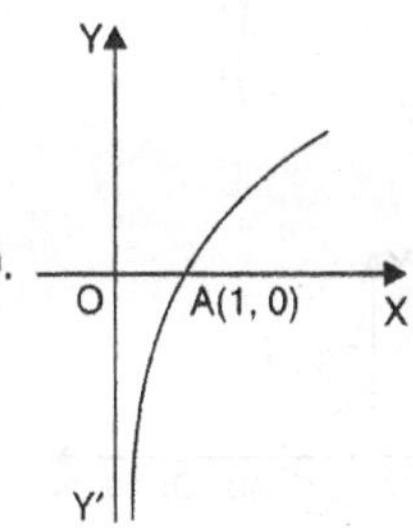

3. $\dfrac{8}{3} a^2$. **4.** πab.

5. 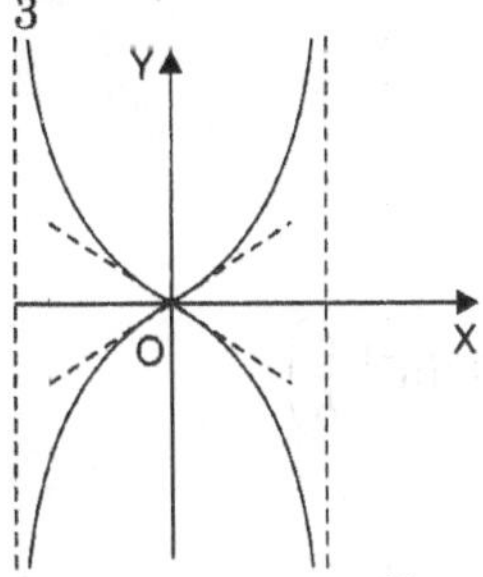

Area = $4a^2$.

6. 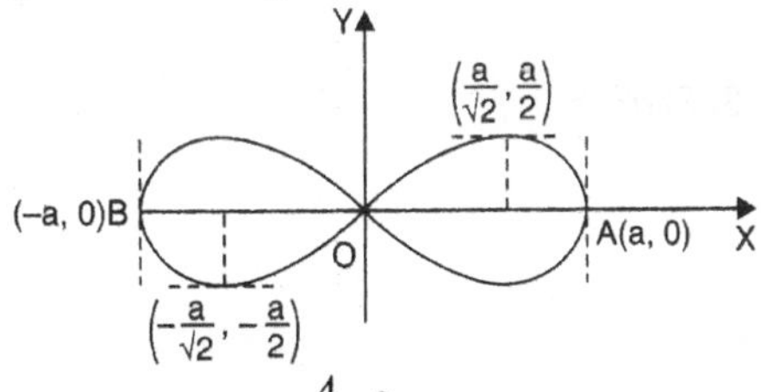

Area = $\dfrac{4}{3} a^2$.

7. 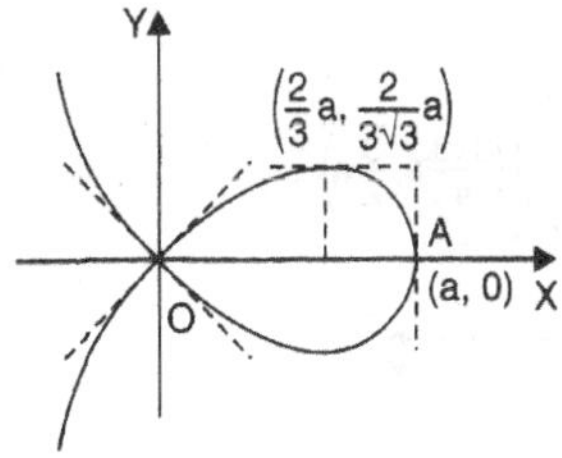

8. (i) 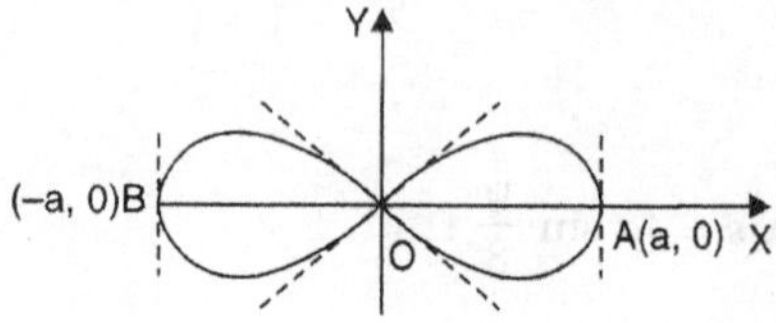

(ii) 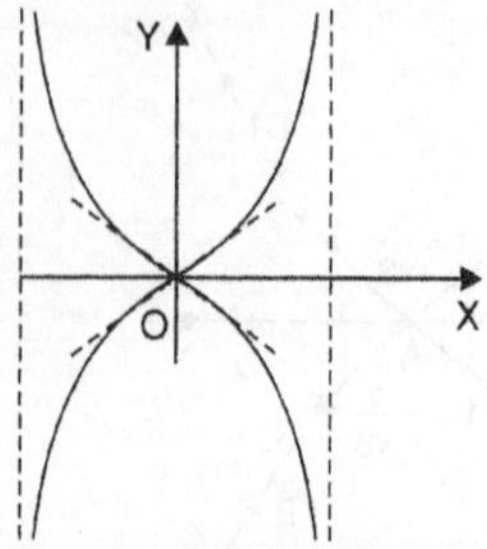

Area = $a^2(\pi - 2)$. Area = $a^2(\pi + 2)$.

9. 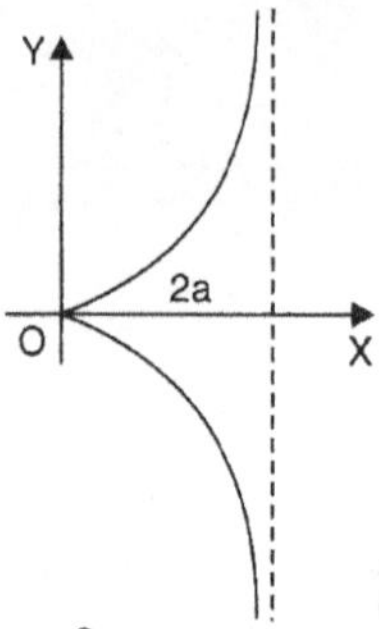

Area = $3\pi a^2$.

10. 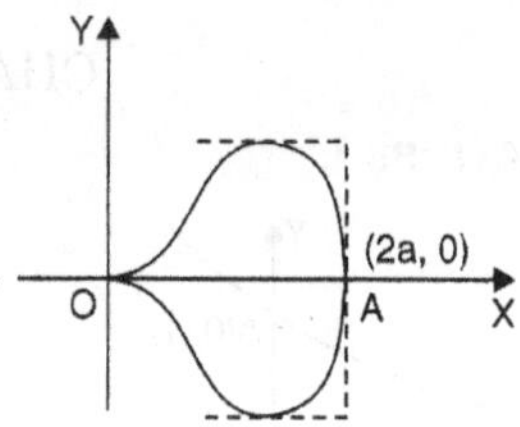

11. $3\pi a^2$.

12. 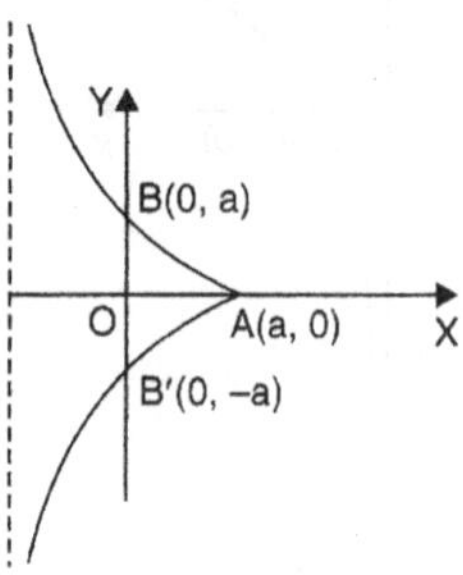

13. 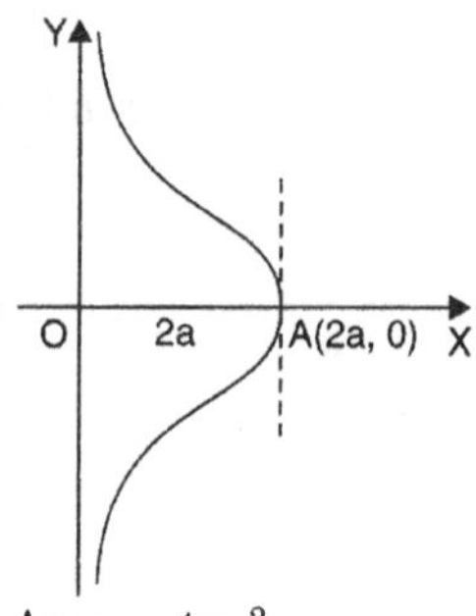

Area = $4\pi a^2$.

Pages 256—257. Art. 96-II.

1. $\dfrac{\pi ab}{2} - \left(\dfrac{ac}{b}\sqrt{b^2 - c^2} + ab\sin^{-1}\dfrac{c}{b}\right), \dfrac{\pi ab}{2} + \left(\dfrac{ac}{b}\sqrt{b^2 - c^2} + ab\sin^{-1}\dfrac{c}{b}\right)$

2. 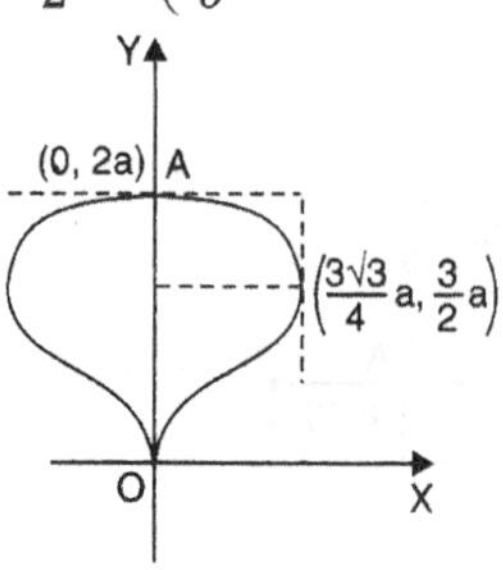

3. $3\pi a^2$.

Page 258.

Ex. 1. $\frac{16}{3}a^2$. **2.** $2a^2\left(\pi - \frac{2}{3}\right)$.

Pages 261—263. Art. 98.

1. (*b*) (*i*) $\dfrac{a^2}{2}\left(\dfrac{1}{\alpha} - \dfrac{1}{\beta}\right)$. (*ii*) $\dfrac{a}{2}\log\dfrac{\beta}{\alpha}$.

(*iii*) $\dfrac{a^2}{6}\left(\dfrac{1}{\alpha^3} - \dfrac{1}{\beta^3}\right)$. (*iv*) $\dfrac{a^2}{4}\tan\alpha \,.\, e^{2\beta\cot\alpha}\,(e^{2\gamma\cot\alpha} - 1)$.

2. (α) $\dfrac{\pi a^2}{8}$. (β) $\dfrac{\pi a^2}{16}$.

(γ) $\dfrac{\pi a^2}{4n}$. Total area : n a + ve odd integer, $\dfrac{\pi a^2}{4}$; n a + ve even integer, $\dfrac{\pi a^2}{2}$.

3. n a + ve odd integer, $\dfrac{\pi a^2}{4}$; n a + ve even integer, $\dfrac{\pi a^2}{2}$.

5. $\dfrac{\pi}{12}(a^2 + b^2)$. **6.** (*i*) $\dfrac{3}{2}\pi a^2$. (*ii*) $\dfrac{3}{2}\pi a^2$.

8. 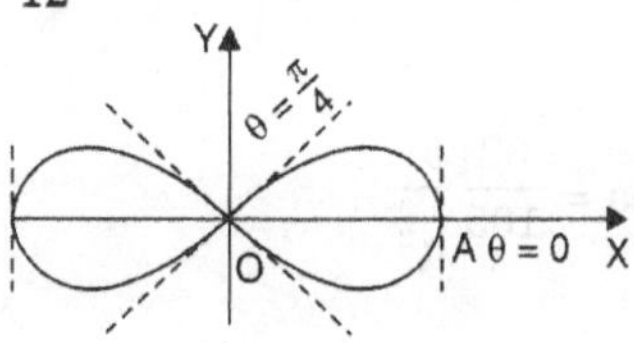

Area = a^2. **9.** $\dfrac{\pi}{2}(a^2 + b^2)$.

10. $\dfrac{\pi a^2}{16}\left(\dfrac{\pi^2}{6} - 1\right)$.

12. (*a*) 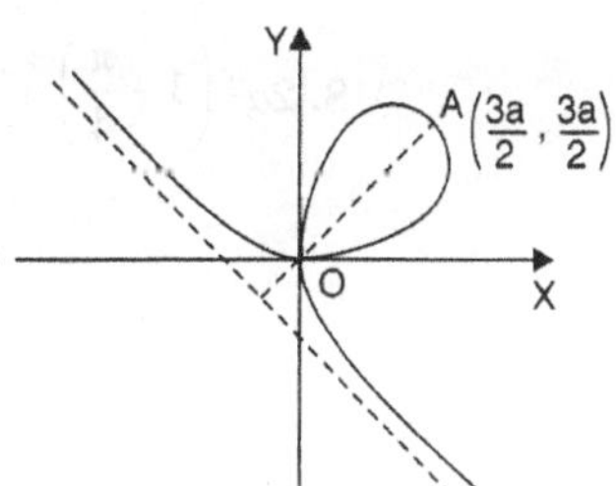

Area = $\dfrac{3a^2}{2}$. (*b*) $\dfrac{3a^2}{2}$. **13.** $\dfrac{\pi a^2}{4}$.

Page 263. Art. 99.

Ex. $\frac{5}{4}\pi a^2$.

Page 268.

Ex. (*i*) $\dfrac{\pi c^2}{2ab}(a^2 + b^2)$. (*ii*) $\dfrac{1}{3m^4n^4}(an^3 + bm^3)$.

Pages 271—272.

1. .693. **2.** 1.299. **3.** 7.8. **4.** 710 sq. ft.
5. 3.141. **6.** .693.

MISCELLANEOUS EXAMPLES ON CHAPTER X

Pages 273—278.

1. (*ii*) $\dfrac{n}{n+1} a^{\frac{n-1}{n}} h^{\frac{n+1}{n}}$.

4. Area $= 2a^2\left(1-\dfrac{\pi}{4}\right)$.

[It will be found that area $= 2a^2\left(\dfrac{\pi}{4}-1\right) = -2a^2\left(1-\dfrac{\pi}{4}\right)$.

$\therefore$ changing the sign, area (in magnitude) $= 2a^2\left(1-\dfrac{\pi}{4}\right)$.]

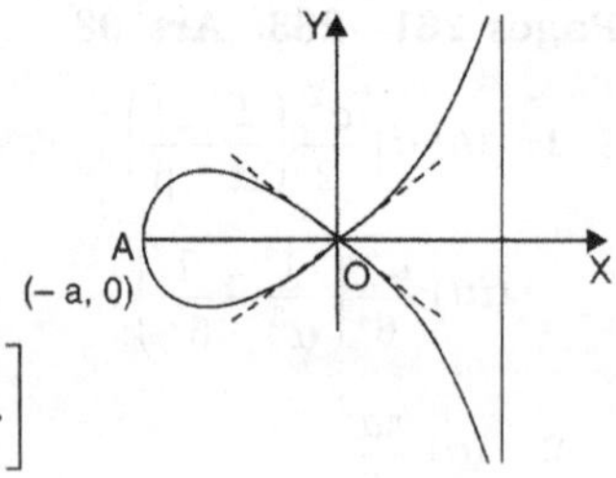

5.

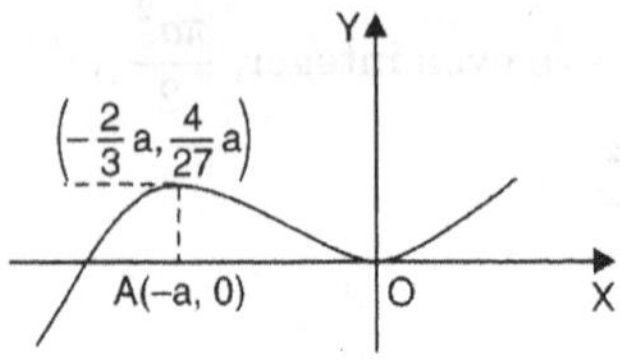

[There is a pt. of inflexion at

$\left(-\dfrac{a}{3}, \dfrac{2}{27}a\right)$.]

6.

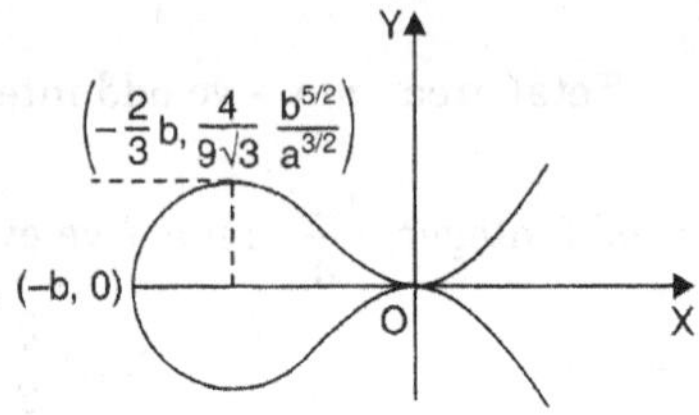

Area $= \dfrac{32}{105}\dfrac{b^{\frac{7}{2}}}{a^{\frac{3}{2}}}$.

7. Area $= \pi a^2$.

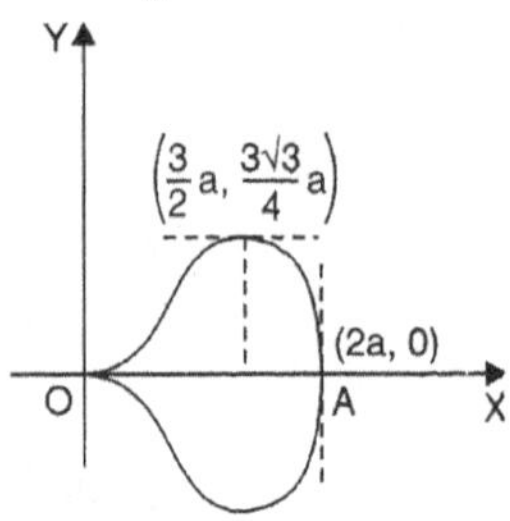

8. $2a^2\left(1-\dfrac{\pi}{4}\right)$.

10.

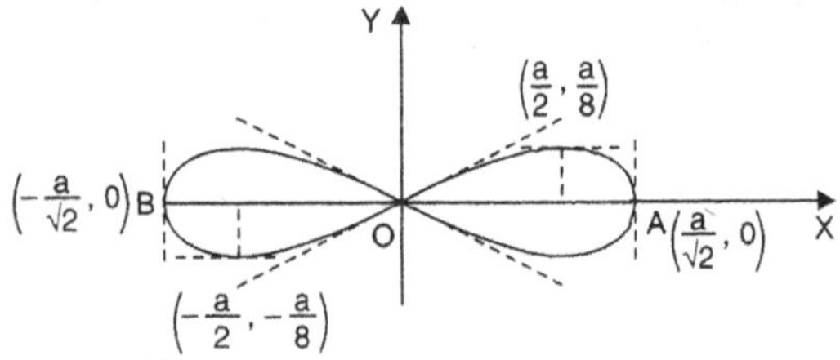

11.

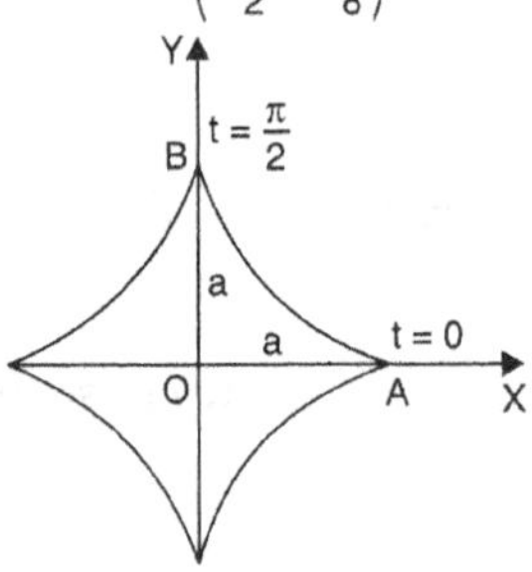

Area $= \frac{3}{8}\pi a^2$. **12.** Area $= \frac{3}{8}\pi a^2$.

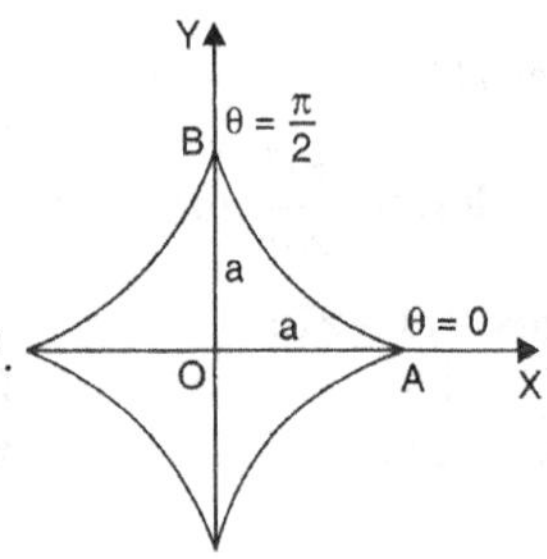

13. 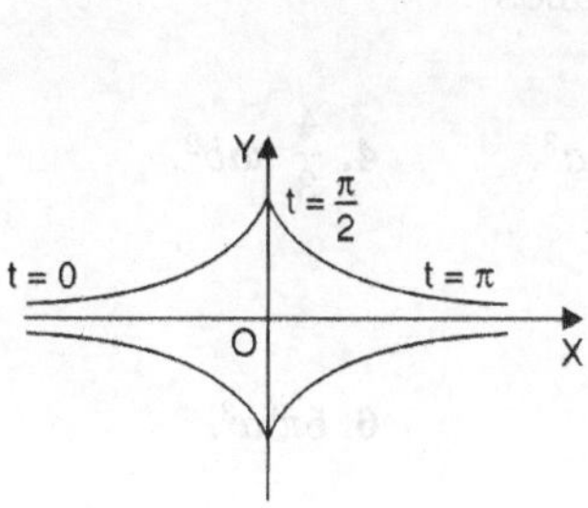

14. 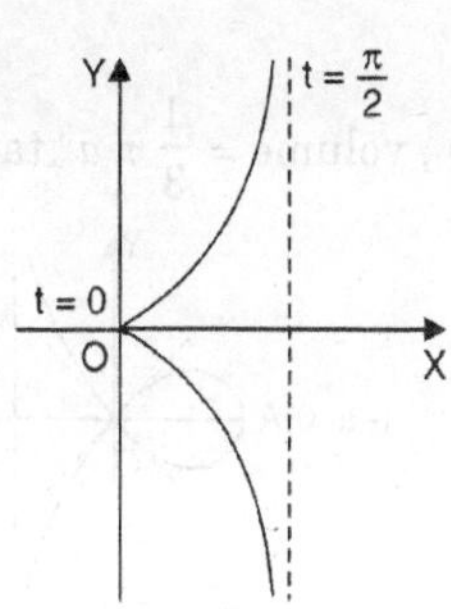

15. 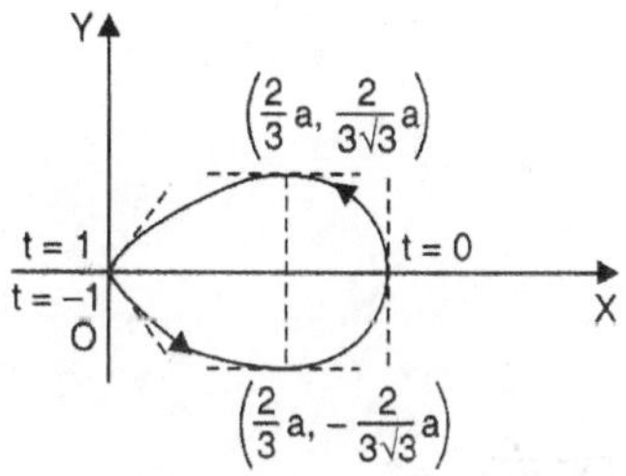

16. 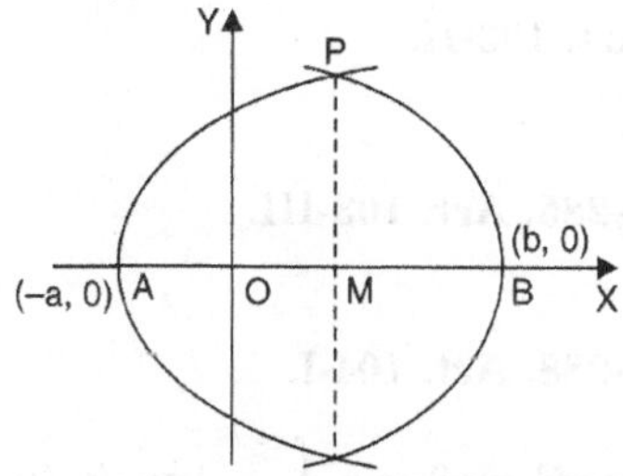

[Required area = 2. area ABP
= 2 (area AMP + area PMB)]

17. 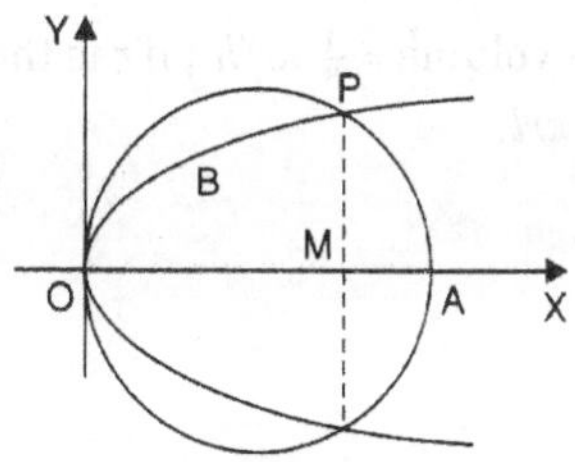

$(3\sqrt{3} + \frac{4}{3}\pi)\,a^2$.
[Common area = 2. area OAPB
= 2 (area OMPB + area PMA)]

19. $\dfrac{1}{3}$.

22. $a^2\left(\dfrac{10\pi}{3} + 3\sqrt{3}\right)$.

24. $4c^2\sqrt{2}\tan^{-1}\dfrac{1}{\sqrt{2}}$.

26. 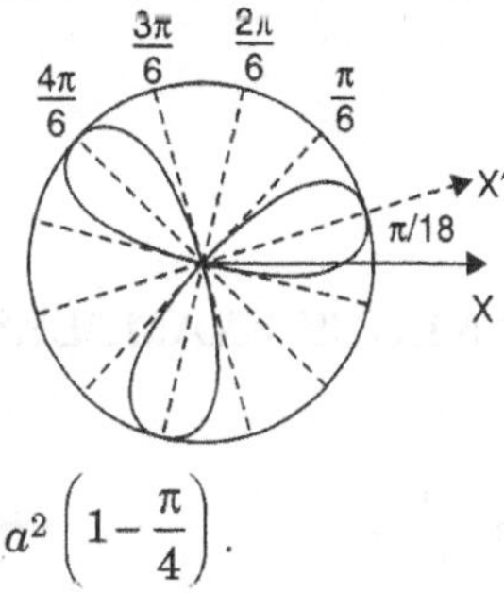

Area of loop $= \dfrac{\pi}{3}$.

27. $a^2\left(\dfrac{15\sqrt{3}}{16} - \dfrac{\pi}{2}\right)$

28. $a^2\left(1 - \dfrac{\pi}{4}\right)$.

32. $a^2(\pi - 1)$.

33. $(9\pi + 16) : (9\pi - 16)$.

34. $\frac{1}{4}\tan\alpha \,.\, (r_2^{\,2} - r_1^{\,2})$.

CHAPTER XI

Pages 281—282.

1. (*i*) $\dfrac{1}{(2n+1)\,a^{2n-2}}\,\pi\,c^{2n+1}$. (*ii*) $\dfrac{n}{n+2}\,\pi\,a^{\frac{2n-2}{n}}\,c^{\frac{n+2}{n}}$.

2. Area = $\frac{a^2}{2}\tan\theta$; volume = $\frac{1}{3}\pi a^3 \tan^2\theta$. **3.** $\frac{4}{3}\pi a^3$. **4.** $\frac{4}{3}\pi ab^2$.

5. $2\pi a^3\left(\log 2 - \frac{2}{3}\right)$.

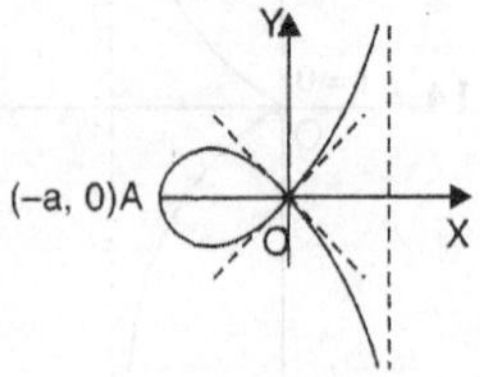

6. $5\pi^2 a^3$.

Page 283. Art. 103-II.

1. $\frac{4}{5}\pi a^3$.

Pages 283—285. Art. 103-III.

3. $2\pi^2 a^3$. **4.** $\frac{1}{2}\pi^2 a^3$.

Pages 287—288. Art. 104-I.

1. (*i*) Curved surface = $\int_{x=a}^{x=b} 2\pi y\, ds$, (*ii*) volume = $\int_a^b \pi y^2\, dx$.
2. (*i*) $4\pi a^2$; (*ii*) $4\pi a^2$.
3. If r is the base-radius and h the height of the cone, then volume = $\frac{1}{3}\pi r^2 h$; if r is the base-radius and l the slant side of the cone, then surface = πrl.
4. Surface = $\frac{32}{3}\pi a^2$; volume = $\pi^2 a^3$. **5.** $\frac{64}{3}\pi a^2$.
6. Volume = $\frac{32}{105}\pi a^3$; surface = $\frac{12}{5}\pi a^2$. **7.** $\frac{12}{5}\pi a^2$.

Page 288. Art. 104-II.

Ex. (*i*) $\pi a^2 [3\sqrt{2} - \log(\sqrt{2}+1)]$. (*ii*) = $\frac{4}{5}\pi a^3$.

Pages 291—292.

1. $2\pi^2 a^2 b$, where a, b are the semi-major and semi-minor axes of the ellipse.
2. $\frac{8}{15}\pi a^3\sqrt{2}$.

Pages 294—295. Art. 106.

2. $\frac{1}{12}\pi^2 a^3$.

MISCELLANEOUS EXAMPLES ON CHAPTER XI

Pages 295—299.

1. 90π. **2.** $\frac{2}{15}\pi$. **5.** $\frac{1536}{5}\pi$ cu. in. **7.** $2\pi^2$, π^2.

8. Area = 4π ;

volume = $4\pi^2$.

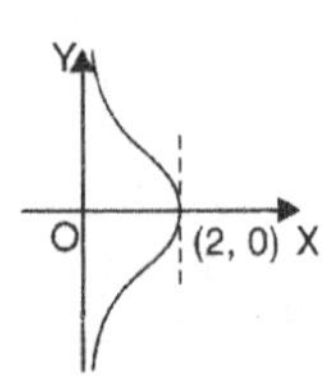

9. $2\pi a^2 b\left[2\frac{b}{a} - \sin^{-1}\frac{b}{a} - \frac{b}{a^2}\sqrt{a^2-b^2} - \frac{1}{3}\frac{b^3}{a^3}\right]$.

14. $\frac{8}{3}\pi a^2 (2\sqrt{2} - 1)$.

16. $8\pi\left(1+\frac{4\pi}{3\sqrt{3}}\right)$. **17.** $\pi a^2\left(\sqrt{6}-1-\frac{1}{\sqrt{2}}\log\frac{2+\sqrt{3}}{\sqrt{2}+1}\right)$.

19. Volume = $\frac{\pi c^2}{2}\left[x+\frac{c}{2}\sinh\frac{2x}{c}\right]_{x_1}^{x_2}$, surface = $\pi c\left[x+\frac{c}{2}\sinh\frac{2x}{c}\right]_{x_1}^{x_2}$

23. $\frac{48\sqrt{2}}{5}\pi a^2$. **25.** $4\pi a^2\left(1-\frac{1}{\sqrt{2}}\right)$. **27.** $\frac{8}{3}\pi a^3$. **29.** $\frac{\pi^2 a^3}{4\sqrt{2}}$.

CHAPTER XII

Pages 303—304.

1. (*i*) $(a+y)(a-x)=c$. (*ii*) $y\sqrt{1-x^2}+x\sqrt{1-y^2}=c$.
2. (*i*) $\log xy+x-y=c$. (*ii*) $\log\frac{x}{y}-\frac{y+x}{xy}=c$.
3. (*i*) $\tan x\tan y=c$. (*ii*) $\tan y=c(1-e^x)^3$.
4. $y=c(1-ay)(a+x)$.
5. (*i*) $y-a\tan^{-1}\frac{x+y}{a}=c$. (*ii*) $y-\frac{a}{2}\log\frac{x-y-a}{x-y+a}=c$.
6. (*i*) $\frac{1}{2}\tan^{-1}\frac{4y+y+1}{2}=x+c$. (*ii*) $y-\log(x+y+2)=c$.

Page 306.

1. (*i*) $\frac{x}{y}=\log x+c$. (*ii*) $\tan^{-1}\frac{y}{x}+\frac{1}{2}\log(x^2+y^2)=c$.
2. (*i*) $m\tan^{-1}\frac{y}{x}-\frac{1}{2}\log(x^2+y^2)=c$. (*ii*) $-\frac{x^3}{3y^3}+\log y=c$.
3. (*i*) $\frac{y}{x}-\log y=c$. (*ii*) $xy^2=c(x+2y)$. **4.** $y+\sqrt{x^2+y^2}=cx$.
5. (*i*) $x^2+y^2=cx$. (*ii*) $xy\cos\frac{y}{x}=c$.

Pages 308—310.

2. $\log(x^2+y^2+x+y+\frac{1}{2})=2\tan^{-1}\frac{2y+1}{2x+1}+c$.
3. (*i*) $7(x-2y)+3\log(14x+21y+22)=c$. (*ii*) $3(x-2y)-\log(3x+3y+4)=c$.

Pages 312—313.

1. (*i*) $y\,x^n=ax+c$. (*ii*) $y(x^2+1)^2=\tan^{-1}x+c$.
2. (*i*) $y(x^2+1)=4\frac{x^3}{3}+c$. (*ii*) $y=(x+1)^n(e^x+c)$.
3. (*i*) $y(x+\sqrt{a^2+x^2})=a^2\log\frac{x+\sqrt{a^2+x^2}}{a}+c$.

   (*ii*) $y\,e^{2\sqrt{x}}=2\sqrt{x}+c$.

4. (*i*) $y = x^2 (1 + ce^{\frac{1}{x}})$. (*ii*) $y = \tan x - 1 + c\, e^{-\tan x}$.

5. $y = ax + cx\sqrt{1-x^2}$. **6.** $x = \tan^{-1} y - 1 + c\, e^{-\tan^{-1} y}$.

7. $y = \cos x - 2\cos^2 x$.

Pages 315—316. Art. 113.

1. (*i*) $y^{\frac{1}{2}} = -\frac{1}{3}(1-x^2) + c(1-x^2)^{\frac{1}{4}}$. (*ii*) $y^{1-n} = \dfrac{2}{1-n}[(1-n)\sin x - 1] + c\, e^{-(1-n)\sin x}$.

2. (*i*) $y^{-2} = x^2 + 1 + c\, e^{x^2}$. (*ii*) $y^{-1} = \dfrac{1}{x} + \dfrac{c}{\sqrt{x}}$.

3. $x^{-1} = 2 - y^2 + ce^{-\frac{1}{2}y^2}$.

4. (*i*) $y^3 (x+1)^2 = \dfrac{x^6}{6} + \dfrac{2}{5}x^5 + \dfrac{x^4}{4} + c$. (*ii*) $y^2 = \dfrac{2a}{4b^2+1}(2b\cos x + \sin x) + c\, e^{-2bx}$.

5. (*i*) $y^3 = ax + cx\sqrt{1-x^2}$. (*ii*) $e^y (x+1) = 2x + c$.

6. $\tan y = \frac{1}{2}(x^2 - 1) + ce^{-x^2}$.

MISCELLANEOUS EXAMPLES ON CHAPTER XII

Pages 316—318.

1. (*i*) $\log x - \dfrac{x^2}{2} = -\dfrac{y^2}{2} - 2y - 2\log(1-y) + c$.

(*ii*) $y \sin y = x^2 \log x + c$. **2.** $y - \tan \dfrac{x+y}{2} = c$.

3. (*i*) $\log(y-x) - \dfrac{x}{y-x} = c$. (*ii*) $\dfrac{2xy}{(x+y)^2} + \log(x+y) = c$.

4. (*i*) $x^2 + y^2 = c(x+y)$. (*ii*) $(x^2+y^2)^2 = c(x^2-y^2)$.

5. (*i*) $y + \sqrt{x^2+y^2} = cx\, e^x$. (*ii*) $4(2y - x) + \log(4x + 8y + 5) = c$.

6. (*i*) $y\dfrac{1+\sqrt{x}}{1-\sqrt{x}} = x + \frac{2}{3}x^{\frac{3}{2}} + c$. (*ii*) $y = \sqrt{1-x^2} + c(1-x^2)$.

7. (*i*) $y(x-1) = x^2(x^2 - x + c)$. (*ii*) $x^2 y = -x^2\cos x + 2(x\sin x + \cos x) + c$.

8. (*i*) $y = \sin x + c\cos x$. (*ii*) $y(\sec x + \tan x) = \sec x + \tan x - x + c$.

9. (*i*) $x^2 y = \dfrac{x^4}{4}\log x - \dfrac{x^4}{16} + c$. (*ii*) $y \log x = (\log x)^2 + c$.

10. (*i*) $x = y(y^2 + c)$. (*ii*) $x\, e^{\tan^{-1} y} = \tan^{-1} y + c$.

11. (*i*) $xy = \frac{1}{4}(x^4 + 3)$. (*ii*) $y(1+x^2) = 4\dfrac{x^3}{3} + c$; $3y(1+x^2) = 4x^3$.

12. (i) $y^{-1} = 1 + \dfrac{c}{\sqrt{1-x^2}}$. (ii) $y^{-1} \sec^2 x = -\dfrac{\tan^3 x}{3} + c$.

13. (i) $y^{-2} e^{x^2} = 2x + c$. (ii) $y^{-1} = \sin x + c \cos x$.

14. (i) $y^2 = x^2 (-\frac{2}{3} e^{-\frac{1}{x^3}} + c)$. (ii) $y^2 = x(-\log x + c)$.

15. (i) $\sin y = (1 + x)(e^x + c)$. (ii) $e^y = e^x - 1 + c\, e^{-e^x}$.

CHAPTER XIII

Pages 327—328.

1. (i) $y = c_1 e^{3x} + c_2 e^{4x}$. (ii) $y = e^{3x}(c_1 + c_2 x)$.
2. (ii) $y = c_1 \cos \mu x + c_2 \sin \mu x$, $y = c_3 e^{\mu x} + c_4 e^{-\mu x}$.
3. (i) $y = c_1 e^x + c_2 e^{2x} + c_3 e^{3x}$. (ii) $y = e^x (c_1 + c_2 x) + c_3 e^{2x}$.
   (iii) $y = c_1 e^{2x} + c_2 \cos 2x + c_3 \sin 2x$.
5. (i) $y = c_1 \cos 2x + c_2 \sin 2x + c_3 \cos 3x + c_4 \sin 3x$.
   (ii) $y = \cos 2x.(c_1 + c_2 x) + \sin 2x.(c_3 + c_4 x)$.
6. $y = e^{-x}(c_1 \cos x + c_2 \sin x) + e^x (c_3 \cos x + c_4 \sin x)$.
7. $x = 0$.

Pages 339—340.

1. (i) $y = c_1 e^x + c_2 e^{2x} + \frac{1}{12} e^{5x}$. (ii) $y = e^{-x}(c_1 + c_2 x) + \frac{2}{9} e^{2x}$.
2. (i) $y = c_1 e^{-x} + e^{2x}(c_2 + c_3 x) + \frac{1}{4} e^{3x}$. (ii) $y = c_1 e^{2x} + c_2 e^{-2x} - \frac{8}{17} \sin \frac{1}{2} x$.
3. (i) $y = e^{-\frac{1}{2}x}\left(c_1 \cos \dfrac{\sqrt{3}}{2} x + c_2 \sin \dfrac{\sqrt{3}}{2} x\right) - \frac{1}{13}(2 \cos 2x + 3 \sin 2x)$.

   (ii) $y = c_1 e^{-x} + e^{\frac{1}{2}x}\left(c_2 \cos \dfrac{\sqrt{3}}{2} x + c_3 \sin \dfrac{\sqrt{3}}{2} x\right) + \dfrac{1}{730}(\sin 3x + 27 \cos 3x) - \dfrac{1}{2} - \dfrac{1}{4}(\cos x - \sin x)$.
4. (i) $y = \cos nx.(c_1 + c_2 x) + \sin nx.(c_3 + c_4 x) + \dfrac{1}{(n^2 - m^2)^2} \cos mx$.

   (ii) $y = c_1 e^{4x} + c_2 e^{5x} + x + \frac{9}{20}$.
5. (i) $y = c_1 e^{2x} + c_2 e^{3x} + \frac{1}{6}(x + \frac{5}{6}) + \dfrac{1}{m^2 - 5m + 6} e^{mx}$.

   (ii) $y = c_1 e^x + c_2 e^{4x} + c_3 e^{-2x} + \frac{1}{8}(x + \frac{3}{4})$.
6. (i) $y = c_1 e^x + c_2 e^{-x} - 2 - 3x$. (ii) $y = c_1 e^x + c_2 e^{-4x} + c_3 e^{3x} + \frac{1}{12}(x + \frac{13}{12})$.
7. (i) $y = c_1 \cos 2x + c_2 \sin 2x - \frac{1}{5} \sin 3x + \frac{1}{5} e^x + \frac{1}{4}(x^2 - \frac{1}{2})$.

   (ii) $y = c_1 + e^{-x}(c_2 + c_3 x) + \dfrac{1}{18} e^{2x} + \dfrac{x^3}{3} - \dfrac{3}{2} x^2 + 4x$.
8. (i) $y = c_1 e^{ax} + c_2 e^{-ax} + c_3 \cos ax + c_4 \sin ax - \dfrac{1}{a^4}\left(x^4 + \dfrac{24}{a^4}\right)$.

(*ii*) $y = c_1 + c_2x + e^x (c_3 + c_4x) + x^3 + 6x^2$.

9. (*i*) $y = c_1 \cos x + c_2 \sin x + \frac{1}{5} e^{2x} (x - \frac{4}{5})$.

(*ii*) $y = c_1 e^{-x} + c_2e^{-2x} + c_3e^{3x} - \frac{1}{12} e^{2x} (\frac{17}{12} + x)$.

10. (*i*) $y = e^x (c_1 \cos \sqrt{3}\, x + c_2 \sin \sqrt{3} x) + \frac{1}{2} e^x \cos x$.

(*ii*) $y = c_1 e^x + c_2 e^{-x} + c_3 \cos x + c_4 \sin x - \frac{1}{5} e^x \cos x$.

Pages 344—345. Art. 123.

1. (*i*) $y = c_1e^{ax} + c_2 e^{-ax} + \dfrac{x}{2a} e^{ax} + \dfrac{1}{n^2 - a^2} e^{nx}$.

(*ii*) $y = c_1 e^x + c_2e^{-x} + \dfrac{x}{2} \sinh x$.

2. (*i*) $y = c_1 e^x + e^{-\frac{1}{2}x} \left(c_2 \cos \dfrac{\sqrt{3}}{2} x + c_3 \sin \dfrac{\sqrt{3}}{2} x \right) + \dfrac{1}{7} e^{2x} + \frac{2}{3} x e^x - 1$.

(*ii*) $y = e^{-x} (c_1 + c_2x + c_3 x^2) + \dfrac{x^3}{6} e^{-x}$.

3. (*i*) $y = c_1 e^x + e^x (c_2 \cos x + c_3 \sin x) + x e^x + \frac{1}{10} (3 \sin x + \cos x)$.

(*ii*) $y = c_1 \cos 2x + c_2 \sin 2x + \dfrac{1}{5} e^x - \dfrac{x}{4} \cos 2x$.

MISCELLANEOUS EXAMPLES ON CHAPTER XIII

Pages 345—348.

1. (*i*) $y = c_1 e^{-ax} + c_2 e^{-bx}$. (*ii*) $y = e^{2x} (c_1 + c_2 x)$.

(*iii*) $y = 2 (\cos x - \sin x)$.

2. (*i*) $y = e^{-\frac{m}{\sqrt{2}}x} \left(c_1 \cos \dfrac{m}{\sqrt{2}} x + c_2 \sin \dfrac{m}{\sqrt{2}} x \right) + e^{\frac{m}{\sqrt{2}}x} \left(c_3 \cos \dfrac{m}{\sqrt{2}} x + c_4 \sin \dfrac{m}{\sqrt{2}} x \right)$.

(*ii*) $y = e^{-\frac{1}{2}x} \left[\cos \dfrac{\sqrt{3}}{2} x \,.\, (c_1 + c_2x) + \sin \dfrac{\sqrt{3}}{2} x \,.\, (c_3 + c_4\, x) \right]$.

3. (*i*) $y = e^{-px} (c_1 \cos qx + c_2 \sin qx) + \dfrac{1}{(a + p)^2 + q^2} e^{ax}$.

(*ii*) $y = c_1 e^{2x} + c_2 e^{-2x} - \frac{1}{3} e^x - \frac{1}{8} \sin 2x$.

4. (*i*) $y = c_1 e^{(4 + \sqrt{7})x} + c_2 e^{(4 - \sqrt{7})x} - \frac{5}{29} (2 \sin 5x - 5 \cos 5x)$.

(*ii*) $y = c_1 e^x + c_2 e^{2x} - \frac{1}{130} (7 \cos 3x + 9 \sin 3x)$.

(*iii*) $y = e^{-kx} (c_1 + c_2\, x) + \mathrm{A}\, \dfrac{(k^2 - p^2) \cos px + 2kp \sin px}{(k^2 + p^2)^2}$.

5. (*i*) $y = c_1e^x + c_2 e^{12x} + \frac{1}{12} (x + \frac{13}{12})$.